797,885 Books
are available to read at

Forgotten Books

www.ForgottenBooks.com

Forgotten Books' App
Available for mobile, tablet & eReader

ISBN 978-1-332-59785-7
PIBN 10012551

This book is a reproduction of an important historical work. Forgotten Books uses
state-of-the-art technology to digitally reconstruct the work, preserving the original format
whilst repairing imperfections present in the aged copy. In rare cases, an imperfection in
the original, such as a blemish or missing page, may be replicated in our edition. We do,
however, repair the vast majority of imperfections successfully; any imperfections that
remain are intentionally left to preserve the state of such historical works.

Forgotten Books is a registered trademark of FB &c Ltd.
Copyright © 2015 FB &c Ltd.
FB &c Ltd, Dalton House, 60 Windsor Avenue, London, SW19 2RR.
Company number 08720141. Registered in England and Wales.

For support please visit www.forgottenbooks.com

1 MONTH OF FREE READING

at

www.ForgottenBooks.com

By purchasing this book you are eligible for one month membership to ForgottenBooks.com, giving you unlimited access to our entire collection of over 700,000 titles via our web site and mobile apps.

To claim your free month visit:

www.forgottenbooks.com/free12551

* Offer is valid for 45 days from date of purchase. Terms and conditions apply.

English
Français
Deutsche
Italiano
Español
Português

www.forgottenbooks.com

Mythology Photography **Fiction**
Fishing Christianity **Art** Cooking
Essays Buddhism Freemasonry
Medicine **Biology** Music **Ancient Egypt** Evolution Carpentry Physics
Dance Geology **Mathematics** Fitness
Shakespeare **Folklore** Yoga Marketing
Confidence Immortality Biographies
Poetry **Psychology** Witchcraft
Electronics Chemistry History **Law**
Accounting **Philosophy** Anthropology
Alchemy Drama Quantum Mechanics
Atheism Sexual Health **Ancient History**
Entrepreneurship Languages Sport
Paleontology Needlework Islam
Metaphysics Investment Archaeology
Parenting Statistics Criminology
Motivational

SCHOOL OF EDUCATION
LIBRARY

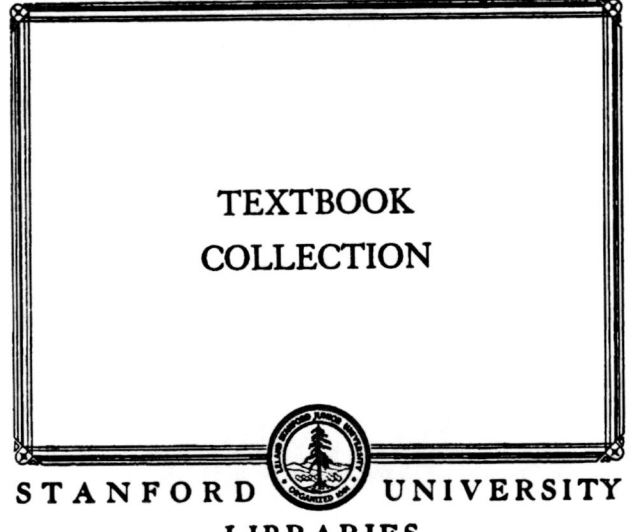

TEXTBOOK
COLLECTION

STANFORD UNIVERSITY
LIBRARIES

C.A.Young

LABORATORY ASTRONOMY

BY

ROBERT WHEELER WILLSON, Ph.D.
PROFESSOR OF ASTRONOMY IN HARVARD UNIVERSITY

626077

COPYRIGHT, 1900, 1905
BY ROBERT W. WILLSON

ALL RIGHTS RESERVED

66.1

The Athenæum Press
GINN & COMPANY · PROPRIETORS · BOSTON · U.S.A.

PREFACE

The subjects treated in elementary text-books of astronomy which are most difficult and discouraging to the beginner are those which deal with the diurnal motion of the heavens and the apparent motions of the sun, moon, and planets among the stars. A clear conception of these fundamental facts is, however, necessary to a proper understanding of many of the striking phenomena to which the study of astronomy owes its hold upon the intellect and the imagination.

No adequate notion of those subjects which involve the ideas of force and mass can be given to the average student who has not mastered the elements of mechanics; but to explain the motions of the heavenly bodies, the knowledge of a few principles of solid geometry and of the properties of the ellipse will suffice, — no more, indeed, than may be easily explained in the pages of the text-book itself.

Most of the difficulties which arise at the outset of the study may be satisfactorily met by methods which require the student to make and discuss simple observations and to solve simple problems. This necessity is recognized in many recent text-books which introduce such methods to a greater or less extent, — in all cases to great advantage and in some with marked success. I have gathered in this book some of those which I have found practicable, intending that they should explain in natural sequence those phenomena which depend on the diurnal motion, the moon's motion in her orbit and the change in position of that orbit, the motion of the sun in the ecliptic, and the geocentric motions of the planets.

The methods chosen may be carried out with fair-sized classes and do not require a place of observation favored with an extensive view of the heavens. The gnomon-pin, the hemisphere, the cross-staff, a simple apparatus for measuring altitude and azimuth which may be converted into an equatorial by inclining it at the proper angle, together with a few maps and diagrams, form an outfit so inexpensive that it may be supplied to each pupil, and much work may be done at home. It is obvious that the possibility thus offered of utilizing favorable opportunities for observation is especially valuable in a study which is so much dependent on the weather. All members of the class, too, will be doing the same or similar work at the same time, — a principle of cardinal importance in elementary laboratory work with large classes.

The meridian work of Chapter VI is added for the sake of logical completeness, to explain the determination of the zero of right ascensions, — a subject which is usually neglected in the text-books and would not be included in an ordinary course.

Nothing has been directly planned for teaching the names of the constellations and the use of star maps. The work of Chapters II, III, and IV, covering a period of some months, results in a very good acquaintance with the principal stars and asterisms. It may be assumed, too, that the teacher is familiar with the heavens and will gather the class as early as possible to introduce them at least to the polar constellations.

The book is intended primarily for teachers, but much of it is suitable for use as a text-book, in spite of its rather condensed form. It is meant to be used in connection with one of the many admirable text-books on descriptive astronomy adapted to high-school pupils.

The first six chapters were printed in 1900, and various changes and additions might now be made, notably an improvement in the protractor for laying off altitudes on the hemisphere, which is now so constructed that it may be used as a ruler for the accurate drawing of great circles. This permits a much simpler determination of the pole of a small circle than that described in the first chapter.

ROBERT W. WILLSON

HARVARD UNIVERSITY
STUDENTS' ASTRONOMICAL LABORATORY
December, 1905

TABLE OF CONTENTS

CHAPTER I

THE SUN'S DIURNAL MOTION

	PAGE
Path of the Shadow of a Pin-head cast by the Sun upon a Horizontal Plane	1
Altitude and Bearing	4
Representation of the Celestial Sphere upon a Spherical Surface	5
The Sun's Diurnal Path upon the Hemisphere is a Circle — a Small Circle except about March 20 and September 21	8
Determination of the Pole of the Circle	9
Bearing of the Points of Sunrise and Sunset	11
The Meridian — the Cardinal Points	11
Magnetic Declination	12
Azimuth	12
The Equinoctial	14
Position of the Pole as seen from Different Places of Observation	15
Latitude equals Elevation of Pole	16
Hour-angle of the Sun	17
Uniform Increase of the Sun's Hour-angle — Apparent Solar Time	18
Declination of the Sun — its Daily Change	20

CHAPTER II

THE MOON'S PATH AMONG THE STARS

Position of the Moon by its Configuration with Neighboring Stars	21
Plotting the Position of the Moon upon a Star Map	24
Position of the Moon by Measures of Distance from Neighboring Stars	25
The Cross-staff	25
Length of the Month	29
Node of the Moon's Orbit	30
Errors of the Cross-staff	31

CHAPTER III

THE DIURNAL MOTION OF THE STARS

	PAGE
Instrument for measuring Altitude and Azimuth	34
Adjustment of the Altazimuth	35
Determination of Meridian by Observations of the Sun	37
Determination of Apparent Noon by Equal Altitudes of the Sun	39
Meridian Mark	40
Selection of Stars — Magnitudes	41
Plotting Diurnal Paths of Stars on the Hemisphere	42
Paths of Stars compared with that of the Sun	42
Drawing of Hemisphere with its Circles	42
Rotation of the Sphere as a Whole	43
Declinations of Stars do not change like that of the Sun	43
Equable Description of Hour-angle by Stars	43
Hour-angle and Declination fix the Position of a Heavenly Body as well as Altitude and Azimuth — Comparison of the Two Systems of Coördinates	44
Equatorial Instrument for measuring Hour-angle and Declination	45
Universal Equatorial — Advantages of the Equatorial Mounting	45

CHAPTER IV

THE COMPLETE SPHERE OF THE HEAVENS

Rotation of the Heavens about an Axis passing through the Pole explains Diurnal Motions of Sun, Moon, and Stars	47
Relative Position of Two Stars determined by their Declinations and the Difference of their Hour-angles	48
Use of Equatorial to determine Positions of Stars	49
Use of a Timepiece to improve the Foregoing Method	50
Map of Stars by Comparison with a Fundamental Star	53
Extension of Use of Timepiece to reduce Labor of Observation	54
The Vernal Equinox to replace the Fundamental Star — Right Ascension	56
Sidereal Time — Sidereal Clock	57
Right Ascension of a Star is the Sidereal Time of its Passage across the Meridian	58
Right Ascension of any Body plus its Hour-angle at any Instant is Sidereal Time at that Instant	58
Finding Stars by the Use of a Sidereal Clock and the Circles of the Equatorial Instrument	59
The Clock Correction	60
List of Stars for determining Clock Error	61

TABLE OF CONTENTS

CHAPTER V

MOTION OF THE MOON AND SUN AMONG THE STARS

	PAGE
Plotting Stars upon a Globe in their Proper Relative Positions	63
Plotting Positions of the Moon upon Map and Globe by Observations of Declination, and Difference of Right Ascension from Neighboring Stars	64
Variable Rate of Motion of the Moon	65
Variable Semi-diameter of the Moon	65
Position of Greatest Semi-diameter and of Greatest Angular Motion	65
Plotting Moon's Path on an Ecliptic Map	65
Observations of Sun's Place in Reference to a Fundamental Star by Equatorial and Sidereal Clock	66
Sun's Place referred to Stars by Comparison with the Moon or Venus	68
Plotting the Sun's Path upon the Globe — the Ecliptic	70

CHAPTER VI

MERIDIAN OBSERVATIONS

Use of the Altazimuth or Equatorial in the Meridian	72
The Meridian Circle	73
Adjustments of the Meridian Circle	74
Level	74
Collimation	78
Azimuth	78
Determination of Declinations	80
Determination of the Polar Point	81
Absolute Determination of Declination	81
Determination of the Equinox	83
Absolute Right Ascensions	84
Autumnal Equinox of 1899	85
Autumnal Equinox of 1900	87
Length of the Year	88

CHAPTER VII

THE NAUTICAL ALMANAC

Mean Time	91
The Equation of Time	92
Standard Time	93

TABLE OF CONTENTS

	PAGE
The Calendar Pages	94
Examination of the Several Columns	99
Data for the Planets and Stars	102
Comparison of Observations with the Ephemeris	102
Observations of the Moon with the Cross-staff; Length of the Month	103
Observation at Apparent Noon	104
Observations of the Planets. Observations of the Moon with Equatorial	105
Observations of the Sun's Place	106
Determination of the Equinox	107

CHAPTER VIII

THE CELESTIAL GLOBE

Description of the Globe	108
Rectifying the Globe for a Given Place and Time	111
The Sun's Place on the Globe	112
The Altitude Arc	113
Problems which do not require Rectification of the Globe	114
Problems which require Rectification of the Globe for a Given Time	117
Finding an Hour-angle by the Globe	119
Reduction to the Equator	121

CHAPTER IX

EXAMPLES OF THE USE OF THE GLOBE

Problems which require Rectification of the Globe for a Given Place	122
Rising and Setting of Stars	122
Sunrise	124
Altitude and Azimuth; Hour-angle	125
Finding the Time from the Sun's Altitude	126
Identifying a Heavenly Body by its Altitude and Azimuth at a Given Time	129
Aspect of the Planets at a Given Time	130
Rising and Setting of the Moon	131
Twilight	133
Orientation of Building by Sun Observation	134
Latitudes in which Southern Cross is Visible	135
The Midnight Sun; the Harvest Moon	136
Change of Azimuth at Rising and Setting	137
Graduating a Horizontal Sundial	137
Graduating a Vertical Sundial	138
Determining Path of Shadow by Globe	139
The Hour-index	141

TABLE OF CONTENTS ix

CHAPTER X

THE MOTION OF THE PLANETS

	PAGE
Elliptic Orbits a Result of the Law of Gravitation	143
Properties of the Ellipse	144
To draw an Ellipse from Given Data	145
Mean and True Place of a Planet; Equable Description of Areas	146
The Equation of Center	148
Measurement of Angles in Radians	149
The Diagram of Curtate Orbits	151
To find the Elements of an Orbit from the Diagram	154
Place of the Planet in its Orbit	156
To find the True Heliocentric Longitude of a Planet	157
To find the Heliocentric Latitude	161
Geocentric Longitude of a Planet	161
The Sun's Longitude and the Equation of Time	162
Geocentric Latitude	163
Perturbations; Precession	166
The Julian Day	167
Right Ascensions and Declinations of the Planets	167
Configurations of the Planets	168
The Path of Mars among the Stars in 1907	169

LABORATORY ASTRONOMY

PART I

CHAPTER I

THE DIURNAL MOTION OF THE SUN

THE most obvious and important astronomical phenomenon that men observe is the succession of day and night, and the motion of the sun which causes this succession is naturally the first object of astronomical study. Every one knows that the sun rises in the east and sets in the west, but very many educated people know little more of the course of the sun than this. The first task of the beginner in astronomy should be to observe, as carefully as possible, the motion of the sun for a day. What is to be observed then? A little thought shows that it can only be the direction in which we have to look to see it at different times; that is, toward what point of the compass — how far above the ground. All astronomical observation. indeed, comes down ultimately to this — the direction in which we see things. The strong light of the sun enables us to make use of a very simple method depending on the principle that the shadow of a body lies in the same straight line with the body and the source of light.

Path of the Shadow of a Pin-head. — If we place a pin upright on a horizontal plane in the sunlight and mark the position of the shadow of its head at any time, we thus fix the position of the sun at that time, since it is in the prolongation of the line drawn from the shadow to the pin-head. In order to carry out systematic observations by this method in such a form that the results may be easily discussed, it will be convenient to have the following apparatus: (1) A firm *table* in such a position as to receive sunlight for as long a period as possible. It is better that it should be in the open air, in which case it may be made by driving small posts into

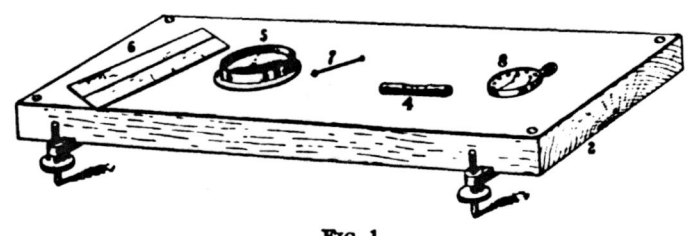

Fig. 1

the ground and securely fastening a stout plank about 18 inches square as a top. (2) A *board*, 18 inches long and 8 inches broad, furnished with leveling screws and smoothly covered with white paper fastened down by (3) *thumb tacks*. (4) A *level* for leveling the board. (5) A *compass*. (6) A *glass plate*, 6 inches long and 2 inches broad, along the median line of which a straight black line is drawn. (7) A *pin*, 5 cm. long, with a spherical head and an accurately turned base for setting it vertical. (8) A *timepiece*.

Draw a straight pencil line across the center of the paper as

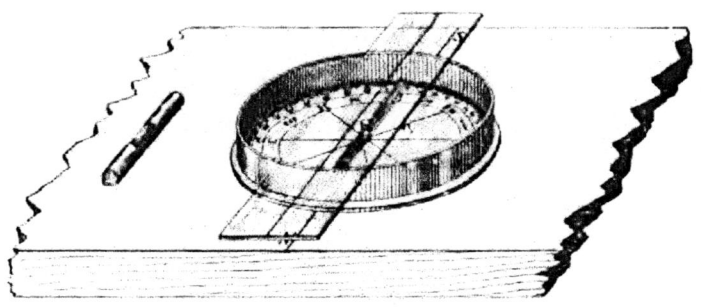

Fig. 2

nearly as possible perpendicular to the length of the board. Place the board upon the table and level approximately. Put the compass on the middle of the pencil line and put the glass plate on the compass with its central line over the center of the needle; turn the plate till its median line is parallel to the pencil line (Fig. 2), and swing the whole board horizontally, till the needle is parallel to the two lines, which are then said to be in the magnetic meridian. Press the leveling screws firmly into the table, and thus make dents by which the board may at any future time be placed in the same position without the renewed use of the compass. Level the board

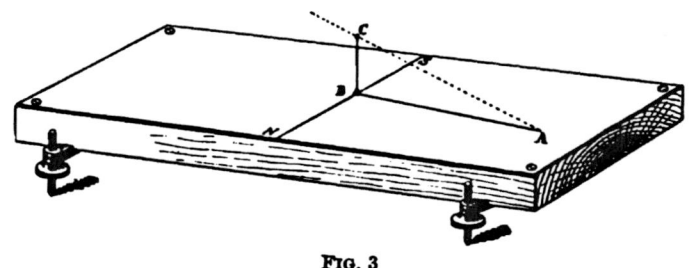

FIG. 3

carefully, placing the level first east and west, then north and south. Place the pin in the pencil line, — in the center if the observation is made between March 20 and September 20, but near the southern edge of the board at any other time of the year, — pressing it firmly down till the base is close to the paper, so that the pin is perpendicular to the paper. Mark with a hard pencil the estimated center of the shadow of the pin-head, A (Fig. 3), noting the time by the watch to the nearest minute, affix a number or letter, and affix the same number to the recorded time of the observation in the note-book. It is a good plan to use pencil for notes made while

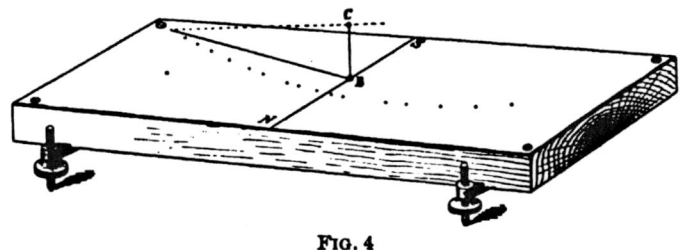

FIG. 4

observing, and ink for computations or notes added afterward in discussing them. Repeat at hourly, or better half-hourly, intervals, thus fixing a set of points (Fig. 4), through which a continuous curve may be drawn showing the path of the shadow for several hours. The same observation should be repeated two weeks later.

ALTITUDE AND BEARING

By the foregoing process we obtain a diagram on which is shown the position of the pin point, a magnetic meridian line through this point, and a series of numbered points showing the position of the shadow of the pin-head at different times; the height of the pin is known and also the fact that its head was in the same vertical line with its point.

In the discussion of these results, it will be convenient to proceed as follows:

Remove the pin and draw with a hard pencil a fine line, *AB* (Fig. 5), through the pinhole and the point marked at the first observation. This line is called a line of bearing, and the angle which

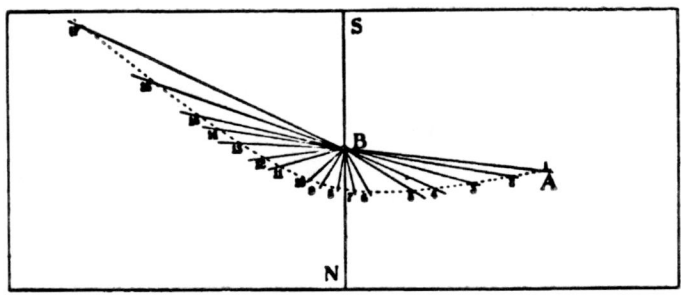

FIG. 5

it makes with the magnetic meridian is called the magnetic bearing of the line. This angle, which may be directly measured on the diagram by a protractor, fixes the position of the vertical plane which contains the observed point and passes also through the center of the pin-head and the sun. If this point bears N.W. from the pin, the sun evidently bears S.E.

Imagine a line, *AC* (Fig. 3), connecting the observed point with the sun's center and passing also through the center of the pin-head. The position of the sun *in* the vertical plane is evidently fixed by this line. The angle between the line of bearing and this line, *BAC*, is called the altitude of the sun; it measures, by the ordinary convention of solid geometry, the angle between the sun's direction and the plane of the horizon.

To determine this angle, lay off the line $B'C'$ (Fig. 6), equal in length to the pin, 5 cm., draw a perpendicular through B', and by means of a pair of compasses or scale laid between the two points A and B (Fig. 5), lay off the line $A'B'$ on the perpendicular, draw $A'C'$, and measure the angle $B'A'C'$ by a protractor. We now have the bearing and altitude of the sun at the time of the first observation, the bearing of the sun from the pin being opposite to that of the point from the pin. In like manner the altitude and bearing are determined for each observed point upon the path of the shadow, and noted against the corresponding time, in the note-book (to avoid confusion, it is convenient to make a separate figure for the morning and afternoon observations, as shown in Fig. 6). We have thus obtained a series of values which will enable us to study more easily the path of the sun upon the concave of the sky.

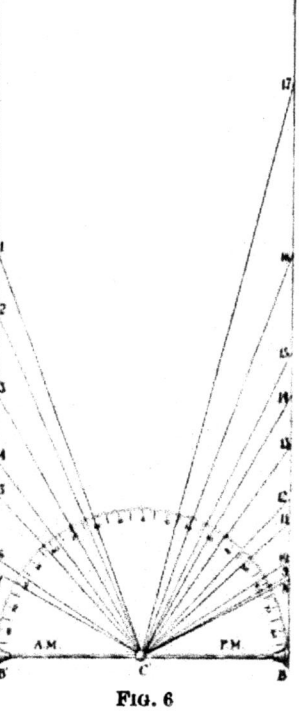

FIG. 6

Plotting the Sun's Path on a Spherical Surface. — Probably the most evident method of accomplishing this object would be to

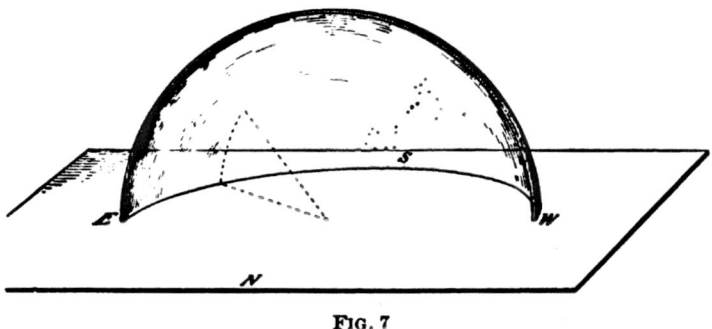

FIG. 7

construct a small concave portion of a sphere, as in the accompanying figure, which suggests how the position of the sun might be referred to the inside of a glass shell.

But the hollow surface offers difficulty in construction and manipulation, and it requires but little stretch of the imagination to pass to the convex surface as follows. The glass shell, as seen from the other side, would appear thus:

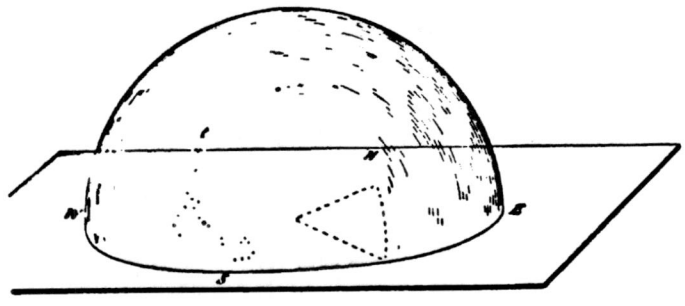

Fig. 8

and we can more readily get at it to measure it, and moreover can more easily recognize the properties of the lines which we shall come to draw upon it, since we are used to looking upon spheres from the outside rather than from the inside, except in the case of the celestial sphere.

On both Figs. 7 and 8 is shown a group of dots which have nearly the configuration of a group of stars conspicuous in the southern heavens in midsummer and called the constellation of Scorpio. It is evident that the constellation has the same shape in both cases, except that in Fig. 8 it is turned right for left or semi-inverted, as is the image of an object seen in a mirror. This property obviously belongs to all figures drawn on the concave surface as seen from the center, when they are looked at from the outside directly toward the center.

So also the diurnal motion of the sun, which as we see it from the center is from left to right, would be from right to left as viewed from the outside of such a surface. This latter is so slight an inconvenience that it is customary to represent the motions of the heavenly bodies in the sky upon an opaque globe, and to determine the angles which these bodies describe about the center, by measuring the corresponding arcs upon the convex surface.

THE DIURNAL MOTION OF THE SUN 7

Plotting on a Hemisphere. — The apparatus required for plotting the sun's path consists of: a *hemisphere*, a, 4½ inches in diameter; a *circular protractor*, b, a *quadrantal protractor*, c, of 2¼ inches

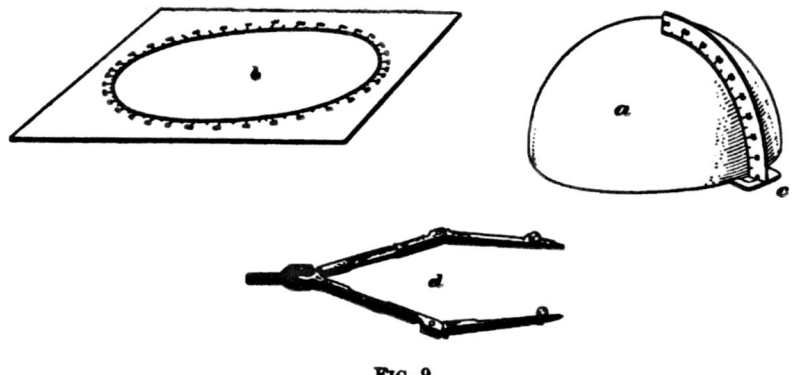

FIG. 9

radius, and a *pair of compasses*, d, whose legs may be bent and one of which carries a hard pencil point.

Determine by trial with the compasses the center of the base of the hemisphere, and mark two diameters by drawing straight lines upon the base at right angles through the center. Prolong these by marks about ⅛ inch in length upon the convex surface. Place the

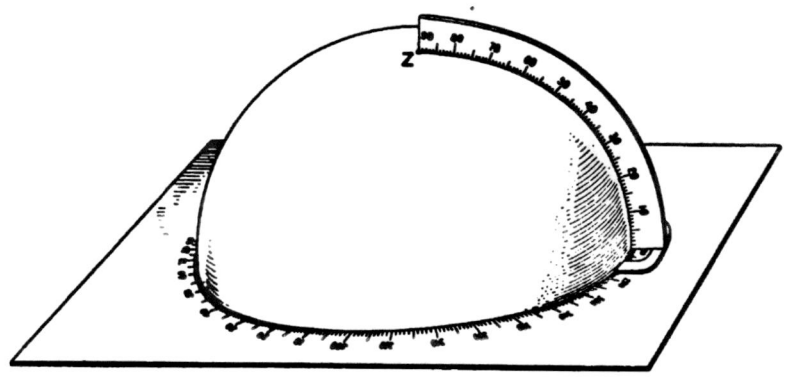

FIG. 10

hemisphere exactly central upon the circular protractor, by bringing the marked ends of one of the diameters upon those divisions of the protractor which are numbered 0° and 180°, and the other on the divisions numbered 90° and 270°. Determine and mark the highest point of the hemisphere by placing the quadrant with its base upon the circular protractor, and its arc closely against the sphere, and marking the end of the scale (Fig. 10). Repeat this with the arc in four positions, 90° apart on the base. The points thus determined should coincide; if they do not, estimate and mark the center of the four points thus obtained. This point represents the highest point of the dome of the heavens — the point directly overhead, called the zenith, and the zero and 180° points on the base protractor may be taken as representing the south and north points respectively of the magnetic meridian.

The Sun's Path a Circle. — To plot the altitude and bearing of the first observation, place the foot of the quadrant or altitude arc close against the sphere, the foot of its graduated face on the degree of the protractor which corresponds to the bearing. Mark a fine point on the sphere at that degree of the altitude arc corresponding to the altitude at the first observation. This point fixes the direction in which the sun would have been seen from the center of the hemisphere at the time of observation if the zero line had been truly in the magnetic meridian. Proceed in the same manner with the other observations of bearing and altitude, and thus obtain

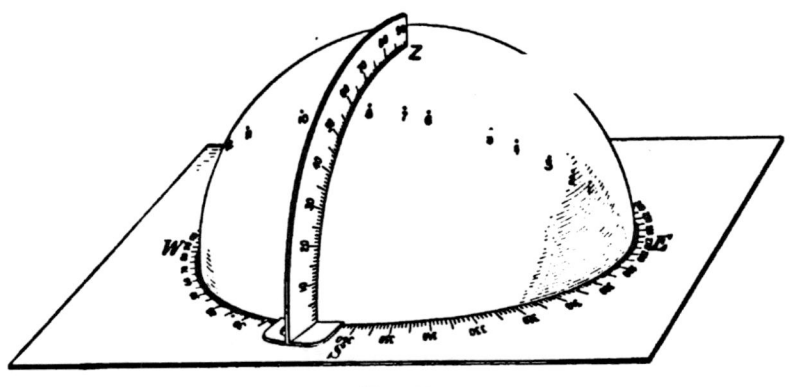

FIG 11

a series of points (Fig. 11), through which may be drawn a continuous line representing the sun's path upon that day.

It will appear at once that the arcs between the successive points are of nearly equal length if the times of observation were equidistant, and otherwise are proportional to the intervals of time between the corresponding observations — a property which does not at all belong to the shadow curve from which the points are derived. We thus have a noteworthy simplification in referring our observations to the sphere. It will also appear that a sheet of

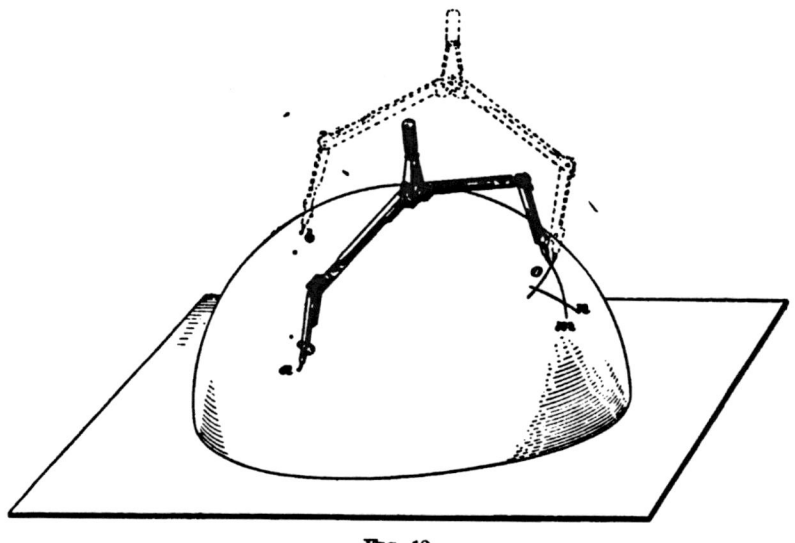

FIG. 12

stiff paper or cardboard may be held edgewise between the hemisphere and the eye, so as to cover all the points; that is, they all lie in the same plane. This fact shows that the sun's path is a circle on the sphere. It is shown by the principles of solid geometry that all sections of the sphere by a plane are circles. If the plane of the circle passes through the center, it is the largest possible, its radius being equal to that of the sphere; it is then called a great circle. Near the 20th of March and 22d of September it will be found that the path of the shadow is nearly a straight line on the diagram, and that the path of the sun is nearly a great circle; that is, the plane of this circle passes nearly through the center of the sphere. In general, the shadow path is a curve, with its concave side toward the pin in summer and its convex side toward it in winter, while the path on the sphere is a small circle, that is, its plane does not pass through the center of the sphere.

Determining the Pole of the Circle. — It is proved by solid geometry that all points of any circle on the sphere are equidistant from two points on the sphere, called the poles of the circle. It is important to determine the pole of the sun's diurnal path.

Estimate as closely as possible the position on the sphere of a point which is at the same distance from all the observed points of the sun's path and open the compasses to nearly this distance. For a closer approximation to the position of the pole, place the steel point of the compasses at the point on the hemisphere corresponding to the first observation, a, and with the other (pencil) point draw a short arc, m (Fig. 12), near the estimated pole. Draw the arc n from the point of the last observation, c, and join these two arcs by a third drawn from an observed point, b, as near as possible to the middle of the path; the pole of the sun's diurnal circle will lie nearly on the great circle drawn from b to the middle point o of the arc last drawn. Place the steel point at o, and the pencil point at b, and try the distance of the pencil point from the sun's path at either extremity. If the pencil point lies above (or below) the path at both extremities, the compasses must be opened (or closed) slightly and the assumed pole shifted directly away from (or toward) the middle of the path.

FIG. 13

The proper opening of the compasses is thus quickly determined as well as a close approximation to the position of the pole. Place the steel point at this new position, p, the pencil point at b, and again test the extreme points. If the west end of the path is below the pencil point (Fig. 13), the latter should be brought directly down

THE DIURNAL MOTION OF THE SUN

to the path by shifting the steel point on the sphere in the plane of the compass legs, that is, along the great circle from p to s.

From the point thus found a circle can be described with the compasses so as to pass approximately through all the observed points; that is, this point is the pole of the sun's path, and when it is fixed as exactly as possible a circle is to be drawn from horizon to horizon which will represent the sun's path from the point of sunrise to that of sunset, and passing very nearly through all the observed points. The bearing of the points of sunrise and sunset may then be read off on the horizontal circle.

THE MERIDIAN

The pole as thus determined marks a very interesting and important point in the heavens. We will draw a great circle through the zenith and the pole. To do this, place the altitude arc against the sphere, as if to measure the altitude of the pole; and

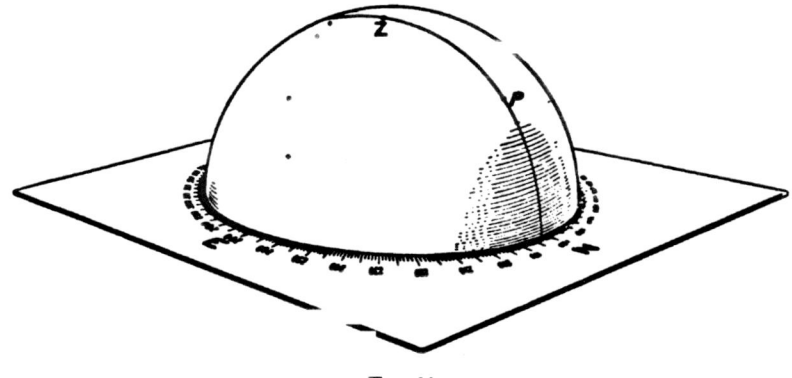

FIG. 14

using it as a guide, draw the northern quadrant of the vertical circle through the zenith and the pole. Note the bearing of this vertical circle. Place the altitude arc at the opposite bearing, and draw another or southern quadrant of the same great circle till it meets the south horizon. This great circle (Fig. 14) is called the meridian of the place of observation, and its plane is called the plane of the meridian of the place of observation, — sometimes the true meridian, to distinguish it from the magnetic meridian.

The line in which it cuts the base of the hemisphere represents the meridian line or true meridian line, just as the line first drawn represents the line of the magnetic meridian. If the observations are made in the United States, near a line drawn from Detroit to Savannah, it will be found that the true meridian coincides very nearly with the magnetic meridian. East of the line joining these cities, the north end of the magnet points to the west of the true meridian by the amounts given in the following table:

    21° at the extreme N.E. boundary of Maine.
    15 at Portland.
    10 at Albany and New Haven.
    5 at Washington and Buffalo.

While on the west the declination, as it is called, is to the east of the true meridian.

    5° at St. Louis and New Orleans.
    10 at Omaha and El Paso.
    15 at Deadwood and Los Angeles.
    20 at Helena, Montana, and C. Blanco.
    23 at the extreme N.W. boundary of the United States.

By drawing these lines on the map, as in Fig. 15, it is easy to estimate the declinations at intermediate points within one or two degrees, — at the present time west declinations in the United States are increasing and east declinations decreasing by about 1° in fifteen years.

A great circle perpendicular to the meridian may be drawn by placing the altitude protractor at readings 90° and 270° from the meridian reading and drawing arcs to the zenith in each case. This circle is the prime vertical, and intersects the horizon in the east and west points; thus all the cardinal points are fixed by the meridian determined from our plotting of the sun's path.

Azimuth. — Place the hemisphere upon the circular protractor in such a position that the line of the *true meridian* on the hemisphere coincides with the zero line of the protractor.

Place the altitude arc so as to measure the altitude at any part of the sun's path west of the meridian (Fig. 16). The reading of the foot of the arc will give the angle between the true meridian and the vertical plane containing the sun at that point of its diurnal circle. This angle is its *true* bearing and differs from its magnetic

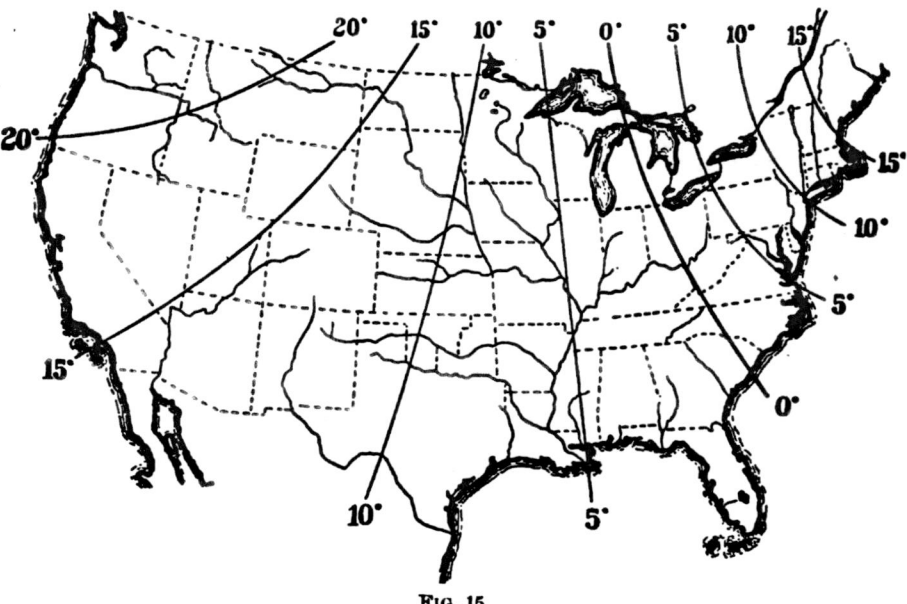

Fig. 15

bearing by the declination of the compass, being evidently less than the magnetic bearing, if the declination is west of north. It is also called the azimuth of the sun's vertical circle, or, briefly, of the sun.

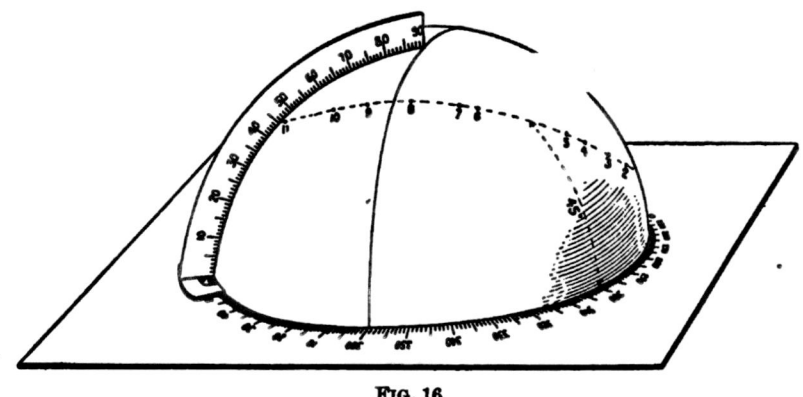

Fig. 16

Formerly azimuth was usually reckoned from north through the west or east, to 180° at the south point. It is now customary to measure it from south through west up to 360°, so that the azimuth of a body when east of the meridian lies between 180° and 360°. The present method is more convenient because the given angle fixes the position of the vertical circle without the addition of the letters E. and W. It is worthy of notice that with this notation the azimuth of the sun as seen in northern latitudes outside of the tropics always increases with the time; and indeed this is true of most of the bodies we shall have occasion to observe.

Now place the altitude quadrant so that its foot is at a point on the circular protractor where the reading is 360° minus the azimuth of the point just measured; the sun at this point of its path is just as far east of the meridian as it was west of the meridian at the point last considered, and it will be found that the altitude of the two points is the same. On the path shown in Fig. 16 the altitude is 45° at the points whose azimuths are 60° and 300° (60 E. of S.).

This fact, that equal altitudes of the sun correspond to equal azimuths east and west of the true meridian, is an important one, and will presently be made use of to enable us to determine the position of the true meridian with a greater degree of precision.

THE EQUINOCTIAL

We shall find it convenient to draw upon the hemisphere another line, which plays an important rôle in astronomy, the great circle 90° from the pole. Placing the steel point of the compasses at the zenith, open the legs until the pencil point just comes to the horizon plane where the spherical surface meets it, so that if it were revolved about the zenith, the pencil point would move in the horizon. The compass points now span an arc of 90° upon the hemisphere. Place the steel point at the pole, and draw as much of a great circle as can be described on the sphere above the horizon. This will be just one-half of the great circle, and will cut the horizon in the east and west points. The new circle is called the equinoctial or celestial equator (Fig. 17).

We have seen that the path of the sun over the dome of the heavens appears to be a small circle described from east to west about a fixed point in the dome as a pole. The ancient explanation of this fact was that the sun is fixed in a transparent spherical shell of immense size revolving daily about an axis, the earth being a plane in the center of unknown extent, but whose known regions are so small compared to the shell that from points even widely separated on the earth the appearance is the same; just as the

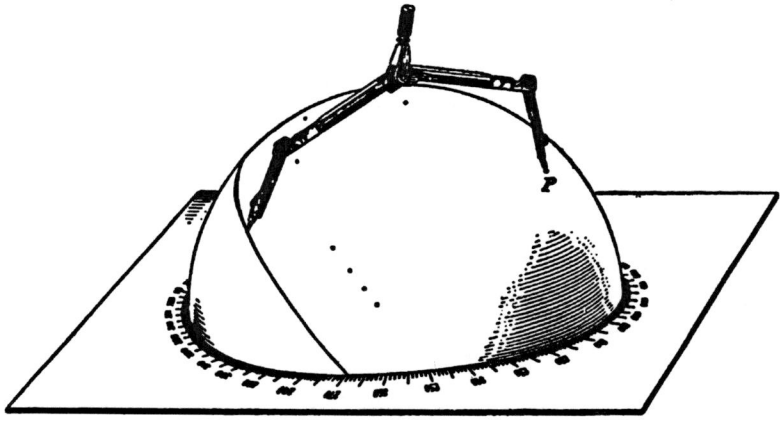

Fig. 17

apparent direction and motion of the sun would be practically the same on our hemisphere to a microscopic observer at the center, and to another anywhere within one-hundredth of an inch of the center. When observations were made, however, at points some hundreds of miles apart on the same meridian, very perceptible differences were found, whose nature will be understood from a comparison of the hemisphere (Fig. 18 a), plotted from

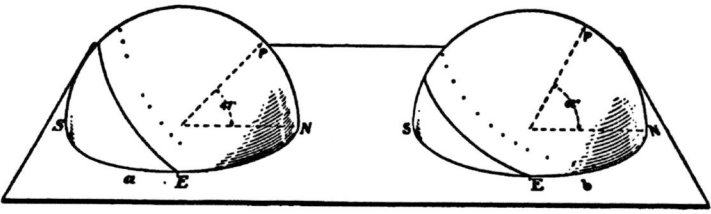

Fig. 18

observations made Aug. 8, 1897, at a point in Canada, not far from Quebec, with a second hemisphere (Fig. 18 b), on which is shown the path of the sun on the same date derived from observation of the shadow of a pin-head at Polfos in Norway. It appears on comparison that the distance of the pole above the north horizon is considerably greater in the latter, while the equator is just as much nearer the southern horizon; the sun is at the same distance from the equator in each case. This fact cannot be explained on the supposition that the horizon planes of the two places are the same, for in that case we should have the spherical shell which contains the sun revolving at the same time about two different fixed axes, which is impossible. It is not, however, improbable that the earth's surface should be curved, if we can admit as a possibility that the direction of gravity, which is perpendicular to a horizontal plane, may be different at different places. That the earth's surface in the east and west direction is curved, we know; for men have traversed it from east to west and returned to the starting point, so that we have good reason to believe that its surface is everywhere curved. Long before this conclusive proof was obtained, however, the globular form of the earth was inferred on good grounds.

It was early suggested (regarding the fact that, if the sun is fixed in a shell, that shell is of enormous size as compared with the earth) that it is inherently more probable that the apparent motion of the sun is due to a rotation of the spherical earth about an axis passing through the earth's center and the poles of the sun's circle. This argument is greatly strengthened when we investigate the apparent motion of the stars in connection with their size and distance, and it is now beyond a doubt that this is the true explanation of the apparent diurnal motion of the sun.

LATITUDE EQUALS ELEVATION OF THE POLE

This subject is treated in all text-books on descriptive astronomy, and it is pointed out that the pole of the sun's path is the point where the line of the earth's axis of rotation cuts the sky, and the equinoctial or celestial equator is the great circle in which the plane of the earth's equator cuts the sky. The fact is proved also that the elevation of the pole above the horizon at any place is equal to the latitude of the place.

This angle, as measured on the hemisphere shown in Fig. 18 *a*, is 47°, and on the hemisphere of Fig. 18 *b* is 62°. The latitudes of

Quebec and Polfos as determined by more accurate measures are 46° 50' and 61° 57'.

It is easy to see that the arc of the meridian from the zenith to the equinoctial is also equal to the latitude, while the arc from the south point of the horizon to the equator and that from the zenith to the pole are each equal to 90° minus the latitude, or, as it is usually called, the co-latitude.

It will be well here, as in all our measurements, to form some idea of the accuracy of our results. As one degree on our hemisphere is quite exactly equal to 1^{mm}, a quantity easily measured by ordinary means, it is not difficult with ordinary care to determine the

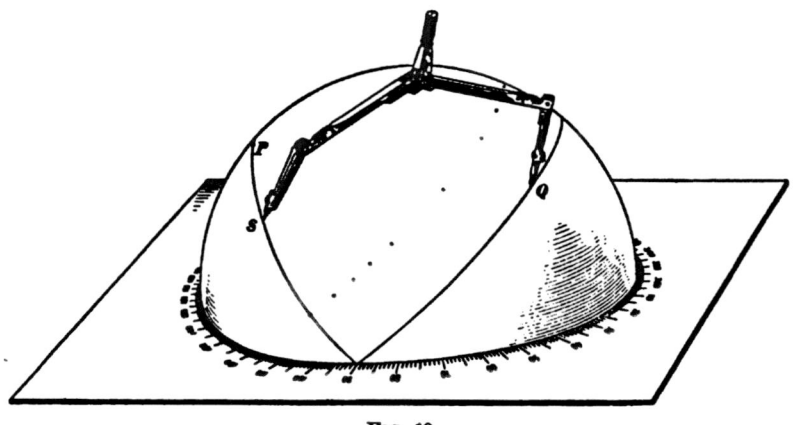

FIG. 19

pole of the sun's path so closely that no observed point lies more than a degree from the path. The pole is then fixed within one degree unless the length of the path is very short; usually if the path is more than 90° in length the pole may be placed within less than a degree of its true place and the latitude measured with an error of less than one degree.

HOUR-ANGLE OF THE SUN

Open the dividers as before (see p. 14) so as to draw a great circle. Place the steel point upon the place of the sun, S, on its diurnal circle at the time of the last observation in the afternoon (Fig. 19), and with the pencil point strike a small arc cutting the equator at Q.

Place the steel point where this arc cuts the equator, and draw a great circle which will pass through the sun's place and the pole; notice that it also cuts the equator at right angles. Such a circle is called an hour-circle. It is the intersection of the surface of the sphere with a plane that passes through the poles and the place of the sun. The number of degrees in the arc of the equator, included between the meridian and the hour-circle which passes through the sun, is called the hour-angle of the sun. By the ordinary convention of solid geometry it measures the wedge angle between the plane of the hour-circle and the plane of the meridian. If a book be placed with its back in the line from the pole to the center of the sphere, and with its title-page to the west, and the western cover opened till it is in the plane of the hour-circle, while the title-page is in the plane of the meridian, the wedge angle between the title-page and the cover will be the hour-angle and will be measured by the arc of the equator indicated above. It is reckoned as increasing from the meridian towards the west in the direction in which the cover is opened. If the hour-circle of the first morning observation is determined in the same way, the hour-angle measured in the opposite direction from the meridian is sometimes called the hour-angle east of the meridian; but more commonly by astronomers this value is subtracted from 360°, and the angle thus obtained is called the hour-angle, this being more convenient because the hour-angle of the sun thus measured constantly increases with the time as the sun pursues its course; being 0° at noon, 180° at midnight, 360° at the next noon, etc.

UNIFORM INCREASE OF HOUR-ANGLE

Let us now examine more carefully the truth of the surmise previously made, that the arc of the sun's path between two successive observations is proportional to the interval of time between the observations. Draw the hour-circles of the sun at each point of observation (Fig. 20); measure the arc on the equator between the first and the last hour-circles; divide by the number of minutes between the two times. This will give the average increase of hour-angle per minute. Multiply this increase by the difference in

THE DIURNAL MOTION OF THE SUN 19

minutes of each of the observed times from the time of the first observation, and compare with the progressive increase of the hour-angle as measured off on the equator by means of the graduated quadrant. They will be found to be nearly the same in each case. It is thus shown that the hour-angle of the sun increases uniformly with the time. The rate is nearly a quarter of a degree per minute, since 360° are described in 24 hours. Notice that when the hour-angle is zero, the actual time by the watch is not very far from 12 o'clock (in extreme cases it may be 45 minutes, if the clock is keeping standard time), and that if the hour-angle in degrees (west of the meridian) is divided by 15, the number of hours differs from the

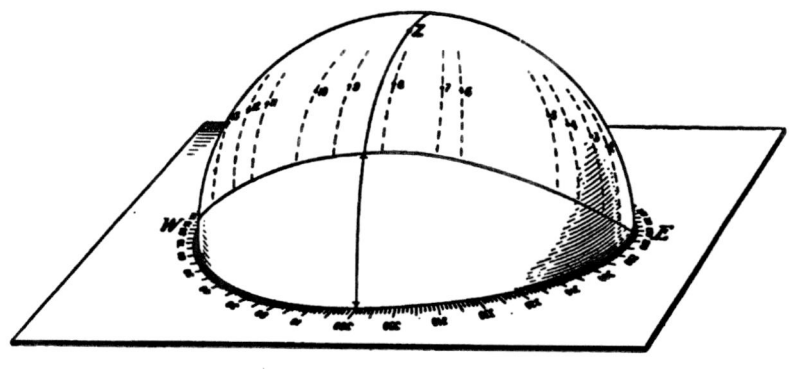

FIG. 20

watch time just as much as the time of meridian passage differs from 12 hours. In fact, the hour-angle of the sun measures what is called apparent solar time, $i.e.$, when H.A. = 15°, it is 1 o'clock; H.A. = 75°, it is 5 o'clock; H.A. = 150°, 10 o'clock, etc.; those angles east of the meridian lying between 180° and 360°, $i.e.$, between 12^h and 24^h, so that 12 hours must be subtracted to give the correct hours by the ordinary clock, which divides the day into two periods of 24 hours each; for instance, if H.A. = 270°, it is 18^h past noon or 6 A.M. of the next day. Astronomical clocks usually show the hours continuously from 0 to 24, thus avoiding the necessity of using A.M. and P.M. to discriminate the period from noon to midnight and from midnight to noon.

DECLINATION OF THE SUN

The distance of the sun's path from the celestial equator, measured along the arc of an hour-circle, is called its declination, and will be found appreciably the same at all points. It requires more delicate observation than ours to find that it changes during the few hours covered by our observation. If, however, the observation be repeated after an interval, say, of two weeks at any time except for a month before or after the 20th of June or December, it will be found that although the sun at the second observation describes a circle, this circle is not in the same position with regard to the equator — that its declination has changed (between March 13 and 27, for instance, by about $5°.5$). The inference to be drawn is that even during the period of our observation the sun's path is not exactly parallel to the equator, although our observations are not delicate enough to show that fact.

It is true in general, as in this case, that the first rude measurements applied to the heavenly bodies give results which when tested by those covering a longer time, or made with more delicate instruments, are found to require correction.

CHAPTER II

THE MOON'S PATH AMONG THE STARS

NEXT to the diurnal motion of the sun the most conspicuous phenomenon is the similar motion of the stars and the moon; this will form the subject of a future chapter.

The study of the moon, however, discloses a new and interesting motion of that body. It partakes indeed of the daily motion of the heavenly bodies from east to west, but it moves less rapidly, requiring nearly 25 hours to complete its circuit instead of 24, as do the sun and stars, and returning to the meridian therefore about an hour later on each successive night.

In consequence of this motion it continually changes its place with reference to the stars, moving toward the east among them so rapidly that the observation of a few hours is sufficient to show the fact. At the same time its declination changes like that of the sun, but much more rapidly.

We should begin early to study this motion, and it will be found interesting to continue it at least for some months at the same time that other observations are in progress — a very few minutes each evening will give in the course of time valuable results.

POSITION BY ALIGNMENT WITH STARS

The first method to be used consists in noting the moon's place with reference to neighboring stars at different times. Some sort of star map is necessary upon which the places of the moon may be laid down so that its path among the stars may be studied. As the configurations that offer themselves at different times are of great variety, it will be well to give a few examples of actual observations of the moon's place by this method.

Dec. 12, 1899, at 12^h 0^m P.M., the moon was seen to be near three unknown stars, making with them the following configuration, which was noted on a slip of paper as shown in Fig. 21. The relative size of the stars is indicated by the size of the dots. (The *original* papers on which the observations are made should be carefully preserved; indeed, this should always be the practice in all observations.)

Fig. 21

At the same time, for purposes of identification, it was noted that the group of stars formed, with Capella and the brightest star in Orion, both of which were known to the observer, a nearly equilateral triangle. It was also noted that the moon was about 6° from the farthest star, this being estimated by comparison with the known distance between the "pointers" in the "Dipper" (about 5°). With these data it was easily found by the map that these stars were the brightest stars in Aries, and the moon was plotted in its proper place on the map (page 24).

December 13, at 5^h 35^m P.M., the moon was $\frac{1}{4}$° (half its diameter) below (south of) a line drawn from Aldebaran (identified by its position with reference to Capella and Orion and by the letter V of stars in which it lies, the Hyades) to the faintest of the three reference stars of December 12. It was also about $\frac{3}{4}$° west of a line between two unknown stars identified later as Algol (equidistant from Capella and Aldebaran) and γ Ceti (at first supposed on reference to the map to be α Ceti, but afterward correctly identified by comparing the map with the heavens). The original observation is given below (Fig. 22) of about one-half the size of the drawing, all except the underscored names being in pencil. The underscored names are in ink and made after the stars were identified. This is a useful practice when additions are made to an original, so that subsequent work may not be given the appearance of notes made at the time of observation. It is well to give on the sketch map several stars in the neighborhood of those used for alignment, to facilitate identification.

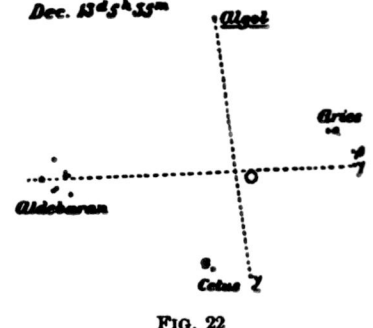

Fig. 22

The alignment was tested by holding a straight stick at arm's length parallel to the line joining the stars.

December 14, $6^h\ 30^m$ P.M. Moon on a line from Algol through the Pleiades (known) about $2\frac{1}{2}°$ (5 diameters of moon) beyond the latter, which were very faint in the strong moonlight. *No figure.*

December 15, $5^h\ 10^m$ P.M. Moon in a line between Capella and Aldebaran. Line from Pleiades to moon bisects line from Aldebaran to β Tauri (identified by relation to Aldebaran and Capella).

$9^h\ 25^m$ P.M. Moon in line from β Aurigæ to Aldebaran (Fig. 23).

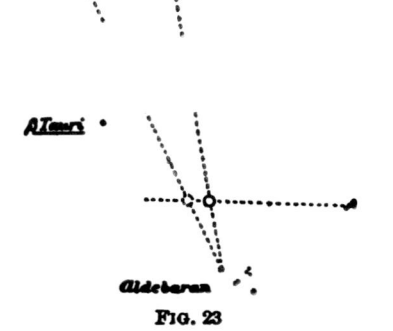

FIG. 23

(NOTE. — Henceforth details of identification are omitted.)

December 16, $7^h\ 40^m$ P.M. Moon almost totally eclipsed $2\frac{1}{4}°$ east of line from β Aurigæ to γ Orionis; same distance from β Tauri as ζ Tauri (revised estimate about $\frac{1}{2}°$ nearer β Tauri than is ζ Tauri) (Fig. 24).

December 18, $10^h\ 30^m$ P.M. Observation snatched between clouds. Moon's western edge tangent to line from α Geminorum to Procyon and about $1°$ north of center of that line.

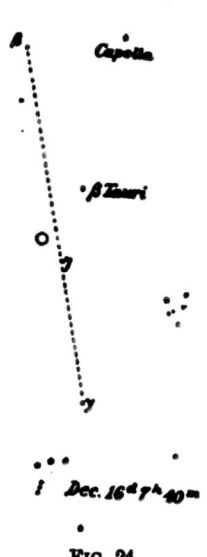

FIG. 24

In the sketch maps above no great accuracy is attempted in placing the stars, but in the final plotting on the map the directions of the notes are carefully followed. The plotting should be done as soon as possible after the observation is made, for even a hasty comparison with the map will often show that stars have been misidentified or that there is some obvious error in the notes, which may be rectified at once if there is an opportunity to repeat the observation. Such a case occurs in the observations of December 13 recorded above, where γ Ceti was mistaken for α.

PLOTTING POSITIONS OF THE MOON ON A STAR MAP

Figure 25 shows the positions of the moon plotted from the foregoing observations, together with the lines of construction from which they were determined.

A drawing should be made of the shape of the illuminated portion of the moon at each observation, and the direction among the stars

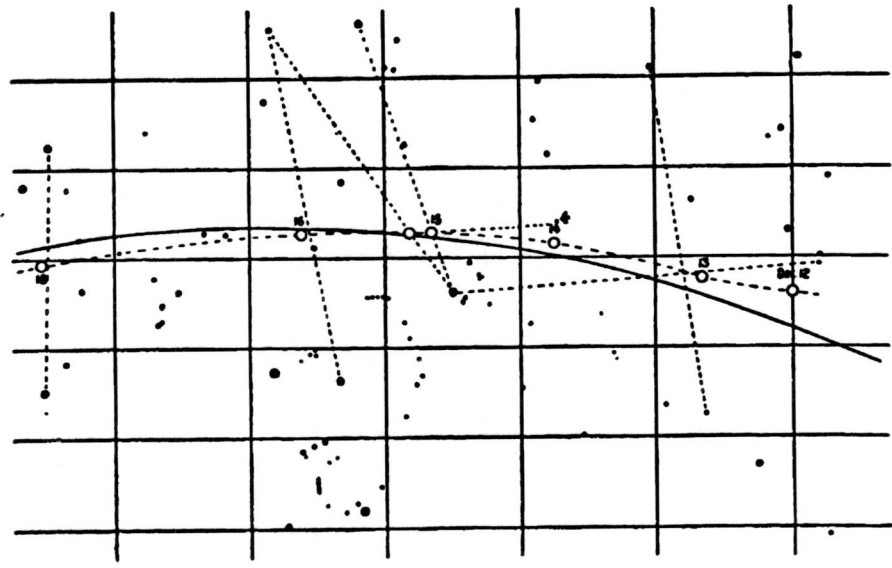

FIG. 25

of the line joining the points of the horns (cusps) for future study of the cause of the moon's changes of phase.

If the star map accompanying this book is used, the identification of the stars consists in determining which of the dots represents the star of reference; the name may be determined by reference to the list; thus the two stars near the line XXIV on the upper portion of the map are "α Andromedæ $0^h\ 5^m + 29°$" and "γ Pegasi $0^h\ 8^m + 14°$." The meaning which attaches to these numbers is given in Chapter IV. It is a good plan to keep a copy of the map on which to note the names for reference as the stars are learned; most of the conspicuous ones will soon be remembered as they are used.

THE MOON'S PLACE FIXED BY ITS DISTANCE FROM NEIGHBORING STARS

One month's observation by this method will show that the moon's path is at all points near to the curved line drawn on the map, which is called the ecliptic and which is explained on page 70. To establish more accurately its relations to this line it will be advisable in the later months to adopt a more accurate means of observation, although when the moon is very near a bright star, its position may be quite accurately fixed by the means that we have indicated; and if it chances to pass in front of a bright star and produce an occultation, the moon's position is very accurately fixed indeed, as accurately as by any method. But such opportunities are rare, and for continuous accurate observation we should have a means of measuring the distance of the moon from stars that are at a considerable distance from it. An instrument sufficiently accurate for our purpose is the cross-staff described below. It should be mentioned that, on account of the distortion of the map, the place of the moon is usually more accurately given by distances from the comparison stars than by alignment. The sextant may be used instead of the cross-staff, but is less convenient and also more accurate than is necessary.

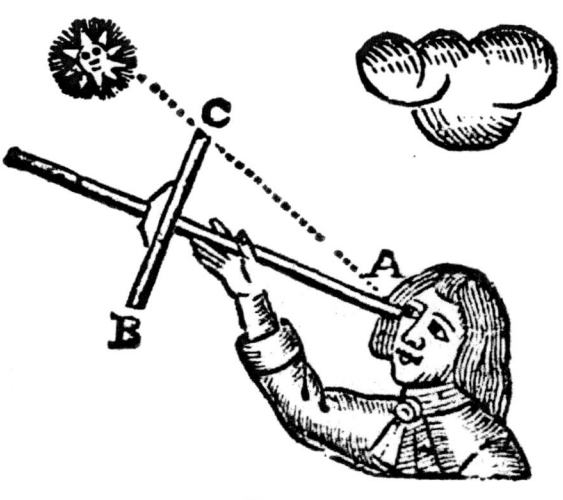

FIG. 26

The Cross-staff. — The cross-staff (Fig. 26) consists of a straight graduated rod upon which slides a "transversal" or "cross" perpendicular to the rod; one end of the staff is placed at the eye and the "cross" is moved to such a place that it just fills the angle from one object to another; its length is then the chord of an arc equal to the angle between the objects as seen from that end of the staff at which the eye is placed. The figure, which is taken from an old book on navigation, illustrates the use of this instrument for measuring the sun's altitude above the sea horizon; the rod in the position shown indicates that the sun's altitude is about 40°.

Obviously a given position of the cross corresponds to a definite angle at the end of the rod, and the rod may be graduated to give this angle directly by inspection, or a table may be constructed by which the angle corresponding to any division of the rod may be found; such a table is given on page 27. For our purpose an instrument of convenient dimensions is made by using a cross 20 cm. in length, sliding on a rod divided into millimeters (Fig. 27); this may be used for measuring angles up to 30°, which is enough for our

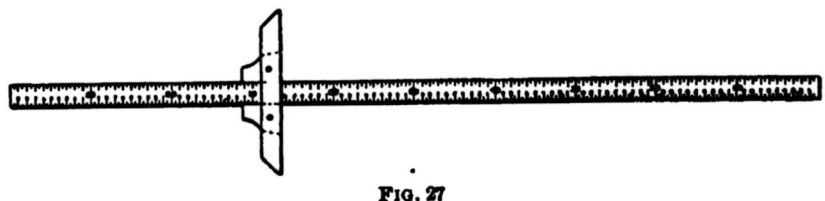

Fig. 27

purpose. The smallest angle that can be measured is about 12°, which corresponds to a chord of ⅓ of the radius; but by making a part of the cross only 10 cm. long, as shown in the figure, we may measure angles from 6° upwards, and for smaller angles may use the thickness of the cross, which is 5 cm., and thus measure angles as small as 3°; the longer cross will not give good results above 30°, as a slight variation of the eye from the exact end of the rod makes a perceptible difference in the value of the angles greater than 30°.

Measures with the Cross-staff. — As an example of the use of the cross-staff, the following observations are given: They were made with a staff about 3 feet in length, graduated by marking the point for each degree at the proper distance in millimeters from the eye end of the staff, as given by Table II on page 27. After the points were marked a straight line was drawn through each entirely across the rod, using the cross itself as a ruler; graduations were thus made on one side for use with the 20 cm. cross, on the other for the

Table I — Angle subtended by Crosses

Distance from Eye	Length of Cross			Distance from Eye	Length of Cross		
	20 cm.	10 cm.	5 cm.		20 cm.	10 cm.	5 cm.
100 cm	11°.4	5°.7	2°.9	62 cm	18°.3	9°.2	4°.6
99	11.5	5.8	2.9	61	18.6	9.4	4.7
98	11.6	5.8	2.9	60	18.9	9.5	4.8
97	11.8	5.9	3.0	59	19.2	9.7	4.9
96	11.9	6.0	3.0	58	19.6	9.9	4.9
95	12.0	6.0	3.0	57	19.9	10.0	5.0
94	12.1	6.1	3.0	56	20.2	10.2	5.0
93	12.3	6.2	3.1	55	20.6	10.4	5.2
92	12.4	6.2	3.1	54	21.0	10.6	5.3
91	12.5	6.3	3.1	53	21.4	10.8	5.4
90	12.7	6.4	3.2	52	21.8	11.0	5.5
89	12.8	6.4	3.2	51	22.2	11.2	5.6
88	13.0	6.5	3.3	50	22.6	11.4	5.7
87	13.1	6.6	3.3	49	23.1	11.6	5.8
86	13.3	6.7	3.3	48	23.5	11.9	6.0
85	13.4	6.7	3.4	47	24.0	12.1	6.1
84	13.6	6.8	3.4	46	24.5	12.4	6.2
83	13.7	6.9	3.5	45	25.1	12.7	6.4
82	13.9	7.0	3.5	44	25.6	13.0	6.5
81	14.1	7.1	3.5	43	26.2	13.3	6.7
80	14.3	7.2	3.6	42	26.8	13.6	6.8
79	14.4	7.2	3.6	41	27.4	13.9	7.0
78	14.6	7.3	3.7	40	28.1	14.3	7.2
77	14.8	7.4	3.7	39	28.8	14.6	7.3
76	15.0	7.5	3.8	38	29.5	15.0	7.5
75	15.2	7.6	3.8	37	30.2	15.4	7.7
74	15.4	7.7	3.9	36	31.0	15.8	7.9
73	15.6	7.8	3.9	35	31.9	16.3	8.2
72	15.8	7.9	4.0	34	32.8	16.7	8.4
71	16.0	8.1	4.0	33	33.7	17.2	8.7
70	16.3	8.2	4.1	32	34.7	17.7	8.9
69	16.5	8.3	4.2	31	35.8	18.3	9.2
68	16.7	8.4	4.2	30	36.9	18.9	9.5
67	17.0	8.5	4.3	29	38.1	19.6	9.9
66	17.2	8.7	4.3	28	39.3	20.2	10.2
65	17.5	8.8	4.4	27	40.6	21.0	10.6
64	17.7	8.9	4.5	26	42.1	21.8	11.0
63	18.0	9.1	4.5	25	43.6	22.6	11.4

Table II

Angle subtended by 20 cm. Cross	
12°	951 mm
13	878
14	814
15	760
16	711
17	669
18	631
19	598
20	567
21	540
22	514
23	491
24	470
25	451
26	433
27	416
28	401
29	387
30	373
31	361
32	349
33	338
34	327
35	317
36	308
37	299
38	290
39	282
40	275

10 cm. cross, and on one edge for the thickness of the cross. By means of these graduations the angle subtended by the cross in any position is read directly from the scale, quarters or thirds of a degree being estimated and recorded in minutes of arc.

The observations are:

1900. January 2. 5ʰ 15ᵐ.
    Moon to ε Pegasi,   35° 45′
    " " Altair,   26 30
    " " Fomalhaut,   41 40

January 3. 6ʰ 0ᵐ.
    Moon to ε Pegasi,   23° 30′
    " " Altair,   29 20
    " " β Aquarii,   8 20

January 4. 5ʰ 20ᵐ.
    Moon to ε Pegasi,   17° 40′
    " " β Aquarii,   8 30
    " " δ Capricorni,   9 45

January 6. 5ʰ 50ᵐ.
    Moon to γ Pegasi,   12° 0′
    " " α Pegasi,   16 40
    " " ε Pegasi,   33 30

January 7. 5ʰ 45ᵐ.
    Moon to γ Pegasi,   9° 40′
    " " β Arietis,   19 45
    " " α Andromedæ, 21 15
    " " β Ceti,   27 30

January 8. 6ʰ 0ᵐ.
    Moon to α Arietis,   11° 0′
    " " γ Pegasi,   21 30

January 9. 10ʰ 0ᵐ.
    Moon to α Arietis,   9° 45′
    " " Alcyone,   16 0
    " " α Ceti,   15 30

To represent these observations on the star map, open the compasses until the distance of the pencil point from the steel point is equal to the measured distance — making use for this purpose of the scale of degrees in the margin, and then with the steel point

THE MOON'S PATH AMONG THE STARS

carefully centered on the comparison star, strike a short arc with the pencil point near the estimated position of the moon; the intersection of any two of these arcs fixes the position of the moon. If the different stars give different points, those nearest the moon may

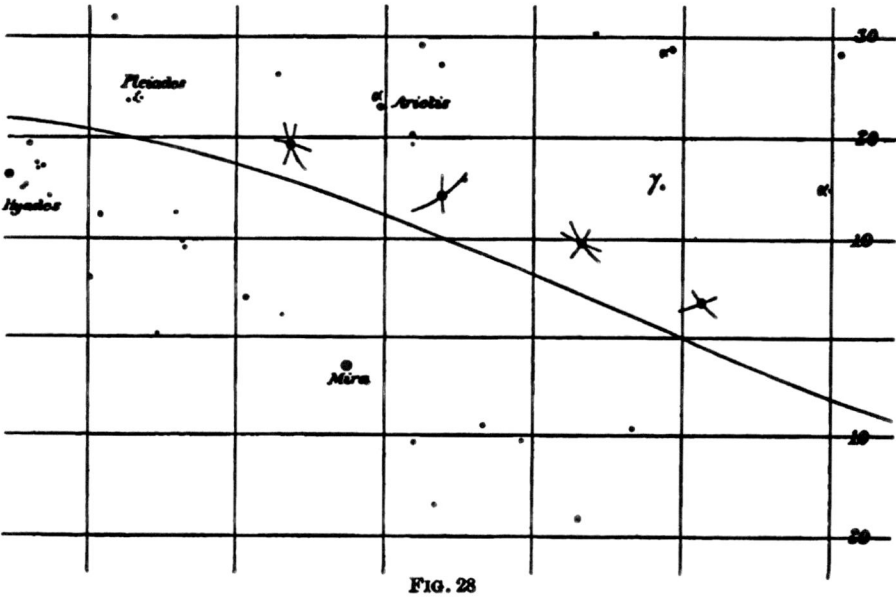

FIG. 28

be assumed to give results nearer the truth. Fig. 28 shows the positions of the moon January 6 to January 9 as plotted from the above measures.

Length of the Month. — If it happens that one of the positions observed in the second month falls between the places obtained on two successive days of the first month, or *vice versa*, a determination of the moon's sidereal period may be made by interpolation. Thus, on plotting the observation of December 12 (p. 22), which places the moon between the two observations on January 8^d 6^h 0^m and January 9^d 10^h 0^m, its distance from the former is $6°.0$ and from the latter $10°.0$, while the interval is 28^h; the moon's place on December 12 at 12^h 0^m is therefore the same as on January 8 at $6^h + \frac{6}{16} \times 28^h$, or January 8^d $16^h.5$, that is, January 9 at 4^h 30^m A.M., and the interval between these two times is 27^d 4^h 30^m, which is the time required for the moon to make a complete circuit among the stars or the length of the sidereal month. This is a fairly close approximation; the observation of December 12 having been made under favorable circumstances, the configuration being well defined and the stars near, so that the position on that date by alignment is nearly as accurate as those determined by the measures on January 8 and 9.

After three months the moon comes nearly to the same position at about the same time in the evening, so that it is convenient to determine its period without interpolation by observing the time when the moon comes into the same star line as at the previous observation; moreover, the interval being three months, an error of an hour in the observed interval causes an error of only 20^m in the length of the month.

THE MOON'S NODE

When a sufficiently large number of observations have been plotted to give a general idea of the moon's path among the stars, a smooth curve is to be drawn as nearly as possible through all the points and this curve should be compared with the ecliptic, as shown on the map. Its greatest distance from the ecliptic and the place where it crosses the ecliptic — the position of the node — should be estimated with all possible precision. For this purpose, only the more accurate positions obtained by the cross-staff should be used.

After a few observations of alignment are made, the student will desire to use the more accurate method at once, but it is better to have at least one month's observation by the first method (even if the cross-staff is also used) for comparison with later observations by alignment for the purpose of determining the length of the month, as suggested above, without any instrumental aid whatever.

The records of the positions of the node should be preserved by the teacher for comparison from year to year to show the motion of this point along the ecliptic. The node, as determined by the observations above given, was nearly at the point where the ecliptic crosses the line from γ Orionis to Capella. Observations made in November, 1897, by the method of Chapter IV, gave its place on the ecliptic at a point where the latter intersects a line drawn through Castor and Pollux, thus indicating a motion of about 40° in the interval.

Observations made with the cross-staff are sufficiently accurate to show that the motion of the moon is not uniform, but as the distortion of the map complicates the treatment of this subject, we shall defer its consideration until the method of Chapter V has been introduced.

It will be well, however, as soon as measures with the cross-staff are begun, to devote a few minutes each evening to measures of the moon's diameter with an instrument measuring to $10''$, such as a good sextant; or, better, a telescope provided with a micrometer, in order to show the variations of the moon's apparent size at different parts of its orbit. The relative distances of the moon from the earth as inferred from these measures should be compared with the variations of her angular motion as read off from the chart; although on account of the distortion referred to above, it will not be possible to show more than the fact that when the moon is nearest, her angular motion about the earth is greatest, and *vice versa.*

The sextant or micrometer may henceforward be used also for observations of the sun's diameter, which should be measured as often as once a week for a considerable period.

When the moon's diameter is measured, a rough estimate of her altitude should be made in order to make the correction for augmentation in a future more accurate discussion of the measures for determining the eccentricity of her orbit.

DETERMINING THE ERRORS OF THE CROSS-STAFF

Observations with the cross-staff are most easily made just before the end of twilight or in full moonlight, so that the cross may be seen dark against a dimly lighted background. When used for measuring the distance of stars in full darkness, it is convenient to have a light so placed behind the observer that, while invisible to him, it shall dimly illuminate the arms of the cross.

As the angles which are determined by the cross-staff, especially if large, are affected by the observer's habit of placing the eye too near to or too far from the end of the staff, it is a good plan to measure certain known distances and thus determine a set of corrections to be applied, if necessary, to all measures made with that instrument.

The following table gives the distances between certain stars always conveniently placed for observation in the United States, together with the results of measures made upon them with a cross-staff held in the hands without support, and indicates fairly the accuracy which may be obtained with this instrument. The back of the observer was toward the window of a well-lighted room, and the cross was plainly visible by this illumination.

Stars	True Distance	Measured Distances			Mean	Correction
α Ursæ Majoris to β Ursæ Majoris	5°.4	5°.8	—	—	5°.8	− 0°.4
α " " " γ " "	10 .0	10 .5	10°.6	10°.6	10 .6	− 0 .6
α " " " ε " "	15 .2	15 .6	15 .5	15 .7	15 .6	− 0 .4
β " " " ζ " "	19 .9	20 .0	20 .3	20 .2	20 .2	− 0 .3
α " " " η " "	25 .7	26 .6	26 .0	26 .1	26 .2	− 0 .5
α " " " Polaris	28 .5	29 .0	29 .2	28 .9	29 .0	− 0 .5
β " " " "	33 .9	35 .1	34 .7	34 .4	34 .7	− 0 .8
η " " " "	41 .2	42 .2	42 .0	42 .0	42 .1	− 0 .9

The measured distances are about one-half degree too large, and if a correction of this amount is applied to all angles measured by this instrument up to 30°, the corrected values will seldom be so much as half a degree in error, and the mean of three readings will probably be correct within a quarter of a degree.

CHAPTER III

THE DIURNAL MOTION OF THE STARS

As the observations of the moon require but a few minutes each evening, observations may be made on the same nights upon the stars. The first object is to obtain the diurnal paths of some of the brighter stars, and as they cast no shadow we must have recourse to a new method of observation to determine their positions in the sky at hourly intervals.

A simple apparatus for this purpose is represented in Fig. 29. A paper circle is fastened to the leveling board used in the sun

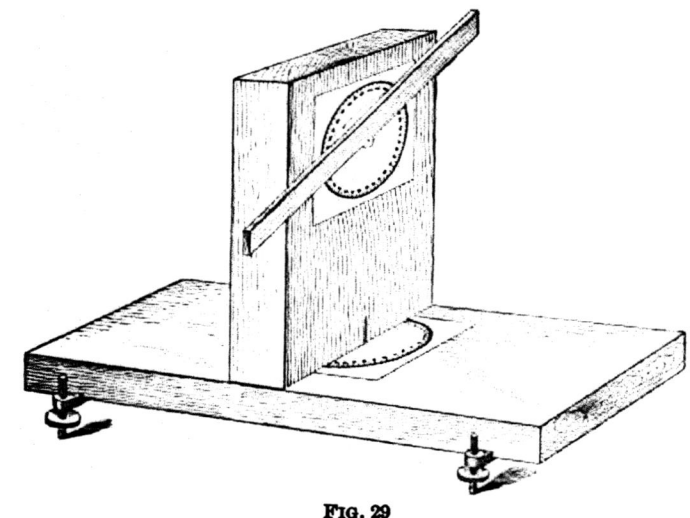

FIG. 29

observations so that the zero of its graduation lies as nearly as possible in the meridian, and a pin with its head removed is placed upright through the center of the circle.

A carefully squared rectangular block about 10 inches by 8 inches by 2 inches is placed against the pin so that the angle which its face makes with the meridian may be read off upon the horizontal circle. A second paper circle is attached to the face of the block with the zero of its graduations parallel to the lower edge; a light ruler is fastened to the block by a pin through the center of its circle; the ruler may be pointed at any star by moving the block about a vertical axis till its plane passes through the star, and then moving the ruler in the vertical plane till it points at the star; a lantern is necessary for reading the circles and for illumination of the block and ruler in full darkness; it should be so shaded that its direct light may not fall on the observer's eye. Sights attached to the ruler make the observation slightly more accurate, but also rather more difficult, and without them the ruler may be pointed within half a degree, which is about as closely as the angles can be determined by the circles.

THE ALTAZIMUTH

An inexpensive form of instrument for measuring altitude and azimuth is shown in Fig. 30. Here the ruler provided with sights A, B is movable about d, the center of the semicircle E. This semicircle is movable about an axis perpendicular to the horizontal circle F, and its position on that circle is read off by the pointer g, which reads zero when the plane of E is in the meridian. The circle F is mounted on a tripod provided with leveling screws. If the circle is so placed that the pointer reads zero when the sight-bar is in the magnetic meridian, then its reading when the sights are pointed at any star will give the magnetic bearing of the star. It will, however, be more convenient to adjust the instrument so that the pointer reads zero when the sight-bar is in the true meridian.

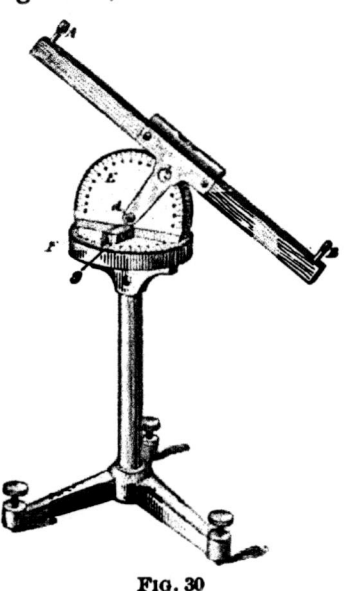

FIG. 30

To insure the verticality of the standard a level is attached to the sight-bar, and by the leveling screws the instrument must be adjusted so that the circle E may be revolved without causing the level bubble to move. (See page 36.)

A more convenient and not very expensive instrument is the altazimuth or universal instrument shown in Fig. 31, which contains some additional parts by the use of which it may be converted into an equatorial instrument. (See page 45.) It consists of a horizontal plate carrying a pointer and revolving on an upright axis which passes through the center of a horizontal circle graduated continuously from 0° to 360°. The plate carries a frame supporting the axis of a graduated circle; this axis is perpendicular to the upright axis, and the circle is graduated from 0° to 90° in opposite directions. Attached to the circle is a telescope whose optical axis is in the plane of the circle. The circle is read by a pointer which is fixed to the frame carrying its axis and reads 0° when the optical axis of the telescope is perpendicular to the upright axis. A level is attached to the telescope so that the bubble is in the center of its tube when the telescope is horizontal. In what follows, all these adjustments are supposed to be properly made by the maker.

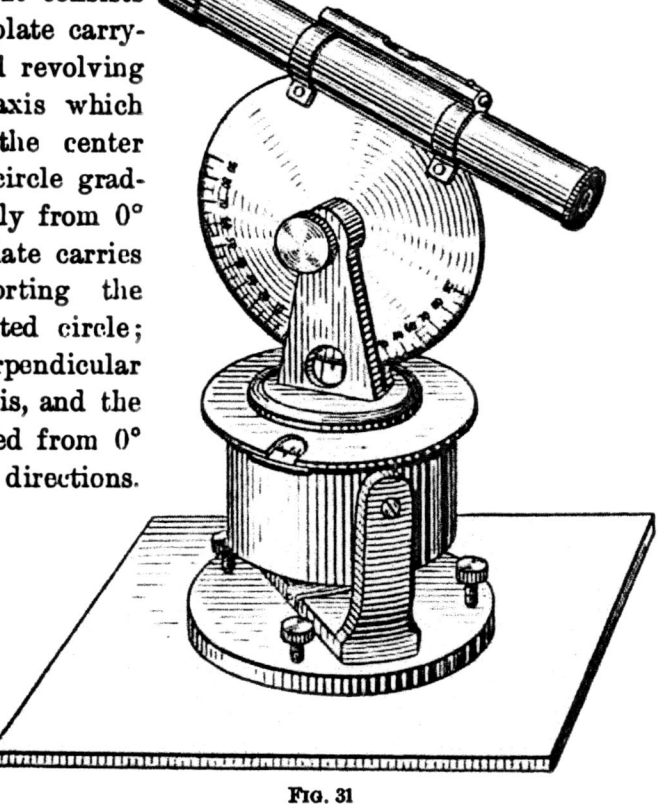

FIG. 31

ADJUSTMENT OF THE ALTAZIMUTH

If the altazimuth is so adjusted that the upright axis is exactly vertical, and if we know the reading of the horizontal circle when the vertical circle lies in the meridian, we may determine the position of a heavenly body at any time by pointing the telescope upon it and reading the two circles. The difference between the reading of the horizontal circle and its meridian reading is the azimuth, and the reading of the vertical circle is the altitude of the body. Before proceeding to the observation of stars, it will be well to repeat our observations on the sun, using this instrument, and making them in such a manner that we may at the same time get a very exact determination of the meridian reading by the method suggested on page 14.

Place the instrument upon the table used for the sun observation; bring the reading of each circle to 0°; and turn the whole instrument in a horizontal plane until the telescope points approximately south, using the meridian determination obtained from the shadow observations. One leveling screw will then be nearly in the meridian of the center of the instrument, while the two others will lie in an east and west line. Bring the level bubble to the middle of its tube by turning the north leveling screw; then set the telescope pointing east; and "set" the level by turning the east and west screws in opposite directions. Be careful to turn them equally; this can be done by taking one leveling screw between the

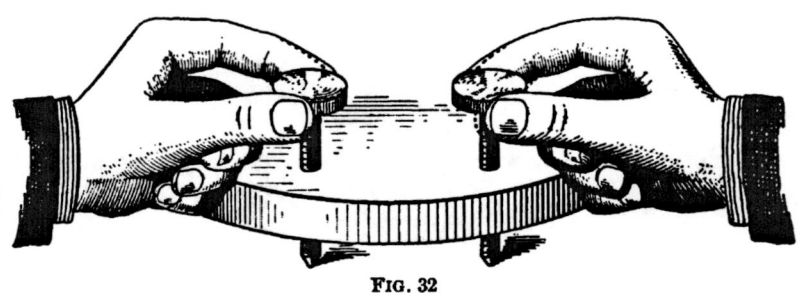

FIG. 32

finger and thumb of each hand, holding them firmly, and turning them in opposite directions by moving the elbows to or from the body by the same amount. Turn the telescope north, and the bubble should remain in place; if it does not, adjust the north screw. The instrument is very easily and quickly adjusted by this method. The upright axis is vertical when the telescope can be turned about it into any position without displacing the bubble.

Determination of the Meridian and Time of Apparent Noon. — After completing the adjustment of the instrument, the reading of the circle

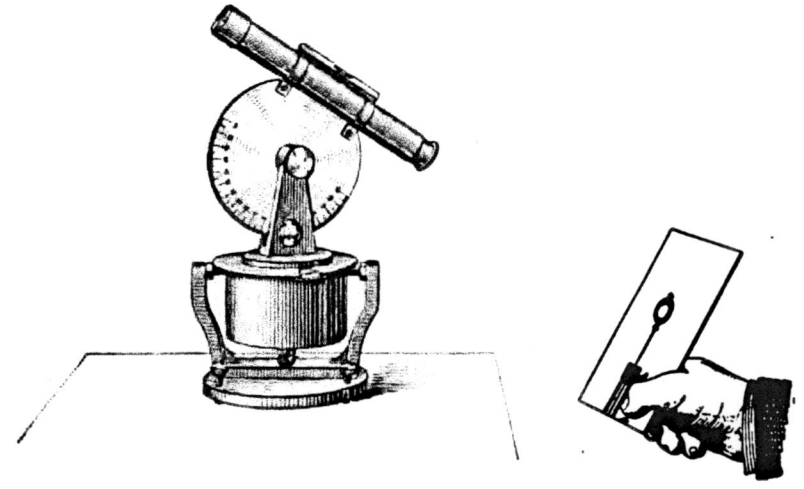

FIG. 33

when the telescope is in the meridian is determined as follows: Point the telescope upon the sun approximately. Place a sheet of paper or a card behind it, and turn the telescope about the vertical axis until the shadow of the vertical circle is reduced to its smallest dimensions and appears as a broad straight line. By moving the telescope about the horizontal axis, bring the shadow of the tube to the form of a circle; in this circle will appear a blurred disk of light. Draw the card about 10 inches back from the eyepiece, and pull out the latter nearly $\frac{1}{8}$ of an inch from its position when focused on distant objects and the disk of light becomes nearly sharp; complete the focusing of this image of the sun by moving the card to or from the eyepiece. The distance of the card and the drawing out of the eyepiece should be such that the sun's image shall be about $\frac{1}{2}$ to $\frac{3}{4}$ of an inch in diameter. Now move the telescope until the image is centered in the shadow of the telescope tube, note the time, and read both circles; this observation fixes the altitude and azimuth of the sun. For determining the meridian it is not necessary that the time should be noted, but it will be convenient to use these observations for a repetition of the determination of the sun's path, determining the altitudes and azimuths by this more accurate method.

This observation should be made at least as early as 9 A.M. Now increase the reading of the vertical circle to the next exact number of degrees, and follow the sun by moving the telescope about the vertical axis. After a few minutes the sun will be again centered by this process. Note the time, and read the horizontal circle. Increase the reading of the vertical circle again by one degree to make another observation, and so on for half an hour. Observations may be made at one-half degree intervals of altitude, but those upon exact divisions will evidently be more accurate. If circumstances admit, observations may be made, during the period of two hours before and after noon, for the purpose of plotting the sun's path; but, owing to the slow change of altitude in that time, the corresponding azimuths are not well determined, and they will be nearly useless for placing the instrument in the meridian.

Some time in the afternoon, as the descending sun approaches the altitude last observed in the forenoon, set the vertical circle upon the reading corresponding to that observation, and repeat the series in inverse order; that is, decrease the readings of altitude by one degree each time, and note the time and the reading of the horizontal circle when the sun is in the axis of the telescope at each successive altitude.

Since equal altitudes correspond to equal azimuths (see page 14), east and west of the meridian, the difference of the horizontal readings is twice the azimuth at either of the two corresponding observations (360° must be added to the western reading, if, as will generally be the case, the 0° point lies between the two readings). Therefore, one-half this difference added to the lesser or subtracted from the greater reading gives the meridian reading. The same value is more easily found by taking half the sum of the two readings. In the same way one-half the interval of time between the two observations added to the time of the first reading gives the watch time of the sun's meridian passage, or apparent noon, as it is called.

THE DIURNAL MOTION OF THE STARS

Each pair of observations gives the value of the meridian reading and of the watch time of apparent noon; their accordance will give an idea of the accuracy of the observations.

The following observations of the sun were made March 8, 1900, with an instrument similar to that shown in Fig. 33.

	Time	Altitude	Horizontal Circle		Time	Altitude	Horizontal Circle
1	8h 54m 37s	27°.5	307°.6	9	2h 32m 10s	31°.0	47°.7
2	8 58 10	28 .0	308 .4	10	2 36 30	30 .5	48 .7
3	9 1 42	28 .5	309 .2	11	2 39 45	30 .0	49 .45
4	9 4 51	29 .0	310 .0	12	2 43 27	29 .5	50 .35
5	9 9 5	29 .5	310 .9	13	2 47 0	29 .0	51 .15
6	9 12 20	30 .0	311 .8	14	2 50 17	28 .5	51 .95
7	9 15 35	30 .5	312 .6	15	2 54 7	28 .0	52 .85
8	9 19 37	31 .0	313 .45	16	2 57 33	27 .5	53 .6

The 1st and 16th of these observations give for the meridian reading $\frac{1}{2}[307.6 + (53.60 + 360)] = 360°.60$, and for the corresponding watch time $\frac{1}{2}[8\ 54\ 37 + (2\ 57\ 33 + 12^h)] = 11^h\ 56^m\ 5^s$.

Taking the corresponding A.M. and P.M. observations in this manner, we find for the eight pairs of observations above the following values.

Altitude	Meridian Reading	Watch Time of Noon
27°.5	360°.6	11h 56m 5.0s
28 .0	360 .625	56 8.5
28 .5	360 .575	55 59.5
29 .0	360 .575	55 55.5
29 .5	360 .625	56 16.0
30 .0	360 .625	56 2.5
30 .5	360 .65	56 2.5
31 .0	360 .575	55 53.5
mean	360.61	11 56 2.9

The agreement of these results is closer than will usually be obtained, the observations being made by a skilled observer and the angles carefully read by means of a pocket lens, which in many cases enabled readings to be made to 0°.05; any reading such as that of the 8th observation, where the value was estimated to lie between two tenths, being recorded as lying halfway between them. This practice adds little to the accuracy if several observations are made, and is not to be recommended to beginners.

MERIDIAN MARK

It will be convenient to fix a meridian mark for future use. This may be done by fixing the telescope at the meridian reading, turning it down to the horizontal position, and placing some object (as a stake) at as great a distance as possible, so that it may mark the line of the axis of the telescope when in the meridian. A mark on a fence or building will serve if at a greater distance than 50 feet, though a still greater distance is desirable. For setting the telescope upon the mark, it is convenient to have two wires crossing in the center of the field of view, but the setting may be made within $0°.1$ without this aid. Having established such a mark, set the horizontal circle at $0°$, and move the whole base of the instrument until the telescope points upon the meridian mark. Level carefully; then set the telescope again, if the operation of leveling has caused it to move from the meridian mark; level again, and by repeating this process adjust the instrument so that it is level and that the telescope is in the meridian. Then press hard on the leveling screws, and make dents by which the instrument can be brought into the same position at any future time.

After the A.M. and P.M. observations recorded above, the telescope was pointed upon a meridian mark established by observations made with the shadow of a pin, and the reading of the horizontal circle was $359°.8$. The mark was then shifted about a foot toward the west, and the telescope again pointed upon it. As the reading of the circle was then $360°.6$, it may be assumed that the mark was now very nearly in the meridian.

If circumstances are such that no point of reference in the meridian is available, it will be necessary, after determining the meridian readings by the sun, to set the telescope upon some well-defined object in or near the horizontal plane and read the circle. The difference between this reading and the meridian reading will be the azimuth of the object. Set the pointer of the horizontal circle to this value, and set the telescope upon the reference mark by moving the whole base as before. If the pointer of the circle is now brought to $0°$, the telescope will evidently be in the meridian; and the position is to be fixed by making dents with the leveling screws as before.

CHOICE OF STARS

We are now ready to begin observations of the stars.

The most familiar group of stars in the heavens is, no doubt, that part of the Great Bear which is variously called the Dipper, Charles's Wain, or the Plough.

At the beginning of October, at 8 o'clock in the evening, an observer anywhere in the United States will see the Dipper at an altitude between 10° and 30° above the N.W. horizon. Set the telescope upon that star which is nearest the north point of the horizon; read both circles to determine its altitude and azimuth, and note the time. Even if the telescope is provided with crosshairs, the illumination of the light of the sky will not be sufficient to render them visible; but sufficient accuracy in pointing is obtained by placing the star at the estimated center of the field. Observe in succession the altitude and azimuth of the other six stars forming the Dipper, noting the time in each case.

Using the Dipper as a starting point, we will now identify and observe a few other stars.* The total length of the Dipper is about 25°. Following approximately a line drawn joining the last two stars of the handle of the Dipper, at a distance of about 30°, we come to a bright star of a strong red color, much the brightest in that portion of the heavens; this is Arcturus. Observe its altitude and azimuth, and note the time as before. Almost directly overhead, too high to be conveniently observed at this time, is a brilliant white star, Vega (α Lyræ). A little east of south from Vega, at an altitude of about 60°, is a group of three stars forming a line about 5° in length. The central and brightest star of the three is Altair (α Aquilæ), and its position should be observed.

Diurnal Paths of the Stars. — Proceed in this way for about an hour, observing also, if time permits, the group of five stars whose middle is at azimuth 220° and altitude 35°. This is the constellation of Cassiopeia. Another interesting asterism will be found — supposing that by this time it is 9 o'clock — at azimuth 270° and altitude 45°, consisting of four stars of about equal magnitude,

---

* Many of the latest text-books on astronomy contain small star maps which are valuable aids in the identification of the less conspicuous groups.

placed at the corners of a quadrilateral whose sides are about 15° in length, and forming what is called the Square of Pegasus.

It is convenient as an aid in identification to note in each case the magnitude of the star observed. As a rough standard of comparison, it may be remembered that the six bright stars of the Dipper are of about the second magnitude; that at the junction of the handle and bowl is of the fourth. The three stars in Aquila are of the first, third, and fourth magnitudes. Vega and Arcturus are each larger than an average first magnitude star. The brightest stars in the constellation Cassiopeia and in the Square of Pegasus range from the second to the third magnitude.

The little quadrilateral of fourth magnitude stars about 15° east of Altair and known as Delphinus, or vulgarly as Job's Coffin, may be observed.

At the expiration of an hour, set again upon the Dipper stars and repeat the series, going through the same list in the same order. Arcturus will have sunk so low in a couple of hours as to be beyond the reach of observation, even if the place of observation affords a clear view of the horizon. Vega, however, will be less difficult to observe, and may be now added to the list. We should not omit to make an observation of the pole star, which, as its name indicates, may be found near the pole and can be easily found, since the azimuth of the pole is 180°, and its altitude is equal to the latitude of the place.

From the observed values of altitude and azimuth plot the successive places on the hemisphere exactly as in the case of the sun, and thus represent upon the hemisphere the paths of a number of stars in various parts of the heavens. It will be found that these paths are all circles of various dimensions, and that the circles are all parallel to the equator, as determined from the sun observations, that is, they have the same pole as the diurnal circles of the sun.

At this stage it is a good plan to devote some attention to the representation of the various results as shown on the hemisphere, by means of figures on a plane surface, that is, to make careful freehand drawings of the hemisphere and the circles which have now been drawn upon it as seen from various points of view. This is an important aid to the understanding of the diagrams by which it is necessary to explain the statement and solution of astronomical problems; with this purpose in view the drawings should be lettered and the definitions of the various points and lines written under them.

ROTATION OF THE SPHERE AS A WHOLE

So far the result of our observations is to show that the heavenly bodies appear to move as they would if they were all attached in some way to the same spherical shell surrounding the earth, and were carried about by a common revolution, as if the shell rotated on a fixed axis, passing through the point of observation. The sun may be conceived as carried by the same shell, but observations at different dates show that its place on the shell must slowly change, since its declination changes slightly from day to day.

If these observations on the stars are repeated ten days or one hundred days later, we shall find that the declinations determined from them are the same; that is, the declinations of the diurnal paths of the stars do not change like that of the sun. It will appear also that, as in the case of the sun, equal arcs of the diurnal circle and consequently equal hour-angles are described in equal times. It follows from this, of course, that stars nearer the pole will appear to move more slowly, since they describe paths which are shorter when measured in degrees of a great circle, as may be shown by measuring the diurnal circles on the hemisphere by a flexible millimeter scale, 1 mm. being equal to $1°$ of a great circle on our hemisphere.

If the field of view of our telescope is $5°$, a star on the equinoctial will be carried across its center by the diurnal motion in 20 minutes, while a star at a declination of $60°$ will remain in the field for twice that time, since its diurnal circle is only half as large as the equinoctial and an angular motion of $10°$ of its diurnal circle is only $5°$ of great circle. Since the declinations of the stars do not change, it is unnecessary to make our observations of the stars on the same night; or, rather, observations made on different nights may be plotted as if made on the same night. We may thus obtain extensions of the diurnal circles by working early on one evening and at later hours of the night on following occasions.

POSITIONS FIXED BY HOUR-ANGLE AND DECLINATION; THE EQUATORIAL

It is evident that we have, in the hour-angles and declinations of the stars, another system of coördinates on the celestial sphere by means of which their position may be fixed. The altitude and azimuth refer the position of the star to the meridian and to the horizon; while the hour-angle and declination refer its position to the meridian and the equator. We have hitherto found it more convenient to deal with the first set of coördinates, but it is often desirable to determine the hour-angle and declination of a body by direct observation, and this may be done by means of an instrument similar to the altazimuth but with the upright axis pointed to the pole of the heavens, so that the horizontal circle lies in the plane of the equator. With this instrument the angles read off on the circle which is directly attached to the telescope measure distances along the hour-circle, perpendicular to the equator, *i.e.*, declinations, while an angle read off on the other circle measures the angle between the meridian and the hour-circle of the star at which the telescope points, and is therefore the star's hour-angle. The two circles are therefore appropriately called the declination circle and the hour-circle of the instrument. As these terms are used with another meaning as applied to circles on the celestial sphere, it would seem that there might be confusion from their use in this sense, but in practice it is never doubtful whether "circle" means the graduated circle of an instrument or a geometrical circle on the surface of the sphere.

It is here supposed that the instrument has been so adjusted that both circles read 0° when the telescope is in the plane of the meridian and points at the equator. An instrument so mounted is called an equatorial instrument. Our altazimuth is adapted to this purpose by constructing the base so that it may be revolved about a horizontal axis perpendicular to the plane in which the altitude circle lies when the azimuth circle reads 0°. If, then, it has been placed in the meridian by the observation of equal altitudes as before described, it may be inclined about this latter axis through an angle equal to the complement of the latitude, and thus brought into the proper position for observing declination and hour-angle directly. An instrument so constructed is called a "universal" equatorial. To adjust the universal equatorial so that the axis points to the pole, adjust it as an altazimuth with both circles reading 0° and level it with the telescope in the meridian pointing south. Depress the telescope till the reading of the vertical circle equals the co-latitude. Tip the whole instrument so as to incline the vertical axis toward the north till the bubble plays and the telescope is horizontal; to do this the vertical axis must have been tipped back through an angle equal to the co-latitude, and it will be in proper adjustment directed toward a point in the meridian whose altitude is equal to the latitude. (Fig. 34 shows the instrument adjusted for latitude 45°.)

FIG. 34

A notch should be cut in the iron arc at the bottom of the counterpoise, into which the spring-catch may slip when the adjustment is correct, so that the instrument may be quickly changed from one position to the other. If the notch is not quite correctly placed, the final adjustment may be made by a slight motion of the north leveling screw to bring the level exactly into the horizontal position, the vertical circle having been set to the co-latitude for this purpose.

The proper adjustment of the altazimuth is simpler, since it depends only on the use of the level, while to place an equatorial instrument in position we must know the latitude as well. On comparing the two systems of coördinates, it is clear that, while the altitude and azimuth both change continuously, but not uniformly with the time, the hour-angle changes uniformly with the time, and the declination remains the same. One advantage of the latter system of coördinates is that in repeating our observations on the same star after the lapse of an hour, we need only set the declination circle to the previously observed declination, and set the hour-circle at a reading obtained by adding to the former setting the elapsed time in hours reduced to degrees by multiplying by 15; we shall then pick up the star without difficulty. This is an important aid in identifying stars, which has no counterpart in the use of the altazimuth, and we shall henceforth use this method of observation in preference to the other.

CHAPTER IV

THE COMPLETE SPHERE OF THE HEAVENS

THE study of the motions of the sun, moon, and stars has thus far led to the conclusion that their courses above the plane of the horizon can be perfectly represented by assuming the daily rotation from east to west of a sphere to which they are attached, or a rotation of the earth itself from west to east about an axis lying in the meridian and inclined to the horizon at an angle equal to the latitude of the place of observation, while the sun moves slowly to and from the equator, and the moon, like the sun, changes its declination continually, and has also a motion toward the east on the sphere at a rate of about 13° in each 24 hours. The combination of the two motions of the moon causes it to describe a path which will be more fully discussed later. We shall now begin to observe the sun, to see if its motion among the stars resembles that of the moon in having an east and west component in addition to its motion in declination.

The motion of the moon can be directly referred to the stars, since both are visible at the same time, although the illumination of the dust of our atmosphere, by strong moonlight, cuts us off from the use of the smaller stars, which cannot be seen except when contrasted with a perfectly dark background.

The illumination produced by the sun, however, is so strong that it completely blots out even the brightest stars, so that we cannot apply either of the methods that we have employed in observing the moon.

We are only able to see the stars, of course, when they are above the plane of the horizon, but it is natural to suppose that they continue the same course below the horizon from their points of setting to those of their rising. This inference is confirmed by the fact that some of the bright stars which set within a few degrees of the north point of the horizon, and which we infer complete their course below the horizon, may be seen actually to do so by an observer at a point on the earth some degrees farther north, from which they may be observed throughout the whole of their courses. In the case of the sun, the following facts lead to the same conclusion. Immediately after sunset a twilight glow is seen in the west whose intensity is greatest at the point where the sun has just set. This glow appears to pass along the horizon towards the north, and its point of greatest intensity is observed to be directly over the position which the sun would occupy in the continuation of its path below the horizon, on the assumption that it continues to move uniformly in that path. In high latitudes this change of position in the twilight arch can be followed completely around from the point of sunset to the point of sunrise, the highest point being due north at midnight. It is impossible not to believe that the sun is actually there, though concealed from our sight by the intervening earth. (Of course, too, it is now generally known that in very high latitudes the sun at midsummer is visible throughout its diurnal course.) As the sun sinks farther, the light of the sky decreases, the brighter stars begin to appear, and it is clearly impossible to resist the conclusion that they have been in position during the daylight, but simply blotted out by the overwhelming light of the sun.

OBSERVATIONS WITH THE EQUATORIAL

When we have fixed the idea that the heavenly sphere revolves as a whole, carrying with it in a general sense all the bodies that we observe, the next step is to devise some means of locating the different bodies in their proper relative positions on the sphere. For this purpose the equatorial instrument furnishes us with an admirable means of observation. The relative position of two stars is completely fixed when we know the position of their parallels of declination and their hour-circles, since the place of each star is at the intersection of these two circles.

Since an observation with the equatorial gives directly the declination and hour-angle of a star, the method of fixing the relative position of two stars, A and B, is as follows:

Point the telescope at A, and read the circles; then set on B, and read the circles; then again on A, and read the circles, taking care that the interval between the first and second observations shall be as nearly as possible equal to the interval between the second and third. Obviously the mean of the two readings of the hour-circle at the pointings upon A gives the hour-angle of A at the time when B was observed, since the star's hour-angle changes uniformly. The difference between this mean and the reading of the hour-circle when the pointing was made upon B is, therefore, the difference between the hour-angles of the stars at the time of that observation; and this fixes the relative position of their hour-circles, since this difference is the arc of the equator included between them; their declinations are given by the readings of the declination circles, and thus the relative position of the two stars is completely known.

As an illustration of this method, we may take the following example:

With the telescope pointed at A, the readings of the hour-circle and declination circle were $68°.2$ and $15°.1$, respectively. The telescope was then pointed at B, and the circles read $85°.9$, $28°.1$, and finally upon A, the readings being $69°.1$, $15°.1$; the intervals were nearly the same, as will usually be the case, unless there is some difficulty in finding the second star. Of course the first star can be *re*-found by the readings at the first observation; indeed, if the intervals are plainly unequal, a repetition of the observation may always be made at equal intervals by setting the circles for each star so that no time is lost in finding.

From the above observations we infer that when the hour-angle of B was $85°.9$, that of A was $68°.65$; and, therefore, that the hour-circles of the two stars cut the equator at points $17°.25$ apart; the hour-circle of B being to the west of that of A, so that B comes to the meridian earlier, or "precedes" A.

It may be noted that the observations apparently occupied a little less than 4 minutes, since in the whole interval the hour-angle of A changed by $0°.9$.

USE OF A CLOCK WITH THE EQUATORIAL

If the intervals between the observations are not exactly equal, it will still be easy to fix the hour-angle of A at the time of the observation on B if the *ratio* of the intervals is known; if, for instance, the first observation of A gives an hour-angle of 25°.3, and the later observation an hour-angle of 26°.3, while the intervals are 1^m between the first and second observations, and 3^m between the second and third, the hour-angle of A at the second observation was obviously 25°.3 + 0°.25. We may thus obtain by "interpolation" the hour-angle of A at any known fraction of the interval. Plainly it is an advantage to note the time of each observation for this purpose, as in the following observations, which were made Feb. 5, 1900, for the purpose of determining the relative positions of the stars forming the Square of Pegasus.

STAR	WATCH TIME	DECL. CIRCLE	HOUR-CIRCLE
1 γ Pegasi	7^h 14^m 0^s	+ 15°.2	66°.3
2 α Pegasi	15 0	+ 15 .2	83 .6
3 β Pegasi	16 15	+ 28 .1	84 .1
4 α Andromedæ	17 10	+ 29 .0	68 .3
5 γ Pegasi	18 30	+ 15 .2	67 .6
6 γ Pegasi	21 30	+ 15 .1	(69 .2)
7 α Andromedæ	22 30	+ 29 .1	69 .6
8 β Pegasi	23 30	+ 28 .1	85 .9
9 α Pegasi	24 20	+ 15 .3	86 .0
10 γ Pegasi	25 30	+ 15 .1	69 .1
11 γ Pegasi	27 30	+ 15 .1	69 .6

The observations here follow each other rapidly. They were made by an experienced observer, and the arrangement of the stars is such that, after setting γ Pegasi, α Pegasi is brought into the field by moving the telescope about the hour-axis only; we pass to β Pegasi by motion around the declination axis only, to α Andromedæ by motion about the hour-axis, and back to γ Pegasi by rotation about the declination axis; so that the stars are found more quickly than if both axes must be altered in position at each change; in observations 6 to 10 the series is observed in reversed order.

THE COMPLETE SPHERE OF THE HEAVENS

If the instrument was correctly adjusted, the declination of the four stars was as follows: γ Pegasi + 15°.14, α Pegasi 15°.25, β Pegasi 28°.1, α Andromedæ 29°.05, each being determined as the mean of all the observations made upon the star.

The first advantage of the recorded times is to show that the reading of the hour-circle in 6 was an error, probably for 68°.2, as we see by comparison with the other values of the hour-angle of γ Pegasi, which increase uniformly about 1° in each 4 minutes. It will be better, however, to reject the observation entirely, as it is not necessary to use it for the first set of observations 1 to 5, which we will now discuss.

By interpolation between 1 and 5 we find that the hour-angle of γ Pegasi at $7^h 15^m 0^s$ was $\frac{2}{9}$ of 1°.3 greater than 66°.3, or 66°.59; at $7^h 16^m 15^s$ it was $\frac{1}{2}$ of 1°.3 greater than 66°.3, or 66°.95; and at $7^h 17^m 10^s$ it was $\frac{80}{270}$ of 1°.3 less than 67°.6, or 67°.21. As the hour-angles of the other stars were observed at these times, we can at once find the differences of their hour-angles from that of γ Pegasi, which are as follows: α Pegasi, 17°.01; β Pegasi, 17°.15; α Andromedæ, 1°.09. All the hour-angles are greater than those of γ Pegasi, so that all the stars precede γ Pegasi. By using all the observations we may presumably obtain more accurate results, and it will be well, as in all cases when a considerable number of observations must be dealt with, to arrange the reductions in a more systematic manner.

In the table on the following page the difference of hour-angle is obtained by subtracting the observed hour-angle in each case from the hour-angle of γ Pegasi, so that its value is negative, if, as in the results given above, the stars precede γ Pegasi, and positive if they follow it. An observation of Venus, made on the same occasion, is added to the list, and an additional observation of α Pegasi is included; the erroneous observation of γ Pegasi at $7^h 21^m 30^s$ is excluded.

The values of the hour-angle of γ Pegasi at the successive times, as given in column 6, are computed from the following considerations, the proof of which is left to the student. If a quantity changes uniformly, and its values at several different times are known, the mean of these values is the same as the value which

52 LABORATORY ASTRONOMY

STAR	TIME	DECL.	H.A.	H.A. OF γ PEG.	STAR FOLLOWS γ PEG.
1 Venus	7ʰ 12ᵐ 0ˢ	+ 4°.0	75°.5	65°.86	− 9°.64
2 α Peg.	13 0	+ 15 .1	83 .1	66 .10	− 17 .00
3 γ Peg.	14 0	+ 15 .2	66 .3	66 .35	+ 0 .05
4 α Peg.	15 0	+ 15 .2	83 .6	66 .59	− 17 .01
5 β Peg.	16 15	+ 28 .1	84 .1	66 .89	− 17 .21
6 α Androm.	17 10	+ 29 .0	68 .3	67 .12	− 1 .18
7 γ Peg.	18 30	+ 15 .2	67 .6	67 .44	− .16
8 α Androm.	22 30	+ 29 .1	69 .6	68 .44	− 1 .16
9 β Peg.	23 30	+ 28 .1	85 .9	68 .88	− 17 .22
10 α Peg.	24 30	+ 15 .3	86 .0	68 .93	− 17 .07
11 γ Peg.	25 30	+ 15 .1	69 .1	69 .17	
12 γ Peg.	27 30	+ 15 .1	69 .6	69 .65	+ 0 .05

the quantity has at the mean of the times. Using this principle, we find the hour-angle of γ Pegasi at 7ʰ 21ᵐ 22ˢ was 68°.15.

Between observations 3 and 12 it changed 3°.3 in 13½ᵐ, or 0°.244 per minute. Assuming this rate of change, it is easy, though laborious, to compute the hour-angle at any one of the given times; for example, at 7ʰ 12ᵐ 0ˢ the hour-angle was 68°.15 − (9$\frac{22}{30}$ times 0°.244), or 65°.86. Labor will be saved by making a table of the values at the even minutes by successive additions of 0°.244, from which the values at the observed times are rapidly interpolated. The sixth column contains the number of degrees by which the hour-circle of the star follows that of γ Pegasi. The mean values for each star obtained from this column are as follows.

STAR	DECL.	DIFF. H.A.
γ Pegasi	+ 15°.15	0°.00
Venus	− 4 .0	− 9 .64
α Pegasi	+ 15 .20	− 17 .03
β Pegasi	+ 28 .10	− 17 .22
α Androm.	+ 29 .05	− 1 .17

The true values of the declinations of these stars as determined by many years of observations are for γ Pegasi 14°.63, α Pegasi 14°.67, β Pegasi 27°.55, α Andromedæ 28°.53. The values from our

THE COMPLETE SPHERE OF THE HEAVENS

observations are 15°.15, 15°.20, 28°.10, 29°.05, so that the latter require corrections of − 0°.52, − 0°.53, − 0°.55, and − 0°.52, respectively. This is due to a faulty adjustment of the instrument, but the error from this cause evidently affects all the observations by nearly the same amount, 0°.53, so that the relative positions are given quite accurately; our observations placing the whole constellation about ¼° too far north.

Since Venus is in the near neighborhood of γ Pegasi, we may assume that the observations of that planet are subject to the same corrections, that she preceded γ Pegasi by 9°.64, and that her true declination was − 4°.0 − 0°.53, or − 4°.53. The correction is applied algebraically with the same sign as to the other stars, since it must be so applied as to make the true place farther south than the observed place.

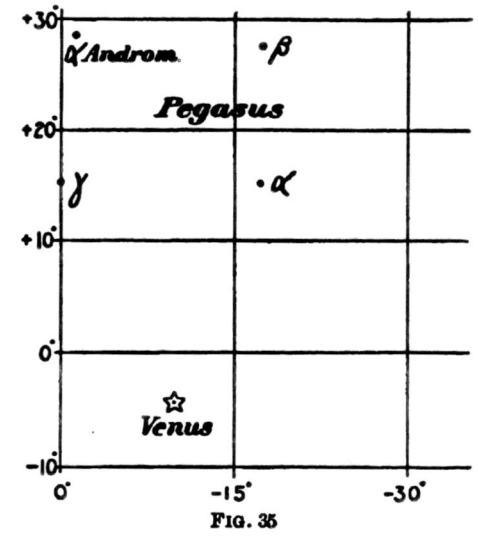

Fig. 35

The places of the Square of Pegasus and the planet Venus, as seen in the sky Feb. 5, 1900, are shown in Fig. 35.

Before plotting the stars on the hemisphere from the above data, it must be prepared by drawing upon it in their proper positions the meridian, zenith, pole, and equator. Draw the hour-circle of γ Pegasi (see Fig. 19, p. 17) at the proper hour-angle from the meridian, to give its position at the time of the last observation, which may be determined by making it intersect the equator at the proper point 69°.6 west of the meridian, and place the star upon it at a distance from the equator equal to the observed declination, 15°.14. The hour-angle of α Pegasi should be drawn in the same manner to cut the equator at 86°.66 from the meridian, and the star placed upon it at the observed declination, 15°.20. Of course on the scale of so small a hemisphere the nearest half degree is sufficiently accurate. Remember that the configurations on the hemisphere and on the map are semi-inverted.

CLOCK REGULATED TO SHOW THE HOUR-ANGLE OF THE FUNDAMENTAL STAR

The method of calculating the hour-angles of γ Pegasi in the last example shows that if the reading of the watch can be relied upon, the observations of that star need only be made at the beginning and at the end of the period of observation, the hour-angle at any time being determined by its uniform increase; or even from a single observation at the beginning of the period, since at the time of observation of any star the hour-angle of γ Pegasi can be inferred from that at its first observation by adding the number of degrees which it would have described in the time elapsed, obtained by multiplying the number of hours by 15, or, what gives the same results, dividing the minutes by 4. Moreover, if the rate of the watch is such that it completes its 24 hours in the time in which the stars complete their daily revolution, and if its hands are so set as to read 12 hours when γ Pegasi is on the meridian, the difference of hour-angle at any time will be equal to the reading taken directly from the hands of the watch reduced as above to degrees, for when the star is on the meridian and its hour-angle therefore zero, the watch marks 0^h 0^m 0^s. Four minutes later by the watch the hour-angle of the star has increased by the diurnal revolution to 1°; in four minutes more to 2°; when the watch indicates 1 hour, the star's hour-angle has increased to 15°, and so on, till 24 hours have elapsed, when the star will again be on the meridian and the cycle recommences.

The rate of an ordinary watch is sufficiently near to that of the stars to allow of its use for this purpose for periods of an hour without causing any error in our observations.

In the use of this method we may regard the observation of the fundamental or zero star as a means of finding out whether the clock is set to the right time: thus, in the following set of observations the first observation gives the hour-angle of γ Pegasi 67°.6 at 7^h 15^m 10^s, but as 67°.6 equals 4^h 30^m 24^s, we may regard the clock as 2^h 44^m 46^s fast; and by applying this correction to all the observed times, may write down at once under the title "corrected time" what the readings would have been if the clock had been set

THE COMPLETE SPHERE OF THE HEAVENS 55

to show 0 hours, when the star's hour-angle was 0°. Multiplying these by 15 we have the hour-angle in degrees given in column 4.

The following observations were undertaken for determining the configuration of the stars in Orion and its neighborhood, Feb. 6, 1900.

STAR	OBS. TIME	CORRECTED TIME	H.A. OF γ PEG.	OBSERVED H.A. OF STAR	DECL.	FOLLOWS γ PEG.
γ Pegasi	7h 15m 10s	4h 30m 24s	67°.6	67°.6	+ 15°.5	
a	7 20 0	4 35 14	68 .8	348 .5	− 1 .4	80°.3
b	7 22 0	4 37 14	69 .3	349 .95	− 0 .6	79 .35
c	7 23 30	4 38 44	69 .7	348 .1	− 2 .2	81 .8
d	7 25 20	4 40 34	70 .1	344 .8	+ 7 .1	85 .3
e	7 27 10	4 42 24	70 .6	353 .0	+ 6 .1	77 .6
f	7 28 45	4 43 59	71 .0	347 .5	− 9 .9	83 .5
g	7 30 20	4 45 34	71 .4	356 .4	− 8 .2	75 .0
h	7 32 0	4 47 14	71 .8	351 .6	− 5 .5	80 .2
i	7 34 0	4 47 14	72 .3	334 .7	− 16 .9	97 .6
j	7 35 45	4 50 59	72 .7	321 .3	+ 5 .05	111 .4
a	7 37 45	4 52 59	73 .2	352 .9	− 1 .4	80 .3
γ Pegasi	7 39 50	4 55 4	73 .8	73 .9	+ 15 .4	
Moon	7 42 0	4 57 14	74 .3	27 .6	+ 20 .4	46 .7

The results of columns 6 and 7 enable us to map the constellation as in Fig. 36.

One or two constellations may be plotted in this manner both on the map, which shows the constellation as seen in the sky, and on

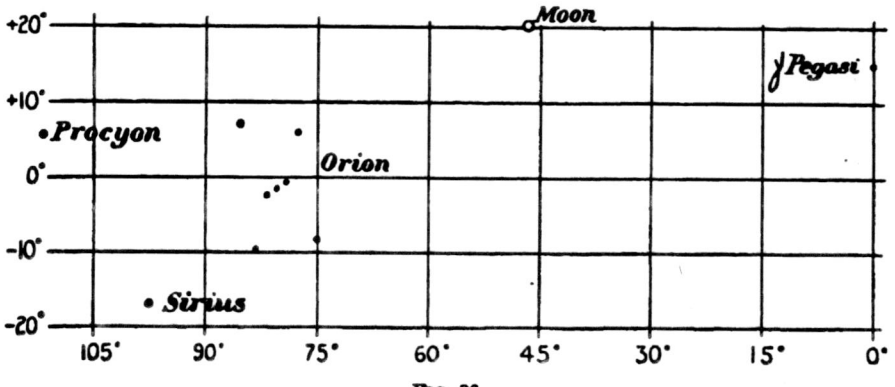

FIG. 36.

the hemisphere, where it is semi-inverted. It will be advisable, however, before much work has been done in this way, to introduce a slight modification.

THE VERNAL EQUINOX — RIGHT ASCENSION

The precession of the equinoxes causes a change in the position of the equator, which slowly changes the declinations of all the stars. For this reason it is found more convenient to select, instead of γ Pegasi as a zero star, the point upon the equator at which the sun crosses it from south to north about March 21 of each year. This point, which is called the vernal equinox, is not fixed, but its motion, due to precession, is simpler than that of any star which might be selected as a zero point; it precedes the hour-circle of γ Pegasi at present by about 8 minutes of time, or 2° of arc, and it was because of this proximity that we first selected that star.

Instead, therefore, of adjusting our clock so that it reads $0^h\ 0^m\ 0^s$ when γ Pegasi is on the meridian, we set it to that time when the vernal equinox is in that plane; its readings then give the hour-angle of the vernal equinox, and the difference between the hour-angles of that point and of the star may be directly obtained from our observations. The distance by which a star follows the vernal equinox is called its right ascension; more carefully defined, it is the arc of the equator intercepted between the hour-circle of the star and the hour-circle of the vernal equinox (which measures the wedge angle between the planes of these circles); it is also the angle between the tangents drawn to these two circles where they intersect at the pole. Since any star which is east of the vernal equinox follows it, the right ascensions of different stars increase toward the east, that is, toward the left in the sky as we face south, but toward the right on the solid hemisphere as we look down from the outside upon its southern face.

Hereafter we shall fix the positions of the stars by their right ascensions and declinations. We may make use of the observations already reduced with very little additional labor. Since γ Pegasi follows the vernal equinox by 2°, we need only add that amount to the quantities given in column 7 on page 55 to know the right ascension of the different stars. If we learn later that on February 6 the right ascension of γ Pegasi was more exactly $0^h\ 8^m\ 5^s.64$, we may further correct by adding 5^s, or even $5^s.64$, if the accuracy of the observations warrants it. The method of determining the exact position of the zero star with reference to the vernal equinox is given in Chapter VI.

Formerly right ascensions were measured altogether in degrees, but owing to the modern use of clocks, it has long been customary to give them in hours; for this reason the hour-circle of instruments mounted as equatorials is graduated to read hours and minutes directly. Since our universal equatorial is intended to serve also as an altazimuth, its circles are both graduated to degrees.

SIDEREAL TIME

In the last section right ascension has been defined as the angle between the hour-circle passing through a star and the great circle passing through the pole and the vernal equinox. The latter circle is called the equinoctial colure. We have also suggested the use of a clock set to read $0^h\ 0^m\ 0^s$ at the time when the vernal equinox is on the meridian; so that the hour-angle of the vernal equinox at any time will be given directly by the reading of the face of the watch in hours, minutes, and seconds, from which the angle in degrees is found by multiplying by 15. A clock set in this manner, and running at such a rate that it completes 24 hours in the time that the star completes its revolution from any given hour-angle to the same hour-angle again, is said to keep sidereal time. We shall find later that a clock so regulated gains about 4 minutes a day on a clock keeping mean time, thus gaining 24 hours on an ordinary clock in the course of a year, and agreeing evidently with a clock keeping apparent time, as defined on page 19, at that time when the sun is at the vernal equinox and crosses the meridian at the same time with the latter.

Let us suppose now that the vernal equinox has passed the meridian by one hour, then its hour-angle is 1^h, or $15°$; and our sidereal clock indicates exactly $1^h\ 0^m\ 0^s$. Any star which is at this time on the meridian, that is, whose hour-angle is $0°$, must therefore follow the vernal equinox by 1^h, or $15°$, while at the same instant the time by our sidereal clock is $1^h\ 0^m\ 0^s$. By our definition of right ascension, since the star follows the vernal equinox by 1^h, its right ascension is 1^h; in this case, therefore, the right ascension of the star in hours, minutes, and seconds has the same value as the time given by the hands of the clock. In the same way, if the vernal equinox has passed the meridian so far that its hour-angle is $2^h\ 15^m$, the face of the clock will show $2^h\ 15^m$; and any star then upon the meridian follows the vernal equinox by $2^h\ 15^m$. The same relation holds here; namely, that the right ascension of the star is equal to the time by the sidereal clock when the star is upon the meridian. This might have been given as a definition of the term "right ascension"; and, indeed, so closely are the two connected in the mind of the practical astronomer that if the right ascension of a star is given, he at once thinks of this number as representing the time of its meridian passage.

RIGHT ASCENSION PLUS HOUR-ANGLE EQUALS SIDEREAL TIME

We may here give an explanation of a general principle of very frequent application, and of which this is simply a particular case. Suppose the vernal equinox, represented by the symbol ♈ (Fig. 37), to have passed the meridian by $5^h\ 10^m$. Then a star, S, whose right ascension is $2^h\ 15^m$, since it follows the vernal equinox by that amount, will have passed the meridian by $2^h\ 55^m$; and its hour-angle will be $2^h\ 55^m$. The arc of the equator between the meridian and the vernal equinox may be considered as made up of two parts: the right ascension of the star, which is measured by the arc eastward from the vernal equinox to the hour-circle of the star, and the hour-angle of the star, which extends from the meridian westward to the hour-circle of the star. Since this is true of any star, or,

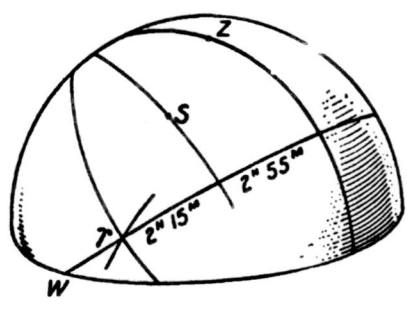

FIG. 37

indeed, of any heavenly body, we may make the following general statement: The right ascension of any body plus its hour-angle at any instant will be equal to the sidereal time at that instant; or, as it is sometimes written: R.A. + H.A. = Sidereal Time. If the body is a point on the meridian, its H.A. = zero; hence the R.A. of a star on the meridian, or briefly, R.A. of the meridian = Sidereal Time, as we have before shown.

From this relation we may most simply determine the right ascension of any heavenly body by observing its hour-angle with the equatorial instrument, and at the same time noting the sidereal time, since R.A. = Sidereal Time − H.A. It is by this method that we shall now proceed to make a somewhat extended catalogue of stars from which we may plot their positions upon the globe.

We will here notice some of the important uses to which this principle may be put. If by any other means the right ascension of a body is known, we may find its hour-angle at any given sidereal time by the equation, Sidereal Time − R.A. = H.A. This gives us an easy way to point upon any object whose right ascension and declination are known, if we have a clock keeping sidereal time; and this is the usual way in which the astronomer finds the objects which he wishes to observe, since they are generally so faint that they cannot be seen by the naked eye. For example, to point the telescope at the great nebula in Orion, whose right ascension is $5^h\ 28^m$, and declination 6° S., we first set the declination circle to −6°, and if the sidereal time is $7^h\ 30^m$ we set the hour-circle to $2^h\ 2^m$, then the telescope will be pointed upon the star. If the sidereal time is $4^h\ 30^m$, in which case the star evidently has not reached the meridian by nearly an hour, we must add 24 hours to the sidereal time; then the expression, H.A. = Sidereal Time − R.A. will become H.A. = $28^h\ 30^m - 5^h\ 28^m$, or $23^h\ 2^m$, the hour-angle being reckoned, as before stated, from 0^h up to 24^h. If then the hour-circle is brought to the reading $345\frac{1}{2}° = 15° \times 23\frac{2}{60}$, we shall find the star in the field.

THE CLOCK CORRECTION

The same principle enables us to set our clock correctly to sidereal time by observing the hour-angle of *any* star whose right ascension is known. For example, the right ascension of Sirius being $6^h 40^m$, or 100°, and its hour-angle being observed to be 330°, or 22^h, the sidereal time is R.A. + H.A., that is, 430°, or, subtracting 360°, is 70°, corresponding to $4^h 40^m$; and a clock may be set to agree; or, by subtracting the time which it then indicates, we determine a correction to be applied to its reading to give the true sidereal time. If, for instance, at the observation above, the clock time is $4^h 41^m 10^s$, the clock correction is $-1^m 10^s$. In this case the clock is $1^m 10^s$ fast, the time which it indicates is greater than the true time, and its "error" is said to be $+1^m 10^s$. On the other hand, when the clock is slow the correction to true time is positive, while the "error" is negative.

It is customary to observe this distinction between the terms "error" and "correction"; the former is the amount by which the observed value of a quantity exceeds its true value, while the correction is the quantity which must be added to the observed to obtain the true value. They are thus numerically equal but of opposite sign.

The error of the declination circle determined by the observations of page 53 was $+0°.53$, while the correction was $-0°.53$.

For the constantly occurring "clock correction," we shall use the symbol Δt, the value of which is positive if the clock is slow and negative if it is fast.

If, as is often desirable, we wish to observe a body of known right ascension upon the meridian, we have only to observe it when the time by the sidereal clock is equal to its right ascension.

We may of course find the right ascension of the moon by a direct comparison with the neighboring stars, just as we have determined the difference in right ascension of α Pegasi, from that of γ Pegasi, for the brighter stars can be easily observed at the same time as the moon; but no star is so bright that it can be readily observed by our small instrument when the sun is above the horizon,* and we have therefore no means of making a direct comparison between

---

* See, however, page 69.

THE COMPLETE SPHERE OF THE HEAVENS

a star and the sun. But by means of our clock and our new method of observation this becomes easy; and the sun is to be added to the list of bodies whose right ascension we are to observe regularly. It is only necessary that we should be provided with a clock which keeps correct sidereal time. (See page 67.)

We have already spoken of the means of setting the clock; now a few words as to how the regularity of its rate may be determined. It is only necessary to observe the watch time at which any star is at a given hour-angle on successive nights. If the rate of the clock is such that the interval between the observations is greater than 24 hours, the watch is gaining; if the amount is less than half a minute a day, the watch may be assumed for our purposes to be keeping correct sidereal time, its actual error at any time being checked, as before described, by the observation of the hour-angle of bodies of known right ascension.

LIST OF STARS

Our first care will be to observe a number of bright stars not very far from the equator which will serve for setting the clock or determining its error, selecting them so that several shall always be above the horizon and may at any time be used for this purpose. Several of those already observed will be found in the list given on the following page, which contains the approximate places of a number of conspicuous stars.

By repeated comparisons of these stars with each other and with γ Pegasi, their right ascensions may easily be fixed within 30^s, and they may then be used for determining the clock error at any time when they are visible. The observations of each evening should be reduced as soon as possible and maps made of the various constellations similar to those of Figs. 35 and 36; it is, however, impossible to represent any large portion of the sphere satisfactorily on a plane surface, and, in order to have a proper idea of the relative positions of the various constellations, we must plot our results on a globe — a proceeding still more necessary when we come to study the motion of the sun and moon among the stars by the method of the following chapter.

A globe 6 inches in diameter is sufficiently large for our purpose; it should be so mounted that it may be turned about its axis on a firm support, and upon it should be traced 24 hour-circles 15° apart, and small circles (parallels of declination) parallel to the equator and 10° apart; its surface should be smooth and white, and of such a texture as to take a lead-pencil mark easily, but permit of erasure.

TIME STARS

Star	Mag.	R.A.	δ	Star	Mag.	R.A.	δ
γ Pegasi	3	0^h.1	+ 15°	Denebola	2	11^h.7	+ 15°
β Ceti	2	0 .6	− 19	δ^2 Corvi	3	12 .4	− 16
β Andromedæ	2	1 .1	+ 35	Spica	1	13 .3	− 11
α Arietis	2	2 .0	+ 23	Arcturus	1	14 .2	+ 20
α Ceti	2½	3 .0	+ 4	α^2 Libræ	3	14 .8	+ 16
Alcyone	3	3 .7	+ 24	α Serpentis	3	15 .7	+ 7
Aldebaran	1	4 .5	+ 16	Antares	1	16 .4	− 26
Capella	1	5 .2	+ 45	α Ophiuchi	2	17 .5	+ 13
Rigel	1	5 .2	− 8	γ^2 Sagittarii	3	18 .0	− 30
ε Orionis	2	5 .5	− 1	Vega	1	18 .6	+ 39
Betelgeuze	1	5 .8	+ 7	Altair	1	19 .8	+ 9
Sirius	1	6 .7	− 17	α^2 Capricorni	4	20 .2	− 13
Castor	2	7 .5	+ 32	α Delphini	4	20 .6	+ 16
Procyon	1	7 .6	+ 5	ε Pegasi	2½	21 .7	+ 9
Pollux	1	7 .7	+ 28	α Aquarii	3	22 .0	− 1
α Hydræ	2	9 .4	− 8	α Pegasi	2½	23 .0	+ 15
Regulus	1	10 .1	+ 12				

The number attached to the Greek letter indicates that the star to be observed is the following of two neighboring stars.

CHAPTER V

MOTION OF THE MOON AND SUN AMONG THE STARS

For plotting the stars on the globe in their proper places, as given by their right ascensions and declinations, it is convenient to have the equator graduated into spaces of 10^m each; this may be done by laying the edge of a piece of paper along the equator, and marking off the points of intersection of the equator with two consecutive hour-circles; laying the paper upon a flat surface, bisect the space between the two lines with the dividers, and trisect each of these spaces by trial, testing the equality of the spacing by the dividers; this may be satisfactorily done by two or three trials, and the short scale thus obtained may be easily transferred to the arcs on the equator between each two hour-circles. It may be found convenient to bisect each of the spaces on the scale, thus dividing the equator into spaces of 5^m each.

A strip of parchment or parchment paper about 8 inches long and ¼ inch wide, of the shape shown in Fig. 38, and graduated to degrees, completes the apparatus necessary for plotting. The hole being placed over the axis of the globe, the graduated edge of

FIG. 38

the strip may be made to coincide with the hour-circle of any star by causing it to intersect the equator at a point corresponding to the star's right ascension, taking care that the edge lies in a great circle of the sphere; the graduated edge gives at once the proper declination for plotting the star upon its hour-circle, and the point may be marked with a well-sharpened, hard lead pencil; the latter should be carefully kept, and used for purposes of plotting only. With this simple apparatus the stars may be rapidly and accurately placed upon the globe.

An attempt should be made to represent the magnitudes of the stars by the size of the dots which indicate their places.

THE MOON'S PATH ON THE SPHERE

The moon should be placed on the list of objects for regular observation, the observations being made in precisely the same manner as those of the stars, and its place should be plotted upon the globe at each observation and marked by a number, giving the date of the month. This method of fixing the moon's place is much more accurate than those made use of in Chapter II, and, as the places are plotted upon a globe, we may study to better advantage those peculiarities of her motion which are masked by the distortion of the map referred to in Chapter II.

The position of the node may now be fixed with such a degree of accuracy that its regression is shown by the observations of two or three months, if some care is taken to observe as nearly as possible at the same altitude in the successive months, so that the corrections for parallax may be nearly the same; indeed, a very few months will force upon the notice of the observer the fact that the moon's path does not lie in one plane, just as observations a few days apart show that the sun's diurnal path is not really a small circle lying in one plane.

We also study the variable motion of the moon by applying dividers between the successive plotted places and then placing the dividers against the parchment scale to measure the distance in degrees traversed in the plane of the orbit. The scale must lie along an hour-circle so as to conform to the curvature of the sphere.

The average rate being about 13° a day, the points on the orbit should be determined as nearly as possible at which the motion is greater and less than this amount, and the point of most rapid motion fixed as closely as possible; this point is most simply fixed by its distance in degrees from the ascending node of the moon's orbit. Since the latter point, however, is continually changing, it is customary to reckon the so-called "longitude in the orbit" of the point by measuring from the vernal equinox along the ecliptic to the node, and adding the angle measured along the orbit from the node to the point.

The variations of the moon's angular diameter and the point of the orbit where the diameter is greatest should be compared with

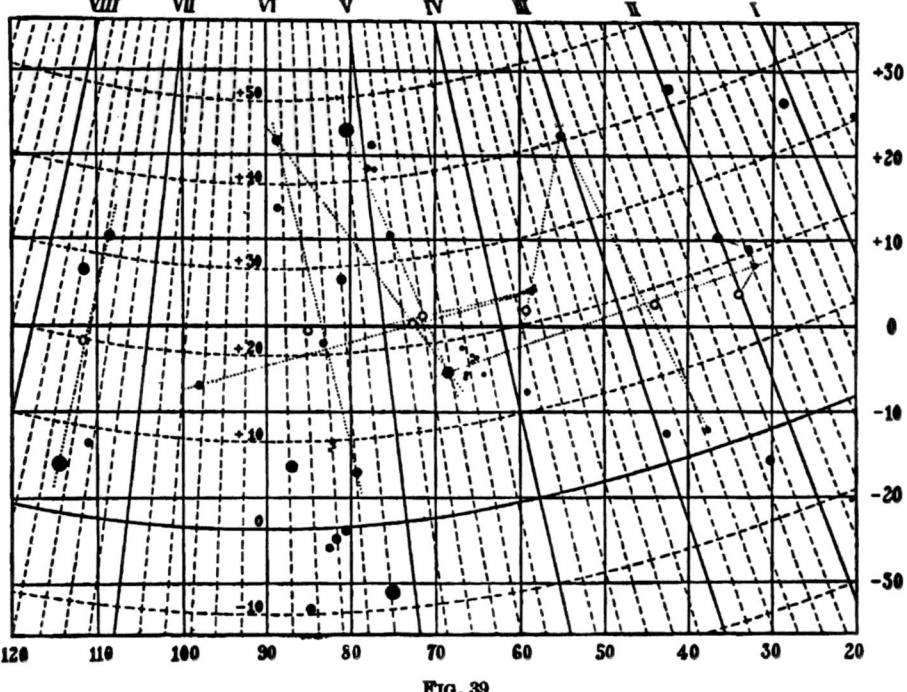

FIG. 39

the results obtained from the investigation of the angular velocity in the orbit, since we thus gain some knowledge of the moon's relative distances from us at different points of its orbit, and of the relation between its distance and its rate of motion about the earth.

The scale of the 6-inch globe is too small to do justice to the accuracy of our observations, which are accurate to a quarter or a tenth of a degree, and it will be interesting to plot these observations on a map constructed on a larger scale, and on a plan which reduces the distortion to very small limits in the region of the ecliptic; such a map is shown in Fig. 39 on a reduced scale; the ecliptic is here taken as a straight horizontal line, as the equator is in the star map previously used; the latitude, or angular distance of a point from the ecliptic measured on a great circle perpendicular to the latter, serves as the coördinate corresponding to the declination on our former map, while right ascension is replaced by longitude, or distance along the ecliptic measured from the vernal equinox up to 360°. The same map will serve also for plotting the paths of the planets in our later study.

For convenience in plotting, the parallels of declination and the hour-circles are printed in broken lines upon the map. The observations of the moon shown in the figure are those of December, 1899, already plotted on the map of Fig. 25.

THE SUN'S PLACE AMONG THE STARS

By means of the equatorial we may also determine the place of the sun among the stars, although the method of direct comparison with stars we have used in the case of the moon is not applicable, since the stars are not visible when the sun is above the horizon; the most obvious method which is capable of any degree of accuracy involves the use of a clock regulated to sidereal time.

To determine the place of the sun, point upon it with the equatorial about two hours before sunset; note the time, and read the circles; as soon as possible after sunset observe a star in the same manner, with the instrument as near as may be to its position at the sun observation. It is evident that if the circumstances were fortunately such that the telescope did not have to be moved between the observations, the difference in right ascension of the sun and the star would be the difference in time noted by the sidereal clock, while the declinations of the sun and star would be the same. The nearer the star is to the position in which the sun was observed, the less will be the errors arising from imperfect adjustment and orientation of the instrument; while the shorter the interval between the observations, the smaller will be the error due to the uncertainty in the rate of the clock. As the condition of not moving the telescope can seldom be fulfilled, however, we must treat the observation as follows:

Let R.A., H.A., t, and Δt be the right ascension, hour-angle, clock time, and clock correction at the time of the star observation, and R.A.', H.A.', t', and Δt, the corresponding quantities at the sun observation. The equation

$$\text{R.A.} + \text{H.A.} = \text{Sidereal Time} = t + \Delta t$$

determines the value of Δt, which substituted in the equation

$$\text{Sidereal Time} = t' + \Delta t = \text{R.A.}' + \text{H.A.}'$$

determines the value of R.A.', the sun's right ascension at the moment of observation.

The value of Δt, as determined from the first equation, will be negative if the clock is fast, and positive if the clock is slow; and it must always be applied to the observed time with the proper sign. The declination of the sun is, of course, given directly by the reading of the declination circle.

The following example illustrates the method:

March 29, 1899, an observation of the sun with an equatorial telescope, and a clock keeping sidereal time, gave the following values:

Observed time $= 5^h\ 36^m\ 26^s$; H.A. $= 75°.7 = 5^h\ 2^m\ 48^s$; $\delta = + 4°.1$. About an hour after sunset an observation of α Ceti was made in nearly the same position of the instrument, which gave the following values:

Observed time $= 7^h\ 53^m\ 43^s$; H.A. $= 74°.1 = 4^h\ 56^m\ 24^s$; $\delta = + 4°.2$. This latter gives, from the known right ascension of α Ceti,

$$2^h\ 57^m\ 0^s + 4^h\ 56^m\ 24^s = \text{Sidereal Time} = 7^h\ 53^m\ 43^s + \Delta t,$$

and hence $\Delta t = -19^s$; and, applying the same equation to the sun observation,

$$\text{Sun's R.A.} + 5^h\ 2^m\ 48^s = 5^h\ 36^m\ 26^s - 19^s = 5^h\ 36^m\ 7^s;$$

hence the sun's right ascension at the time of the first observation was $0^h\ 33^m\ 19^s$. This is liable to an error equal to the uncertainty of the circle readings, which may be at least one-twentieth of a degree, or 12^s of time, and to an error equal to the uncertainty of the gain or loss of the clock during the interval of $2\frac{1}{4}$ hours between the two observations, probably five or ten seconds of time. We may assume that the errors arising from defective adjustment of the instrument were the same for both objects, and may be neglected, since the position of the instrument was very nearly the same for both observations.

DIFFERENTIAL OBSERVATIONS

The declination of α Ceti, as read from the circles, was $+4°.2$, while its known declination was $+3°.7$. The correction necessary to reduce the circle reading to the true value is, therefore, $-0°.5$, and, applying this quantity to the reading on the sun, we have for the true value of the sun's declination $+4°.1 - 0°.5 = +3°.6$. It is worthy of note that the correction is about the same as that determined from the observations discussed on page 53, which were made with the same instrument in nearly the same adjustment, but from a different place of observation. These results indicate an inherent defect in the instrument, which is at least in great part neutralized by the method of observation. It is a very important thing, even with the most delicate instruments, to avail ourselves of methods which accomplish this object, and surprisingly good work may be done with poor instruments by paying proper attention to the details of observation for this purpose.

Methods by which an unknown body is thus compared with a known body under circumstances as nearly alike as possible are called "differential methods."

INDIRECT COMPARISON OF THE SUN WITH STARS

It is often possible to determine the difference of right ascension of the sun and some well-known star by using the moon as an intermediary, determining the difference of right ascension of the sun and moon during the daytime and comparing the moon and a star as soon as possible after sunset, the motion of the moon during the interval being allowed for. The irregularity of the moon's motion may, however, introduce a greater error than that arising from uncertainty in the rate of the clock. A better method is offered on those not infrequent occasions when the planet Venus is at its greatest brilliancy, when it may be easily observed in full daylight; the motion of Venus in the interval is much smaller and more nearly uniform, and, therefore, more accurately determined; and by this method the interval between the observations connecting the sun with Venus and Venus with the star may be reduced to a very few minutes, or even seconds, so that the error due to the clock may be regarded as negligible.

The following observations illustrate the method.

1900		Watch Time	H.A.	δ
April 19.3.	Procyon	$8^h\ 17^m\ 45^s$	15°.4	+ 5°.5
	Venus	8 19 33	56 .1	+26 .1
	Procyon	8 21 45	16 .4	+ 5 .55
	Venus	8 23 0	57 .0	+26 .05
	Procyon	8 24 53	17 .2	+ 5 .3
April 20.0.	Sun	1 28 45	358 .5	+11 .6
	Venus	1 31 35	313 .2	+25 .35
	Sun	1 36 10	0 .15	+11 .6
	Venus	1 38 33	315 .0	+25 .3
	Sun	1 41 21	1 .4	+11 .6
April 20.3.	Procyon	9 31 27	33 .25	+ 5 .45
	Venus	9 32 30	72 .9	+25 .9
	Procyon	9 33 28	33 .9	+ 5 .4
	Venus	9 35 0	73 .3	+25 .9
	Procyon	9 36 0	34 .25	+ 5 .4

The observations April 19.3, that is, April 19 about 7 P.M., give for the hour-angle of Venus 56°.55 at the watch time $8^h\ 21^m\ 17^s$, and for that of Procyon 16°.33 at $8^h\ 21^m\ 28^s$; hence at $8^h\ 21^m$ Procyon followed Venus 40°.22.

In the same way we find that April 20.3 Procyon followed Venus 39°.3, the change of the right ascension of Venus being 0°.92 in 25.2 hours. A simple interpolation shows that April 20.0 Procyon followed Venus 39°.59, and the observations at that time show that Venus followed the sun 45°.92, so that Procyon followed the sun 39°.59 + 45°.92 = 85°.51, and the difference of right ascension between Procyon and the sun at noon on April 20 was, therefore, $5^h\ 42^m\ 2^s$.

ADVANTAGES OF THE EQUATORIAL INSTRUMENT

Observation with the equatorial we shall find especially useful in getting exact positions of the moon, since it is available at any time when the moon is above the horizon, and after sunset we can always find some bright star sufficiently near to afford a fairly accurate value of its place.

It is often inconvenient to observe the moon by the more accurate method which is described in Chapter VI, that of meridian observations, which is confined to a short interval of one or two minutes each day, and is often interfered with by clouds passing at the critical moment, although nine-tenths of the whole day may be suitable for observations made out of the meridian. Moreover, until the moon is several days old, it is too faint for observation at its meridian passage. It is, therefore, upon the equatorial that we shall mainly rely for the determination of the moon's motion, as well as for many observations of the planets out of the meridian.

Although it is far more convenient to find the right ascension and declination of the sun by the method of the following chapter, at least a few positions should be found by observations with the equatorial and plotted on the globe. The result will be to show that the path of the sun is very exactly a great circle fixed on the sphere or so nearly fixed that some years of observation with the most refined instruments are necessary to detect any change in its position among the stars, although a much shorter time even would serve to show the slow change of its intersection with the celestial equator due to precession.

This great circle is called the ecliptic, and its position is shown on the map which we have used for plotting our first moon observation.

Three months will give a sufficient arc of this circle to enable us to determine with some accuracy its position with respect to the equator, its inclination to the latter, and their points of intersection;

if possible, observations should, however, be continued throughout the year which the sun requires to complete its circuit, so that the variability of its motion may be observed, most of the work, however, being done with the meridian circle.

The sun's diameter should occasionally be measured to determine the points at which it is nearest to and farthest from the earth.

CHAPTER VI

MERIDIAN OBSERVATIONS

WE have now arrived at a point where we can see what are the desirable conditions for making observations as accurately as possible of the position of a heavenly body. To adjust the equatorial instrument so that its axis lies in the meridian and at the proper inclination, and to keep it so adjusted, is a matter of some difficulty. In the last chapter we have shown how, by observing an unknown body in a certain fixed position of the instrument, and later a body whose right ascension and declination are known in as nearly as possible the same position of the instrument, we lessen the effect of the instrumental errors. We made our observation of the sun shortly before sunset, so that the interval between this observation and that of the comparison star should be as short as possible. If, however, the rate of the clock can be relied upon, there is no reason why the observation should not be made when the sun is on the meridian, the interval of time required to connect it with stars in that case being not necessarily more than eight or nine hours in the most extreme case; and the comparative ease with which an instrument may be constructed so that it shall be at all observations exactly in the meridian, and the possibility of constructing very accurate timepieces, has determined the use of such instruments for all the more precise observations in astronomy, such as fix the positions of the fundamental stars and the vernal equinox on the celestial sphere.

The equatorial instrument may be used for this purpose by clamping it in such a position that the reading of the hour-circle is 0°, in which case the declination axis is horizontal east and west, and when the telescope is moved about its axis it always lies in the plane of the meridian. If, with the instrument so adjusted, we observe the sun at the time of its meridian passage, we may find its declination by reading the declination circle, and its right ascension by noting the interval which elapses before the meridian transit of some known star after nightfall, free from any error involved in reading the hour-circle. As before, a star should be chosen at nearly the same declination, so that the interval of time may be very nearly equal to the difference in right ascension between the sun and the star, even if the instrument is not very exactly in the meridian. Observation of several different stars will enable us to determine whether the instrument actually does describe the plane of the meridian as it is rotated about the horizontal axis (see Chapter VIII); and by the observation of stars near the pole, as described on page 81, we may determine whether the declination circle reads exactly 0° when the telescope points to the equator, as should be the case.

THE MERIDIAN CIRCLE

An instrument which is to be used in this manner, however, is not usually so constructed that it can be pointed at any point in the heavens. Thus, it is unnecessary that it should consist of so many moving parts as the equatorial instrument, and steadiness, strength, and ease of manipulation are very much increased by constructing it as shown in Fig. 40, which represents a very small instrument built on the plan of the meridian circle of the fixed observatory. The strong horizontal axis revolves in two Y's, which are set in strong supports in an east and west line. The axis is enlarged towards the center, and through the center passes at right angles the telescope tube. The axis carries at one end a graduated circle

FIG. 40

perpendicular to the axis of rotation. If the axis of the telescope is perpendicular to the axis of rotation, and if the latter is adjusted horizontally east and west, the telescope may be brought into any position of the meridian plane, but must always be directed to some point of the latter. A pointer attached to the support marks the zero of the vertical circle when the telescope points to the zenith, and if the telescope be pointed to a star at the time of its meridian passage, the angle as read off on the circle is the zenith distance of the star; while the time of the star's meridian passage by a clock giving true sidereal time is its right ascension. If the latitude of the place of observation is known, the star's declination is determined by the fact that the zenith distance plus the declination of any body equals the latitude (see page 81). At first the latitude may be used as determined by the sun observation of Chapter I, or from a good map showing the place of observation, but ultimately its value should be determined with the meridian circle itself.

LEVEL ADJUSTMENT

We will now proceed to show how to make the necessary adjustments for placing the telescope so that it may move in the plane of the meridian.

Place the instrument on its pier and bring the Y's as nearly as possible into an east and west line. If the pier is the same that has been used in the previous work, this may be done by bringing the telescope into the meridian which has been determined by the method of equal altitudes.

The axis must first be brought into a horizontal line, making use for this purpose of the striding level (Fig. 41), which is a necessary auxiliary of this instrument. This is a glass tube nearly but not quite cylindrical, ground inside to such a shape that a plane passing through its axis, CD, cuts the wall in an arc, AB, of a circle whose center is at O. In this tube is hermetically sealed a very mobile liquid in sufficient quantity nearly but not quite to fill it — the space remaining, called the "bubble," always occupying the top of the tube. When CD is horizontal, the bubble rests in the middle of the tube with its ends, of course, at equal distances from the middle; the tube is graduated so that this distance may be measured, the numbering of the graduations usually increasing in both directions from the center of the tube. If the radius of the arc is 14.3 feet, a length of 3 inches of this arc will be equal to about 1°, since the arc of 1° in any circle is about $\frac{1}{57.3}$ of the radius; 1 inch of the arc will then be about 20', and 0.05 inch 1'. These are about the actual values for the level used with the instrument

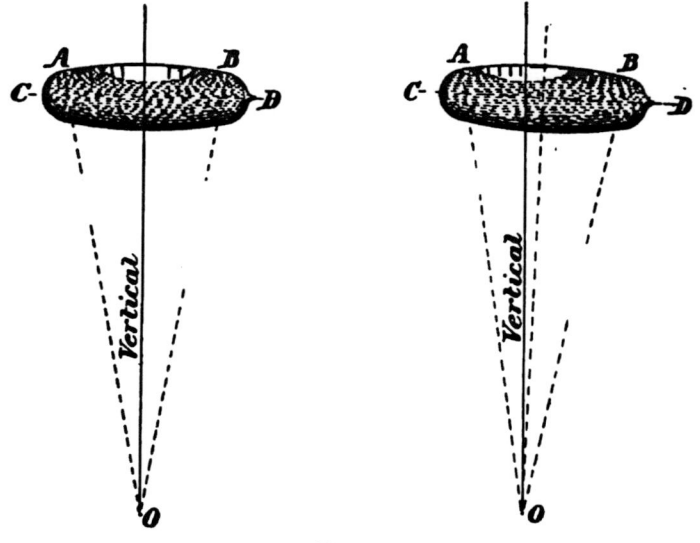

Fig. 41

shown in Fig. 40, the scale divisions being about $\frac{1}{20}$ of an inch apart and therefore corresponding to an arc of 1'.

If the line CD is inclined at an angle of 1' to the horizontal line by raising the end A, the center of curvature will be displaced toward the left, and the level will have the same inclination as if the whole tube had been turned to the right about the point O through an angle of 1'; and the highest point of the arc, which is always directly above O, is now $\frac{1}{20}$ of an inch from the middle toward A. Since the bubble always rests at the highest point of the arc, it follows that its ends will each be moved toward A by one division; if, for instance, the readings of the ends are 5 and 5 when CD is horizontal, they will be 6 and 4 when CD is inclined by 1', and evidently 7 and 3 when CD is inclined 2', etc., the inclination in minutes of arc being one-half the difference of the readings of the ends of the bubble, or $\frac{A-B}{2}$ if A and B represent the readings of the ends of the bubble in each case. If the reading of B is greater, the end A is depressed by one-half the difference of the readings; and the above expression applies to both cases if we agree that it shall always denote the elevation of A, a negative value of $\frac{A-B}{2}$ indicating depression of A.

REVERSAL OF THE LEVEL

The level tube is attached to a frame (Fig. 40) resting on two stiff legs terminating in Y's, which are of the same shape and size as those in which the axis of the meridian circle rests, the axis of the level tube being adjusted as nearly as possible parallel to the line joining the Y's. It is difficult to insure this condition, but if it is not exactly fulfilled, the horizontality of the axis may still be determined by placing the level on the axis, and determining the value $\frac{A-B}{2}$, and then turning it end for end, and again reading the value; for if the end A is high by the same amount in each case, the axis is obviously horizontal, and the measured angle of inclination is due to the fact that the leg of the level adjacent to A is longer than the other leg. The practical rule is to read the west and east ends in each position. If these readings are $W_1 E_1 W_2 E_2$, $\frac{W_1 - E_1}{2}$ is the elevation of the west end according to the first observation, and $\frac{W_2 - E_2}{2}$ at the second. If the leg which is west at the first observation is too long, the first observation gives a value for the elevation of the west end too great, and the second a value too small *by the same amount*; and the average of the two values $\frac{W_1 - E_1}{2}$ and $\frac{W_2 - E_2}{2}$ gives the true value of the inclination of the axis.

It is usual to write this $\frac{(W_1 + W_2) - (E_1 + E_2)}{4}$ and to record the observations in the following form:

$$\begin{array}{cc} W_1 & E_1 \\ W_2 & E_2 \\ \hline W_1 + W_2 & E_1 + E_2 \end{array}$$

Subtract the second sum from the first and divide by 4. This gives a positive value if the west end is high, and the axis may be made horizontal by turning the leveling screw so as to make the level bubble move through the proper number of divisions. The level should be again determined in the same way, and the axis is level when

$$(W_1 + W_2) - (E_1 + E_2) = 0.$$

The following record of level observation made Feb. 26.3, 1900, conforms to the above scheme:

$$\begin{array}{cc} W & E \\ 1\tfrac{1}{2} & 2\tfrac{1}{2} \\ 2 & 2 \\ \hline 3\tfrac{1}{2} & 4\tfrac{1}{2} \end{array}$$

-1

$-\tfrac{1}{4}$ division $= 15''$

The west end being too low, the screw was turned so as to raise it enough to move the bubble $\tfrac{1}{4}$ division toward the west, the level remaining on the axis during the adjustment and watched as the screw was turned; the readings were then as follows:

$$\begin{array}{cc} 2\tfrac{1}{4} & 1\tfrac{3}{4} \\ 1\tfrac{3}{4} & 2\tfrac{1}{4} \\ \hline 4 & 4 \end{array}$$

$$0$$

And the axis was truly level, since $(W_1 + W_2) - (E_1 + E_2) = 0.$

COLLIMATION ADJUSTMENT

The line of collimation of the telescope is the line drawn from the center of the lens to the wires that cross in the center of the field. When the telescope is "pointed" or "set" upon a star, the image of the star falls upon the point where these wires cross, and when the instrument is correctly adjusted the line of collimation is perpendicular to the axis of rotation, so that the line of collimation cuts the celestial sphere in a great circle as the telescope turns upon its axis.

To make this adjustment, point the telescope exactly upon any well-defined distant point, — the meridian mark will, of course, be chosen if it has been located, — then remove the axis from its Y's and replace it after turning it end for end; if the telescope is still set on the mark in the second position, the adjustment is correct; otherwise move the wire halfway toward the mark by means of the screws a, a (Fig. 40). Set again upon the mark by moving the screws in the eyepiece tube; reverse the axis again, and thus continue until the telescope points exactly upon the mark in both positions of the axis.

If the adjustments for level and collimation are properly made, the intersection of the wires in the center of the field of view will appear to describe a vertical circle, that is, a great circle through the zenith, as the instrument is turned on its axis. The final adjustment consists in bringing this circle to coincide with the meridian, but for this we must have recourse to observations of stars.

AZIMUTH ADJUSTMENT

The simplest method is to observe the time of transit by a sidereal clock of a circumpolar star at its upper transit, and again, 12 hours later, at its transit below the pole; if the interval is exactly 12 hours, the adjustment is correct; if the interval is less than 12 hours, the telescope evidently points west of the pole, and the west end of the rotation axis must be moved toward the north. This is done by the screws a, a (Fig. 40), the fraction of a turn being noted;

the observation is repeated upon the following night, and by comparing the change which has been produced by moving the screws, the further alteration required is readily estimated. On Feb. 26, 1900, the lower transit of ε Ursæ Minoris was observed at $4^h\ 58^m\ 12^s$, and the upper transit at $16^h\ 53^m\ 32^s$; the times were taken by a sidereal clock and have been corrected for its error; the interval being $11^h\ 55^m\ 40^s$, it is evident that the telescope pointed to the east of the meridian, the arc of the star's diurnal path between the lower and upper transits lying to the east of the meridian and being less than 12^h by $4^m\ 20^s$ or 260^s.

To correct the error, the west end of the axis was moved toward the south by turning the adjusting screws through one-quarter of a turn. On the following day the observations were repeated as follows:

Feb. 27.25, lower transit $4^h\ 54^m\ 45^s$; Feb. 27.75, upper transit $16^h\ 54^m\ 28^s$; the eastern arc was still too small, but the error had been reduced to 17^s, and required a further correction of $\frac{17}{243}$ of a quarter turn of the screws, which were therefore turned through about $6°$ in the same direction as before, and the instrument was thus brought very closely into the meridian.

This method can only be used with small instruments when the night is more than 12 hours long; but it is the only independent method; it requires that the rate of the clock shall be known between the two observations, and it requires observations at inconvenient times. A more convenient method is always used in practice, but requires an accurate knowledge of the right ascensions of a considerable number of stars in the neighborhood of the pole.

It has been stated that it is often inconvenient to observe the moon when on the meridian, but with this exception all the fundamental observations of astronomy are now made with meridian instruments on account of the simplicity and permanence of the necessary adjustments. A body observed on the meridian is also at its greatest altitude and least affected by atmospheric disturbances, which often interfere with the observation of bodies near the horizon.

DETERMINATION OF DECLINATIONS WITH THE MERIDIAN CIRCLE

The circle of the meridian instrument may be used to determine the declination of a star in two ways, of which that now described is perhaps the most obvious, but also the least convenient.

If the reading of the circle is known when the telescope is pointed at the pole, the angle through which the telescope must be moved to point upon any star, that is, the polar distance of the star, is the difference between this value and the circle reading when the telescope is pointed at the star; this angle is 90°—the star's declination; if the star is on the equator, the angle is 90°; and if the star is south of the equator, the angle is greater than 90° by an amount equal to the declination of the star; if we consider the declination a negative quantity for a star south of the equator, the value $90° - \delta$ represents the polar distance in all cases.

To determine the reading of the "polar point" we may set the telescope upon a circumpolar star at its "upper culmination" and read the circle, and again, 12 hours later, set on the same star at its "lower culmination," the mean of the two readings is the reading of the polar point. The effect of refraction may be neglected with our small instruments without causing an error of $\frac{1}{40}$ of a degree at any place in the United States if we restrict ourselves to stars within 10° of the pole, or the circle readings may be corrected by a refraction table. Immediately after making this determination it is advisable to make a setting on the meridian mark and note the reading; this point may thereafter be used as a reference point from which the reading of the polar point may be at any time determined if the meridian mark has not in the mean time changed its position.

Better still, the observation of the polar point may be combined with a determination of the circle reading when the telescope points at the zenith, by one of the methods to be described later; the difference of the readings in this case is obviously equal to the co-latitude, and such an observation constitutes an "absolute determination of the latitude," that is, a determination made without reference to observations made at any other place. When the latitude has once been satisfactorily determined, the observations of the declinations of stars can be made to depend upon determinations of the zenith point by means of the fact that for a body on the meridian

Declination = Latitude − Zenith Distance,

latitude and declinations being reckoned positive northward from the plane of the equator, and zenith distance positive southward from the zenith. The proof of this relation is left to the student as well as the interpretation of the result when the observation is made at the transit below the pole.

At the time of observing the transits of ε Ursæ Minoris described on page 79 the following readings of the circle were made when the star was in the center of the field. Each of these observations consists of two readings: one of the index A on the south end of a horizontal bar fixed to the supports of the axis, and the other of an index B at the other extremity of the bar, as nearly as possible half a circumference from A. An angle given by the mean of two readings made in this manner is free from the "error of eccentricity," which affects readings by a single index in case the center of the graduated circle does not exactly coincide with the axis about which it is turned between the two observations.

DATE	A	B	MEAN
February 26.25	55°.45	55°.35	55°.40
26.75	39 .95	39 .85	39 .90
27.25	55 .45	55 .35	55 .40
27.75	39 .95	39 .85	39 .90

Hence the reading when the instrument was pointed at the pole was $\dfrac{55°.40 + 39°.90}{2} = 47°.65$.

Evidently the polar distance of the star was $\dfrac{55°.40 - 39°.90}{2} = 7°.75$, and its declination 82°.25; and we have thus obtained an "independent" or "absolute" determination of the declination of ε Ursæ Minoris; that is, a determination independent of the work of other observers, and only dependent on the accuracy of our circle and of our observations.

The circle was known to be adjusted so that the reading of the zenith was very exactly zero, hence the latitude of the place of observation was 42°.35. The exact agreement of these observations indicates that the magnifying power of the telescope was such that it could be set more accurately than the circle could be read, and not that the results are reliable to a hundredth of a degree.

For convenience in recovering the zenith reading, in case the adjustment of the circle should be disturbed, the zenith distance of a meridian mark was measured repeatedly, the result showing that its polar distance was 137°.47, and this was used to check the polar reading in later observations upon stars when it was impossible to get observations of the same star above and below the pole.

Another method of making absolute determinations of the latitude with the meridian circle is to observe the zenith distance of the sun at the solstices; the mean of these values being the zenith distance of the equator, which is equal to the latitude. This observation, however, is subject to considerable uncertainty on account of the difference in atmospheric conditions at the summer and winter solstice, and to great inconvenience on account of the lapse of time; it is, however, of course, the means upon which we must rely for the accurate determination of the obliquity of the ecliptic, one of the fundamental quantities of astronomy.

For the use that we shall make of the meridian circle, it will probably be most convenient to make a careful determination of the polar distance of the meridian mark, and use this habitually as a point for reference.

PROGRAM OF WORK WITH THE MERIDIAN CIRCLE

Work with the meridian circle should at first consist of reobservation of all the stars which have been previously observed with the equatorial, except those which are west of the meridian after nightfall and cannot be observed for six months. Attention should be given to gathering a list of stars within 15° or 20° of the pole for the purpose of quickly setting the instrument in the meridian by the methods of page 79. The sun should be observed at least once a week and its place plotted on the globe, and many stars in the neighborhood of the moon's path to form a basis for finding the moon's place by differential observations, of course, also the moon itself, the planets and a comet, if any of sufficient brightness appears. In this way, by observing a few stars each night, a great amount of material may be stored for future use.

Especial attention should be given to getting a good number of observations of stars near the equator, so that fairly accurate values of their differences of right ascension may be obtained, and at the first opportunity the absolute right ascension of one of their number must be determined in order that thus the places of all may be known. The results may be best recorded by making a list of their right ascensions referred to an assumed vernal equinox. Thus, the observations discussed on page 52 show that α Pegasi precedes γ Pegasi by $17°.03 = 1^h\ 8^m\ 7^s$, or, in other words, follows it by $22^h\ 51^m\ 53^s$; and if the right ascension of γ Pegasi referred to the assumed equinox is $0^h\ 8^m$, that of α Pegasi is $22^h\ 59^m\ 53^s$. If in the course of the year observation shows that the true right ascension of γ Pegasi is $0^h\ 8^m\ 5^s$, it is evident that the true value for α Pegasi is $22^h\ 59^m\ 58^s$, and that the right ascension of all stars referred to the assumed equinox by comparison with γ Pegasi must be increased by 5^s.

DETERMINATION OF THE EQUINOX

An opportunity for observing the absolute right ascension of the zero star, which is often called a "determination of the equinox," occurs about the middle of March and September.

If the course is begun in September, it will be well to make this determination with the help of more experienced observers, even before the nature and object of the measures are understood.

The observation consists in determining the difference of right ascension of some star from the sun at the instant when the latter crosses the equator, for at that time it is either at the vernal or autumnal equinox, and its right ascension is in the one case 0 hours and in the other 12 hours.

If a meridian observation of the sun's altitude shows that the sun is exactly on the equator at meridian passage, and the time of transit is noted by a sidereal clock, and as soon as it is sufficiently dark the transit of a star is observed, the difference of the times is the absolute right ascension of the star if the observation is made at the vernal equinox, or equals the right ascension of the star minus 12^h if the observation is made at the autumnal equinox.

Inasmuch as the meridian of the observer will rarely be that one on which the sun happens to be as it crosses the equator, we must make observations on the day before and the day after the equinox, thus getting the difference of right ascension of the star from the sun at noon on both days. The declination of the sun being also measured at these two times, a simple interpolation gives the time at which the sun crossed the equator, and this time being known, another simple interpolation between the differences of right ascension at the two noons gives the difference of right ascension of the sun and star at the time when the sun was at the equinox, which is the star's absolute right ascension.

The first interpolation assumes that the sun's declination changes uniformly with the time, and the second that its right ascension changes uniformly with the time.

Observations should extend over a period of a week before and a week after the equinox to test the truth of these assumptions.

In observing the sun, a shade of colored or smoked glass may be placed over the eyepiece, or the eyepiece may be drawn out as in the method of observation described on page 37, and the screen held in such a position that the cross-wires are sharply focused upon it. As the image of the sun enters the field it should be adjusted by moving the telescope slightly north or south till the horizontal wire passes through the center of the disk, and as the latter advances, the time should be noted when the preceding and following limbs cross the vertical wire, as well as the time when the vertical wire bisects the disk; at the instant of transit the disk should be neatly divided into four equal divisions, a very small deviation from this condition being quite perceptible to the eye.

THE AUTUMNAL EQUINOX OF 1899

The following table gives the details of observations taken at the autumnal equinox of 1899 for the purpose of determining the equinox.

The latitude of the place of observation was 42°.5, and the declinations given in the last column are calculated by subtracting the zenith distance in each case from this quantity, as explained on page 81.

Date	Object	Time of Transit	Zen. Dist.	Decl.
Sept. 22	Sun	12^h 0^m 2^s.0	S 42°.2	+ 0°.3
	η Serpentis	18 18 22.6	45 .4	− 2 .9
	λ Sagittarii	18 24 2.4	67 .95	− 25 .45
	Vega	18 35 44.5	3 .87	+ 38 .63
	Altair	19 48 7.6	33 .98	+ 8 .52
Sept. 23	Sun	12 3 45.1	42 .62	− 0 .12
	η Serpentis	18 18 20.1	45 .4	− 2 .9
	λ Sagittarii	18 23 57.3	67 .97	− 25 .47
	Vega	18 35 42.6	3 .85	+ 38 .65
	Altair	19 48 1.5		

The intervals between the observed times of transit of each star on the two different dates range from 23^h 59^m 53^s.9 to 23^h 59^m 58^s.1, showing that the clock was losing about 4^s daily, a quantity so small that for our purpose it may be neglected.

Observations of the sun made on different dates between September 18 and September 23, but not here recorded, showed that its right ascension and declination were changing uniformly at the rate of about 3^m 45^s and 0°.39 per day. The table above shows that from September 22 to September 23 the rates were 3^m 43^s.1 (or, allowing for clock rate, about 3^m 39^s) and 0°.42 per day, and the latter value we shall use to determine the time of the equinox, as follows:

At noon September 22, or September 22^d.0, as it is expressed by astronomers, the sun's declination was + 0°.3, and September 23.0

its declination was $-0°.12$. Hence its declination was $0°$ September $22\frac{3}{4}\varrho$, or September $22^d.714$. It was at that time, as exactly as our observations can show, at the autumnal equinox, and its right ascension was $12^h\ 0^m\ 0^s$.

Since η Serpentis followed it to the meridian $6^h\ 18^m\ 20^s.6$, that quantity is the difference between the right ascension of the star and that of the sun September 22.0. Similarly the difference of right ascension of sun and star September 23.0 was $6^h\ 14^m\ 35^s.0$; that is, it was $3^m\ 45^s.6$ less than at the previous date. Assuming this change to be uniform, the difference of right ascension of sun and star at the moment of the equinox on September $22^d.714$ was $0.714 \times 3^m\ 45^s.6$, or $2^m\ 41^s.1$ less than on September 22.0; that is, it was $6^h\ 15^m\ 39^s.5$, and since the right ascension of the sun September 22.714 was $12^h\ 0^m\ 0^s$, the right ascension of η Serpentis was $18^h\ 15^m\ 39^s.5$.

The following table gives the data from which the "absolute right ascensions" of the four stars are thus determined. In the last column are the declinations, which are the means obtained from several observations between September 14 and September 23.

STAR	R.A. OF STAR MINUS R.A. OF SUN			STAR'S	
	SEPT. 22.0	SEPT. 23.0	SEPT. 22.714	R.A.	DECL.
η Serpentis	$6^h\ 18^m\ 20^s.6$	$6^h\ 14^m\ 35^s.0$	$6^h\ 15^m\ 39^s.5$	$18^h\ 15^m\ 39^s.5$	$-2°.89$
λ Sagittarii	6 24 0.4	6 20 12.2	6 21 17.4	18 21 17.4	-25.48
Vega	6 35 42.5	6 31 57.5	6 33 1.8	18 33 1.8	$+38.65$
Altair	7 48 5.6	7 44 16.4	7 45 21.9	19 45 21.9	$+8.59$

The measurements upon which the above results depend are of two kinds: observed clock times, which are liable to errors of a very few seconds, so that the *differences* of right ascension may be assumed to be correct within perhaps 4^s; and measures of the sun's declination, which with the greatest care may be in error at least $0°.05$ on any given date.

It is quite within the bounds of probability, for instance, that the sun's declination was $+0°.25$ on September 22.0 and $-0°.17$

on September 23.0; and recomputing with these values, the date of the equinox was September 22$\frac{24}{41}$, or September 22^d.595, and the right ascensions of the stars 18^h 16^m 6^s.4, 18^h 21^m 44^s.6, 18^h 33^m 28^s.6, 18^h 45^m 49^s.2; that is, the uncertainty of the equinox is 0.12 days and of the right ascensions about 27^s, although the relative right ascension is altered only by a fraction of a second in each case. It is thus evident that the accuracy of the right ascensions depends chiefly upon the accuracy with which the sun's declination can be measured.

THE AUTUMNAL EQUINOX OF 1900

In order to increase the accuracy of determination of declination, a new circle reading to minutes of arc was substituted for that used for the observations of the equinox in 1899, and the observations were repeated at the same place in 1900. The weather conditions were unfavorable, so that only the following observations could be made.

Date	Object	Time of Transit	Zen. Dist.	Decl.
Sept. 22	Sun	11^h 58^m 44^s.8	S42° 11′.5	+ 0° 18′.5
	Vega	18 35 27 .0	3 49 .0	+38 41 .0
	Altair	19 47 49 .0	33 51 .0	+ 8 39 .0
Sept. 23	Sun	12 3 1 .5	42 33 .1	− 0 3 .1
	Altair	19 48 35 .0	33 54 .0	+ 8 36 .0

From these data, by the same method as before, the date of the equinox is found to be September 22$\frac{18.5}{21.6}$, or September 22.8565. If each declination of the sun is accurate to 1′, the result *may* be in error by $\frac{2}{21.6}$ days, or about .09 day; the actual error is probably less than half this amount, and the concluded right ascensions probably within 10^s of the true values.

The observed times of Altair on the two dates show that the clock was gaining 46^s daily, since the true sidereal time of transit, being equal to the star's right ascension, is the same on both nights. This rate is so large that it cannot be neglected as in the discussion of the result for 1899.

If the clock correction Δt (see page 60) at the time of the sun's transit, September 22, be assumed 0^s and the gaining rate 46^s per day, or $1^s.916$ per hour, the corrections for Vega and Altair September 22 were $-12^s.6$ and -14.9, and for the sun and Altair September 23 were -45.9 and $-61^s.0$. The times obtained by applying these corrections are said to be "corrected for rate of the clock to the epoch September 22.0."

In this manner the times, as they would have been observed with a clock having an exact sidereal rate, are found to be:

	September 22	September 23
Time of transit of the Sun . . .	11^h 58^m $44^s.8$	12^h 2^m $15^s.6$
" " " Vega	18 35 15.4	
" " " Altair	19 47 34.1	19 47 34.0

Hence Altair followed the sun

September 22.0	7^h 48^m $49^s.3$
" 23.0	7 45 18.4
" 22.856	7 45 48.8

and the right ascension of Altair was 19^h 45^m $48^s.8$; since Vega precedes Altair by 1^h 22^m $18^s.7$, its right ascension was 18^h 33^m $30^s.1$.

In 1899 the difference of right ascension of the two stars was 1^h 22^m $20^s.1$, but the right ascensions of 1900 are greater by $28^s.3$ and $26^s.7$ than those of 1899.

If we assume the later determination to be absolutely correct, we must regard the earlier as having placed the equinox farther toward the east among the stars than its true place, so that right ascensions referred to the equinox observed in 1899 are too small. We may say that the observations of 1900 indicate a correction of $-27^s.5$ to the "equinox of our little catalogue of four stars"; that is, a correction of $+27^s.5$ to all their right ascensions as determined in 1899.

Applying these corrections, their right ascensions become for

η Serpentis	18ʰ	16ᵐ	7ˢ.0
λ Sagittarii	18	21	44 .9
Vega	18	33	29 .3
Altair	19	45	49 .4

Since the later observations were made with an instrument giving more accurate values of the declination, it is probable that their results are more nearly correct. The clock rate was neglected in the first observations, and the effects of precession, parallax, and refraction in both series, following out the principle that no corrections will be made until observations shall show their necessity.

The effect of refraction is to delay the autumnal equinox about an hour, and hence to decrease the right ascensions of the stars by about 10ˢ. At the vernal equinox, however, refraction hastens the equinox an hour and increases the right ascensions by 10ˢ; its effect may be shown by observations at the two equinoxes of the same year and eliminated by their combination. Parallax hastens the autumnal and delays the vernal equinox by about 8ᵐ, thus affecting right ascensions by a little more than 1ˢ, the mean of observations at the two equinoxes being free from error from this source. The effect of precession will be manifest in less than ten years with an instrument like that used in the above observations of 1900.

By comparing the equinox of September 22.714 ± 0.12, 1899, and September 22.856 ± .09, 1900, the length of the tropical year is found to be 365°.142, but may lie between 364.93 and 365.35 as far as our observations can surely determine. Since refraction delays the vernal and hastens the autumnal equinox by nearly the same amount (about an hour) in each case, it has no effect upon the length of the year. As the greatest error to be feared with our improved instrument is less than 0.1 day, the length of ten or one hundred years may be determined with less than twice that error, in those periods the length of the year may be determined within 0.02 and .002 day, respectively.

With the best modern instrument used to the greatest advantage, the sun's declination may be determined near the equinox within

$0''.5$, and hence the time of the equinox within 30^s and right ascensions within $0^s.08$. A single tropical year may be measured with an error of less than 1^m.

We have now explained the methods by which it is possible to fix the places of the sun, moon, and stars at different times and thus to obtain data from which their apparent motions about the earth may be studied and theories formed from which their future places may be predicted. More or less complete accounts of these theories are to be found in all works on descriptive astronomy, and the predictions derived from them are published for three years in advance by several governments for the use of navigators and astronomers. Such a publication is the American Ephemeris and Nautical Almanac, of which it will be convenient to give some account before taking up the motions of the planets.

The apparent motions of the planets are less simple than those of the sun, moon, and stars, which at all times seem to move about the earth as a center with approximately uniform velocities. The planets, it is true, in the long run continually move like the sun and moon around the heavenly sphere toward the east, but their velocities are variable within wide limits and at certain times are even reversed, so that they move in the opposite direction or "retrograde" among the stars.

For this reason a longer period of observation is necessary to determine their motions than can be given by the individual student. We may, however, regard the nautical almanacs of past years as predictions that have been verified, and they stand for us as an accredited set of exceptionally accurate observations from which we may draw material to combine with the results of our own observations.

CHAPTER VII

THE NAUTICAL ALMANAC

THE American Ephemeris and Nautical Almanac consists of two parts,—the Nautical Almanac proper, which is published separately and contains data especially useful in navigation, and a second part, which contains additional tables adapted to the use of astronomers. The Nautical Almanac will suffice for most of our purposes, but the complete work is convenient for a few references.

The tables contain data for the sun, moon, and planets, for successive equidistant points of Greenwich mean time, so near together that the values at any intermediate time may be obtained by interpolation with a degree of accuracy greater than can be obtained by a single observation made with the most refined instruments. The dates are given in astronomical time, each day beginning at noon of the corresponding civil date.

At this point a few words are necessary in explanation of the term "mean time."

We have already defined apparent solar time as the hour-angle of the sun, and sidereal time as the hour-angle of the vernal equinox. Owing to the fact that the sun moves at a varying angular rate and in a path inclined to the equinoctial, the hour-angle of the sun does not increase uniformly, and the hours of apparent time are, therefore, of unequal length.

We have not yet obtained material for a complete discussion of the relation between apparent and mean solar time, and for this we must refer to the text-books of descriptive astronomy. It will be convenient to explain one simple statement of this relation which is not always explicitly given.

The time required by the sun to complete its circuit of the heavens, from one passage through the vernal equinox to another, is 365.2422 days. As it describes 360° of longitude in that time, its average daily motion in longitude is 0°.985647.

To establish a convenient measure of time not greatly different from apparent solar time, a fictitious body is imagined to start with the sun at perihelion and to move along the ecliptic with a uniform daily motion in longitude of 0°.98565. Its longitude at any time is called, appropriately enough, the "mean longitude of the sun."

When this body reaches the vernal equinox, a second fictitious body, called the "mean sun," is supposed to start out from that point eastward along the equator, moving with a uniform velocity equal to the mean daily motion of the sun in the ecliptic.

The mean sun, therefore, continually increases its *right ascension* by 0°.98565 per day; and since both fictitious suns are at the vernal equinox in longitude zero at the same instant and move at the same rate, one in the ecliptic and the other in the equator, it is obvious that at all times the right ascension of the mean sun is equal to the sun's mean longitude.

The hour-angle of the mean sun is equal to the mean solar time, just as the hour-angle of the true sun is equal to the apparent solar time.

A clock, properly regulated and set so that it shows $0^h\ 0^m\ 0^s$ at each successive passage of the mean sun over the meridian of a given place, is said to keep the local mean time of that place. When the hour-angle of the mean sun is 10°, 20°, 30°, the local mean time is $0^h\ 40^m$, $1^h\ 20^m$, 2^h, respectively.

It is of course true of the mean sun as of any other heavenly body (see page 58) that its H.A. + R.A. = Sid. T. We may therefore write:

H.A. of mean sun + R.A. of mean sun = Sid. T.
H.A. of sun + R.A. of sun = Sid. T.

And from these equations, remembering the definitions of mean and apparent time, we derive the following:

Mean T. = App. T. + (R.A. of sun − R.A. of mean sun).

The quantity in the parenthesis, which must be added to App. T. to give the corresponding Mean T., is called the equation of time.

The equation of time is the difference between mean time and apparent time, and when *positive* must be *added* to apparent time to give the corresponding mean time, or *subtracted* from mean time to find the corresponding apparent time.

Standard Time. — It is now usual to regulate the clocks over large sections of country to the mean time of a neighboring meridian. Thus, clocks in the central part of the United States are set to show $0^h\ 0^m\ 0^s$ when the sun is in the meridian whose longitude is 90° west of Greenwich, and they are said to keep Central standard time; which is, therefore, 6 hours slow of Greenwich time. Other sections use the mean time of the 75th, 105th, and 120th meridians, 5, 7, and 8 hours slow of Greenwich, respectively. More than one half the people of the United States use Central standard time.

The fact that our watches are set to standard time is a convenience in using the Almanac, since the watch time gives us Greenwich mean time by applying so simple a correction, the minutes and seconds being unchanged and the hours increased by a small constant number.

THE CALENDAR

About four-fifths of the Nautical Almanac consists of data regarding the sun and moon, eighteen successive pages being devoted to each month, and the corresponding pages of the different months numbered with the Roman numerals from I to XVIII. These pages, which form the Calendar, we will now consider in detail. The reading matter of the Explanation which follows the tables should be carefully read in connection with the following paragraphs: reduced facsimiles of several pages are shown at page 176, to which reference may be made.

The positions are given as they would appear to an observer at the earth's center, and the times are, as stated at the head of each page, Greenwich mean time. We pass at once to page II, which, rather than the very similar page I, it will be always more convenient to use when, as in most of our observations, the Greenwich time is known for which the data are required.

Page II. — The first and second columns give the day of the week and month. The third column contains the sun's apparent right ascension at Gr. Mean Noon, — that is, its right ascension as affected by the annual aberration (which makes it appear to be about 20″ behind its true place in its orbit) and measured from the actual equinox of the date. Column 4 contains the hourly difference, or the amount by which the right ascension is changing per hour.

To illustrate the use of this column, let it be required to find the right ascension of the sun at the time of the first observation recorded on page 39 at $8^h 54^m 37^s$ A.M., Eastern standard time, March 8, 1900.

We must first notice that the corresponding astronomical time, which is reckoned from noon to noon, is $20^h 54^m 37^s$ after noon of the preceding day, — that is, the local date was March $7^d 20^h 54^m 37^s$; adding 5^h to change E. Std. T. to G.M.T., we have March $7^d 25^h 54^m 37^s$, or March $8^d 1^h 54^m 37^s$, G.M.T.

The sun's right ascension, March 8, at Greenwich mean noon, is given as $23^h 13^m 57^s.68$. To this, since the sun's right ascension is always increasing, must be added the change in $1^h 54^m 37^s (= 1^h.91)$, the time elapsed since noon, which is obtained by multiplying the hourly difference found in column 4 by 1.91; this gives the correction to be added to the tabular right ascension as $1.91 \times 9^s.237$, or $17^s.64$, and the right ascension at the time of observation was therefore $23^h 13^m 57^s.68 + 17^s.64$, or $23^h 14^m 15^s.32$.

This simple process, which is fully illustrated in the Explanation, will never give a value more than $0^s.4$ in error. A method of interpolation by which an accuracy of $0^s.01$ may be attained is given in the Explanation. The error of the simple method arises from the fact that the hourly difference is not constant, as will appear at once from inspection of the values in the fourth column.

Columns 5 and 6 give the sun's apparent declination and its hourly difference. The value at any time may be found by interpolation in the manner just explained.

North declinations are regarded as positive, and south declinations negative, and in accordance with this convention the hourly difference is marked + when the change of declination is toward the north and − when toward the south, so that the true declination is found by applying the correction algebraically: thus, to find the declinations at 4 P.M., G.M.T., on the following dates, we have:

1900	δ AT MEAN NOON	H. DIFF.	CORR. FOR 4ʰ	δ AT 4ʰ G.M.T.
Jan. 10	− 21° 59′ 4″.0	+ 22″.25	+ 4 × 22″.25 = + 89″.0	− 21° 57′ 35″.0
April 10	+ 7 53 3 .7	+ 55 .48	+ 4 × 55 .48 = + 221 .9	+ 7 56 45 .6
Aug. 10	+ 15 38 18 .2	− 43 .73	− 4 × 43 .73 = − 174 .9	+ 15 35 23 .3
Nov. 10	− 17 6 18 .2	− 42 .31	− 4 × 42 .31 = − 169 .2	− 17 9 7 .4

The error in a declination determined by a simple interpolation from the preceding mean noon can never exceed 12″. By the more accurate method given in the Explanation, it is always less than 0″.1.

To make sure that the correction has been applied with the proper sign, it is sufficient to notice that the computed value must lie between the values for the including dates.

Columns 7 and 8 contain the equation of time and its hourly difference. The correction to be applied is obtained, as in the preceding examples, by multiplying the hourly difference by the number of hours elapsed since Greenwich mean noon, and must either be added or subtracted so as to give a value between the values of the including dates.

The heading of the column indicates whether the equation of time is to be added to or subtracted from mean time to give apparent time. Of course when it is additive to mean time it must be subtracted from apparent time to give mean time, as will appear on comparing the corresponding column of page I.

Example. What is the equation of time January 10, 1900, at $3^h 45^m$, Central standard time?

The corresponding G.M.T. is $9^h 45^m = 9^h.75$
Eq. of T. at Gr. Mean Noon . $+ 7^m 39^s.87$ H. Diff. $= 1^s.014$
Change in $9^h.75 = 9.75 \times 1^s.014$ 9 .88 $\times 9.75$
Eq. of T. at $3^h 45^m$, Cent. T. . $+ 7$ 49 .75 Corr. $= 9^s.88$

The correction $9^s.88$ is added because the value of the equation January 11 is seen to be $8^m 3^s.90$, and the correction must be applied so as to increase numerically the value on January 10.

The ninth column contains the right ascension of the mean sun. Since at mean noon the mean sun is on the meridian and since (p. 59) the right ascension of a body which is on the meridian at a given instant equals the sidereal time at that instant, the right ascension of the mean sun at Greenwich mean noon equals the Greenwich sidereal time at Greenwich mean noon, and this explains the alternative heading which appears at the top of the column.

Since the right ascension of the mean sun increases uniformly, the constant hourly difference requires no special column, but is given at the foot of the page. For interpolation it is most convenient to use Table III, which occupies three of the last pages of the Almanac, and gives directly the multiples of $9^s.8565$ by each hour and minute up to 24 hours, thus saving the reduction of minutes to decimals of an hour.

Example. Right ascension of mean sun, January 15, 1900, at $4^h\,44^m\,30^s$.

R.A. mean sun, Gr. Mean Noon	$19^h\,37^m\,55^s.26$
Add $4^h\,44^m\,30^s \times 9^s.8565$ (Table III)	46.74
R.A. mean sun at $4^h\,44^m\,30^s$	19 38 42 .00

This is obviously the sidereal time of mean noon at a place in longitude $4^h\,44^m\,30^s$ west, and if desired a table of this quantity may be computed for such a place by adding $46^s.74$ to the values given each day in the Almanac for Greenwich.

Page I.—The quantities on page I are only used for reducing meridian observations of the sun, which are made, of course, at local apparent noon. This page is convenient when the Greenwich mean time has not been noted, for the time elapsed since the preceding Greenwich apparent noon is equal to the west longitude of the place of observation. This is the quantity, therefore, by which the hourly difference must be multiplied to give the correction. An example of the use of this page is given on page 104.

All the quantities given on page I may be found more easily from page II if we know the G.M.T. for which they are required. The only quantity for which we are obliged to consult page I is the semi-diameter, and this never differs by so much as $0''.01$ from its value at mean noon.

Page III. — Column 2 gives the day of the year corresponding to the given date, and is convenient for finding the number of days intervening between dates. Thus, January 15, 1900, is the 15th day of the year and September 25 is the 268th; hence from noon, January 15, to noon, September 25, is 268 − 15, or 253 days.

Column 3 contains the sun's longitude measured from the vernal equinox of the given date. For some purposes it is more convenient to measure from the mean equinox of the beginning of the fictitious year, an epoch much used in astronomical calculations but of no intrinsic interest. The minutes and seconds of the longitude as thus measured are found in column 4. The longitude of column 3 is measured from the actual place of the equinox at the given date as affected by precession and nutation.

Column 6 gives the sun's latitude, which is always nearly but not exactly zero, as will be explained further on in this chapter.

Column 7 gives the logarithm of the earth's distance from the sun in astronomical units. An astronomical unit is equal to the semi-axis major of the earth's orbit, — about 93,000,000 miles. For those unacquainted with logarithms the following table will make it easy to find by interpolation the approximate distance corresponding to a given logarithm.

Logarithm 9.9925000 corresponds to 0.9829 astronomical units.
" 9.9950000 " " 0.9886 " "
" 9.9975000 " " 0.9943 " "
" 0.0000000 " " 1.0000 "
" 0.0025000 " " 1.0058 "
" 0.0050000 " " 1.0116 "
" 0.0075000 " " 1.0174 "

Example. January 19, 1900, log radius vector = 9.99299, which is very nearly $\frac{1}{4}$ of the way from 9.9925 to 9.9950; hence on that date the distance of the earth from the sun is $\frac{1}{4}$ of the way between 0.9829 and 0.9886, or 0.9840 astronomical units. The value can be obtained within less than $\frac{1}{300}$ of its amount without interpolation by taking the nearest value of the logarithm given in the table.

Column 9 gives the mean time at which the vernal equinox is on the meridian of Greenwich (when the number of hours is greater than 12 the time is after midnight, and therefore during the morning hours of the next civil date). This quantity is sometimes used in converting sidereal to mean time, but its use may be easily avoided and is sufficiently treated in the Explanation.

Page IV. — The quantities on page IV relate to the moon. They are given for each 12 hours of Greenwich mean time, and seem to call for no explanation, except perhaps the symbol ☌, signifying conjunction, which occurs once (and occasionally twice) upon each page, on the day before or after that of new moon. Since successive transits follow each other nearly 25 hours apart, in general one date in each month would be left blank, the moon crossing the meridian during the hour preceding noon of one date, and during the hour following noon of the succeeding date. The symbol ☌ occupies the vacant space and marks the date of new moon.

Pages V to XII contain the right ascension and declination of the moon for every hour of G.M.T., together with their differences for each minute of time. The rapid motion of the moon makes it necessary to give these quantities at shorter intervals than suffice for the sun, in order that an equal accuracy may be attained in interpolation.

These are of course places as seen from the earth's center, and it is to be remembered that at any point on the earth's surface the moon may be displaced by parallax a little more than 1°.

On page XII are given the exact dates to the nearest hour of G.M.T. of the moon's phases and the times of perigee and apogee.

Pages XIII to XVIII contain tables of "lunar distances," — that is, distances for each three hours of Greenwich mean time between the moon's center and certain bright stars and planets not far from the plane of its motion; the sun is included in the list, as the moon is often visible in full daylight, so that its distance from the sun may be easily measured.

This table is used in determining longitude; the local time being known, the G.M.T. may be found by the method of lunar distances, as follows: The distance from moon to star or sun being measured is found to lie between two distances given in the table; the G.M.T. of the observation then lies between the hours corresponding to the two tabular distances, and its exact value may be determined by interpolation. The difference between this time and the known local time of the observation is the longitude.

The method requires accurate observations, and troublesome computations are necessary to correct the measured distance for the effects of refraction and parallax so as to find the distance from moon to star as seen from the earth's center.

Data for the Planets, Eclipses. — Following the calendar pages of the Nautical Almanac are thirty pages giving the right ascension and declination and the time of meridian passage of the five planets which are visible to the naked eye, and three pages containing the right ascensions and declinations of 150 of the brighter fixed stars.

A few pages are devoted to the eclipses of the year, with maps from which may be obtained the approximate times of the successive phases of the solar eclipses as seen from any given point of observation on the earth.

EXAMINATION OF THE SEVERAL COLUMNS

Having given this general summary of the contents of the tables, we will now call attention to some of the interesting facts and relations that appear on running through the various columns throughout the whole year.

The date of the solstices may be determined as the days on which the sun's declination has its maximum northern and southern values.

The date of the equinoxes may be found, from either the right ascension or declination columns, as the date on which the declination changes sign, and the right ascension is either 0^h or 12^h; the exact time may be found by interpolation. (See page 107.)

The number of days between the equinoxes may be determined by using the column of days, page III. It will be found that the sun is for some days more than half the year in that part of its orbit which lies in the northern hemisphere.

The column of hourly difference shows that the declination is changing slowly at the solstices and most rapidly at the equinoxes; moreover, the change at the latter dates is nearly uniform both in right ascension and declination, as stated on page 85. If a right triangle be drawn with the difference in right ascension for the date of the equinox as base and difference in declination as altitude, the angle between the base and the hypotenuse measured by a protractor will be found to be $23\frac{1}{2}°$. It obviously equals the angle between the equator and the ecliptic.

Notice that the equation of time is the difference between right ascension of mean and true sun, as stated on page 92, thus:

From the Almanac for 1900 (p. II), we have the following values: January 21, Sun's R.A. = $20^h 13^m 2^s.79$; R.A. Mean Sun = $20^h 1^m 34^s.61$. Subtracting the latter from the former, we have for the equation of time $+11^m 28^s.18$. This is the value given on page II; the positive sign indicates that it is to be added to apparent time to find mean time, or subtracted from mean time to find apparent time.

The dates on which the equation of time is 0 and dates and values of greatest and least equations should be noticed; also that on those dates for which the equation is 0 the values of the sun's right ascension and declination, etc., on pages I and II, are the same, since apparent noon and mean noon coincide. For 1900 the civil dates are as follows:

	EQ. OF T.
February 11	$+ 14^m 27^s.28$
April 15	0
May 15	$- 3^m 49^s.40$
June 14	0
July 27	$+ 6^m 17^s.22$
September 1	0
November 3	$- 16^m 20^s.40$
December 25	0

The hourly difference of the right ascension of the mean sun has the same integers as the mean daily motion of the sun in longitude, 0.98565; for $0°.98565$ per day $= \dfrac{0°.98565}{24}$, or $\dfrac{0.98565 \times 3600''}{24}$, per hour, and reducing this to seconds of time by dividing by 15, we find the motion of the mean sun to be $9^s.8565$ per hour. This illustrates the fact that the mean motion of the sun in longitude ($0°.98565$ per day) is the same as that of the mean sun in right ascension ($9^s.8565$ per hour), page 92.

The column which gives the sun's latitude will repay an investigation. It appears at a glance that there is a small but regular change, from south to north and return, with a period of about 27 or 28 days.

The principal cause of this is that it is not the earth, but the center of gravity of the earth and moon, which describes an orbit in the plane of the ecliptic; and by the known properties of the center of gravity, when the moon is above the ecliptic the earth must be below. It is not very difficult to show that from this cause the latitude may be 0".67 greater or less than when both bodies are in the ecliptic, that is, when the moon is at one of her nodes.

The attractions of Venus and Jupiter also draw the earth out of the ecliptic by an amount which may reach 0".5. In January, 1900, this "planetary perturbation" was about $+$ 0".13. The total range of latitude during the month (see page 178) was from $+$ 0".68 to $-$ 0".48. The moon was at her nodes January 12.33 and January 26.85.

From the radius vector column (p. III) we may find the sun's distance at any date by the table on page 97. By comparing this with the semi-diameter column (p. I), it is shown that the sun's distance is inversely proportional to its angular semi-diameter. Thus, January 1, 1904:

Log r = 9.9926540, Dist. = 0.9832, Semi-diam. = 16' 17".90

and July 1, 1904:

Log r = 0.0072095, Dist. = 1.0167, Semi-diam. = 15' 45".67

and

0.9832 : 1.0167 = 945".67 : 977".90,

as appears on multiplying the means and extremes and comparing the products.

The dates of the moon's perigee and apogee may be determined from the greatest and least semi-diameter, page IV, column 2, or from the greatest and least parallax in column 4. Since both semi-diameter and parallax are inversely proportional to the moon's distance from the earth, the latter may be determined by multiplying the former by a constant quantity. This constant is 3.6625, and it is not difficult to show that it is the ratio of the earth's equatorial radius to that of the moon.

Compare the last two columns, noting that at new moon the moon comes to the meridian with the sun at noon and that at full moon (age 15 days) it comes to the meridian at midnight.

TABLES OF THE PLANETS AND STARS

The data for the planets which follow the calendar pages illustrate many facts which are explained in the text-books on descriptive astronomy.

Retrograde motion, for example, is shown by negative hourly differences in right ascension; the stationary points occur on those dates on which the hourly difference changes sign; opposition takes place when the time of transit is 12^h; conjunction, when it is 0^h; the retrograde motion is a maximum at opposition.

By means of the right ascensions and declinations the path for the year may be plotted on a star map, for which purpose an ecliptic map (see page 65) is especially adapted.

The time of passing the node may be found from the point where the path cuts the ecliptic, and the sidereal period from the interval between two passages of the same node.

A series of Almanacs covering some years is useful in following the outer planets as well as for comparison of the calendar pages to show the repetition of the solar data after four years.

The table of star places contains columns of annual variation, — that is, the sum of the precession and proper motion (the latter always a very small quantity), — which are useful in showing the effects of precession on the right ascensions and declinations of stars in different parts of the heavens. Compare in this respect δ Draconis, β Ursæ Minoris, Polaris, γ Pegasi, η Geminorum, and λ Sagitarii.

COMPARISONS OF OBSERVATIONS WITH THE EPHEMERIS

Many of the facts which we have obtained by observation in former chapters may be found in the columns of the Almanac, and after a thorough comprehension of the methods has been acquired much time may be saved by employing these data; but it is to be remembered that facts thus obtained are not so thoroughly grasped or so easily retained. With this caution, we may compare some of the results of our previous work with the tables, to give an idea of the methods of using the latter. Following are comparisons of a few of the observations of the preceding chapters with the values given by the Ephemeris:

Observations of the Moon. — From careful measurement of the map on page 29, the moon's declination on January 9, 1900, at 10 P.M., was $+19°.3$, and its right ascension was $2^h 38^m$. The place of observation was $4^h 44^m.5$ west of Greenwich, and the time used was Eastern standard time, which is 5 hours slow of Greenwich; the G.M.T. was therefore $15^h 0^m$, at which time the moon's declination and right ascension are given in the Ephemeris (p. 180) as $+18° 48'$ and $2^h 39^m$. The difference between the observed and calculated places is about $\frac{1}{2}°$ in declination and 1^m in right ascension, mainly due to error of observation with the cross-staff.

Length of the Month. — We may use the Ephemeris to find the length of the month by seeking the next date at which the moon's right ascension and declination are the same, which is February 5, at about 21 hours, G.M.T., as will be seen from page VI for February. This gives $27^d 6^h$ as the period of the moon's revolution among the stars.

Passing to page V for December, we find that the right ascension was again $2^h 39^m$ on December 3 at 19 hours, at which time the declination was $17° 19'$. This shows that the moon's orbit had shifted during this time so that it did not pass through exactly the same points of the heavens in these two months, its December path in the neighborhood of right ascension $2^h 39^m$ being $1\frac{1}{2}°$ south of the corresponding point of its path in January.

By column 2 of page III, January 9 is the 9th day of the year and December 3 is the 337th; hence the moon completed an integral number of revolutions in $337^d 19^h - 9^d 15^h$, or $328^d 4^h$.

The period having been determined as $27\frac{1}{4}$ days approximately and $328 \div 27\frac{1}{4}$ being nearly 12, it is evident that the number of complete revolutions between these dates is 12. Dividing $328^d 4^h$ by 12, we have $27^d 8^h$ as a closer approximation to the sidereal month.

Taking the length of the successive months during the year, it is interesting to note how very considerable is the difference in length of the successive sidereal months due to the "perturbations" of the moon's motion.

Observations at Apparent Noon. — The observations recorded on page 39 were made at Cambridge, in longitude $4^h 44^m.5$ west of Greenwich, and the watch time of apparent noon was $11^h 56^m 2^s.9$.

By the use of the Almanac, we find the correction of the watch to standard time as follows:

Since the observation was made at local apparent noon, it will be better to use page I of the Almanac, which gives for March 8, at Greenwich apparent noon, equation of time $11^m 1^s.46$, to be added to apparent time, and hourly difference $0^s.619$.

The time of observation was $4^h 44^m.5$, or nearly $4^h.75$ later, and the change of the equation of time in this interval was $4.75 \times 0^s.619 = 2^s.93$. As the equation of time was decreasing, its value at the time of observation was $10^m 58^s.53$. Since no sign is appended to the hourly difference, we check this result by noting that it falls between the values tabulated for March 8 and 9. Hence:

Camb. App. T.	12^h	0^m	0^s
Eq. of T. (add)		10	58.53
Camb. M.T.	12	10	58.53
Corr. for Long.	4	44	30
G.M.T. of observation	16	55	28.53
Subtracting	5	0	0
Eastern Std. T. of observation	11	55	28.53
Observed watch time	11	56	2.9
Corr. of watch to Std. T.		−	34.37

The correction for longitude to give G.M.T. is added, because at any given instant the local time of any place is greater than that of a place to the westward, since the sun passes its meridian earlier and always has a greater hour-angle than at the western place.

Remembering that Cambridge is $15^m 30^s$ east of the meridian from which Eastern standard time is reckoned, we may find the watch correction more simply, thus:

Camb. M.T.	12^h	10^m	$58^s.53$
Reduction for Long. (subtract)		− 15	30
Eastern Std. T.	11	55	28.53
Watch time	11	56	2.9
Δt		−	34.37

Observations of the Planets. — The data on page 52 show that on February 5, 1900, at $7^h 12^m$ (the watch keeping Eastern standard time), the right ascension of Venus was $9°.64 = 38^m 33^s.6$ less than that of γ Pegasi, which from the Ephemeris was $0^h 8^m 5^s.69$; hence from this differential observation the right ascension of Venus was $23^h 29^m 32^s.09$.

The G.M.T. of the observation was $12^h 12^m = 12^h.2$.
The tables for Venus (p. 224) give:

FEBRUARY 5	R.A. OF VENUS	DECL.	H. DIFFS.	
At Gr. M. noon . .	$23^h 25^m 38^s.14$	$-4° 55' 46''$	$+11^s.106$	$+77''.31$
Diff. for $12^h.2$ . .	$+2\ 15.49$	$+15\ 43$	$\times 12.2$	$\times 12.2$
At $12^h 12^m$, G.M.T. .	$23\ 27\ 53.63$	$-4\ 40\ 3$	135.49	943.2
Observed values (p. 53)	$23\ 29\ 32.1$	$-4\ 32$	$2^m 15^s.49$	$15' 43''.2$

The observation differs from the Ephemeris by $1^m 38^s$ in right ascension and $8'$ in declination, although the method *should* give angles within $0°.2$. The discrepancy is much greater than usually occurs, and this observation of Venus is affected by some unexplained error; it depends on a single reading of the hour-angle. To exhibit the usual accuracy, we may compare with the following observations, made February 6:

	WATCH TIME	H.A.	DECL.
γ Pegasi	$7^h 3^m 10^s$	$64°.6$	$+15°.45$
Venus	$.5\ 10$	74.05	-3.4
γ Pegasi	$7\ 10$	65.7	$+15.5$

Hence Venus preceded γ Pegasi $8°.90 = 35^m 36^s$, Decl. $= -3°.4 - 0°.53 = -3°.93$; and since the right ascension of γ Pegasi was $0^h 8^m 6^s$, our observation gives for the place of Venus at $12^h 5^m$ G.M.T., R.A. $= 23^h 32^m 30^s$, and $\delta = -3° 56'$. The Ephemeris gives R.A. $= 23^h 32^m 17^s.2$, and $\delta = -4° 9' 14''.5$.

Observations of the Moon's Place. — The data given on page 55 show that on February 6, 1900, the moon followed γ Pegasi $46°.7 = 3^h 6^m 48^s$. The right ascension of γ Pegasi was $0^h 8^m 6^s$; hence the moon's right ascension was $3^h 14^m 54^s$, while its declination, given directly by the circle, was $+20°.4$. The Eastern standard time was $7^h 42^m$, corresponding to $12^h 42^m$ G.M.T.

The Ephemeris gives:

	MOON'S R.A.	DECL.	DIFFS. FOR 1ᵐ	
At 12ʰ G.M.T.	3ʰ 14ᵐ 34ˢ	+ 20° 25′	+ 2ˢ.33	+ 6″.2
Diff. for 42ᵐ	+ 1 38	+ 4	× 42	× 42
At time of observation	3 16 12	+ 20 29	97.9	260.4
Observed values	3 14 54	+ 20 24	1ᵐ 37ˢ.9	+ 4′ 20″

The agreement here is satisfactory considering that the moon is more than 45° from the star with which it is compared. Part of the difference is due to parallax.

Observations of the Sun's Place. — By the observation treated on page 67, the sun's right ascension and declination at 5ʰ 36ᵐ 26ˢ, Cambridge sidereal time, March 29, 1899, by comparison with α Ceti, were found to be 0ʰ 33ᵐ 19ˢ and + 3°.6. To compare this with the Ephemeris of the sun, we must first find the Greenwich mean time corresponding to 5ʰ 36ᵐ 26ˢ, Cambridge sidereal time. Heretofore we have had given either local apparent time or standard time of observations, and the Greenwich mean time has been found by adding the equation of time and longitude in one case or an integral number of hours in the other. In this case we have given the local sidereal time, to find the corresponding Greenwich mean time.

The first step is to find the Greenwich sidereal time by adding the longitude west of Greenwich, after which G.M.T. is found as follows:

Gr. Sid. T. = 5ʰ 36ᵐ 26ˢ + 4ʰ 44ᵐ 30ˢ . . . . . . = 10ʰ 20ᵐ 56ˢ	
March 29, Gr. Sid. T. of Gr. M. noon	0 30 37.57
Hence the sidereal interval elapsed since Gr. M. noon is	9 50 18.43
And, by Table II, the quantity to be subtracted from this to give the equivalent mean interval is . . .	1 36.71
Hence the corresponding mean time interval is . . .	9 48 41.72

This is the mean time interval since Greenwich mean noon, which of course is the required G.M.T.

We may now determine the sun's place at 9ʰ 48ᵐ, or 9ʰ.8, G.M.T., by means of page II of the Ephemeris, as follows: .

	SUN'S R.A.	DECL.	H. DIFFS.	
At Gr. M. noon	0ʰ 31ᵐ 33ˢ	+ 3° 24′.4	+ 9ˢ.1	+ 58″
Diff. for 9ʰ.8	+ 1 29	+ 9.5	× 9.8	× 9.8
At time of observation	0 33 2	+ 3 34.9	89	568
Observed values (p. 67)	· 0 33 19	+ 3 36	1ᵐ 29ˢ	9″.5

Determination of the Equinox. — The following data from the Almanacs of 1899 and 1900 may be compared with the results of page 89:

AT GR. APP. NOON	SUN'S DECL.	DIFF.	DATE OF EQUINOX BY INTERPOLATION
1899. Sept. 22.0	$+ 0° 18' 8''.7$	$23' 22''.6$	$22^d + \dfrac{18' 8''.7}{23' 22''.6}$ days $=$ Sept. 22.77620
23.0	$- 0 5 13.9$		
1900. Sept. 23.0	$+ 0 0 26.4$	$23 23.9$	$23^d + \dfrac{0' 26''.4}{23' 23''.9}$ days $=$ Sept. 23.01880
24.0	$- 0 22 57.5$		

The longitude of the place of observation was $4^h\ 48^m\ 40^s$ W.

$$= \frac{4^h\ 48^m\ 40^s}{24^h} = \frac{4.81111}{24} \text{ days} = 0^d.20046.$$

Hence the local dates of the equinoxes were September 22.57574, 1899, and September 22.81834, 1900, and the length of the tropical year was 365.24260 days, as compared with the observed values

$$\begin{cases} \text{September 22.714, 1899,} \\ \text{September 22.856, 1900.} \\ \text{365.14 days.} \end{cases}$$

Observations of Star Places. — The right ascensions and declinations of the stars given on pages 86 and 89 may be compared with the mean places given in the Nautical Almanac for 1899 and 1900, or, better, with the apparent places given in Part II of the American Ephemeris. From the latter we find for September 22, 1900:

	R.A.	DECL.
η Serpentis	$18^h\ 16^m\ 11^s.4$	$-\ 2°\ 55'.3$
λ Sagittarii	$18\ \ 21\ \ 51.9$	$-\ 25\ \ 28\ .5$
Vega	$18\ \ 33\ \ 35.5$	$+\ 38\ \ 41\ .8$
Altair	$19\ \ 45\ \ 57.9$	$+\ 8\ \ 36\ .6$

which are in close agreement with the results of observation.

CHAPTER VIII

THE CELESTIAL GLOBE

When a globe such as that described on page 63 has had a number of constellations plotted on it in their proper positions, and the sun's path added, showing the positions occupied by the sun at different times of the year, it becomes a very useful apparatus for many purposes.

If, for instance, it is so placed that its axis points to the pole, and is turned about the axis until the place of the sun as marked on the globe for a certain date is on the under side and in a vertical plane through the center, the sphere will represent the heavens as seen at midnight on the given date.

When the globe has been so adjusted, if a straight line is drawn from the center to any star on the surface of the globe, the prolongation of this line will lead to the real star at the point which it occupies on the sphere of the heavens. Thus used, such a globe is helpful to a beginner in identifying the constellations. Obviously the plane of the sun's path on the globe, if extended to the heavens, will mark out the ecliptic, and all the hour-circles and parallels of declination will mark the corresponding circles in the sky.

If the globe is turned slowly about its axis so that a point on the equator moves from east to west through 15° per hour, we have a sort of working model of the moving sphere of the heavens on which we may measure off arcs and angles and thus solve approximately many problems suggesting themselves to one beginning to study the apparent motions of the heavens. Such an apparatus has from very early times been an important aid to astronomers and students of astronomy, and no aid is so useful in arriving easily at correct ideas on the subject. Especially was it useful and appropriate in those days when the mechanism of the heavens was believed to correspond closely to that of the model and the globe was regarded as being a fair representation of their actual construction, — in fact, a representation of the eighth or outer sphere which carried the fixed stars, turning about a material axis somehow fixed in the "Primum Mobile." The planets moving inside, each in its crystal sphere, were treated by projecting them each on to its proper place on the outside sphere for any particular time to solve a given problem. For the beginner, who stands to a certain extent in the place of the early astronomers, it is still most important in studying many problems. Usually the diagrams by which we illustrate our statements of astronomical problems are drawn as if the celestial sphere were seen from the outside as we see the globe. This is because it is impossible to represent on a plane any large part of a spherical surface as seen from the inside.

As usually constructed for demonstration and the solution of problems, the celestial globe is made by building up layers of strong paper laid in glue upon a solid wooden sphere so as to cover it with a light but stiff shell, which is then cut through along a great circle, so that the core may be taken out. The two halves of the shell are fastened together by gluing on a strip of thin, strong cloth, and after passing an axis of stiff wire through the center, several layers of a mixture of glue and whiting are applied to the surface, each being smoothed before drying. The whole is then turned so as to form a very light and accurate spherical shell. Upon the surface are pasted gores of paper, on which the circles and principal stars are printed in such a manner as to lie in their proper places on the globe. The outlines of the constellations are shown on the plates, and the conventional figures which have been ascribed to them. A small circular piece centered on the pole completes the map. The figures are colored by hand, and the whole is then covered with a hard, transparent varnish.

Both equinoctial and ecliptic are graduated to degrees, and the hours of right ascension on the former are marked by Roman numerals. The places of the sun are usually indicated on the ecliptic at dates five days apart. Since the circuit of the sun is completed in $365\frac{1}{4}$ days, while the length of the year is sometimes 365 and sometimes 366 days, an average position of the sun must be chosen, which is done with sufficient accuracy by plotting its place for the second year after leap year.

The axis of the globe is supported by a stiff brass circle, so that the center of the sphere lies exactly in the plane of one of its faces, and this face is graduated into degrees, one semicircle near the outer edge from 0° at either pole to 90° at the equator, and the other semicircle near the inner edge from 0° at the equator to 90° at either pole. The inner graduation is used for measuring the angular distance from the equator to any point on the globe, that is, the declination of any point. The graduation on the outer edge is used for placing the axis at the proper angle to the horizon in rectifying the globe, as explained on page 111. This graduated circle which supports the axis is called the "brass meridian." It is mounted in two slots in a somewhat larger wooden circle called the "horizon," in such a manner that it is perpendicular to the latter and that its center lies in the plane of the upper surface of the wooden circle.

The horizon is graduated on its inner edge, and each quadrant has two sets of numbers, one of which reads from 0° at the prime vertical to 90° at the meridian, and the other from 0° at the meridian to 90° at the prime vertical. These numbers serve for the direct reading of amplitude and bearing respectively, which are easily translated into azimuth, remembering that W. is 90°, N. 180°, and E. 270°, if azimuth is measured from the south point toward the west from 0° to 360°. The brass meridian may be turned in its own plane, sliding easily in the slots so that the axis of the globe shall make any desired angle with the horizon.

If the globe is accurately made and mounted, its center will coincide with the common center of the graduated face of the brass meridian and the upper surface of the horizon, whatever may be the inclination of the axis. No irregularities should appear in the small space between these circles and the surface of the globe when the latter is whirled rapidly on its axis. Some idea of the correct placing of the circles on the globe may be obtained by noting whether all points of the equator and parallels come under the proper divisions of the brass meridian, whether all points of the equator pass through the east and west points of the horizon 90° from the graduated face of the brass meridian, and whether the points of the equator which lie in the east and west points of the horizon are twelve hours apart whatever the inclination of the axis.

It is desirable to have a means of fixing a point on the globe by some mark that may be afterward removed without injuring the surface. *Gummed* paper should not be used: small pieces of unglazed paper when well moistened will adhere long enough for ordinary purposes.

A good mark may be made with water-color paint mixed with glycerine so as to be very thick and applied with a rubber point or soft pen point. Such a mark may easily be removed with a moistened finger even after several weeks.

Ink suitable for fountain pens is usually safe if removed within an hour or two.

TO RECTIFY THE GLOBE

In order that the globe shall represent the heavens at any particular place, the axis must be inclined to the horizon by an angle equal to the latitude. This may be accomplished by rotating the brass meridian in its plane and measuring the angle of elevation of the pole by the outside graduation, which reads from 0° at the pole to 90° at the equator. This process is called " rectifying " the globe for a given *place*.

Having been rectified for a given place, the globe may be rectified for a given *time* by bringing it to such a position that a line drawn from its center to any star is parallel to the line drawn from the given place to the actual place of the star in the heavens at the given time. For this purpose, the pole being elevated to the proper inclination, that is, the latitude, the whole apparatus is turned on its base until the brass meridian is in the meridian of the place, and the globe is turned on the polar axis until some one point is known to be in the proper position; then all points of the globe will be in their proper positions.

The point chosen for this purpose will vary with circumstances. If the local sidereal time is given, it is only necessary to place the globe so that the hour-angle of the vernal equinox equals the given sidereal time. (See page 57.) This is easily done by the graduation of the equator on the globe. When the hour-angle of the vernal equinox is 1^h, 2^h, 3^h, the reading of the equinoctial under the brass meridian is 1^h, 2^h, 3^h, etc., and the globe is therefore rectified to a given sidereal time by turning it about the polar axis until the given sidereal time is brought to the graduated face of the brass meridian. The vernal equinox will then be at the proper hour-angle and all points on the globe will be properly related to the corresponding points on the sky.

If the apparent time is given, the globe may be rectified by the following process. Mark the place of the sun in the ecliptic for the given day. Bring this point to the meridian, which rectifies the globe for apparent noon; then, to rectify it for the given apparent time, it is necessary to turn the globe until the hour-angle of the sun is equal to the given apparent time. This may be done by using the graduations of the equator as follows. Rectify for apparent noon and read the hours and minutes of the graduation on the equinoctial which comes under the brass meridian (this is the sidereal time of apparent noon). Add to this reading the given apparent time, and the sum will be the hours and minutes of the equatorial graduation that must be brought to the meridian to place the sun at the proper hour-angle.

If local mean time is given, the apparent time may be obtained by applying the correction for the equation of time for the given date, and the globe may then be rectified for apparent time, as described in the last paragraph.

If, as will generally be the case, standard time is given, this may be reduced to local mean time by applying the correction for longitude, and we may then proceed as before.

We may here remark that in rectifying the globe for solar time we make use of the sun's place as marked on the ecliptic for the given date; and that this place may be inaccurate by as much as half a degree is obvious from the following consideration. Suppose the place of the sun on the globe to be exact for any one year on February 28. It will be exact on March 1 or about 1° in error, according as the year has not or has the date February 29. The following table of the sun's longitude shows more clearly the nature of the facts.

Year	February 20	March 2	September 23
1901	331°.2	341°.2	179°.8
1902	330 .9	341 .0	179 .5
1903	330 .7	340 .7	179 .3
1904	330 .5	341 .5	180 .0
Average	330 .8	341 .1	179 .7

The values nearly repeat themselves after four years.

It is obvious that by assuming an average value of the longitude for February 20, March 2, and September 23, we should sometimes be in error by about $\frac{1}{2}°$ in the sun's place, though never more, and by some such compromise the places must be selected for the position of the sun upon a globe for general use. The error that thus arises may amount to 2^m in the determination of the sun's right ascension from the globe.

An indispensable attachment for the celestial globe is a thin flexible strip of brass graduated to degrees and so constructed that it may be attached to the brass meridian at its highest point by a pivot, about which it can be turned so as to be brought to coincide with any vertical circle; its graduated edge may then be brought over any point on the globe and the azimuth of the point fixed by noting the place where the arc meets the graduations on the horizon. The altitude of the point may be directly read on the flexible arc, which is graduated from 0° at the horizon to 90° at the place where it is fixed to the brass meridian. The graduations are continued below the horizon from 0° to 18° for the purpose of determining the end of twilight (page 133). The flexible arc is usually called the "altitude arc."

The globe thus equipped may be used for the approximate solution of all problems which arise from the diurnal motion, some of which we will now discuss. These approximate solutions are not only sufficient for many purposes, but always indicate the proper statement of the problem for purposes of computation, and serve to detect gross errors in the numerical results.

PROBLEMS WHICH DO NOT REQUIRE RECTIFICATION OF THE GLOBE

Many problems are independent of the position of the observer on the earth's surface, and for their solution it is immaterial at what angle the polar axis is inclined. By bringing the axis to the plane of the horizon, any star may be brought to view above the horizon, but unless it is convenient to stand so that one can look down upon the globe from above, it is often better to take a sitting position and place the polar axis nearly vertical. In following the solutions of the examples below, the accompanying figures serve to show whether the globe has been brought to the proper position.

Problem 1. — *To find the right ascension and declination of a star.*

Rotate the globe until the star is in the plane of the brass meridian; note the hours, minutes, and seconds of that graduation of the equinoctial which falls under the brass meridian. This is the right ascension of the star. This value we may call the "meridian reading" of the equator and in future abbreviate to R.A.M. (right ascension of the meridian). The declination of the star equals that degree of the graduation of the meridian under which the star lies.

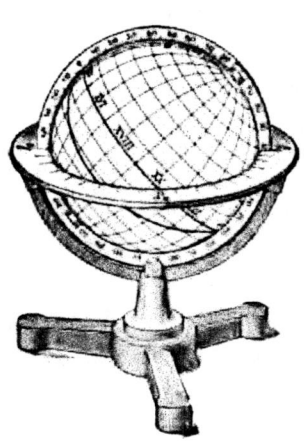

FIG. 42. R.A.M. $13^h 44^m$; Decl. $+49\frac{1}{2}°$

Example 1. The star η Ursæ Majoris in the end of the Dipper handle is brought to the brass meridian (Fig. 42) and is found to lie halfway between the divisions 49 and 50 north of the equator; the declination is therefore $+49°.5$. The meridian reading is $13^h 44^m$, which is the star's right ascension. (For reading the declination the graduations on the inner edge of the brass meridian must be used.)

Problem 2. — *Given the right ascension and declination of a star, to find the star.*

Rotate the globe until the meridian reading (R.A.M.) is equal to the given right ascension, and under the brass meridian at the given declination will be found the star.

Example 2. The right ascension of a certain star is 19ʰ 46ᵐ and its declination + 8¼°. What is the star?

The division on the equator marked 19ʰ 46ᵐ is brought to the brass meridian (Fig. 43), and halfway between the graduations 8 and 9 on the meridian is found Altair, which is the star sought.

Problem 3. — *To find the angular distance between two stars.*

Place the flexible quadrant along the surface of the globe so that its graduated edge passes through both stars, and read the graduation at the points where it touches each star; the difference of the readings is the angular distance between the stars. The graduated edge should lie along the great circle; as this is not always easy to adjust, it is well to repeat the measure with the quadrant in different adjustments and take the smallest value obtained.

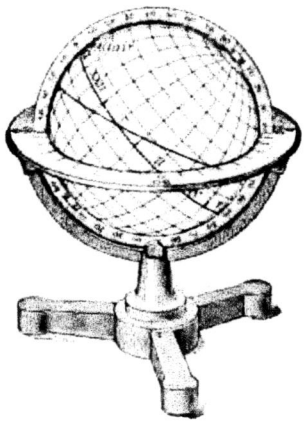

FIG. 43. R.A.M. 19ʰ 46ᵐ; Decl. + 8¼°

An alternative method free from this source of error is to adjust the points of a pair of compasses so that they may just span the distance between the two stars. The compasses may then be applied to the globe with one leg at the vernal equinox (0°); the other leg being brought to the equinoctial its reading will give the angular distance between the stars. To guard against defects in the globe, the second point may be brought to the ecliptic, and the reading should be the same as on the equinoctial; if the readings differ, the mean of the values should be taken.

In the use of the compasses care must be taken not to scratch the surface of the globe.

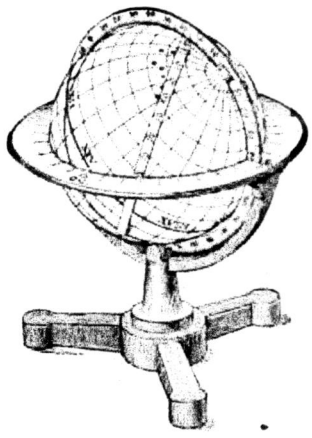

FIG. 44. Length of Dipper 26°

Example 3. The following measures were made to determine the distance between α Ursæ Majoris and η Ursæ Majoris. With the flexible quadrant applied to the globe (Fig. 44) so as to lie as nearly as possible along the great circle between the stars, the readings were:

η URSÆ MAJORIS	α URSÆ MAJORIS	DISTANCE
0.0	26.0	26.0
0.0	26.1	26.1
20.0	46.1	26.1
40.0	66.1	26.1

Here no difficulty was found in laying the arc along the great circle, as the distance is not great, and the value is taken to be 26°.1. Adjusting the points of a pair of compasses to the stars and then placing the compasses with one point at the vernal equinox, the other point was found to reach to 25°.6 of right ascension on the equinoctial and to 25°.6 of longitude on the ecliptic, which gives the distance between the stars as 25°.6.

Problem 4. — *To find the sun's longitude, right ascension, and declination at a given date.*

If the sun's place at different dates is marked on the ecliptic, its longitude may be read off directly on the graduations of the ecliptic. In all old globes, however, and in many modern ones the ecliptic is not thus marked, and the place of the sun must be found by determining the longitude by a table such as that given on page 173, which is nearly correct for the first half of the present century. A substitute for this table is generally to be found in the form of two contiguous concentric circles on the horizon circle, one graduated into degrees of longitude and the other into months and days, so that the line for a given date in the outer circle is found opposite the corresponding degree of the sun's longitude in the inner circle. Commonly also the divisions both of this circle and of the ecliptic are divided into groups of 30°, each corresponding roughly to one month of time. The 30° of Aries reach from the first of Aries on March 20 to the first of Taurus on April 20, and so on in the order of the signs. Thus, opposite May 6 is the fifteenth degree of Taurus, corresponding to longitude 45° in the usual way of reckoning; opposite January 1 is the tenth degree of Capricornus, nine complete signs and 10°, or longitude 280°. In the table on page 173 the equivalents of the degrees of longitude are given in signs and degrees.

By whatever method the sun's place in the ecliptic is fixed, its right ascension and declination are found by the method of Problem 1.

Example 4. What are the sun's right ascension and declination on April 20 ?

The longitude is found by the table to be 29°.5, and on bringing this point of the ecliptic to the meridian (Fig. 45) it is found to be in declination $+11\frac{1}{4}°$, while the reading of the meridian is $1^h\ 50^m$. The sun's right ascension is therefore $1^h\ 50^m$ and its declination is $11\frac{1}{4}°$ north.

PROBLEMS WHICH REQUIRE RECTIFICATION OF THE GLOBE FOR A GIVEN *TIME*

Fig. 45. Sun's R.A. $1^h\ 53^m$; Decl. $+11\frac{1}{4}°$

Such are problems which require a determination of the angle between the meridian and some one of the hour-circles of the globe. They are independent of the latitude of the place of observation, but depend upon the position of the heavenly bodies with respect to the meridian. The brass meridian being taken as the meridian of the place of observation, the only quantities involved are differences of hour-angle and of right ascension, and it will be advisable here to collect the following relations, which have already been explained.

All time is measured by the continually increasing hour-angle of some point of the celestial sphere.

Local sidereal time (Camb. Sid. T.) is the hour-angle of the vernal equinox.

Local apparent (solar) time (Camb. App. T.) is the hour-angle of the sun.

Local mean (solar) time (Camb. M. T.) is the hour-angle of the mean sun.

For example, at $21^h\ 20^m$, Camb. Sid. T., the hour-angle of the vernal equinox at Cambridge is $21^h\ 20^m$; at $10^h\ 30^m$, Chicago apparent time, the hour-angle of the sun at Chicago is $10^h\ 30^m$; at $5^h\ 10^m$, New York mean time, the hour-angle of the mean sun at New York is $5^h\ 10^m$.

The hour-angle is in all cases measured westward from the observer's meridian up to 24^h.

Greenwich mean time (G.M.T.) is the hour-angle of the mean sun measured from the meridian of Greenwich. When we say that a place is a certain number of hours and minutes of longitude west of Greenwich, we mean that the rotation of the earth brings the sun to the meridian of the place just so many hours and minutes after its arrival at the meridian of Greenwich. At local noon, then, its hour-angle, reckoned from the Greenwich meridian, is equal to the difference of longitude between the two meridians. As the sun thereafter moves westward equally from the two meridians, Greenwich time is always greater than that of any place west of it by exactly the difference of their longitudes.

Therefore, to find the G.M.T. corresponding to a given local mean time, we add to the latter the longitude (west) from Greenwich. Standard time is directly obtained from G.M.T. by subtracting 4, 5, 6, 7, 8 hours, respectively, for Colonial, Eastern, Central, Mountain, and Pacific time. Thus, the "reduction for longitude," so called, from Cambridge mean time is $+4^h\ 44^m.5$ to G.M.T. and $+4^h\ 44^m.5 - 5^h$ to Eastern standard time; or, by a single operation, $-15^m.5$ directly to Eastern time. The "reduction for longitude" for San Francisco is $+8^h\ 9^m.7$ to Greenwich and $+8^h\ 9^m.7 - 8^h = +9^m.7$ to Pacific time. Problems, therefore, which involve standard time require a knowledge of the observer's longitude.

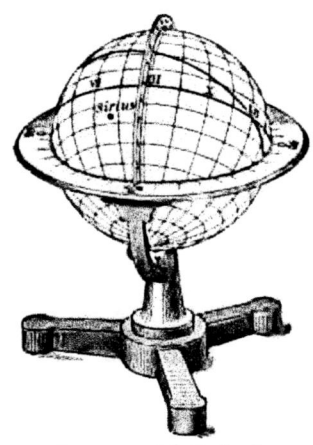

Fig. 46. Sid. T. $7^h\ 50^m$.

Problem 5. — *To rectify the globe for a given sidereal time.*

Rotate the globe till the R.A.M. equals the given sidereal time. This brings the vernal equinox to an hour-angle equal to the given sidereal time, and all points of the sphere into their proper relation to the meridian.

THE CELESTIAL GLOBE

Example 5. To rectify the globe for $7^h\ 50^m$ sidereal time, rotate the globe until R.A.M. is $7^h\ 50^m$ (Fig. 46).

Problem 6. — *The globe being rectified for a given sidereal time, to determine the hour-angle of a body.*

Note the R.A.M. when the globe is in the given position; then bring the body to the meridian and read its right ascension. Subtract the latter reading from the former and the result is the hour-angle of the body.

Since the reading of the meridian is always the sidereal time (page 59), this process exemplifies the equation H.A. = Sid. T. — R.A. It is of course understood that if in adding two times or hour-angles the result is greater than twenty-four hours, that amount is to be subtracted; thus, an hour-angle of $35^h\ 25^m\ 10^s$ corresponds to the same position as an hour-angle of $11^h\ 25^m\ 10^s$. Also, if it is required to subtract a larger from a smaller hour-angle, the latter should be increased by twenty-four hours before performing the subtraction: thus, $6^h\ 41^m - 11^h\ 17^m = 30^h\ 41^m - 11^h\ 17^m = 19^h\ 24^m$.

Example 6. What is the hour-angle of Sirius at (a) $7^h\ 50^m$, sidereal time, and at (b) $4^h\ 20^m$, sidereal time?

(a) Rectifying the globe, as in Problem 5, to $7^h\ 50^m$ Sid. T., the R.A.M. = $7^h\ 50^m$. Bringing Sirius to the meridian (Fig. 47), R.A.M. = $6^h\ 41^m$ = R.A. of Sirius, as in Problem 1. Hence H.A. of Sirius at $7^h\ 50^m$ Sid. T. = $7^h\ 50^m - 6^h\ 41^m = 1^h\ 9^m$ (Fig. 46).

FIG. 47. R.A. of Sirius, $6^h\ 41^m$

(b) Rectifying to $4^h\ 20^m$ Sid. T., R.A.M. = $4^h\ 20^m$, and, as before, H.A. = $4^h\ 20^m - 6^h\ 41^m = 28^h\ 20^m - 6^h\ 41^m = 21^h\ 39^m$.

Problem 7. — *The globe being rectified for a given apparent time, to determine the hour-angle of a body.*

Bring the sun's place to the meridian and take the R.A.M. (this is the sun's right ascension, Problem 4). Rotate the globe through an hour-angle equal to the given apparent time, and the sun is brought to the required hour-angle; the R.A.M. thus becomes H.A.

of the sun + R.A. of the sun, and the globe is properly rectified when this reading of the equator is brought under the meridian.

Since H.A. + R.A. = Sid. T., the rule may be given as follows: Determine the sun's right ascension by the globe (Problem 4). Add the given apparent time. The sum is the sidereal time. For this sidereal time rectify the globe by Problem 5, and find the hour-angle by Problem 6.

Example 7. What is the hour-angle of Sirius at 10 P.M., apparent time, February 13?

Sun's R.A. by globe	21^h	50^m
App. T.	10	0
Sid. T.	7	50
R.A. of Sirius by globe	6	41
H.A. of Sirius	1	9 (Fig. 46)

Problem 8. — *The globe being rectified for a given mean time, to determine the hour-angle of a body.*

Apply the equation of time (with the proper sign) to the given mean time to find the corresponding apparent time, and with this value rectify as in Problem 7.

Example 8. What is the hour-angle of Sirius at 5 A.M., local mean time, July 10?

Equation of time + 5^m (add to App. T.)

July 10, 5 A.M.	= July 9^d 17^h	0^m
Eq. of T. (subtract)		5
App. T.	16	55
Sun's R.A. by globe (add)	7	20 *
Sid. T.	0	15
R.A. of Sirius (Problem 1) (subtract)	6	41
H.A. of Sirius	17	34

Problem 9. — *The globe being rectified for a given standard time, to determine the hour-angle of a body.*

Apply the reduction for longitude to find the corresponding mean time and rectify as in Problem 8.

* The sun's place is marked on the globe for noon of the indicated date. It is therefore more accurate in this problem to make use of the sun's place for July 10 and in general for the nearest noon, which is always that of the civil date.

THE CELESTIAL GLOBE

Example 9. At Chicago (longitude $+ 5^h 50^m$) what is the hour-angle of Sirius at 6.30 P.M., Central standard time, October 30?

Red. for Long. Chicago T. to Central T. $- 10^m$
Eq. of T. $- 16^m$ (subtract from App. T.)

Central standard time	6^h	30^m
Red. of Long. to Chicago M. T.	$+$	10
Chicago M. T.	6	40
Eq. of T. (add to M. T.)	$+$	16
App. T.	6	56
Sun's R.A. by globe (add)	14	23
Chicago Sid. T.	21	19
R.A. of Sirius (subtract)	6	41
H.A. of Sirius (by Problem 5)	14	38

Reduction to the Equator. — In the solution of Example 4, page 117, it was shown that when the sun's longitude is 29°.5 its R.A. is $1^h 50^m$, or 27°.5.

The quantity which must be added to the longitude of a point on the ecliptic to find its R.A. (in this case $- 2°$) is called the "reduction to the equator" and is used in finding the equation of time as explained in Chapter X. Its value for any given point of the ecliptic may be found by the globe as in Example 4.

Following are the results:

Longitude		Red. to Equator	Longitude		Red. to Equator
0° and	180°	0°.0	90° and	270°	0°.0
10	190	$- 0.8$	100	280	$+ 0.9$
20	200	$- 1.5$	110	290	$+ 1.6$
30	210	$- 2.1$	120	300	$+ 2.2$
40	220	$- 2.4$	130	310	$+ 2.4$
50	230	$- 2.4$	140	320	$+ 2.4$
60	240	$- 2.2$	150	330	$+ 2.1$
70	250	$- 1.6$	160	340	$+ 1.5$
80	260	$- 0.9$	170	350	$+ 0.8$
90	270	0.0	180	360	0.0

CHAPTER IX

EXAMPLES OF THE USE OF THE GLOBE

Most of the problems with which we have to deal require that the observer's exact place on the earth shall be known, — that is, his latitude as well as his longitude; and in order that they may be solved it is necessary that the globe should be rectified to the latitude by inclining the axis to the horizon by an angle equal to the latitude.

This chapter contains some typical examples and the methods by which they are solved, with references to the problems of the preceding chapter.* Attention should be paid to the arrangement of the solutions, and all numerical results should be fully labeled so that it may be seen how they are obtained and combined. In all the problems, unless otherwise stated, the globe must be rectified to the latitude of Cambridge, 42°.4 N. The longitude may be assumed $4^h 44^m$ west of Greenwich.

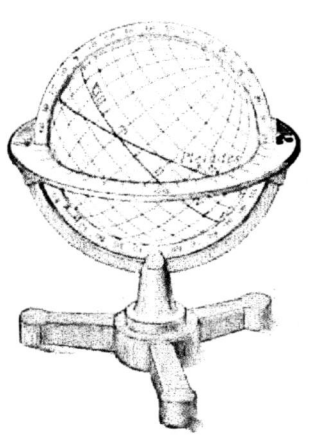

Fig. 48. Rising of Pleiades: $20^h 12^m$ Camb. Sid. T.

Example 10. At what sidereal time do the Pleiades rise at Cambridge?

Rectify the globe by raising the north pole to such an angle that the graduation 42°.4 on the outside edge of the brass meridian coincides with the surface of the horizon. Rotate the globe about the polar axis until the Pleiades are in the plane of the eastern horizon (Fig. 48). The R.A.M. equals the sidereal time sought, — $20^h 12^m$. This result is independent of the longitude. The Pleiades rise at any place in latitude 42°.4 N. at $20^h 12^m$ of local sidereal time.

* These solutions were obtained with a not very accurate globe nine inches in diameter. Better results may be obtained with a larger globe in good condition.

EXAMPLES OF THE USE OF THE GLOBE

Example 11. At what apparent time do the Pleiades rise at Cambridge on October 30?

Determine the sidereal time, as in the last example, $20^h\ 12^m$. The sun's right ascension is determined to be $14^h\ 17^m$ by bringing it to the meridian (Fig. 49), as in Problem 4, and the relation App. T. = Sid. T. − Sun's R.A. gives

$20^h\ 12^m - 14^h\ 17^m = 5^h\ 55^m$ Camb. App. T.

Example 12. At what Cambridge mean time do the Pleiades rise October 30? Eq. of T. $= -16^m$ (subtract from App. T.).

The apparent time being $5^h\ 55^m$ by the last example, the mean time is $5^h\ 55^m - 16^m = 5^h\ 39^m$.

Example 13. At what Eastern standard time do the Pleiades rise at Cambridge October 30?

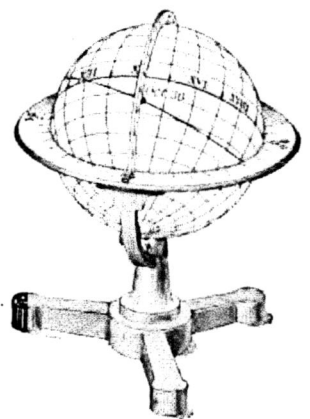

FIG. 49. October 30: Sun's R.A. $14^h\ 17^m$

The arrangement of the work is as follows:

Camb. Sid. T. by globe (Example 10)	$20^h\ 12^m$
Sun's R.A. by globe (Problem 4)	14 17
Camb. App. T. (Example 11)	5 55
Eq. of T. by table	− 16
Camb. M. T. (Example 12)	5 39
Red. to E. Std. T.	− 16
E. Std. T. of rising of Pleiades	5 23

Example 14. At what standard time do the Pleiades set at Cambridge March 1?

Bringing the Pleiades to the western horizon, we have, as in Example 13:

Camb. Sid. T. by globe (Fig. 50)	$11^h\ 15^m$
Sun's R.A. (Problem 4)	22 50
Camb. App. T.	12 25
Eq. of T. by table	+ 13
Camb. M. T.	12 38
Red. to E. Std. T.	− 16
E. Std. T. of setting of Pleiades March 1	12 22

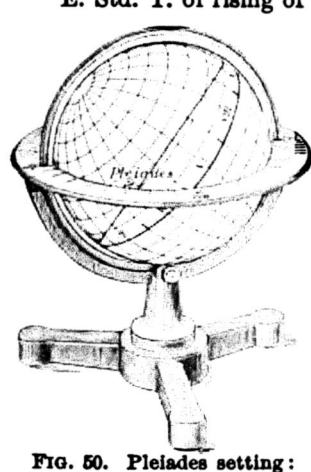

FIG. 50. Pleiades setting: Sid. T. $11^h\ 15^m$

Example 15. What is the standard time of sunrise at Cambridge on May 15?

Mark the place of the sun on the ecliptic for May 15 and bring this point to the plane of the eastern horizon (Fig. 51).

Fig. 51. Sunrise May 15: Sid. T. $20^h\ 28^m$

The R.A.M. gives the Camb. Sid. T. by globe	$20^h\ 28^m$
Sun's R.A. (Problem 4) by globe . . .	3 28
Camb. App. T.	17 00
Eq. of T. by table	− 4
Camb. M. T.	16 56
Red. for Long.	−16
Std. T. of sunrise May 15	16 40

Or, May 16, $4^h\ 40^m$ A.M. But see the note to Problem 8. Since the place of the sun was taken for May 15, the solution gives the time of sunrise for that civil date.

Example 16. What is the azimuth of the sun at Cambridge at sunrise June 21?

The sun's place for June 21, being brought to the horizon as in the preceding problem, was found to be on the division 59 of the graduation which reads from zero at the north point of the horizon to 90° at the east point (Fig. 52); its bearing, therefore, is N. 59° E., and its azimuth reckoned from the south point is 180° + 59°, or 239°.

The graduation on the inner edge of the horizon has a second set of numbers beginning with 0° at the east and west points and running to 90° at the north and south points. By means of this amplitudes may be directly measured. The amplitude of the sun in this case was E. 31° N.

Example 17. At Cambridge, September 10, in the afternoon, the sun's altitude is 20°. What is its azimuth?

For the solution of this problem the altitude arc must be applied to the brass

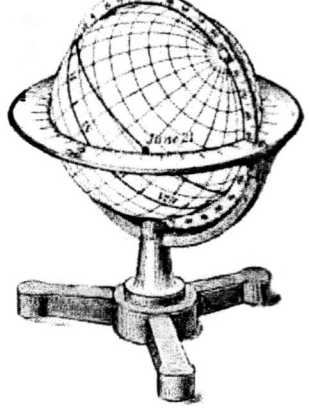

Fig. 52. Sunrise June 21: Sun's Bearing N. 59° E.; Az. 239°

meridian, attaching the clamp so that the 90° mark of the graduation is as exactly as possible under the graduation 42°.4 on the inner edge of the brass meridian; this is at the highest point of the globe, corresponding to the zenith of the sphere in latitude 42°.4 north.

The longitude of the sun for September 10 being found, by the circles printed on the horizon for this purpose, to be 17°.7 in Virgo, or five signs and 17°.7 = 167°.7, this point was brought into the southwest quadrant halfway from the south to the west point and the altitude arc made to pass through it; the altitude was seen to be approximately 40°. The foot of the arc was then moved about 20° toward the west point and the sun's place brought to it; the altitude was now about 30°. The foot of the arc was moved again about 20° farther toward the west point and the sun's place brought to it, the sun's altitude being about 15°. The arc was now moved back a few degrees toward the south and by a few trials a position found (Fig. 53) such that the sun's place coincided exactly with the division marking an altitude of 20°; the zero of the graduated edge of the arc was then halfway between 77° and 78° of the graduation on the inner edge of the horizon circle. The bearing was then S. 77°.5 W. and the azimuth 77°.5.

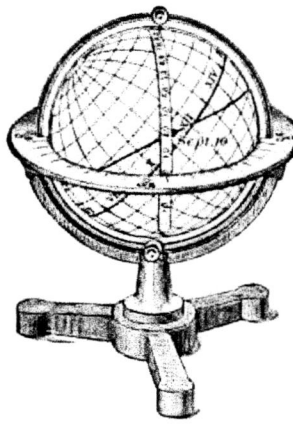

FIG. 53. September 15: Sun's Alt. 20°; Az. 77°.5

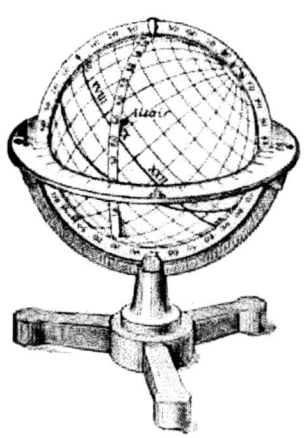

FIG. 54. Alt. of Altair 30°: H.A. 20ʰ 13ᵐ; Az. 287°; Sid. T. 15ʰ 56ᵐ

Example 18. At Cambridge Altair is east of the meridian at an altitude of 30°. Find its azimuth and hour-angle and the sidereal time. Bringing the place of Altair to 30° on the flexible arc, as described in the last problem, the bearing is found to be S. 73° E. Hence the azimuth is 287°. With the same adjustment the R.A.M. is $15^h\ 56^m$, which is the sidereal time. By bringing Altair to the meridian, its right ascension is found to be $19^h\ 43^m$, and, by Problem 5, H.A. = $15^h\ 56^m - 19^h\ 43^m$ = $20^h\ 13^m$.

Example 19. On September 10, at Cambridge, in the forenoon, the sun's altitude is 20°. What is the local mean time?

The sun's longitude being 167°.7, as in Example 17, its place is brought to 20° on the flexible arc in the southeast quadrant (at a bearing S. 78° E., with which compare the result of Problem 17) and the problem solved as follows:

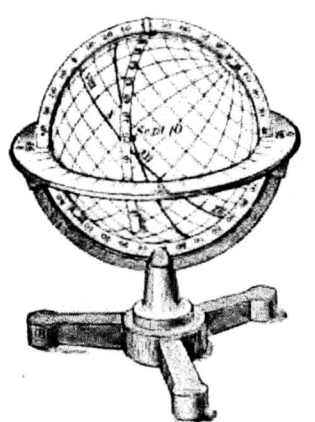

FIG. 55. Sun's Alt. 20°; R.A.M. 6ʰ 42ᵐ

Sun's forenoon Alt. 20°

R.A.M.	6ʰ	42ᵐ
Sun's R.A. (Problem 4)	11	13
App. T.	19	29
Eq. of T. by table	−	3
Camb. M. T.	19	26
Or	7	26 A.M.

It would appear that our result means 7.26 A.M. of the following day. But it is to be remembered that we have used the sun's place for September 10 (the places are marked for noon), and our solution then applies more nearly to the morning of that date. Example 19 is perhaps the most important that we have solved, since it illustrates the method by which the longitude is determined at sea. The sun's altitude is measured by a sextant and its hour-angle computed. From the apparent time thus obtained the local mean time is found as above and compared with G.M.T. kept by a chronometer.

Example 20. On July 10, at Cambridge, what is the sun's hour-angle when it is in the prime vertical? What is the local mean time?

In the summer half of the year the sun is in the prime vertical once in the forenoon and once in the afternoon, so that there will be two solutions of the problem.

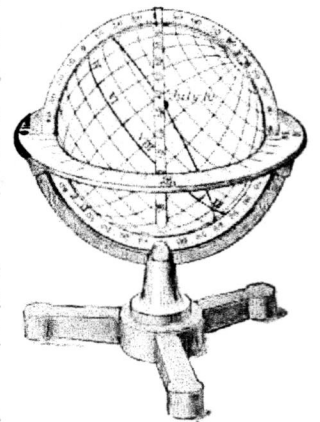

FIG. 56. Sun in Prime Vertical: July 10, forenoon; R.A.M. 3ʰ 3ᵐ

The place of the sun July 10 is found by the table to be in longitude 107°.7. The altitude arc being adjusted with its foot at the east point of the horizon, the sun's place is brought to the graduated edge of the arc and R.A.M. noted. The altitude arc being brought in the same way to coincide with the west quadrant of the prime vertical, the sun's place is brought again to the graduated edge and R.A.M. noted. Then the sun's right ascension is determined, and the results may be recorded and the computation made in the following form:

Sun in prime vertical	A.M.	P.M.
R.A.M.	$3^h\ 3^m$	$11^h\ 36^m$
Sun's R.A. by globe	7 20	7 20
App. T.	19 43	4 16
Eq. of T. by table	+5	+5
Local M. T.	19 48	4 21
Or	7 48 A.M.,	4 21 P.M.

Example 21. At Cambridge, at 0^h sidereal time, what bright stars are seen near the meridian? What are their declinations?

Rectify the globe for latitude $+42°.4$. Rotate the globe until the R.A.M. is 0^h, and the following stars will be found near the meridian: γ Pegasi, Decl. $+14°.0$; α Andromedæ, Decl. $+27°.5$; β Cassiopeiæ, Decl. $+58°$; Polaris, of course, but too near the pole to be seen on the globe; γ Ursæ Majoris, Decl. $54°$; δ Ursæ Majoris, Decl. $58°$. The two latter are below the pole, and to determine their declinations the globe must be rotated $180°$ to bring them under the inner graduations of the meridian.

FIG. 57. Stars on Meridian at Cambridge at 0^h Sidereal Time

Notice that the four first stars lie along the same hour-circle, which is the equinoctial colure, in R.A. 0^h, and that this circle is divided roughly by them into multiples of $15°$, thus: Polaris to β Cassiopeiæ, $30°$; β Cassiopeiæ to α Andromedæ, $30°$; α Andromedæ to γ Pegasi, $15°$.

By continuing the line of stars about $15°$ we arrive at Decl. $=0°$, R.A. $=0°$, that is at the vernal equinox, which though marked by no conspicuous star is easily fixed by this alignment.

Example 22. What is the standard time corresponding to 0^h of sidereal time at Cambridge October 10?

The sidereal time being given, this problem is similar to Examples 13, 14, and 15, and illustrates the general process of passing from sidereal to mean or standard time by means of the globe, thus:

Sid. T.	0^h 0^m
Sun's R. A. by globe	13 4
App. T.	10 56
Eq. of T.	− 13
Camb. M. T.	10 43
Red. for Long. to Std. T.	− 16
Eastern standard time	10 27

Example 23. Find the altitude and azimuth of Arcturus at 8 P.M., standard time, at Cambridge, September 10.

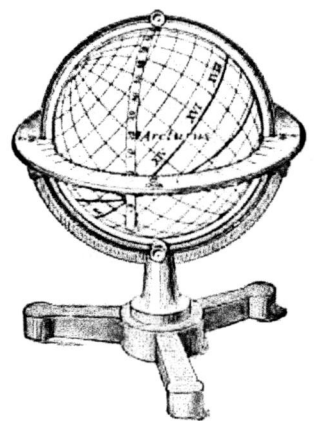

FIG. 58. Arcturus: September 10, 8 P.M., E. Std. T.; Alt. 20°; Az. 98°

This problem requires the globe to be rectified for both latitude and time. The latter adjustment is made as follows:

Std. T.	8^h 0^m
Red. for Long.	+ 16
Camb. M. T.	8 16
Eq. of T. by table (add to M.T.)	+ 3
App. T.	8 19
R. A. Sun by globe	11 15
Camb. Sid. T.	19 34

Rectify for Cambridge, Lat. + 42°.4. Rotate the globe till the R.A.M. is 19^h 34^m. Apply the altitude quadrant so as to pass through Arcturus, and we find its altitude 19°.5, and its bearing N. 80°.5 W.; hence its azimuth is 99°.5.

Example 24. What constellation is rising in the east at 9 P.M., Eastern standard time, at Cambridge, November 10?

As in the preceding problem:

Std. T.	9^h 0^m
Red. for Long. to Camb. M. T.	+ 16
Camb. M. T.	9 16
Eq. of T. by table (subtract from App. T.)	+ 15
Camb. App. T.	9 31
R.A. of Sun by globe	15 3
Sid. T.	0 34

To rectify for time rotate the globe till the R.A.M. is 0^h 34^m. It will be found that the constellation of Orion has just risen above the eastern point of the horizon. Compare the form of this solution with that of Example 13, which is the inverse of this, the rising of a star being given and the standard time sought.

PROBLEMS INVOLVING THE USE OF THE NAUTICAL ALMANAC

Example 25. At Cambridge, November 30, 1904, at 5^h 15^m P.M., standard time, a bright star is seen due southwest about 10° above the horizon. No other stars being visible in the twilight, it is desired to identify the star.

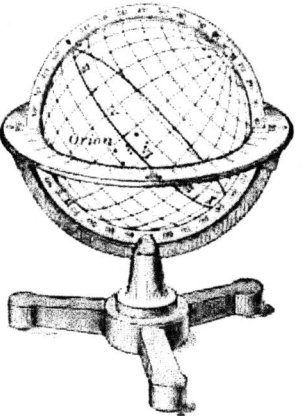

FIG. 59. Orion rising: Cambridge, November 10, 9 P.M., Std. T.

E. Std. T.	5^h	15^m
Red. for Long.	+	16
Camb. M. T.	5	31
Eq. of T. (subtract from App. T.)	+	11
App. T.	5	42
R.A. of Sun	16	26
Camb. Sid. T.	22	8

Rectifying for Cambridge, Lat. + 42°.4, and for 22^h 8^m Sid. T., it is found, by means of the altitude arc (Fig. 60), that there is no star upon the globe at the given altitude and azimuth, the nearest star being σ Centauri, which would not be visible at that altitude

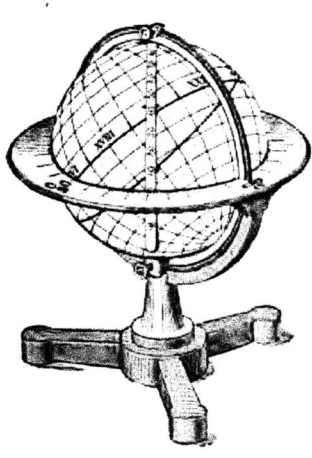

FIG. 60. Star 10° above Southwest Horizon, Cambridge, November 30, 1904; Sid. T. 22^h 7^m

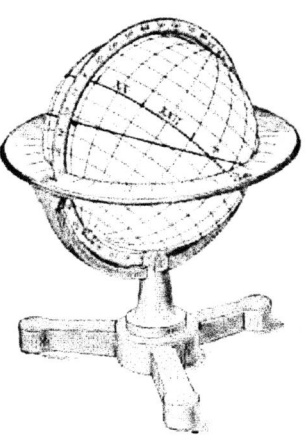

FIG. 61. Star brought to Meridian: R.A 18^h 56^m; Decl. $-23\frac{1}{4}$°

in twilight. The exact point being marked is brought to the meridian and found to be in R.A. $18^h\ 56^m$ and Decl. $-23\frac{1}{2}°$ (Fig. 61). The fact that its position is very near the ecliptic suggests that it may be a planet, and on consulting the Almanac it is found that on November 30 the right ascension of Venus is $19^h\ 4^m$ and its declination $-24°.7$, or within about 2° of the observed place.

Example 26. Which of the planets that are visible to the naked eye are above the horizon at Cambridge at 8 P.M., standard time, October 1, 1904?

From the Nautical Almanac are taken the following data for the given date:

	R.A.	DECL.
Mercury	$11^h\ 25^m$	$+ 5°.1$
Venus	13 55	$- 11 .4$
Mars	10 10	$+ 12 .7$
Jupiter	1 44	$+ 9 .1$
Saturn	21 10	$- 17 .6$

FIG. 62. Planets, October 1, 1904

Marking these places upon the globe and rectifying for the given place and time, it is at once seen that the first three are below the western horizon, while Jupiter is 20° above the east point of the horizon and Saturn approaching the meridian at an altitude of about 30°.

Where only an approximate result is desired, it will often be sufficient to neglect the corrections for longitude and equation of time, the sum of which at Cambridge never amounts to much more than half an hour. This of course assumes standard time to equal apparent time. Thus, in this problem we may bring the sun to the meridian and, noting R.A.M. $= 12^h\ 30^m$ and adding 8^h, we have $20^h\ 30^m\ (\pm 30^m)$ as the R.A.M. corresponding to 8^h apparent time. The general terms in which the answer is given above will apply equally well, and some time is saved where only the general aspect of the heavens is required.

Example 27. At what standard time does Jupiter set at Cambridge December 25, 1904?

By the tables in the Nautical Almanac, we find that on the given date the right ascension of Jupiter is $1^h 17^m$ and its declination $+ 6°.8$. Marking this place on the globe and bringing it to the western horizon, the R.A.M. is $7^h 38^m$, which is the sidereal time. Converting to standard time:

Sid. T.	$7^h 38^m$
Sun's R.A. by globe	18 12
App. T.	13 26
Eq. of T.	0
Camb. M. T.	13 26
Red. for Long.	− 16
Std. T.	13 10
Or	1 10 A.M.

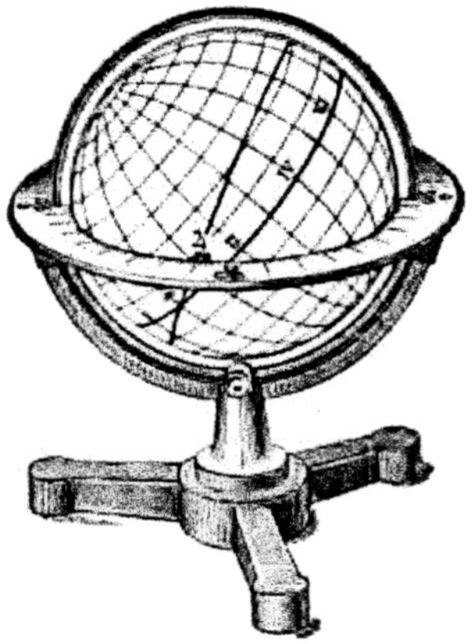

Example 28. At what time does the moon rise at Cambridge December 25, 1904?

If the moon's position were known directly from the Nautical Almanac, the solution of this problem would be similar to the last; but the moon's right ascension and declination are changing so rapidly that we must reach the result by approximation. We may first assume the moon's place at rising to be the same as at standard noon, December 25 (or 5^h, G.M.T.), and at that time the Almanac gives the moon's right ascension $8^h 54^m$, Decl. $+ 14°.9$. Marking this place on the globe and bringing it to the eastern horizon, we find R.A.M. $= 1^h 56^m$, and continue the computation as in the second column of the table below. (See Example 15.)

FIG. 63. Jupiter setting: R.A.M. $7^h 38^m$; E. Std. T. $13^h 10^m$.

G.M.T.	5^h	$12^h 28^m$	$12^h 45^m$
Moon's Place	$8^h 54^m, + 14°.9$	$9^h 11^m, + 13°.9$	$9^h 13^m, + 13°.9$
R.A.M.	$1^h 56^m$	$2^h 13^m$	$2^h 18^m$
R.A. of Sun	18 12	18 12	18 12
App. T.	7 44	8 1	8 6
Eq. of T.	0	0	0
Camb. M. T.	7 44	8 1	8 6
Red. for Long.	− 16	− 16	− 16
E. Std. T.	7 28	7 45	7 50

This gives as the approximate time of moonrise $7^h 28^m$, E. Std. T., or $12^h 28^m$, G.M.T., and finding the moon's place for this time, R.A. $9^h 11^m$, Decl. $+ 13°.9$, we better our result by the computation shown in the third column, which gives $7^h 45^m$, E. Std. T., or $12^h 45^m$, G.M.T. With this value we find the moon's place $9^h 13^m$, $+ 13°.9$, and compute as in the last column, finding E. Std. T. $= 7^h 50^m$.

FIG. 64. Moonrise at Cambridge December 25, 1904: R.A.M. $2^h 18^m$

As this is within ten minutes of the time for which the data were assumed, and since in ten minutes the moon's right ascension, as shown by the difference column, changes by 24^s, — a quantity too small to be surely measured on an ordinary 10-inch globe, — we may regard the last solution as sufficiently accurate.

It would appear that the two last results should be in closer agreement, since the difference in the assumed times is only seventeen minutes; the two first measures, however, were not made with care, as only approximate values were sought.

It is obviously an advantage to estimate the approximate time of moonrise as closely as possible before beginning the solution: this may be done by noting the age of the moon (page IV of the month) and remembering that the moon rises and sets about 48^m, or $0^h.8$, later each night than the night before, and that at new moon sun and moon rise and set together. Assuming that the sun rises at 6 A.M. and sets at 6 P.M., standard time, we shall find an approximate value of the standard time of moonrise or moonset by adding to these times a number of hours equal to eight-tenths of the moon's age in days.

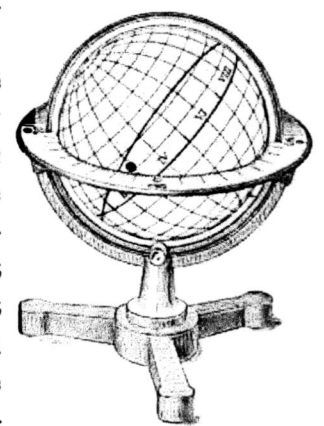

FIG. 65. Moonset at Cambridge December 18, 1904: R.A.M. $9^h 41^m$

Thus, in the preceding problem, the moon's age being eighteen days on December 25, we add $0.8 \times 18^h = 14^h.4$ to 6 A.M. to find the time of moonrise;

this gives 8ʰ.4 P.M. as the approximate time, which is within an hour of the final result.

Example 29. Find the time at which the moon sets at Cambridge December 18, 1904.

The moon's age is found by the Ephemeris to be eleven days; hence we add 9ʰ to 6ʰ P.M., and have as the approximate time of moonset 15ʰ, corresponding to 20ʰ, G.M.T. We may record the successive approximations as follows:

	First Approximation	Second Approximation
Assumed G.M.T.	20ʰ	20ʰ 33ᵐ
Moon's R.A. and Decl.	2ʰ 58ᵐ, + 12°.2	2ʰ 59ᵐ, + 12°.3
R.A.M. at moonset	9ʰ 39ᵐ	9ʰ 41ᵐ
Sun's R.A.	17 46	17 46
App. T.	15 53	15 55
Eq. of T.	− 4	− 20
Red. for Long.	− 16	
Std. T.	15 33	15 35

A single recomputation will always be sufficient if the moon's place is first determined by computing from its age.

MISCELLANEOUS EXAMPLES

Example 30. Find the duration of twilight at Cambridge March 1.

Evening twilight ends when the sun has sunk so far below the horizon that his direct rays can no longer fall upon and be reflected by any particles in that portion of the atmosphere which lies above the plane of the horizon. This is usually assumed to be the case when the sun is 18° below the horizon.

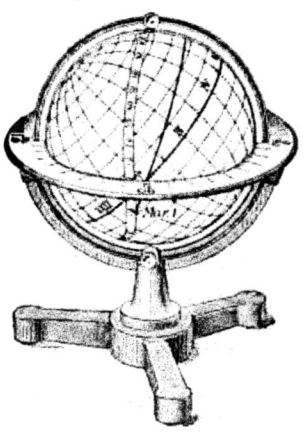

Fig. 66. End of Twilight at Cambridge March 1

Bringing the sun's place for March 1 to the horizon, and then, by means of the extension of the altitude arc, to a point 18° below the horizon (Fig. 66), we have the following values:

R.A.M. at sunset	4ʰ 20ᵐ
R.A.M. at end of twilight	6 0
Difference	1 40

which equals the change in the sun's hour-angle, or the time elapsed between sunset and the end of twilight.

Example 31. At what hour, apparent time, does morning twilight begin at Cambridge June 21?

June 21. Sun's place 18° below E. horizon, R.A.M.	$20^h\ 8^m$
Sun's R.A. by globe	6 0
App. T.	14 8
Or	2 8 a.m.

Example 32. At what point of the horizon does the first glimmer of dawn appear in latitude 42°.4 on June 21?

Bringing the sun's place by trial to the altitude arc at a point 18° below the horizon (Fig. 67), the reading on the horizon at the graduated edge of the altitude arc is E. 57° N. = Az. 213°; and as this is the nearest point of the horizon to the sun when it is 18° below the horizon, it is at this point or a little to the south that the first light will appear.

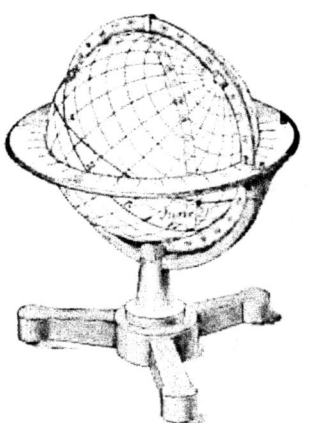

Fig. 67. Dawn at Cambridge June 21, at $2^h\ 8^m$ a.m.: Sun's Az. 213°

Example 33. How many hours can the sun shine into north windows June 21 in latitude 41°?

By the method of Example 15, it is found that the apparent times of sunrise and sunset on June 21 are $4^h\ 30^m$ a.m. and $7^h\ 30^m$ p.m., and by the method of Example 20, that the sun is in the prime vertical at $7^h\ 56^m$ a.m. and $4^h\ 4^m$ p.m. Hence from $4^h\ 30^m$ to $7^h\ 56^m$ a.m. and from $4^h\ 4^m$ to $7^h\ 30^m$ p.m., a total of $6^h\ 52^m$, the sun shines on the north face of an east and west wall. The length of the day is fifteen hours.

Example 34. August 20, in latitude $42\frac{1}{2}°$, longitude $4^h\ 48^m$, at ten minutes past 10 a.m., Eastern standard time, the sun begins to shine upon the front wall of a building. How does the building face?

Since at the given time the sun is in the same vertical plane with the front wall of the building, the problem requires us to determine the direction of this plane by finding the sun's azimuth, which may be done as follows:

EXAMPLES OF THE USE OF THE GLOBE

Rectifying for latitude 42½°, we have:

Std. T. 10ʰ 10ᵐ A.M.	= 22ʰ 10ᵐ
Red. for Long. (from E. Std. T.)	+ 12
Local M.T.	22 22
Subtract Eq. of T. (additive to App. T.)	− 3
App. T.	22 19
Sun's R.A.	10 1
Sid. T.	8 20

Rectifying for this time and bringing the altitude arc to the sun's place for August 20, we find the sun's azimuth to be 315°. Hence the front wall is in a line from southeast to northwest, and the building fronts southwest.

Example 35. What is the greatest northern latitude in which all of the four bright stars of the Southern Cross are visible? What must be the time of year?

Rectifying the globe for the equator, the Southern Cross (about R.A. 12ʰ, Decl. − 60°) is brought to the meridian and the brass meridian is moved in its own plane until the lowest star is brought to the horizon at its south point. The elevation of the pole above the north horizon is then read on the brass meridian and found to be 28°, which is the required latitude. The star being still in the same position, the altitude arc is then

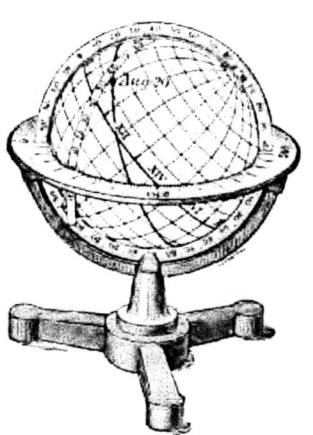

FIG. 68. August 20: Std. T. 10ʰ 10ᵐ; R.A.M. 8ʰ 20ᵐ; Sun's Az. 315°

used to mark the points of the ecliptic which are 18° below the horizon. These are found to be at points occupied by the sun January 2 and May 25, and between these dates, therefore, the whole cross may be above the horizon in latitude 28° in the full darkness of night, the sun being below the twilight limit.

Example 36. What is the latest date at which we can see Sirius in the evening twilight in latitude 42°?

Sirius is visible when the sun is about 10° below the horizon, and cannot be seen later than the day on which he sets at the instant that the sun is 10° below the horizon.

Rectifying for 42° and bringing Sirius to the western horizon, we find that the point of the ecliptic which is 10° below the horizon is the place occupied by the sun on May 15, which is, therefore, the required date.

Example 37. Between what dates is the sun visible at midnight at the North Cape, in latitude 70° north?

Rectifying the globe for 70° north and rotating the globe slowly, it is found that points on the ecliptic in longitudes 58° and 122° can be brought exactly to the north point of the horizon; any point between these may be brought to the meridian below the pole and above the horizon. The dates at which the sun occupies these positions are May 19 and July 25, and between these dates the sun will always come to the meridian at midnight *above* the horizon.

Example 38. Illustrate the "harvest moon" by finding the time of moonrise at Edinburgh, latitude 56°, on successive dates about the time of full moon, September 24, 1904.

As only approximate results are desired, we may take from the Ephemeris the moon's place for 6^h P.M., G.M.T., and solve as follows:

1904	R.A.	Decl.	R.A.M. at Moonrise	Sun's R.A.	Apparent Time
September 22	$22^h\ 30^m$	$-8°$	$17^h\ 22^m$	$12^h\ 0^m$	$5^h\ 22^m$
23	23 22	-4	17 42	12 4	5 38
24	0 7	-1	18 9	12 7	6 2
25	0 52	$+3$	18 28	12 10	6 18
26	1 38	$+7$	18 51	12 14	6 37

And it appears that the moon rises about twenty minutes later each night than it did on the previous night.

Example 39. Find the time of moonrise at Edinburgh on successive nights at full moon, March 31, 1904.

We have, as in Example 38, the moon's place at 6^h P.M., G.M.T.:

1904	R.A.	Decl.	R.A.M. at Moonrise	Sun's R.A.	Apparent Time
March 30 . .	$11^h\ 56^m$	$+1°$	$5^h\ 57^m$	$0^h\ 38^m$	$5^h\ 19^m$
31 . .	12 52	-4	7 20	0 41	6 39
April 1 . . .	13 49	-8	8 42	0 44	7 58

Therefore the full moon at the time of the vernal equinox rises about one hour and twenty minutes later each night. (Notice and explain the difference in the accuracy attained in these two examples.)

Example 40. Find the rate at which δ Orionis is changing its azimuth at rising and setting in latitude 42°.

Rectifying for 42° and bringing δ Orionis to the eastern horizon, we find R.A.M. = 23^h 23^m; Az. = 271°. Increasing the hour-angle half an hour by making R.A.M. = 23^h 53^m, we find, by the altitude arc, Az. = 276°. Bringing the star to the western horizon, we have R.A.M. = 11^h 24^m; Az. = 89½°. Decreasing the hour-angle by making R.A.M. = 10^h 54^m, we find Az. = 84½° half an hour before setting. In both cases the diurnal rotation causes the azimuth to *increase* at the rate of 5° in half an hour.

By solving the same problem for stars in various parts of the heavens, as, for instance, Vega, γ Pegasi, Antares, and α Gruis, it appears that stars of whatever declinations, when near the horizon, are *increasing* their azimuths by about 10° per hour in latitude 42°. (This is the rate at which the plane of the pendulum appears to revolve in Foucault's experiment.)

Example 41. To mark the hour-lines on a horizontal sundial for use in latitude 42°.

The gnomon of an ordinary sundial (Fig. 69) is directed toward the pole, and its shadow at apparent noon falls upon the horizontal dial on the line of XII hours, which, when properly adjusted, lies in the direction of the meridian. The shadow at that time is in a line drawn through the foot of the gnomon toward azimuth 180°. It always passes through the intersection of the gnomon with the dial and, continually shifting toward the east, at any instant lies in the plane containing the sun and the gnomon.

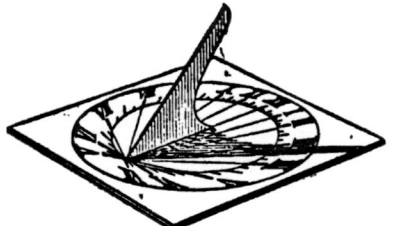

FIG. 69. Horizontal Sundial, Latitude 42°

This plane cuts the celestial sphere in the sun's hour-circle. The shadow, therefore, is a line which passes through the foot of the gnomon and whose azimuth is that of the intersection of the sun's hour-circle with the plane of the horizon. For a given hour-angle the position of this line will be the same whatever the position of the sun upon its circle, and is therefore the same for a given apparent time whatever the time of year.

We may find the azimuth of the intersection of a given hour-circle with the horizon by means of the globe as follows. Rectifying the globe for 42°, the vernal equinox is brought to the meridian, so that the equinoctial colure cuts the horizon at azimuth 180°. In this position R.A.M. is 0^h, and the azimuth of the shadow is 180°. Increasing the hour-angle of the colure by successive increments of 15°, we have the following values for the azimuths of the hour-lines:

FOR THE P.M. HOURS:			AND SIMILARLY FOR THE A.M. HOURS:		
	R.A.M.	Azimuth of Shadow		R.A.M.	Azimuth of Shadow
I	1^h	190	XI	23	170
II	2	201	X	22	159
III	3	214	IX	21	146
IV	4	230	VIII	20	130
V	5	249	VII	19	111
VI	6	270	VI	18	90
VII	7	291	V	17	69

If the hour-circles are shown for each 15° as on most modern globes, it is sufficient to bring one hour-circle to the meridian and note the points where the other circles cut the horizontal plane; Fig. 57 shows the globe rectified to 42° and 0^h Sid. T., and therefore in position for reading the azimuths of the successive hour-lines directly on the horizon.

Example 42. To mark the hour-lines of a vertical sundial for use in latitude 42° N., the bearing of the plane being W. 24° S.

Here the shadow of the gnomon falls upon a vertical plane, and the line for noon is a vertical line through the intersection of the gnomon with the plane.

At any given hour after noon the shadow falls below the gnomon and to the east of the XII line (Fig. 70), since it marks the intersection of the plane of the dial by the sun's hour-circle. It makes an angle with the XII line which may be defined as the "nadir distance" of the line of intersection of the two planes, and this is equal to the zenith distance of that part of the same line which lies above the gnomon.

This problem therefore requires us to find the zenith distance of the intersection of the sun's hour-circle with the vertical plane for a given hour-angle of the sun, and may be solved with the globe as follows:

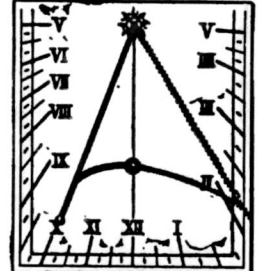

Fig. 70. Vertical Dial, Latitude 42°

Rectify the globe for latitude 42°, and adjust the altitude arc to the zenith with its foot at azimuth 66° on the horizon; its plane then corresponds to that of the dial.

Bringing the vernal equinox to the meridian, R.A.M. = 0^h, the equinoctial colure intersects the altitude arc at zenith distance 0°. Increasing the hour-angle of the colure, as in Example 41, we have successively

Hour-Line	R.A.M.	Zenith Distance of Intersection
I	1^h	13°
II	2	30
III	3	49
IV	4	70
V	5	90

which gives the angles of the afternoon lines from the noon line.

Setting the arc at azimuth 246°, we find in the same way

Hour-Line	R.A.M.	Zenith Distance of Intersection
XI	23^h	11°
X	22	22
IX	21	33
VIII	20	44
VII	19	56
VI	18	70
V	17	90

which gives the morning lines. The A.M. and P.M. divisions will not be symmetrical about the XII line unless the vertical plane faces due south.

Example 43. Find the path of the shadow of a pin head on a horizontal plane at Cambridge March 21, from 8 A.M., apparent time, to 5 P.M., apparent time.

Rectifying the globe for latitude 42°, bringing the sun's place to hour-angles which correspond to the successive hours from 8 A.M. to 5 P.M., and measuring its altitude and azimuth in each position by the altitude arc, we have the following results:

App. Time	R.A.M.	Altitude	Azimuth	Distance
8ʰ A.M.	20ʰ	22°	291°	12.5 cm.
9	21	32	303	8.1
10	22	40	318	6.0
11	23	46	337	4.9
Noon	0	48	0	4.5
1ʰ P.M.	1	46	22	4.9
2	2	40	42	6.0
3	3	32	56	8.0
4	4	22	68	12.6
5	5	11	80	26.5

To construct the curve we must know the length of the pin; assuming this to be 5 cm. long, a point on the paper is chosen to represent the point vertically under the pin head, and through it is drawn a line to represent the meridian, and other lines are

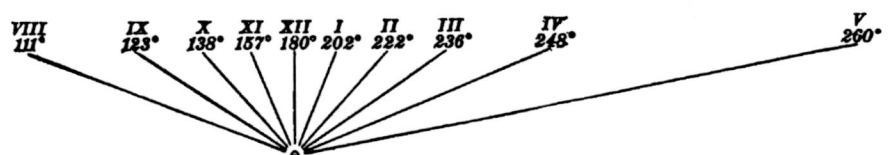

FIG. 71. Azimuth of Shadow

drawn at the azimuths differing by 180° from those given in the above table. (See Fig. 71.) The shadow path will cross these lines at the corresponding hours.

To find the distance of any point of the shadow path from the foot of the pin, we may reverse the process explained on page 5. Drawing a line from C, the center of the base in Fig. 6, through the divisions of the protractor corresponding to any one of the altitudes of the above table and measuring the line $A'B'$, we have the distance in centimeters from the foot of the pin to the point where the shadow falls on the corresponding azimuth line. The last column of the above table gives the distances measured in this manner.

Fig. 72 shows the shadow path as thus constructed, and it is evidently a straight line. This will always be the case on the day of the equinox, when the sun is in the equator and its diurnal path is consequently a great circle.

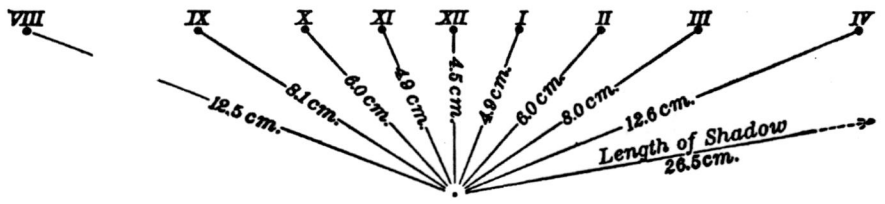

FIG. 72. Path of Shadow

THE HOUR-INDEX

The globe is usually provided with an arrangement by means of which approximate solutions may be made of problems involving time without the use of the graduations of the equinoctial.

This process is so simple that its explanation might well have preceded that of the method of finding the sun's hour-angle given on page 112 and used in Problem 7. It is, however, very inaccurate, and should only be chosen where an error of several minutes is unimportant.

The most convenient form given to the attachment is that of a small pointer fixed to the brass meridian in such a manner that it revolves about the same center as the polar axis, but with sufficient friction to keep it fixed in any position where it may be placed.

This pointer, or "hour-index," lies close to the surface of the globe, which revolves freely under it. The end of the index lies over a small circle on the globe, about 15° from the pole; and this circle is graduated into hours and quarters in two groups of 12 hours each, numbered in the same direction as the graduations of the equinoctial.

The following example illustrates the use of the hour-index, which in this case gives sufficiently good results with less trouble than the method already explained.

Example 44. Find the apparent times, October 1, 1904, of rising and setting of the planets whose places are given on page 130.

Mark the places of the planets and of the sun; bring the latter to the meridian and set the hour-index to read XII noon. Rotate the globe through any angle, and the reading of the index will equal the hour-angle of the sun in its new position, and thus will give directly the corresponding apparent time.

We may, therefore, rapidly determine the apparent time of rising and setting of all the planets by bringing each in turn to the eastern and western horizon and noting the reading of the hour-index.

The hour-index may be adjusted to give local mean time or standard time directly by making it read the local mean time or standard time of apparent noon when the sun is brought to the meridian. Thus, for October 1, at Cambridge, longitude $4^h 44^m$:

App. T. of App. noon	$12^h \ 0^m$
Eq. of T.	$- 10$
Camb. M. T. of App. noon	$11 \ 50$
Red. for Long.	$- 16$
Std. T. of App. noon	$11 \ 34$

And the index should be set to read $11^h 34^m$ when the sun is on the meridian, in order to give Eastern standard time.

CHAPTER X

THE MOTIONS OF THE PLANETS

It has been the aim of the preceding chapters to show how the diurnal motion and the motion of the sun and moon among the stars may be studied in such a manner that the student shall acquire and fix his knowledge in large part by his own observations.

There remains to be considered the motion of the planets, which cannot be studied in the same way because they move so slowly that a long time would be required to obtain a sufficient number of observations on which to base a satisfactory theory. It is of course desirable, however, during the continuance of the observations on the moon and stars to include the planets in order to establish a few fundamental facts, such as that they never appear far from the ecliptic and that in general they move from west to east like the sun and moon, but that when opposite the sun, so that they come to the meridian at midnight, they are moving from east to west among the stars. Their places in the heavens should be occasionally observed, for comparison with the places derived from the theory which forms the subject of the present chapter.

In treating of this theory we shall first assemble the few principles which have been shown to account for the observed motions, and shall then show how these principles may be applied to the graphical solution of problems involving the determination of the place in the heavens of a planet as seen from the earth at any given time. These problems serve to illustrate and explain the phenomena resulting from the planetary motions, as the globe problems of the preceding chapter serve for those resulting from the diurnal rotation of the earth.

Results of the Law of Gravitation. — In consequence of the attraction of the sun, each planet describes an ellipse, having the sun in one focus; this is "Kepler's first law." The mutual attractions of the planets produce "perturbations" of their motion, but in no case are these perturbations sufficient to alter the place of the planet by so much as one degree from its place as determined by the sun's attraction. Jupiter may be displaced about 0°.3 and Saturn nearly 0°.8; but with this exception no displacement of a planet amounts to ¼°. The asteroids are subject to much greater perturbations.

The orbit of each planet is in a plane which remains nearly fixed, and the planes of all the orbits are so nearly coincident with the ecliptic that the projections of their paths on the ecliptic are no more distorted than the roads of a moderately rugged country are distorted in their representations on an ordinary plane map. This fact makes it as easy to determine their motions by an accurate map of their orbits on the plane of the ecliptic as to follow the motion of a traveler over a well-charted country, when his point of departure and rate of travel are known.

PROPERTIES OF THE ELLIPSE

An ellipse may be drawn by putting two pins upright in a board, as in Fig. 73, laying a knotted loop of thread on the board so as to include both pins, and then putting the point of a well-sharpened pencil on the surface inside the loop. Let the pencil be moved out

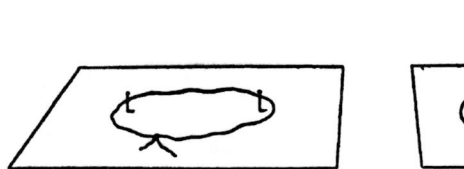

FIG. 73. Drawing an Ellipse

so as to form the loop into a triangle, and then drawn along the surface so as to pass successively through all the points which it can reach without allowing the thread to become slack. The curve which it follows will be an ellipse whose shape and size will depend only on the distance between the pins and the size of the loop.

The form of the curve is shown in Fig. 74.

F_1 and F_2 are the foci, AB the major axis, and C, which bisects both F_1F_2 and AB, is the center of the ellipse. PF_1 is the radius vector from any point P to F_1, and PF_2 the radius vector to F_2. They are usually represented by r_1 and r_2. $r_1 + r_2$ is a constant for all points of the ellipse, being always equal to the length of the thread minus F_1F_2. For the point A

$$r_1 + r_2 = AF_1 + AF_2;$$

and since from the symmetry of the curve

$$AF_1 = BF_2,$$

$$r_1 + r_2 = BF_2 + AF_2 = AB.$$

AC is usually represented by a, and CF_1 or CF_2 by c.

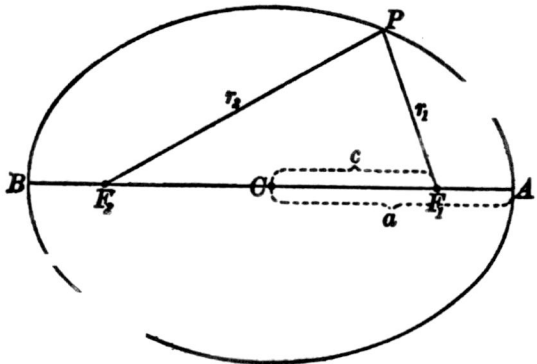

FIG. 74. Fundamental Points and Lines

Since $2c$ equals the distance between the foci, and $2a + 2c$ the length of the thread, the shape and size of the ellipse are completely fixed by the values of a and c. The ratio c/a is called the eccentricity and is represented by e; it is always less than unity. The line along which the major axis lies is called the line of apsides.

To draw a Given Ellipse. — Let it be required to draw an ellipse whose semi-major axis is one inch, and eccentricity $\frac{1}{4}$, with one focus at the point F_1 of Fig. 75, and with its major axis inclined 30° to the horizontal.

Draw the line of apsides MN at the proper angle. Since $e = \frac{1}{4}$, we locate C one-fourth of an inch from F_1 on the line of apsides. Take F_2 at an equal distance beyond C, make the total length of the thread $2\frac{1}{2}$ inches $= 2a + 2c$, and draw the ellipse as shown in the figure.

The dotted line surrounding the ellipse is a circle drawn about C as a center with a radius of one inch (equal to the semi major axis). It is worthy of notice that the ellipse differs but little from this circle, the greatest distance between the two being about $\frac{8}{100}$ of an inch. With a less eccentricity the agreement of the two curves is closer. For $e = 0.10$ the difference is but .005 of the semi major axis, so that an ellipse of that eccentricity whose semi major axis is two inches differs at no point more than $\frac{1}{100}$ of an inch from a circle struck about its center with a radius of two inches. If the orbits of the planets are drawn with their true eccentricities and with a line 0.01 inch in width, and in each case a circle is struck with radius a about the center of the ellipse, and having a width of .01 inch, no white space will be anywhere visible between the two lines unless the diameter of the circle is greater

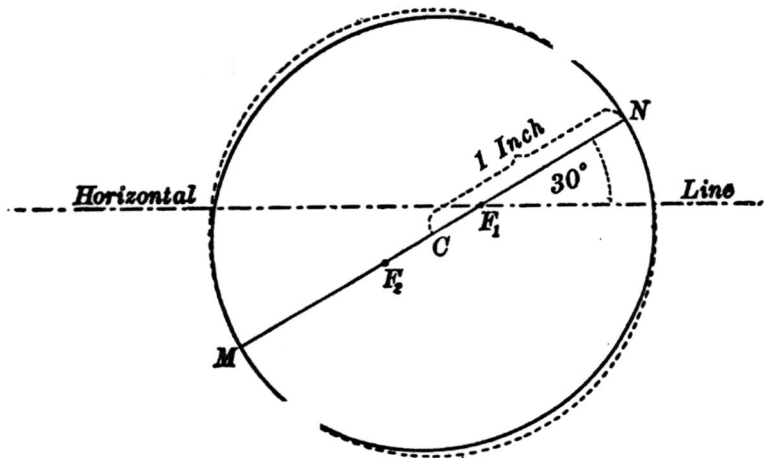

FIG. 75. Ellipse drawn with Given Constants

than about 1 inch for Mercury, 4½ inches for Mars, 17 inches for Jupiter, and 12½ inches for Saturn. For Venus and the earth the circles may be several feet in diameter. The orbits may therefore be represented by such circles with a considerable degree of accuracy.

MEAN AND TRUE PLACE OF A PLANET

Having considered the geometrical properties of the planetary orbits, it is next in order to inquire as to the law which regulates the motions of the planets in their orbits.

Since the sun is at one focus of the orbit, the planet's distance from the sun varies continually. It is nearest the sun at the perihelion point, which is at one extremity of the major axis. Aphelion occurs at the opposite end of the major axis, and the planet is then at its greatest distance.

THE MOTIONS OF THE PLANETS

Kepler's second law states that the planet moves in such a way that its radius vector sweeps over equal areas in equal times. The application of this principle will be evident from the following illustration.

Fig. 76 represents the orbit of Mercury in its true proportions. The period of the revolution of the planet is eighty-eight days, in which time the radius vector sweeps over the whole area of the ellipse. To pass from perihelion to aphelion would require forty-four days, or one-half the period, since the area described is one-half the area of the whole ellipse. It is not difficult to fix very nearly the point reached by the planet twenty-two days after passing through perihelion. It will then have accomplished a quarter of a revolution, and be at such a point P that the area ASP is one-quarter of the ellipse, or one-half of $APBS$, so that APS equals BPS.

It may be shown that this point must be very nearly in the line Pf drawn perpendicular to the major axis through f, the "empty" focus of the orbit, as it is sometimes called.

Assuming P to be on this line, and drawing a perpendicular Sk through the focus

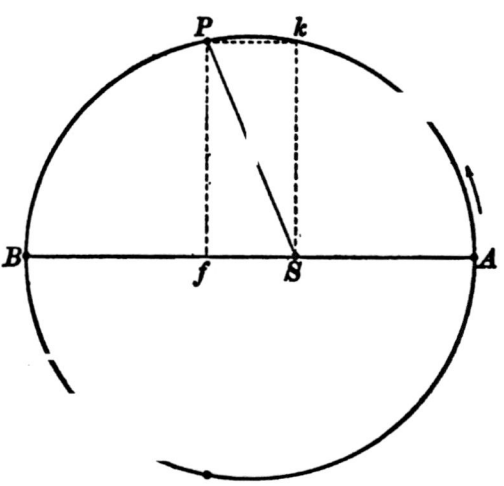

FIG. 76. Equal Areas in the Ellipse

occupied by the sun, and also the radius vector PS, we have from the symmetry of the ellipse, Area ASk equal Area BfP, and the triangle PkS evidently equals the triangle PfS. The difference of the two areas ASP and BSP is therefore the segment of the ellipse cut off by the chord Pk; this segment is so very small that the area ASP is very nearly equal to BSP.

The angle ASP through which the planet has moved about the sun since perihelion is called its "true anomaly." In this case it is about 110°. We may now infer that the true anomalies of Mercury 22, 44, 66, and 88 days after perihelion would be about 110°, 180°, 250°, and 0°, respectively.

It is convenient to refer the motion of the planet to that of a hypothetical planet moving in the orbit in such a way as to be at perihelion with the real planet and describe equal *angles* in equal times; thus the anomaly of the so-called "mean planet" after 22, 44, 66, and 88 days would be 90°, 180°, 270°, and 360°, respectively.

The Equation of Center. — The quantity to be added to the anomaly of the mean planet, or briefly, the "mean anomaly" of the planet, in order to find its true anomaly, is called the "equation of center"; in the cases above given it is for the four positions 0°, + 20°, 0°, and − 20°. It is always positive for values of the mean anomaly between 0° and 180°, and negative for values between 180° and 360°. It appears from Fig. 77, in which P and P' mark the true and mean places of the planet respectively, that at all points from perihelion A to aphelion B, the true anomaly ASP is greater than the mean anomaly ASP', while from aphelion to perihelion ASP is less than ASP'.

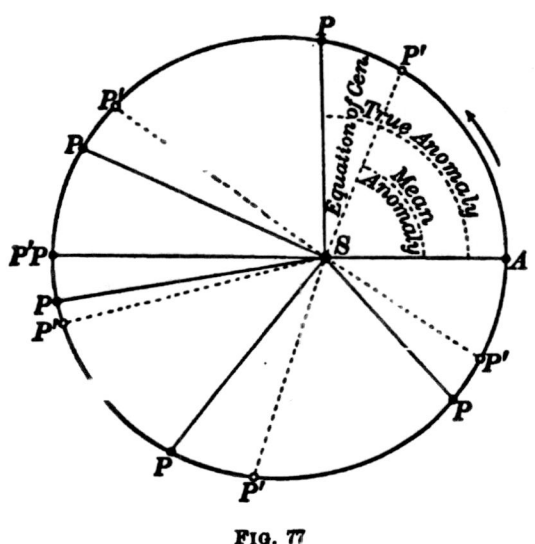

FIG. 77

The value of the mean anomaly being given for any time, its value for any other time is easily found, since it increases uniformly from 0° to 360° in the time required for the planet to make one revolution.

The mean anomaly being known, we may pass to the true anomaly by means of a table of the equation of center (page 174), in which the value of the latter is given for each degree or ten degrees of the planet's mean anomaly.

The computation of these tables lies far beyond our scope, but it is worth while to note that approximate values of the equation of center may be found by a graphical method, which rests upon the principle that in describing equal *areas* about one focus of an ellipse of small eccentricity, a planet describes very nearly equal *angles* about the other focus.

If then the ellipse be carefully constructed on a large scale, say with a major axis of ten inches, and through the empty focus lines be drawn making angles of 10°, 20°, 30°, etc., with the line of

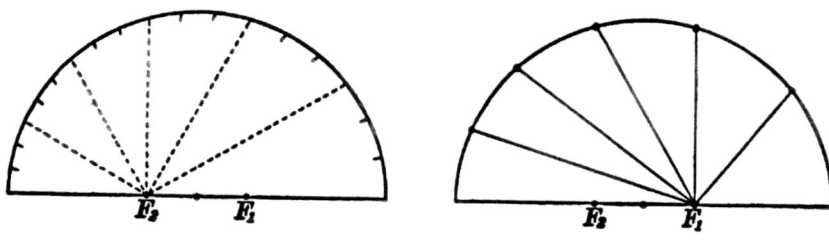

FIG. 78

apsides, these lines will cut the ellipse at the places occupied by the true planet when its mean anomalies are 10°, 20°, 30°, etc. Fig. 78 shows one-half of the orbit of Mercury divided into six equal parts in this manner.

The true places being thus fixed, and lines drawn from each to the sun, the true anomalies may be read off with a protractor; and by comparison with the mean anomalies the equation of center for each ten degrees of mean anomaly may be determined.

MEASUREMENT OF ANGLES IN RADIANS

It has been assumed that the student is familiar with the ordinary method of measuring angles in degrees. For some purposes it is convenient to select a different unit, the "radian."

One radian is the angle subtended by an arc whose length (measured by a flexible scale laid along the curve of the arc) is equal to that of the radius. This angle measured in the ordinary way is found to be $57°.3 = 3438'$, or $206,265''$.

If the length of an arc a is known, and also the radius of the circle r, the angle subtended by the arc is a/r (arc ÷ radius) radians. Thus in a circle two feet in diameter, an arc of one inch subtends an angle of 1/12 radian, — 6 inches of 0.5 radian, 1 foot of 1 radian, etc. Since 1 radian equals $57°.3$, an arc of one inch in the above circle subtends $1/12 \times 57.3°$; and, in general, radians are transformed to degrees, minutes, or seconds of arc by multiplying by 57.3, 3438, and 206,265, respectively; and degrees, minutes, or seconds to radians by dividing by 57.3, 3438, and 206,265, respectively.

The use of the radian is especially convenient in problems involving an angle so small that the corresponding arc nearly equals its chord or the perpendicular drawn from one extremity of the arc to the radius drawn through its other extremity. The method is illustrated by the following instances:

1. The moon's distance is 240,000 miles, and its angular diameter is 31', or 31/3438 radian. Its diameter in miles is given by the equation

$$\frac{\text{arc}}{\text{radius}} = \frac{D}{240,000} = \frac{31}{3438}.$$ Hence $D = 2164$ miles, approximately.

2. The height of a tree is 30 feet, and the length of its shadow is 150 feet. The altitude of the sun is

$$a/r = 30/150 = 0.2 \text{ radian} = 11°.46.$$

The true value obtained by trigonometrical computation is $11°.54$, differing by $.08°$, and this approximate method will give results within $0°.1$ so long as the angle does not exceed this value.

3. By means of a sextant the angle between the water line of a distant war ship (Fig. 79) and the top of its military mast is found

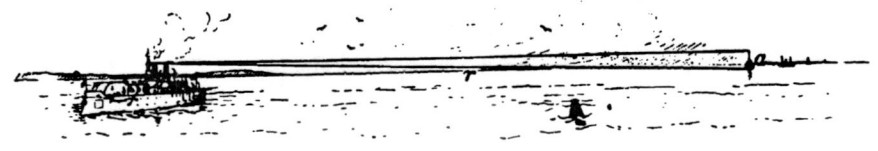

FIG. 79

to be $17' 10''$. The height of the mast is known to be 120 feet. Assuming this height to be equal to the arc subtended by the measured angle, we have

$$17' 10'' = 0.005 \text{ radian} = \frac{a}{r} = \frac{\text{height of mast}}{\text{distance of ship}},$$

and the distance of the ship is about 8000 yards.

DIAGRAM OF CURTATE ORBITS

Fig. 80 represents a diagram of the orbits of the five inner planets projected on the plane of the ecliptic, which serves to solve many problems regarding the planetary motions. The diagram is of convenient size for actual use, if its dimensions are such that one astronomical unit equals about $\frac{3}{4}$ of an inch.

In order to show how small is the distortion of the orbits as projected, we may compare the length of the radius vector to any point in the orbit with that of its projection on the ecliptic, which is called the "curtate" distance from the sun.

Even in the case of the orbit of Mercury, which has the greatest inclination, the curtate distance differs from the true distance at most by $\frac{1}{130}$, in the case of Venus by less than $\frac{1}{800}$, and in the case of all the other planets by less than $\frac{1}{1000}$. If the scale of the diagram is such that one astronomical unit equals $1\frac{1}{2}$ inches, no radius vector drawn in any one of the "curtate" orbits will differ from the corresponding radius vector drawn in the actual orbit by so much as $\frac{1}{130}$ of an inch; and by referring to the data given on page 146 it will be seen that on that scale the elliptic orbits may be represented with considerable accuracy as circles.

The position of the line of apsides is fixed by the longitude of perihelion, page 174; the distance c of the center of the ellipse from the sun is found from the ratio $c/a = e$, and a circle struck about the center with a radius a very closely represents the curtate orbit; the distances c and a are of course to be laid off from the scale of astronomical units.

To draw such a diagram is a useful exercise, and by careful drawing and erasure a single diagram may serve for many problems, but it is convenient to have several printed copies when it is desired to preserve the solutions.

It is also convenient to have diagrams on which an astronomical unit equals $2\frac{1}{4}$, $\frac{3}{4}$, and $\frac{3}{8}$ inches, respectively, the first extending to the orbit of Mars, the second to that of Jupiter, and the third to that of Saturn. The larger scale should be used for problems referring to Mercury and Venus, while the smaller scales are required for the major planets.

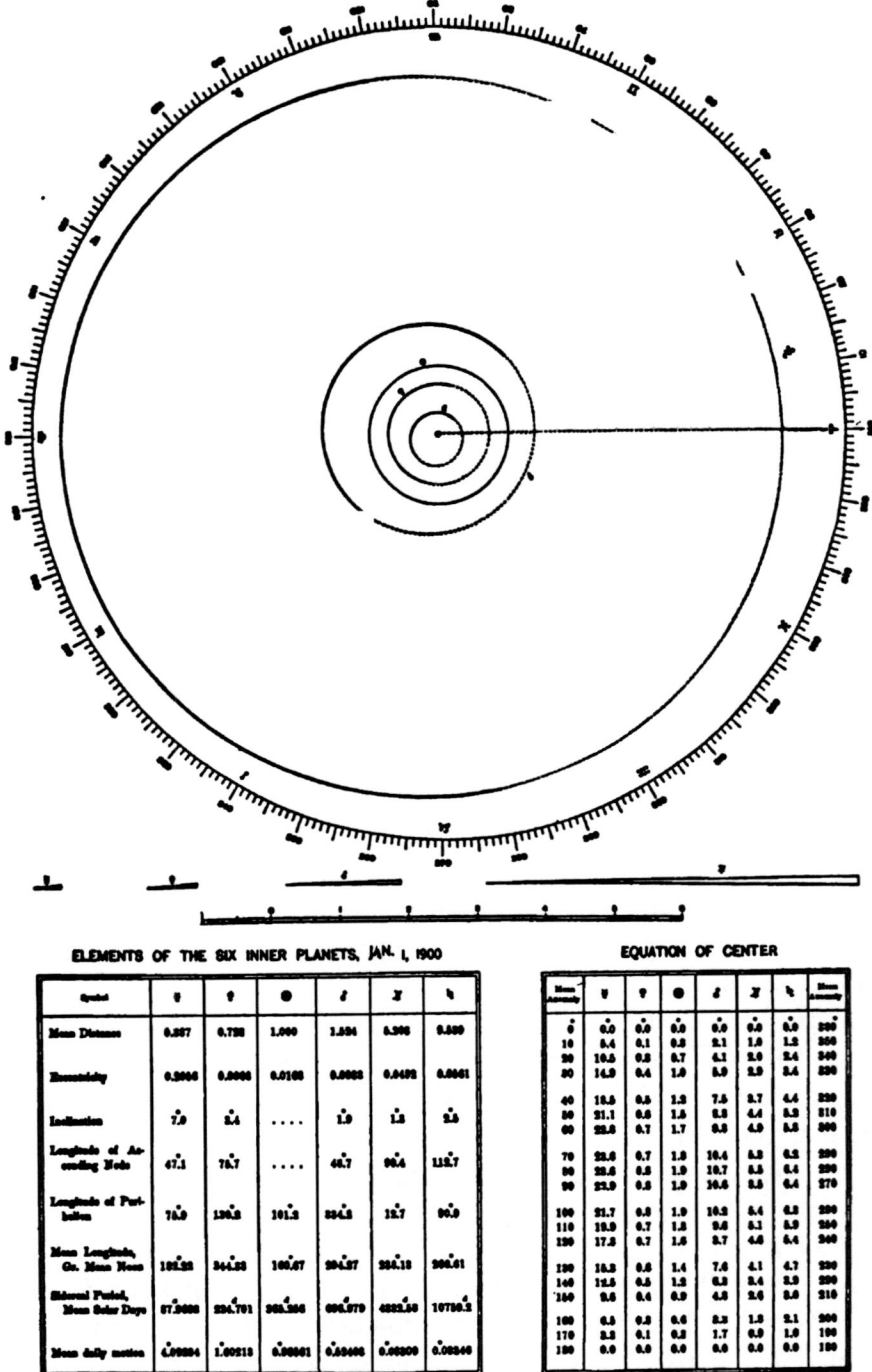

FIG. 80. Diagram of Curtate Orbits

THE MOTIONS OF THE PLANETS 153

On the plan of each orbit the symbol of the planet is placed at the perihelion point, whose position is thus approximately known at a glance.

That part of the orbit which is above the plane of the ecliptic is marked with a full line, and the part below is marked by a broken line. The line of nodes is therefore determined as a line joining the two points where the character of the line changes. This line, of course, passes through the sun.

The inclinations of the orbit planes are shown by the triangles which appear below the diagram, each marked by the symbol of the planet to whose orbit it pertains. A scale of astronomical units is printed at the bottom.

The attached tables (see page 174) give the values of the elements of each orbit and certain other quantities which are required in finding the place of the planet in its orbit at a given time.

Measurements may be made on the diagram between any two points by laying a strip of paper with its straight edge through the points, and marking the edge of the strip opposite each point. By laying the straight edge along the scale the distance in astronomical units is found. Instead of the paper strip a pair of compasses may be used.

The map shows the orbits as they would be seen from the north side of the ecliptic, and the motions of the planets as thus seen are always counter-clockwise about the sun. The plane of the map is that of the ecliptic, and it is so oriented on the paper that horizontal lines drawn from left to right would strike the celestial sphere at the vernal equinox. Therefore the direction which on an ordinary terrestrial map would be east on this map is toward longitude zero; up is toward longitude 90°, down toward longitude 270°; and the direction of any other line on the map is fixed by determining the angle which it makes with the line drawn to the vernal equinox. Thus, the line in Fig. 81 from E to M makes an angle of 45° with the line SR, and is therefore directed toward longitude 45°, and EJ is directed toward longitude 260°. By drawing lines through the sun parallel to EM and EJ, respectively, the longitude may be read off directly on the circle which bounds the diagram.

154   LABORATORY ASTRONOMY

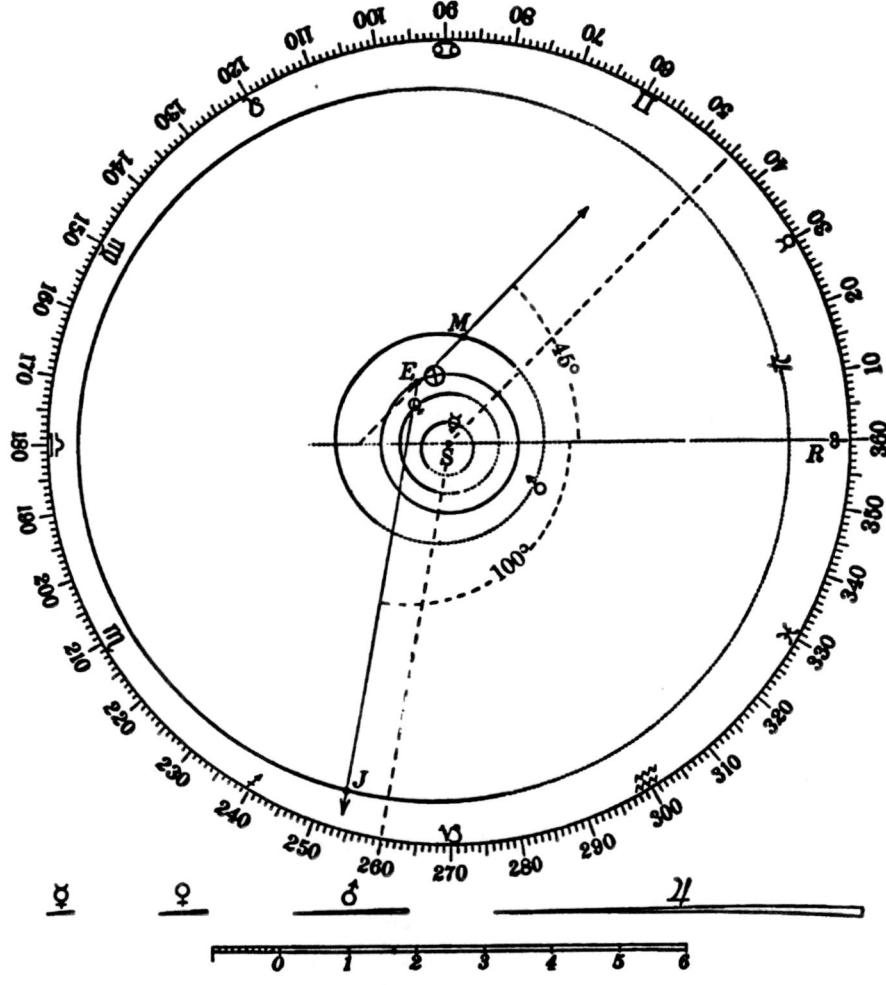

Fig. 81. Direction of a Line fixed by Longitude.

To find the Elements of an Orbit. — The elements of the planetary orbits may be obtained from measurements on the diagram. These elements are as follows:

- a Semi-axis major of the ellipse or mean distance.
- e Eccentricity of the ellipse $= c/a$, where c is distance of focus from center.
- π Heliocentric longitude of perihelion.
- Ω Heliocentric longitude of node.
- i Inclination of plane of orbit to plane of the ecliptic.

To find a draw a straight line from the perihelion point of the orbit through the sun to cut the orbit at the aphelion point. This is the line of apsides. Measure the distance from perihelion to aphelion along the line of apsides in astronomical units. This gives the major axis of the ellipse, one-half of which is the value of a.

To find c, bisect the major axis and thus fix the center of the ellipse. The distance from focus to center may then be measured in astronomical units. This is the value of c; it is not regarded as one of the elements, since it is fixed by the values a and e.

To find e, determine c/a from the above measurements.

To find π, prolong the line of apsides through the perihelion point; the reading at the point where it cuts the graduated circle is the longitude of perihelion.

To find Ω, prolong the line of nodes through the point where the planet moving counter-clockwise passes from the dotted portion of the orbit to the full line. The reading at the point where this line cuts the graduated circle is the longitude of the ascending node.

To find the inclination i, measure the angle of the proper triangle by a protractor; or, more accurately, measure the altitude h and the base b of the triangle; h/b is equal to the inclination in radians. $57°.3\ h/b = i$ in degrees.

The following measurements were made on the orbit of Jupiter:

	Sun to perihelion	4.96
	Sun to aphelion	5.42
	Major axis	10.38
a	Semi-axis a	5.19
	Center to perihelion	5.19
	Focus to perihelion	4.96
c	Center to focus c	0.23

$$\frac{c}{a} = \frac{0.23}{5.19} = 0.044$$

π The line of apsides cuts the circle at 12°.7.
Ω The line to ascending node cuts the circle at 99°.4.
i The altitude of the triangle is 0.13 and the base 5.43; hence
$i = h/b = 0.13/5.43 = 0.024$ radian $= 57°.3 \times 0.024 = 1°.37$.

PLACE OF THE PLANET IN ITS ORBIT

If the heliocentric longitude of a planet is known, it may be plotted at its proper place on the diagram by drawing a line from the sun to that division of the graduated circle which indicates the given longitude; the intersection of this line with the orbit gives the required place. When, for instance, the heliocentric longitude of Jupiter is 280, the intersection falls very close to the descending node. In this particular case the place of the planet is completely known, since it is in the ecliptic. Usually the planet is

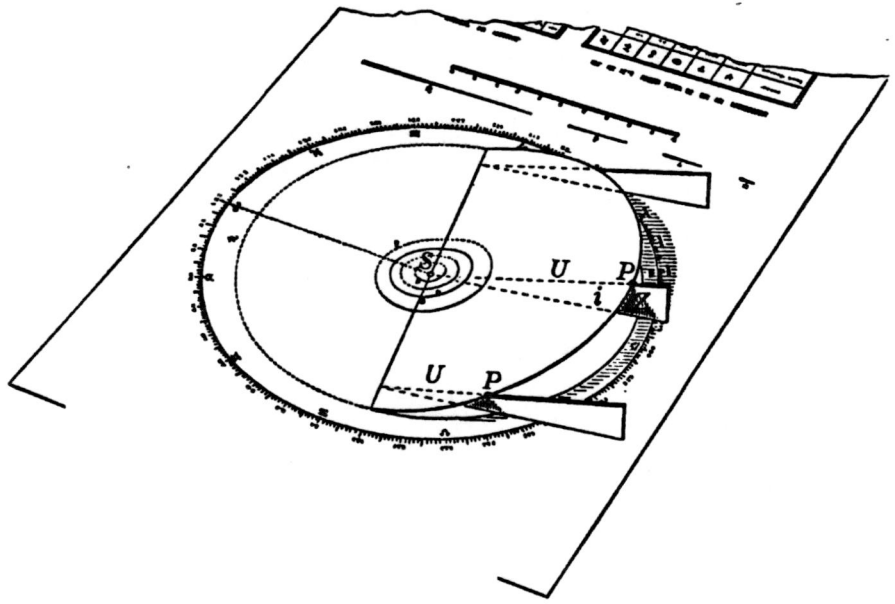

FIG. 82. The Z Coördinate

many millions of miles from the ecliptic, but its exact distance may be easily found by the use of its inclination triangle.

This will appear by consideration of Fig. 82, which represents a diagram in which the orbit of Jupiter has been cut through along the heavy line, and the part of the orbit which is above the ecliptic turned up around the line of nodes so as to be at the proper inclination. The exact angle is insured by supporting it by wedges having the proper angle.

THE MOTIONS OF THE PLANETS

The height of the planet at P above the plane of the ecliptic, which we shall call its "Z coördinate," or simply Z, is evidently the altitude of a right-angled triangle whose small angle is i (the inclination of the orbit), and whose base is the line drawn from the place of the planet on the diagram to the line of nodes. This line (which practically equals the hypotenuse) we will call U.

To find Z, then, it is sufficient to measure U on the diagram and to lay off the same distance along the horizontal side of the inclination triangle. The vertical line drawn to the hypotenuse from the point thus fixed gives the length of Z in astronomical units. A far more accurate method is to make use of the obvious relation $Z/U = i$ in radians, or $57.3\ Z/U = i$ in degrees. Thus, for Jupiter $Z = U \times 1.3/57.3 = 0.023\ U$.

TO FIND THE TRUE HELIOCENTRIC LONGITUDE OF A PLANET

To find the true position of any planet at a given time we must first know its mean anomaly at that time, and then, by applying the equation of center, find the corresponding value of the true anomaly which enables us to place the planet at the proper position in its orbit.

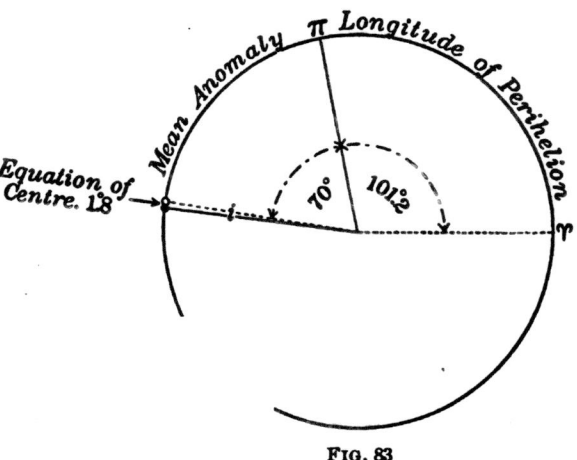

Fig. 83

Thus, if the earth's mean anomaly is 70°, we find by the table, page 174, that the equation of center is $+1°.8$, and hence the true anomaly is 71°.8. Since the longitude of perihelion is 101°.2, the true heliocentric longitude is $101°.2 + 71°.8$, or 173°.0, and this value enables us to plot the earth in its proper place on the diagram, Fig. 83.

We may find the mean anomaly if we know the number of days elapsed since perihelion, and the mean daily motion along the orbit. The fact that the planets move very nearly in the ecliptic, so that the motion in the real and curtate orbit is very nearly the same, makes it easier to proceed in a somewhat different manner, as follows:

In the Table of Elements appended to the chart is given the "mean daily motion" (in heliocentric longitude), which is found by dividing 360° by the period in days. This quantity enables us by a simple multiplication to find the mean motion, or increase in heliocentric longitude of the mean planet in any given number of days.

Knowing the mean (heliocentric) longitude at any given epoch, the mean longitude at any later date is found by addition of the mean motion in the elapsed time. The Table of Elements supplies the necessary "longitude at the epoch" for Greenwich mean noon, January 1, 1900.

We may summarize the process of finding the planet's true heliocentric longitude as follows:

Let
- E be the longitude at the epoch,
- t " " elapsed time in days,
- μ " " mean daily motion,
- π " " longitude of perihelion,
- M " " mean anomaly,
- v " " true anomaly,
- l " " true longitude (heliocentric).

First find the mean anomaly at the time t, as follows:

μt = Mean motion in elapsed time,
$E + \mu t$ = Mean longitude at given date,
$E + \mu t - \pi$ = Mean anomaly.

With this value of the mean anomaly find the equation of center by the table, and since

True anomaly = Mean anomaly + Equation of center,

or $\quad v = E + \mu t - \pi +$ Equation of center,

and True longitude $= v + \pi$, we have directly

True longitude $= E + \mu t +$ Equation of center.

The form of the computation is shown in the solution of the following problem:

THE MOTIONS OF THE PLANETS

Find the true place of Mars and the earth May 8, 1905, at Greenwich, midnight.

The elapsed time may be found as follows:

Gr. Mean Noon. Jan. 1, 1900, to Jan. 1, 1901	365 days
	1902	365
	1903	365
	1904	365
	1905	366
Jan. 1, 1905, to Feb. 1, 1905	31
Mar. 1, 1905	28
Apr. 1, 1905	31
May 1, 1905	30
Noon. May 1 to Midn., May 8, 1905	7.5
Elapsed time =	1953.5 days.

For Mars $\mu t = 0°.52403 \times 1953.5 = 1023°.69$.
For the earth $\mu t = 0°.98561 \times 1953.5 = 1925°.39$.

	MARS	EARTH
Mean longitude Jan. 1, 1900 = E	294°.27	100°.67
Mean motion 1953.5 days = μt	1023 .69	1925 .39
$E + \mu t$	1317 .96	2026 .06
Subtract complete revolutions	1080 .	1800 .
Mean longitude May 8.5, 1905	237 .96	226 .06
Subtract longitude of perihelion π	⎡ 334.2	101.2 ⎤
Mean anomaly M	⎣ 263.8	124.9 ⎦
Equation of center	− 10.36	+ 1.50
True longitude $l = E + \mu t +$ Equation of center	227.60	227.56

It will be noted that in each case the value of $E + \mu t$ has been diminished by an integral number of revolutions: $3 \times 360°$ for Mars and $5 \times 360°$ for the earth. It appears, also, that the numbers inclosed in brackets enter the computation only for the purpose of obtaining the equation of center which is then applied directly to the mean longitude following the equation

$$l = E + \mu t + \text{equation of center.}$$

On plotting the planets it appears that Mars is exactly opposite the sun, as indeed is evident from the fact that the earth and Mars are in the same heliocentric longitude. The Ephemeris gives May 8, 8 P.M., G.M.T., as the time of opposition. The actual distance between Mars and the earth, as measured on the diagram, is 0.56 astronomical units, or fifty-two million miles.

The planet may be plotted with a very fine-pointed, hard pencil, against the edge of a ruler passing through the sun and the point of the graduated circle whose reading equals the planet's true heliocentric longitude. It is quite an advantage to have the ruler of a transparent substance in order that its edge may be correctly placed on the graduations.

A better method, however, is to put a pin through the sun's place firmly into the drawing board or table, and pass around the pin a long loop of smooth black thread. The other end of the loop is

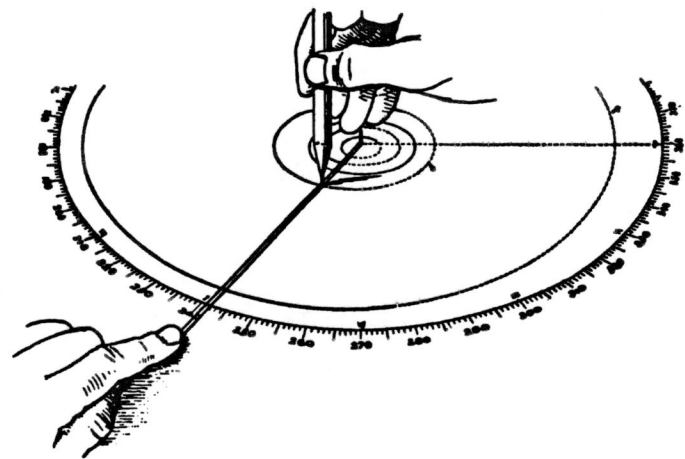

FIG. 84. Plotting with a Loop

held between the thumb and forefinger, with the threads slightly separated (about $\frac{1}{30}$ of an inch). The loop is then drawn taut, and the middle of the white space between the threads may be bisected by the proper point on the graduation; the place of the planet is then marked by putting the point of the pencil exactly midway between the threads where they intersect the orbit (Fig. 84).

The planet having been placed in its true position on the orbit by plotting it as above, so that its curtate radius vector is drawn toward the true heliocentric longitude, its place is completely known if we measure U and find Z, as on page 157. The usual method of fixing the distance of the planet from the ecliptic is to give its heliocentric latitude b, or angular distance from the ecliptic, which may be found thus (Fig. 85):

THE MOTIONS OF THE PLANETS 161

$b = Z/r =$ angular distance (radians) of planet above ecliptic as seen from the sun. Combining this with $Z = U \times i$ (radians), as explained on page 157,

$$b \text{ (radians)} = \frac{U}{r} i \text{ (radians)};$$

and turning each side of the equation into degrees by multiplying by 57.3, we have

$$(57.3\, b)° = \frac{U}{r} \times (57.3\, i)°,$$

or
$$b° = \frac{U}{r} i°.$$

The inclinations are so small that the latitude is always well determined by this method.

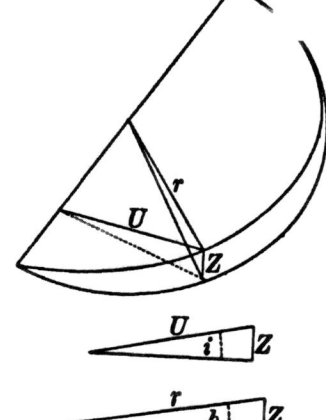

Fig. 85. Heliocentric Latitude

GEOCENTRIC POSITIONS

When a planet has been placed on the diagram by its heliocentric coördinates, we may find its position as seen from the earth; that is, we may find the longitude and latitude of that point of the celestial sphere upon which it is seen projected by an observer upon the earth.

The line drawn *from* the earth *to* the planet is called the "line of sight," and its projection on the ecliptic is the line from the earth to the planet on the diagram. If this line is horizontal, it cuts the celestial sphere at the vernal equinox, and the planet's geocentric longitude is zero.

Geocentric Longitude.—The angle between the (projected) line of sight and the line drawn to the vernal equinox is the planet's geocentric longitude. It is equal to the angle between the line of sight and the line drawn from the sun to the zero of the graduated circle. This angle may be measured in several ways:

1. By prolonging the line of sight, if necessary, till it cuts the line of equinoxes on the diagram, and measuring the angle with a protractor.

2. By drawing a line through the sun parallel to the line of sight, and noting the point where it cuts the graduated circle.

3. The most accurate method is usually the following: Bring a straight edge to pass accurately through the places of earth and planet. Note the points of intersection with the graduated circle.

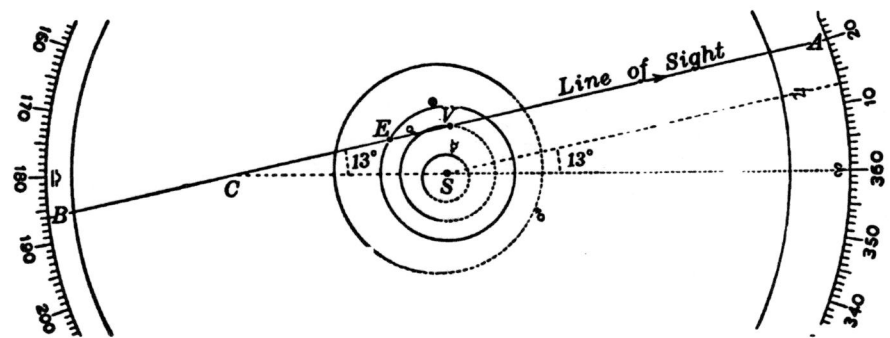

FIG. 86. Geocentric Longitude

Call the reading where the line of sight (*from* earth *to* planet) cuts the circle A, and the other (opposite) reading B. Then the geocentric longitude of the planet is $\frac{A+B}{2} - 90$, if A is less than B; and $\frac{A+B}{2} + 90$, if A is greater than B. This may be proved by the theorem that the angle between two chords of a circle is measured by the half sum or half difference of the included angles, according as they intersect inside or outside the circle.

Better than a straight edge is a fine line on a transparent ruler (celluloid, glass, mica, tracing cloth), or a stretched thread laid over the two points.

Fig. 86 illustrates the three methods, the heliocentric longitudes of the earth and Venus being 150° and 90°, respectively. The angle at C measured by the protractor is 13°, the line through S parallel to AB cuts the graduated circle at 13.0, while the readings at A and B are 20.0 and 186.0, so that $\frac{A+B}{2} - 90° = 13°.0$.

The Sun's Longitude and the Equation of Time.—It is an important fact that, since the line of sight to the sun is drawn to a point whose heliocentric longitude is opposite to that of the earth, the sun's geocentric longitude is always 180° + the earth's heliocentric longitude.

The sun appears to move about the earth in an orbit whose elements are the same as those of the earth about the sun, except that E and π are each greater by 180°.

The sun's mean longitude is therefore $280°.67 + \mu t$ and its mean anomaly is $280°.67 + \mu t - 281°.2$, where t is the number of days since January 1, 1900, and μ is the earth's mean daily motion.

To find the sun's true longitude we add to the mean longitude the equation of center taken from the table for the earth, and from the true longitude we may find the R.A. by adding the reduction to the equator (page 121). We may therefore write:

Sun's R.A. = Sun's mean longitude + Eq. center + Red. to equator.
Sun's R.A. − Sun's mean longitude = Eq. center + Red. to equator.

And since the sun's mean longitude equals the R.A. of the mean sun (page 92),

Sun's R.A. − R.A. of mean sun = Eq. center + Red. to equator.

The first member of the last equation is the equation of time whose approximate value may thus be computed for any date:

$$
\begin{aligned}
\text{Jan. 31, 1900.} \quad \mu t &= 30 \times 0°.9856 = 29°.57 \\
E + \mu t &= 280°.67 + 29°.57 = 310\ .24 \\
E + \mu t - \pi &= 310°.24 - 281°.2 = 29\ .04 \\
\text{Equation of center} &= +0\ .97 \qquad\qquad +0°.97 \\
\text{True longitude} &\quad\ \ \ 311°.21 \\
\text{Red. to equator} &\qquad\qquad\qquad\qquad\quad +2\ .4 \\
\text{Equation of time} &\qquad\qquad\qquad\qquad\quad +3°.37
\end{aligned}
$$

or 13.5 minutes to be added to apparent time.

Geocentric Latitude.—The geocentric latitude β of the planet is the angular distance of the planet from the ecliptic as seen from the earth. It is found by the same method as that used for finding the heliocentric latitude b. (See Fig. 87.)

Draw the line Δ from earth to planet on the diagram. Z/Δ equals the angle β in radians, and $Z = U \times i$ (in radians).

Hence, by reasoning applied on page 161,

$$\beta° = \frac{U}{\Delta} \times i°.$$

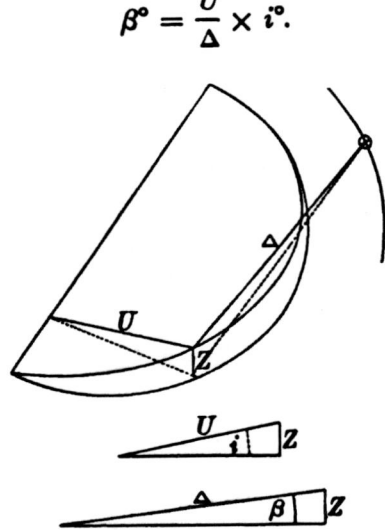

Fig. 87. Geocentric Latitude

The whole process of finding geocentric latitude and longitude is illustrated in the following example:

To find the positions of the five inner planets at Greenwich mean noon, July 6, 1907, the elapsed time from January 1, 1900, being 2742 days (see page 167).

	☿	♀	⊕	♂	♃
E	182°.22	344°.33	100°.67	294°.27	238.13
Add μt	11221 .20	4393 .04	2702 .54	1436 .89	227.83
$E + \mu t$	11403 .42	4737 .37	2803 .21	1731 .16	465.96
Subtr. complete revolutions	11160 .	4680 .	2520 .	1440 .	360.
$E + \mu t$	243 .42	57 .37	283 .21	291 .16	105.96
Subtract π	75 .9	130 .2	101 .2	334 .2	12.7
Mean anomaly	167 .5	287 .2	182 .0	317 .0	93.3
Eq. of center by table .	+ 4 .10	− 0 .73	− 0 .06	− 7 .89	+ 5.47
Heliocentric longitude l .	247 .52	56 .64	283 .15	283 .27	111.43

Plotting the planets on the diagram, we determine the geocentric places by finding the following values:

THE MOTIONS OF THE PLANETS

	☿	♀	♂	♃
From diagram, A	130°.4	80°.1	283°.2	111°.2
" " B	302°.3	266°.9	103°.3	288°.8
" " U	0.14	0.22	1.16	1.09
" " Δ	0.70	1.60	0.40	6.26
From table, i	7°.0	3°.4	1°.9	1°.3
$\dfrac{A+B}{2} \pm 90 = \lambda$	126°.35	83°.5	283°.25	110°.0
$\dfrac{U}{\Delta} \times i = \beta$	$-1°.4$	$-0°.47$	$-5°.51$	$+0°.21$

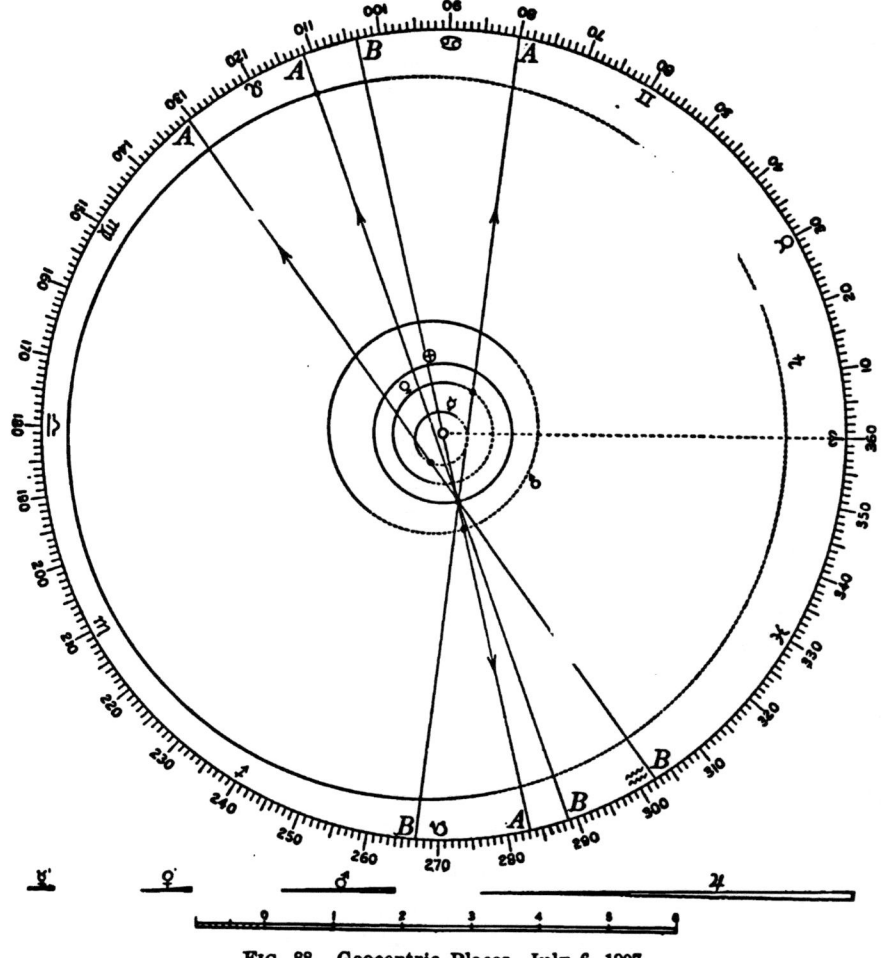

FIG. 88. Geocentric Places, July 6, 1907

The signs attached to the latitudes are fixed by the fact that Jupiter is in the full-line part of its orbit and therefore above the ecliptic, while all the other planets are in the dotted parts of their orbits and therefore in south latitudes.

Since the full line extends from the longitude of the ascending node to that of the descending node, which is 180° greater, we may also fix the sign of β by the following rule:

From the true heliocentric longitude subtract that of the ascending node; if $l - \Omega < 180$, the latitude is positive; if $l - \Omega > 180$, the latitude is negative. Thus, in the above example:

	☿	♀	♂	♃
l	247°.5	56°.6	283°.3	111°.4
Ω	47 .1	75 .7	48 .7	100 .1
$l - \Omega$	200 .4	340 .9	234 .6	11 .3
β	neg.	neg.	neg.	pos.

Perturbations. — The longitudes above obtained are liable to an error of more than a tenth of a degree if the elapsed time exceeds a half century, and the perturbations which are neglected may add somewhat to the error. The effect of the mutual perturbations of Jupiter and Saturn may be approximately corrected by adding to the mean longitudes the following quantities:

♃		♄		♄	
1800–1890	+ 0°.3	1800–1840	− 0°.8	1940–1960	− 0°.4
1890–1950	+ 0 .2	1840–1870	− 0 .7	1960–1980	− 0 .3
1950–1990	+ 0 .1	1870–1910	− 0 .6	1980–1990	− 0 .2
1990–2000	± 0 .0	1910–1940	− 0 .5	1990–2000	− 0 .1

Effect of Precession. — The true longitudes found by the method above described are referred to the equinox of 1900, the point from which the mean longitude of the table is measured.

Since the vernal equinox moves along the ecliptic 50″ per year toward the west, or nearly 6′ in seven years, the longitudes measured from the true equinox of 1907 will be about 0°.1 greater than if measured from the equinox of 1900. This "reduction to the equinox of date" is $50'' \times t$, or $0°.014\,t$, where t is the number of years elapsed since 1900.

The Julian Day. — The process of computing the elapsed time used on page 159 is tedious and liable to error where the elapsed time is considerable. Where the interval between distant dates is to be accurately determined astronomers find it convenient to make use of the number of each day in the Julian period. It is sufficient here to say that January 1, 4713 B.C., was the first day of this period, and the Ephemeris gives each year the number of the Julian day for January 1; thus, the 1st of January, 1900, was No. 2415021 in the cycle. To find the number for any given date, we turn to page III of the corresponding month, add the day of the year (taken from the second column), and subtract 1.

The table on page 175 gives for each year from 1800 to 2000 a number one less than that of the Julian day corresponding to January 1 of the given year. The subsidiary table for months gives for each month a number one less than the day of the year corresponding to the first of the given month.

It is easy to see that by adding together the year number, month number, and day of the month, we get the corresponding Julian day. Thus we compute the interval from January 1, 1900, to July 6, 1907, as follows:

Year number for 1900,	2415020	1907	2417576
Month number for January,	0	July	181
Day of month,	1	6	6
Julian day,	2415021		2417763
			2415021
		Elapsed time,	2742

Right Ascensions and Declinations of the Planets. — By means of the geocentric latitudes and longitudes which we have thus determined the planets may be placed in their respective positions upon the globe.

The proper longitude being found upon the ecliptic of the globe, the latitude is laid off on a strip of paper by placing it along the ecliptic and marking off the proper number of degrees along its edge. The paper is then applied to the globe so as to mark off this distance perpendicular to the ecliptic. The latitude is never so great as 8°, so that no serious error in the place will occur if the strip is not exactly perpendicular to the ecliptic.

The place of the planet being thus marked on the globe, its right ascension and declination may be determined, and problems relating to its diurnal motion, such as its times of rising and setting, may be solved by the methods of Chapters VIII and IX.

CONFIGURATIONS OF THE PLANETS

The elongation of a planet is its distance from the sun along the ecliptic as seen from the earth. It is therefore equal to the difference of the geocentric longitudes of the sun and planet. The elongation is measured either way from the sun up to 180°, at which point the planet is at opposition, or opposite the sun. When the elongation is zero the sun and planet are in the same longitude, and the planet is in conjunction with the sun.

The symbols ☍ and ☌ are used for opposition and conjunction, respectively. When the longitude of the planet is greater than that of the sun it is east of the latter, and follows it in its diurnal revolution. It is therefore above the horizon at sunset and is said to be an "evening star," since it is visible in the twilight after sunset except when near conjunction. On the other hand, all planets whose longitudes are less than that of the sun precede it, and they will be above the horizon at sunrise and therefore visible at dawn, except when very near conjunction. They are then "morning stars," just as stars in eastern elongation are evening stars.

The geocentric longitude of the sun, July 6, 1907, is 103°.2 (since the earth's heliocentric longitude is 283°.2, page 164). The longitude of Jupiter being 110°.2, its elongation is about 7° east, and it is an evening star, though too close to the sun to be visible; it will become a morning star about July 14.

The elongation of Mars is very nearly 180°, and it is at opposition and becoming an evening star. The longitude of Venus is 84°.6; it is 18°.6 west of the sun and is a morning star. On referring to the diagram (Fig. 88), and remembering that it moves more rapidly than the earth, it is evident that it is approaching conjunction *beyond* the sun ("superior" conjunction), after which it will pass to eastern elongations and be an evening star. Mercury's longitude is 126°; it is 23° east of the sun, and referring to the diagram, we see that it is approaching conjunction *between* the earth and sun ("inferior" conjunction), after which it will be a morning star.

The preceding principles enable us to find the place of a planet at any given date, and thus to answer many of the questions which continually suggest themselves to one interested in watching the courses of the planets in the sky.

It is evident, for instance, from the problems solved on pages 159 and 164, that in 1907 the greater proximity of Mars to the earth offers conditions for the study of its surface which are much more favorable than those of the opposition of 1905.

The oppositions of Mars recur at an average interval of about 780 days, which is the synodic period of the earth and Mars, as explained in the text-books of descriptive astronomy.

We may fix the dates of other oppositions approximately, as in August, 1877, September, 1909, November, 1911, December, 1913, etc., and by computing for the first and last days of those months a closer approximation to the day of opposition may quickly be made, and finally a careful computation for the exact date will fix the time within a few hours. The geocentric place and the distance of the planet may then be found.

It appears that favorable oppositions occur in the summer, and that the planet is then quite a distance south of the equator, so that it is far from the zenith of any northern observatory.

The satellites of Mars were discovered in 1877, and in the same year an expedition was sent to the island of Ascension to observe Mars for a determination of the solar parallax.

In conclusion we will consider the motion of Mars during the summer of 1907, to illustrate the form which the computation takes when many places are to be found at comparatively short intervals.

We first carefully determine the mean longitudes of Mars and the earth for March 22 to be 235°.51 and 178°.74, respectively, and then easily form the second column of the following schedule by successive additions of 10°.48 and 19°.71, the mean motions of the two planets in twenty days.

The third column is formed for Mars by writing the longitude of perihelion 334°.2 on the upper edge of a slip of paper and placing it under the numbers of the second column successively, subtracting from each to find the corresponding mean anomaly.

The same result is more easily obtained by adding in the same way 25°.8 (360°−334°.2) to each number in the second column. The third column is checked by noting that the differences of the successive values are 10°.48, which insures the accuracy of both columns. The equation of center is taken from the table and entered in the fourth column, and the true heliocentric longitude found by adding corresponding numbers of the second and fourth columns. The same process gives the earth's true heliocentric longitude.

The labor is by no means proportionate to that required in computing a single place, and the comparison of the successive numbers of each column is an important aid in detecting errors.

Date	\multicolumn{4}{c}{MARS}	\multicolumn{4}{c}{THE EARTH}

Date	$E+\mu t$	$E+\mu t-\pi$	Eq. of Center	l	$E+\mu t$	$E+\mu' t-\pi$	Eq. of Center	l
Mar. 22	235°.51	261°.3	−10°.3	225°.2	178°.74	77°.5	+1°.9	180°.6
April 11	245 .99	271 .8	−10 .6	235 .4	198 .45	97 .2	+1 .9	200 .3
May 1	256 .47	282 .3	−10 .6	245 .9	218 .16	117 .0	+1 .7	219 .9
21	266 .95	292 .8	−10 .3	256 .6	237 .87	136 .6	+1 .3	239 .2
June 10	277 .43	303 .2	− 9 .5	267 .9	257 .58	156 .4	+0 .7	258 .3
30	287 .91	313 .7	− 8 .3	279 .6	277 .29	176 .1	+0 .1	277 .4
July 20	298 .39	324 .2	− 6 .8	291 .6	297 .00	195 .8	−0 .5	296 .5
Aug. 9	308 .87	334 .7	− 5 .0	303 .9	316 .71	215 .5	−1 .1	315 .6
29	319 .35	345 .1	− 3 .1	316 .3	336 .42	235 .2	−1 .5	334 .9
Sept. 18	329 .83	355 .6	− 0 .9	328 .9	356 .13	254 .9	−1 .8	354 .3
Oct. 7	340 .31	6 .1	+ 1 .3	341 .6	15 .84	274 .6	−1 .9	13 .9

The planets were plotted from the above data on a scale of 1.6 inches to the astronomical unit, the boundary circle being 9¼ inches in diameter. The values of A, B, U, and Δ were determined and the geocentric longitudes and latitudes λ and β found as in the following table:

THE MOTIONS OF THE PLANETS 171

	A	B	U	Δ	λ	β
March 22 . .	244°.6	103°.7	+ 0.13	1.10	264°.1	+ 0°.2
April 11 . .	254.8	112.6	− 0.16	0.91	273.7	− 0.3
May 1 . .	263.9	119.2	0.44	0.75	281.5	− 1.2
21 . .	271.4	120.9	0.69	0.60	286.1	− 2.2
June 10 . .	277.8	117.1	0.90	0.50	287.4	− 3.4
30 . .	282.2	108.0	1.10	0.44	285.0	− 4.8
July 20 . .	285.7	94.3	1.25	0.42	280.0	− 5.7
Aug. 9 . .	289.7	84.9	1.36	0.47	277.3	− 5.5
29 . .	296.8	84.1	1.39	0.54	280.4	− 4.9
Sept. 18 . .	305.0	88.2	1.36	0.64	286.6	− 4.0
Oct. 7 . .	316.6	98.4	1.27	0.76	297.5	− 3.2

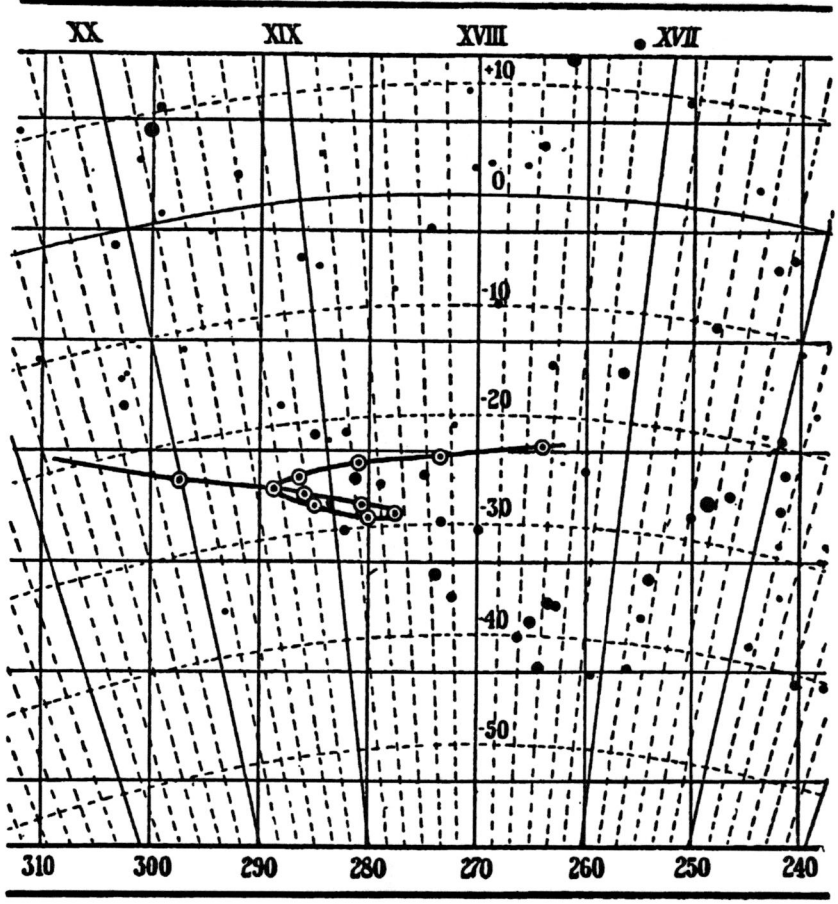

FIG. 89. Path of Mars in the Summer of 1907

In order to form an idea of the path described by the planet among the stars, the positions may be plotted on an ecliptic map, as in Fig. 89, which shows the form of the loop in the constellation of Sagittarius.

During March the motion of the planet is eastward, or in the direction of increasing longitudes, and is said to be "direct." The rate of motion diminishes from one-half degree per day at the outset to half that amount in May, and soon after the beginning of June the planet reaches its first "stationary point" and begins to move slowly in the opposite direction in longitude, or "retrograde." Its continuous motion in latitude toward the south prevents it from exactly retracing its path and causes it to describe a "loop."

Its velocity in the retrograde arc increases to a maximum of a quarter of a degree per day at opposition early in July, and then decreases until the second stationary point is reached about August 9, when the planet resumes its direct motion.

The exact dates of the stationary points may be found by computing a few places in the neighborhood of June 10 and August 9.

The Ephemeris gives the dates as June 5 and August 8.

Table III — Average Values of the Sun's Longitude and the Equation of Time

	Longitude	Longitude	Mean Longitude	Eq. of Time
Jan. 1	280°.3	♑ 10°.3	280°.1	+ 3ᵐ.5
11	290 .5	♑ 20 .5	289 .9	+ 7 .9
21	300 .7	♒ 30 .7	299 .8	+11 .3
31	310 .8	♒ 10 .8	309 .6	+13 .6
Feb. 10	320 .9	♒ 20 .9	319 .5	+14 .4
20	331 .1	♓ 1 .1	329 .4	+14 .0
Mar. 2	341 .3	♓ 11 .3	339 .5	+12 .4
12	351 .4	♓ 21 .4	349 .3	+10 .0
22	1 .3	♈ 1 .3	359 .2	+ 7 .1
April 1	11 .2	♈ 11 .2	9 .0	+ 4 .1
11	21 .0	♈ 21 .0	18 .9	+ 1 .2
21	30 .8	♉ 0 .8	28 .7	− 1 .2
May 1	40 .6	♉ 10 .6	38 .6	− 2 .9
11	50 .3	♉ 20 .3	48 .5	− 3 .7
21	59 .9	♉ 29 .9	58 .3	− 3 .6
31	69 .5	♊ 9 .5	68 .2	− 2 .6
June 10	79 .0	♊ 19 .0	78 .0	− 0 .9
20	88 .6	♊ 28 .6	87 .9	+ 1 .2
30	98 .1	♋ 8 .1	97 .7	+ 3 .3
July 10	107 .7	♋ 17 .7	107 .6	+ 5 .0
20	117 .2	♋ 27 .2	117 .5	+ 6 .1
30	126 .7	♌ 6 .7	127 .3	+ 6 .2
Aug. 9	136 .3	♌ 16 .3	137 .2	+ 5 .4
19	145 .9	♌ 25 .9	147 .0	+ 3 .6
29	155 .6	♍ 5 .6	156 .9	+ 0 .9
Sept. 8	165 .3	♍ 15 .3	166 .7	− 2 .3
18	175 .0	♍ 25 .0	176 .6	− 5 .8
28	184 .8	♎ 4 .8	186 .5	− 9 .2
Oct. 8	194 .6	♎ 14 .6	196 .3	−12 .3
18	204 .5	♎ 24 .5	206 .2	−14 .7
28	214 .5	♏ 4 .5	216 .0	−16 .1
Nov. 7	224 .5	♏ 14 .5	225 .9	−16 .2
17	234 .6	♏ 24 .6	235 .7	−15 .0
27	244 .7	♐ 4 .7	245 .6	−12 .4
Dec. 7	254 .8	♐ 14 .8	255 .4	− 8 .5
17	265 .0	♐ 25 .0	265 .3	− 3 .9
27	275 .2	♑ 5 .2	275 .2	+ 1 .0
Jan. 6	285 .4	♑ 15 .4	285 .0	+ 5 .7

Table IV — Elements of the Six Inner Planets, January 1, 1900

Symbol	☿	♀	⊕	♂	♃	♄
Mean Distance a	0.387	0.723	1.000	1.524	5.203	9.539
Eccentricity e	0.2056	0.0068	0.0168	0.0933	0.0482	0.0561
Inclination i	7°.0	3°.4	...	1°.9	1°.3	2°.5
Longitude of Ascending Node Ω	47°.1	75°.7	...	48°.7	99°.4	112°.7
Longitude of Perihelion π	75°.9	130°.2	101°.2	334°.2	12°.7	90°.9
Mean Longitude E Gr. Mean Noon	182°.22	344°.33	100°.67	294°.27	238°.13	266°.61
Sidereal Period T	87^d.9693	224^d.701	365^d.256	686^d.979	4332^d.58	10759^d.2
Mean Daily Motion μ	4°.09234	1°.60213	0°.98561	0°.52403	0°.08309	0°.03346

Table V — The Long Inequality of Jupiter and Saturn

	♃		♄
1800–1840	+0°.3	1800–1840	−0°.8
1840–1890	+0.2	1840–1870	−0.7
1890–1950	+0.1	1870–1910	−0.6
1950–1990	±0.0	1910–1940	−0.5
		1940–1960	−0.4
		1960–1980	−0.3
		1980–1990	−0.2
		1990–2000	−0.1

Table VI — Equation of Center

Mean Anomaly	☿	♀	⊕	♂	♃	♄	Mean Anomaly
0°	0°.0	0°.0	0°.0	0°.0	0°.0	0°.0	360°
10	5.4	0.1	0.3	2.1	1.0	1.2	350
20	10.5	0.3	0.7	4.1	2.0	2.4	340
30	14.9	0.4	1.0	5.9	2.9	3.4	330
40	18.5	0.5	1.2	7.5	3.7	4.4	320
50	21.1	0.6	1.5	8.8	4.4	5.2	310
60	22.8	0.7	1.7	9.8	4.9	5.8	300
70	23.6	0.7	1.8	10.4	5.3	6.2	290
80	23.6	0.8	1.9	10.7	5.5	6.4	280
90	22.9	0.8	1.9	10.6	5.5	6.4	270
100	21.7	0.8	1.9	10.2	5.4	6.3	260
110	19.9	0.7	1.8	9.6	5.1	5.9	250
120	17.8	0.7	1.6	8.7	4.6	5.4	240
130	15.3	0.6	1.4	7.6	4.1	4.7	230
140	12.5	0.5	1.2	6.3	3.4	3.9	220
150	9.6	0.4	0.9	4.8	2.6	3.0	210
160	6.5	0.3	0.6	3.3	1.8	2.1	200
170	3.3	0.1	0.3	1.7	0.9	1.0	190
180	0.0	0.0	0.0	0.0	0.0	0.0	180

Table VII — The Julian Day

Add together the year number, the month number, and the day of the month.

Year	Number	Year	Number	Year	Number	Year	Number	Year	Number
1800	2378496	1843	2394201	1886	2409907	1929	2425612	1972	2441317
1801	78861	1844	94566	1887	10272	1930	25977	1973	41683
1802	79226	1845	94932	1888	10637	1931	26342	1974	42048
1803	79591	1846	95297	1889	11003	1932	26707	1975	42413
1804	79956	1847	95662	1890	11368	1933	27073	1976	42778
1805	80322	1848	96027	1891	11733	1934	27438	1977	43144
1806	80687	1849	96393	1892	12098	1935	27803	1978	43509
1807	81052	1850	96758	1893	12464	1936	28168	1979	43874
1808	81417	1851	97123	1894	12829	1937	28534	1980	44239
1809	81783	1852	97488	1895	13194	1938	28899	1981	44605
1810	82148	1853	97854	1896	13559	1939	29264	1982	44970
1811	82513	1854	98219	1897	13925	1940	29629	1983	45335
1812	82878	1855	98584	1898	14290	1941	29995	1984	45700
1813	83244	1856	98949	1899	14655	1942	30360	1985	46066
1814	83609	1857	99315	1900	15020	1943	30725	1986	46431
1815	83974	1858	99680	1901	15385	1944	31090	1987	46796
1816	84339	1859	2400045	1902	15750	1945	31456	1988	47161
1817	84705	1860	00410	1903	16115	1946	31821	1989	47527
1818	85070	1861	00776	1904	16480	1947	32186	1990	47892
1819	85435	1862	01141	1905	16846	1948	32551	1991	48257
1820	85800	1863	01506	1906	17211	1949	32917	1992	48622
1821	86166	1864	01871	1907	17576	1950	33282	1993	48988
1822	86531	1865	02237	1908	17941	1951	33647	1994	49353
1823	86896	1866	02602	1909	18307	1952	34012	1995	49718
1824	87261	1867	02967	1910	18672	1953	34378	1996	50083
1825	87627	1868	03332	1911	19037	1954	34743	1997	50449
1826	87992	1869	03698	1912	19402	1955	35108	1998	50814
1827	88357	1870	04063	1913	19768	1956	35473	1999	51179
1828	88722	1871	04428	1914	20133	1957	35839	2000	51544
1829	89088	1872	04793	1915	20498	1958	36204		
1830	89453	1873	05159	1916	20863	1959	36569		
1831	89818	1874	05524	1917	21229	1960	36934		
1832	90183	1875	05889	1918	21594	1961	37300		
1833	90549	1876	06254	1919	21959	1962	37665		
1834	90914	1877	06620	1920	22324	1963	38030		
1835	91279	1878	06985	1921	22690	1964	38395		
1836	91644	1879	07350	1922	23055	1965	38761		
1837	92010	1880	07715	1923	23420	1966	39126		
1838	92375	1881	08081	1924	23785	1967	39491		
1839	92740	1882	08446	1925	24151	1968	39856		
1840	93105	1883	08811	1926	24516	1969	40222		
1841	93471	1884	09176	1927	24881	1970	40587		
1842	93836	1885	09542	1928	25246	1971	40952		

Month Num.	Common	Lp.
Jan.	0	0
Feb.	31	31
March	59	60
April	90	91
May	120	121
June	151	152
July	181	182
Aug.	212	213
Sept.	243	244
Oct.	273	274
Nov.	304	305
Dec.	334	335

MARCH, 1899.

AT GREENWICH MEAN NOON.

THE SUN'S

Day of the Week	Day of the Month	Apparent Right Ascension	Diff. for 1 Hour.	Apparent Declination.	Diff. for 1 Hour.	Equation of Time, to be Subtracted from Mean Time.	Diff. for 1 Hour.	Sidereal Time, or Right Ascension of Mean Sun.
		h m s	s	° ′ ″	″	m s	s	h m s
Wed.	1	22 48 49.18	9.360	S. 7 33 9.5	+56.99	12 31.64	0.496	22 36 17.54
Thur.	2	22 52 33.57	9.340	7 10 18.4	57.26	12 19.48	0.516	22 40 14.09
Frid.	3	22 56 17.49	9.320	6 47 21.0	57.51	12 6.84	0.536	22 44 10.64
Sat.	4	23 0 0.95	9.301	6 24 17.9	+57.75	11 53.75	0.555	22 48 7.20
SUN.	5	23 3 43.98	9.284	6 1 9.3	57.97	11 40.23	0.572	22 52 3.75
Mon.	6	23 7 26.60	9.267	5 37 55.6	58.17	11 26.29	0.589	22 56 0.30
Tues.	7	23 11 8.82	9.251	5 14 37.3	+58.35	11 11.96	0.605	22 59 56.86
Wed.	8	23 14 50.66	9.236	4 51 14.8	58.52	10 57.25	0.620	23 3 53.41
Thur.	9	23 18 32.14	9.221	4 27 48.4	58.67	10 42.17	0.635	23 7 49.96
Frid.	10	23 22 13.27	9.207	4 4 18.6	+58.80	10 26.75	0.649	23 11 46.52
Sat.	11	23 25 54.08	9.194	3 40 45.8	58.92	10 11.01	0.663	23 15 43.07
SUN.	12	23 29 34.57	9.181	3 17 10.3	59.02	9 54.95	0.675	23 19 39.62
Mon.	13	23 33 14.77	9.169	2 53 32.7	+59.10	9 38.60	0.687	23 23 36.17
Tues.	14	23 36 54.69	9.158	2 29 53.2	59.17	9 21.96	0.698	23 27 32.73
Wed.	15	23 40 34.35	9.147	2 6 12.3	59.22	9 5.07	0.709	23 31 29.28
Thur.	16	23 44 13.76	9.137	1 42 30.4	+59.26	8 47.93	0.719	23 35 25.83
Frid.	17	23 47 52.95	9.128	1 18 47.8	59.28	8 30.56	0.728	23 39 22.38
Sat.	18	23 51 31.93	9.120	0 55 4.9	59.28	8 12.99	0.736	23 43 18.94
SUN.	19	23 55 10.72	9.113	0 31 22.2	+59.27	7 55.23	0.744	23 47 15.49
Mon.	20	23 58 49.34	9.106	S. 0 7 39.8	59.25	7 37.30	0.750	23 51 12.04
Tues.	21	0 2 27.82	9.100	N. 0 16 1.7	59.21	7 19.22	0.756	23 55 8.60
Wed.	22	0 6 6.17	9.095	0 39 42.1	+59.15	7 1.02	0.761	23 59 5.15
Thur.	23	0 9 44.41	9.092	1 3 21.0	59.08	6 42.71	0.765	0 3 1.70
Frid.	24	0 13 22.57	9.089	1 26 58.1	59.00	6 24.32	0.767	0 6 58.26
Sat.	25	0 17 0.68	9.087	1 50 33.0	+58.90	6 5.87	0.769	0 10 54.81
SUN.	26	0 20 38.75	9.086	2 14 5.5	58.79	5 47.39	0.770	0 14 51.36
Mon.	27	0 24 16.81	9.086	2 37 35.1	58.67	5 28.90	0.770	0 18 47.91
Tues.	28	0 27 54.88	9.087	3 1 1.6	+58.53	5 10.41	0.769	0 22 44.47
Wed.	29	0 31 32.99	9.089	3 24 24.7	58.38	4 51.97	0.767	0 26 41.02
Thur.	30	0 35 11.16	9.092	3 47 44.0	58.22	4 33.59	0.764	0 30 37.57
Frid.	31	0 38 49.41	9.096	4 10 59.0	58.04	4 15.29	0.760	0 34 34.12
Sat.	32	0 42 27.77	9.101	N. 4 34 9.7	+57.84	3 57.09	0.756	0 38 30.68

NOTE.—The semidiameter for mean noon may be assumed the same as that for apparent noon. The sign + prefixed to the hourly change of declination indicates that south declinations are decreasing, north declinations increasing.

Diff. for 1 Hour, +9ˢ.8565. (Table LII.)

II. JANUARY, 1900.

AT GREENWICH MEAN NOON.

Day of the Week	Day of the Month	THE SUN'S				Equation of Time, to be Subtracted from Mean Time.	Diff. for 1 Hour.	Sidereal Time, or Right Ascension of Mean Sun.
		Apparent Right Ascension.	Diff. for 1 Hour.	Apparent Declination.	Diff for 1 Hour.			
		h m s	s	° ′ ″	″	m s	s	h m s
Mon.	1	18 46 23.63	11.045	S. 23 1 23.1	+12.24	3 40.17	1.190	18 42 43.46
Tues.	2	18 50 48.55	11.031	22 56 15.4	13.39	4 8.53	1.175	18 46 40.02
Wed.	3	18 55 13.12	11.015	22 50 40.4	14.53	4 36.54	1.159	18 50 36.58
Thur.	4	18 59 37.28	10.998	22 44 38.1	+15.66	5 4.15	1.141	18 54 33.13
Frid.	5	19 4 1.04	10.980	22 38 8.7	16.78	5 31.35	1.122	18 58 29.69
Sat.	6	19 8 24.33	10.961	22 31 12.6	17.89	5 58.08	1.103	19 2 26.25
SUN.	7	19 12 47.13	10.940	22 23 49.8	+19.00	6 24.33	1.082	19 6 22.81
Mon.	8	19 17 9.41	10.918	22 16 0.6	20.10	6 50.05	1.060	19 10 19.36
Tues.	9	19 21 31.17	10.895	22 7 45.3	21.18	7 15.25	1.037	19 14 15.92
Wed.	10	19 25 52.36	10.870	21 59 4.0	+22.25	7 39.87	1.014	19 18 12.48
Thur.	11	19 30 12.94	10.845	21 49 57.1	23.31	8 3.90	0.990	19 22 9.04
Frid.	12	19 34 32.93	10.819	21 40 24.8	24.37	8 27.34	0.964	19 26 5.59
Sat.	13	19 38 52.30	10.793	21 30 27.3	+25.41	8 50.14	0.937	19 30 2.15
SUN.	14	19 43 11.01	10.766	21 20 4.9	26.43	9 12.30	0.910	19 33 58.71
Mon.	15	19 47 29.07	10.738	21 9 18.0	27.45	9 33.81	0.882	19 37 55.26
Tues.	16	19 51 46.45	10.710	20 58 6.7	+28.46	9 54.63	0.854	19 41 51.82
Wed.	17	19 56 3.15	10.681	20 46 31.4	29.46	10 14.77	0.825	19 45 48.38
Thur.	18	20 0 19.15	10.651	20 34 32.5	30.44	10 34.21	0.795	19 49 44.94
Frid.	19	20 4 34.42	10.621	20 22 10.1	+31.41	10 52.93	0.765	19 53 41.49
Sat.	20	20 8 48.98	10.591	20 9 24.7	32.37	11 10.94	0.735	19 57 38.05
SUN.	21	20 13 2.79	10.560	19 56 16.6	33.31	11 28.18	0.704	20 1 34.61
Mon.	22	20 17 15.86	10.529	19 42 46.1	+34.23	11 44.70	0.672	20 5 31.16
Tues.	23	20 21 28.17	10.497	19 28 53.5	35.14	12 0.45	0.641	20 9 27.72
Wed.	24	20 25 39.72	10.465	19 14 39.3	36.04	12 15.44	0.609	20 13 24.28
Thur.	25	20 29 50.49	10.433	19 0 3.8	+36.92	12 29.66	0.577	20 17 20.83
Frid.	26	20 34 0.48	10.400	18 45 7.4	37.78	12 43.09	0.544	20 21 17.39
Sat.	27	20 38 9.69	10.367	18 29 50.4	38.62	12 55.74	0.511	20 25 13.94
SUN.	28	20 42 18.10	10.334	18 14 13.3	+39.45	13 7.59	0.478	20 29 10.50
Mon.	29	20 46 25.71	10.300	17 58 16.5	40.27	13 18.64	0.444	20 33 7.06
Tues.	30	20 50 32.49	10.266	17 42 0.4	41.07	13 28.88	0.410	20 37 3.61
Wed.	31	20 54 38.46	10.232	17 25 25.4	41.85	13 38.29	0.376	20 41 0.17
Thur.	32	20 58 43.61	10.198	S. 17 8 31.9	+42.61	13 46.89	0.341	20 44 56.72

NOTE.—The semidiameter for mean noon may be assumed the same as that for apparent noon. The sign + prefixed to the hourly change of declination indicates that south declinations are decreasing.

Diff. for 1 Hour, +9ˢ.8565. (Table III.)

JANUARY, 1900. III.

AT GREENWICH MEAN NOON.

THE SUN'S

Day of the Month.	Day of the Year.	TRUE LONGITUDE. λ	λ'	Diff. for 1 Hour.	LATITUDE.	Logarithm of the Radius Vector of the Earth.	Diff. for 1 Hour.	Mean Time of Sidereal Noon
1	1	280 40 8.7	39 51.2	152.96	+ 0.26	9.9926699	− 0.6	h m s 5 16 24.56
2	2	281 41 20.1	41 2.3	152.96	0.40	9.9926694	+ 0.1	5 12 28.65
3	3	282 42 31.4	42 13.4	152.96	0.50	9.9926706	0.8	5 8 32.74
4	4	283 43 42.4	43 24.3	152.95	+ 0.59	9.9926734	+ 1.5	5 4 36.83
5	5	284 44 53.1	44 34.8	152.94	0.65	9.9926780	2.3	5 0 40.91
6	6	285 46 3.5	45 45.0	152.92	0.67	9.9926845	3.1	4 56 45.00
7	7	286 47 13.3	46 54.7	152.90	+ 0.65	9.9926929	+ 3.9	4 52 49.09
8	8	287 48 22.7	48 3.9	152.88	0.60	9.9927034	4.9	4 48 53.18
9	9	288 49 31.5	49 12.5	152.85	0.51	9.9927163	5.9	4 44 57.27
10	10	289 50 39.7	50 20.6	152.83	+ 0.42	9.9927317	+ 6.9	4 41 1.36
11	11	290 51 47.3	51 28.0	152.81	0.30	9.9927495	8.0	4 37 5.44
12	12	291 52 54.4	52 34.9	152.79	0.17	9.9927699	9.0	4 33 9.53
13	13	292 54 0.8	53 41.2	152.76	+ 0.04	9.9927930	+10.1	4 29 13.62
14	14	293 55 6.7	54 46.9	152.74	− 0.09	9.9928190	11.3	4 25 17.71
15	15	294 56 12.1	55 52.1	152.71	0.20	9.9928476	12.5	4 21 21.80
16	16	295 57 16.9	56 56.7	152.69	− 0.31	9.9928790	+13.6	4 17 25.89
17	17	296 58 21.2	58 0.9	152.67	0.39	9.9929131	14.8	4 13 29.98
18	18	297 59 25.0	59 4.5	152.65	0.45	9.9929500	15.9	4 9 34.07
19	19	299 0 28.3	0 7.7	152.63	− 0.48	9.9929894	+17.0	4 5 38.16
20	20	300 1 31.2	1 10.4	152.61	0.48	9.9930314	18.0	4 1 42.24
21	21	301 2 33.5	2 12.6	152.59	0.46	9.9930759	19.0	3 57 46.33
22	22	302 3 35.4	3 14.3	152.57	− 0.41	9.9931227	+20.0	3 53 50.42
23	23	303 4 36.7	4 15.5	152.54	0.34	9.9931719	20.9	3 49 54.51
24	24	304 5 37.5	5 16.1	152.52	0.23	9.9932232	21.8	3 45 58.60
25	25	305 6 37.8	6 16.2	152.50	− 0.11	9.9932765	+22.6	3 42 2.69
26	26	306 7 37.5	7 15.7	152.48	+ 0.02	9.9933316	23.4	3 38 6.78
27	27	307 8 36.5	8 14.6	152.45	0.15	9.9933886	24.1	3 34 10.87
28	28	308 9 34.8	9 12.8	152.42	+ 0.29	9.9934471	+24.7	3 30 14.96
29	29	309 10 32.4	10 10.2	152.38	0.44	9.9935071	25.3	3 26 19.05
30	30	310 11 29.0	11 6.7	152.34	0.55	9.9935684	25.8	3 22 23.14
31	31	311 12 24.7	12 2.2	152.30	0.63	9.9936309	26.3	3 18 27.23
32	32	312 13 19.2	12 56.4	152.25	+ 0.68	9.9936948	+26.9	3 14 31.32

NOTE.—The numbers in column λ correspond to the true equinox of the date; in column λ' to the mean equinox of January 0.0.

Diff. for 1 Hour, −9ˢ.8296. (Table II.)

IV.   JANUARY, 1900.

GREENWICH MEAN TIME.

THE MOON'S

Day of the Month	SEMIDIAMETER		HORIZONTAL PARALLAX				UPPER TRANSIT		AGE
	Noon.	Midnight.	Noon.	Diff. for 1 Hour	Midnight.	Diff. for 1 Hour.	Meridian of Greenwich.	Diff. for 1 Hour.	Noon.
	′ ″	′ ″	′ ″	″	′ ″	″	h m	m	d
1	16 19.6	16 23.4	59 48.5	+1.33	60 2.8	+1.05	☾		29.5
2	16 26.4	16 28.3	60 13.6	0.75	60 20.7	+0.44	0 57.9	2.43	0.9
3	16 29.2	16 29.2	60 24.1	+0.13	60 23.8	−0.17	1 55.0	2.33	1.9
4	16 28.1	16 26.2	60 20.0	−0.45	60 13.0	−0.70	2 49.6	2.22	2.9
5	16 23.6	16 20.2	60 3.2	0.92	59 50.9	1.10	3 41.9	2.14	3.9
6	16 16.3	16 12.0	59 36.7	1.25	59 20.9	1.36	4 32.8	2.10	4.9
7	16 7.4	16 2.6	59 4.0	−1.44	58 46.4	−1.48	5 23.1	2.10	5.9
8	15 57.7	15 52.8	58 28.4	1.50	58 10.2	1.50	6 13.9	2.13	6.9
9	15 47.9	15 43.1	57 52.2	1.49	57 34.5	1.46	7 5.6	2.18	7.9
10	15 38.4	15 33.8	57 17.2	−1.43	57 0.3	−1.38	7 58.5	2.22	8.9
11	15 29.3	15 25.0	56 44.0	1.34	56 28.2	1.29	8 52.2	2.24	9.9
12	15 20.9	15 16.9	56 13.0	1.24	55 58.4	1.19	9 45.8	2.22	10.9
13	15 13.1	15 9.5	55 44.4	−1.14	55 31.1	−1.09	10 38.4	2.16	11.9
14	15 6.0	15 2.7	55 18.3	1.03	55 6.2	0.98	11 29.1	2.06	12.9
15	14 59.6	14 56.7	54 54.8	0.92	54 44.2	0.85	12 17.2	1.95	13.9
16	14 54.1	14 51.7	54 34.5	−0.77	54 25.7	−0.68	13 2.7	1.85	14.9
17	14 49.6	14 47.9	54 18.1	0.58	54 11.8	0.47	13 46.1	1.76	15.9
18	14 46.5	14 45.6	54 6.9	0.34	54 3.5	−0.21	14 27.8	1.71	16.9
19	14 45.2	14 45.2	54 1.9	−0.06	54 2.1	+0.10	15 8.5	1.69	17.9
20	14 45.9	14 47.1	54 4.4	+0.28	54 8.8	0.46	15 49.3	1.71	18.9
21	14 48.9	14 51.4	54 15.5	0.65	54 24.6	0.85	16 30.9	1.76	19.9
22	14 54.5	14 58.3	54 36.1	+1.06	54 50.1	+1.27	17 14.3	1.86	20.9
23	15 2.8	15 7.9	55 6.5	1.48	55 25.4	1.67	18 0.4	1.99	21.9
24	15 13.7	15 20.0	55 46.5	1.85	56 9.9	2.02	18 49.8	2.14	22.9
25	15 26.9	15 34.2	56 35.1	+2.17	57 1.9	+2.29	19 43.0	2.29	23.9
26	15 41.8	15 49.7	57 30.0	2.37	57 58.8	2.41	20 39.5	2.42	24.9
27	15 57.6	16 5.4	58 27.8	2.40	58 56.4	2.34	21 38.5	2.49	25.9
28	16 12.9	16 19.9	59 23.9	+2.22	59 49.6	+2.04	22 38.3	2.48	26.9
29	16 26.2	16 31.6	60 12.8	1.81	60 32.8	1.52	23 37.3	2.43	27.9
30	16 36.0	16 39.3	60 49.0	1.18	61 0.9	+0.80	☾		28.9
31	16 41.3	16 41.9	61 8.2	+0.40	61 10.6	0.00	0 34.5	2.34	0.4
32	16 41.3	16 39.4	61 8.3	−0.39	61 1.3	−0.75	1 29.8	2.26	1.4

179

JANUARY, 1900.   VII

GREENWICH MEAN TIME.

THE MOON'S RIGHT ASCENSION AND DECLINATION.

Hour.	Right Ascension.	Diff. for 1 Minute.	Declination.	Diff. for 1 Minute.	Hour.	Right Ascension.	Diff. for 1 Minute.	Declination.	Diff. for 1 Minute.
	TUESDAY 9.					**THURSDAY 11.**			
	h m s	s	° ′ ″	″		h m s	s	° ′ ″	″
0	2 5 1.09	2.2544	N.16 38 10.6	9.442	0	3 55 9.21	2.3237	N.22 4 39.6	3.944
1	2 7 16.41	2.2563	16 47 34.2	9.343	1	3 57 28.65	2.3242	22 8 32.5	3.818
2	2 9 31.85	2.2582	16 56 51.8	9.243	2	3 59 48.12	2.3247	22 12 17.8	3.692
3	2 11 47.39	2.2600	17 6 3.4	9.142	3	4 2 7.62	2.3252	22 15 55.5	3.566
4	2 14 3.05	2.2619	17 15 8.9	9.040	4	4 4 27.14	2.3254	22 19 25.8	3.440
5	2 16 18.82	2.2637	17 24 8.2	8.937	5	4 6 46.67	2.3257	22 22 48.4	3.313
6	2 18 34.70	2.2656	17 33 1.4	8.835	6	4 9 6.22	2.3260	22 26 3.3	3.186
7	2 20 50.69	2.2675	17 41 48.4	8.731	7	4 11 25.79	2.3262	22 29 10.7	3.059
8	2 23 6.80	2.2693	17 50 29.1	8.627	8	4 13 45.36	2.3262	22 32 10.4	2.932
9	2 25 23.01	2.2712	17 59 3.6	8.521	9	4 16 4.93	2.3262	22 35 2.5	2.805
10	2 27 39.34	2.2730	18 7 31.6	8.414	10	4 18 24.50	2.3260	22 37 47.0	2.677
11	2 29 55.77	2.2748	18 15 53.3	8.307	11	4 20 44.07	2.3261	22 40 23.8	2.549
12	2 32 12.32	2.2767	18 24 8.5	8.199	12	4 23 3.63	2.3259	22 42 52.9	2.422
13	2 34 28.98	2.2785	18 32 17.2	8.091	13	4 25 23.18	2.3257	22 45 14.4	2.295
14	2 36 45.74	2.2802	18 40 19.4	7.982	14	4 27 42.71	2.3254	22 47 28.3	2.167
15	2 39 2.61	2.2820	18 48 15.1	7.873	15	4 30 2.23	2.3251	22 49 34.5	2.039
16	2 41 19.58	2.2837	18 56 4.2	7.762	16	4 32 21.72	2.3246	22 51 33.0	1.912
17	2 43 36.66	2.2856	19 3 46.6	7.651	17	4 34 41.18	2.3241	22 53 23.9	1.784
18	2 45 53.85	2.2873	19 11 22.4	7.539	18	4 37 0.61	2.3235	22 55 7.1	1.657
19	2 48 11.14	2.2890	19 18 51.4	7.427	19	4 39 20.00	2.3229	22 56 42.7	1.529
20	2 50 28.53	2.2907	19 26 13.7	7.315	20	4 41 39.36	2.3222	22 58 10.6	1.401
21	2 52 46.02	2.2923	19 33 29.2	7.201	21	4 43 58.67	2.3214	22 59 30.8	1.273
22	2 55 3.61	2.2940	19 40 37.8	7.086	22	4 46 17.93	2.3207	23 0 43.4	1.147
23	2 57 21.30	2.2956	N.19 47 39.5	6.971	23	4 48 37.15	2.3197	N.23 1 48.4	1.019
	WEDNESDAY 10.					**FRIDAY 12.**			
0	2 59 39.08	2.2972	N.19 54 34.3	6.856	0	4 50 56.30	2.3187	N.23 2 45.7	0.892
1	3 1 56.96	2.2987	20 1 22.2	6.741	1	4 53 15.40	2.3177	23 3 35.4	0.764
2	3 4 14.93	2.3003	20 8 3.2	6.624	2	4 55 34.43	2.3166	23 4 17.4	0.637
3	3 6 33.00	2.3018	20 14 37.1	6.507	3	4 57 53.39	2.3154	23 4 51.9	0.511
4	3 8 51.15	2.3032	20 21 4.0	6.389	4	5 0 12.28	2.3148	23 5 18.7	0.384
5	3 11 9.39	2.3047	20 27 23.8	6.271	5	5 2 31.09	2.3128	23 5 38.0	0.258
6	3 13 27.72	2.3061	20 33 36.5	6.152	6	5 4 49.82	2.3115	23 5 49.7	0.132
7	3 15 46.12	2.3074	20 39 42.1	6.033	7	5 7 8.47	2.3100	23 5 53.8	+0.006
8	3 18 4.61	2.3087	20 45 40.5	5.913	8	5 9 27.04	2.3087	23 5 50.4	−0.120
9	3 20 23.17	2.3100	20 51 31.7	5.793	9	5 11 45.51	2.3070	23 5 39.4	0.246
10	3 22 41.81	2.3113	20 57 15.7	5.672	10	5 14 3.88	2.3053	23 5 20.9	0.371
11	3 25 0.53	2.3125	21 2 52.4	5.552	11	5 16 22.15	2.3037	23 4 54.9	0.495
12	3 27 19.31	2.3136	21 8 21.9	5.431	12	5 18 40.32	2.3019	23 4 21.5	0.619
13	3 29 38.16	2.3147	21 13 44.1	5.306	13	5 20 58.38	2.3001	23 3 40.6	0.744
14	3 31 57.08	2.3158	21 18 58.9	5.186	14	5 23 16.33	2.2982	23 2 52.2	0.868
15	3 34 16.06	2.3168	21 24 6.4	5.064	15	5 25 34.16	2.2962	23 1 56.4	0.992
16	3 36 35.10	2.3178	21 29 6.6	4.941	16	5 27 51.87	2.2942	23 0 53.2	1.115
17	3 38 54.20	2.3187	21 33 59.3	4.817	17	5 30 9.46	2.2921	22 59 42.6	1.237
18	3 41 13.35	2.3196	21 38 44.6	4.693	18	5 32 26.92	2.2899	22 58 24.7	1.360
19	3 43 32.55	2.3204	21 43 22.5	4.569	19	5 34 44.25	2.2877	22 56 59.4	1.482
20	3 45 51.80	2.3212	21 47 52.9	4.445	20	5 37 1.44	2.2853	22 55 26.8	1.604
21	3 48 11.10	2.3220	21 52 15.9	4.320	21	5 39 18.49	2.2830	22 53 46.9	1.725
22	3 50 30.44	2.3226	21 56 31.3	4.194	22	5 41 35.40	2.2807	22 51 59.8	1.845
23	3 52 49.81	2.3231	22 0 39.2	4.069	23	5 43 52.17	2.2782	22 50 5.5	1.965
24	3 55 9.21	2.3237	N.22 4 39.6	3.944	24	5 46 8.79	2.2757	N.22 48 4.0	2.085

VI. FEBRUARY, 1900.

GREENWICH MEAN TIME.

THE MOON'S RIGHT ASCENSION AND DECLINATION.

Hour.	Right Ascension.	Diff. for 1 Minute.	Declination.	Diff. for 1 Minute.	Hour.	Right Ascension.	Diff. for 1 Minute.	Declination.	Diff. for 1 Minute.
	MONDAY 5.					**WEDNESDAY 7.**			
	h m s	s	° ′ ″	″		h m s	s	° ′ ″	″
0	1 51 20.62	2.2958	N.15 26 59.8	10.261	0	3 42 30.05	2.3083	N.21 30 28.1	4.685
1	1 53 38.36	2.2961	15 37 12.4	10.259	1	3 44 49.75	2.3083	21 35 5.4	4.558
2	1 55 56.16	2.2971	15 47 18.9	10.057	2	3 47 9.45	2.3083	21 39 35.1	4.432
3	1 58 14.01	2.2981	15 57 19.3	9.954	3	3 49 29.15	2.3082	21 43 57.2	4.305
4	2 0 31.93	2.2991	16 7 13.4	9.850	4	3 51 48.84	2.3080	21 48 11.7	4.179
5	2 2 49.90	2.3001	16 17 1.3	9.745	5	3 54 8.51	2.3078	21 52 18.7	4.052
6	2 5 7.94	2.3012	16 26 42.8	9.639	6	3 56 28.17	2.3076	21 56 18.0	3.925
7	2 7 26.04	2.3022	16 36 18.0	9.533	7	3 58 47.82	2.3072	22 0 9.7	3.798
8	2 9 44.19	2.3030	16 45 46.8	9.427	8	4 1 7.44	2.3068	22 3 53.8	3.672
9	2 12 2.40	2.3040	16 55 9.2	9.319	9	4 3 27.04	2.3064	22 7 30.3	3.544
10	2 14 20.67	2.3050	17 4 25.1	9.210	10	4 5 46.61	2.3059	22 10 59.2	3.417
11	2 16 39.00	2.3059	17 13 34.4	9.100	11	4 8 6.15	2.3054	22 14 20.4	3.290
12	2 18 57.38	2.3069	17 22 37.1	8.990	12	4 10 25.66	2.3048	22 17 34.0	3.162
13	2 21 15.83	2.3080	17 31 33.2	8.880	13	4 12 45.13	2.3042	22 20 39.9	3.035
14	2 23 34.34	2.3069	17 40 22.7	8.769	14	4 15 4.57	2.3036	22 23 38.2	2.908
15	2 25 52.90	2.3098	17 49 5.5	8.657	15	4 17 23.96	2.3028	22 26 28.9	2.781
16	2 28 11.52	2.3108	17 57 41.5	8.544	16	4 19 43.31	2.3021	22 29 11.9	2.653
17	2 30 30.20	2.3117	18 6 10.8	8.431	17	4 22 2.61	2.3012	22 31 47.3	2.527
18	2 32 48.93	2.3126	18 14 33.2	8.317	18	4 24 21.86	2.3004	22 34 15.1	2.400
19	2 35 7.71	2.3135	18 22 48.8	8.202	19	4 26 41.06	2.2995	22 36 35.3	2.272
20	2 37 26.55	2.3144	18 30 57.5	8.087	20	4 29 0.20	2.2984	22 38 47.8	2.145
21	2 39 45.44	2.3153	18 38 59.3	7.972	21	4 31 19.27	2.2973	22 40 52.7	2.018
22	2 42 4.39	2.3162	18 46 54.1	7.856	22	4 33 38.28	2.2963	22 42 50.0	1.892
23	2 44 23.39	2.3170	N.18 54 42.1	7.739	23	4 35 57.23	2.2952	N.22 44 39.7	1.765
	TUESDAY 6.					**THURSDAY 8.**			
0	2 46 42.43	2.3177	N.19 2 22.9	7.622	0	4 38 16.10	2.2939	N.22 46 21.8	1.638
1	2 49 1.52	2.3186	19 9 56.7	7.504	1	4 40 34.90	2.2927	22 47 56.3	1.512
2	2 51 20.66	2.3194	19 17 23.4	7.386	2	4 42 53.62	2.3113	22 49 23.2	1.386
3	2 53 39.85	2.3202	19 24 43.0	7.267	3	4 45 12.26	2.3100	22 50 42.6	1.260
4	2 55 59.08	2.3208	19 31 55.5	7.148	4	4 47 30.82	2.3086	22 51 54.4	1.134
5	2 58 18.35	2.3215	19 39 0.8	7.028	5	4 49 49.29	2.3071	22 52 58.7	1.008
6	3 0 37.66	2.3222	19 45 58.9	6.908	6	4 52 7.67	2.3055	22 53 55.4	0.882
7	3 2 57.01	2.3228	19 52 49.8	6.787	7	4 54 25.95	2.3039	22 54 44.6	0.757
8	3 5 16.40	2.3234	19 59 33.4	6.666	8	4 56 44.14	2.3023	22 55 26.3	0.632
9	3 7 35.82	2.3240	20 6 9.7	6.545	9	4 59 2.23	2.3007	22 56 0.5	0.508
10	3 9 55.28	2.3246	20 12 38.8	6.424	10	5 1 20.22	2.2989	22 56 27.3	0.384
11	3 12 14.77	2.3251	20 19 0.6	6.302	11	5 3 38.10	2.2971	22 56 46.6	0.260
12	3 14 34.29	2.3256	20 25 15.0	6.178	12	5 5 55.87	2.2952	22 56 58.5	0.136
13	3 16 53.84	2.3260	20 31 22.0	6.056	13	5 8 13.52	2.2932	22 57 2.9	+ 0.012
14	3 19 13.41	2.3264	20 37 21.7	5.933	14	5 10 31.06	2.2913	22 56 59.9	0.111
15	3 21 33.01	2.3268	20 43 14.0	5.806	15	5 12 48.48	2.2893	22 56 49.6	0.233
16	3 23 52.63	2.3271	20 48 58.8	5.685	16	5 15 5.78	2.2872	22 56 31.9	0.357
17	3 26 12.26	2.3273	20 54 36.2	5.561	17	5 17 22.95	2.2851	22 56 6.8	0.479
18	3 28 31.91	2.3277	21 0 6.1	5.436	18	5 19 39.99	2.2829	22 55 34.4	0.601
19	3 30 51.58	2.3279	21 5 28.5	5.312	19	5 21 56.90	2.2807	22 54 54.7	0.722
20	3 33 11.26	2.3281	21 10 43.5	5.187	20	5 24 13.67	2.2784	22 54 7.7	0.843
21	3 35 30.95	2.3281	21 15 50.9	5.061	21	5 26 30.31	2.2761	22 53 13.5	0.963
22	3 37 50.65	2.3283	21 20 50.8	4.936	22	5 28 46.80	2.2737	22 52 12.1	1.084
23	3 40 10.35	2.3283	21 25 43.2	4.811	23	5 31 3.15	2.2713	22 51 3.4	1.204
24	3 42 30.05	2.3283	N.21 30 28.1	4.685	24	5 33 19.36	2.2689	N.22 49 47.6	1.323

XVIII. FEBRUARY, 1900.

		GREENWICH MEAN TIME.								
		LUNAR DISTANCES.								
Day of the Month.	Name and Direction of Object.		Midnight.	P.L. of Diff.	XVh.	P.L. of Diff.	XVIIIh.	P.L. of Diff.	XXIh.	P.L. of Diff.
18	Pollux	W.	86 46 29	3081	88 15 2	3076	89 43 41	3069	91 12 28	3065
	Regulus	W.	49 49 0	3051	51 18 10	3044	52 47 28	3037	54 16 55	3030
	Antares	E.	50 7 41	3038	48 38 15	3032	47 8 42	3026	45 39 2	3022
	Jupiter	E.	50 57 29	3062	49 28 33	3056	47 59 30	3050	46 30 19	3044
	Saturn	E.	74 45 51	3050	73 16 40	3044	71 47 22	3038	70 17 57	3031
	α Aquilæ	E.	102 48 44	3500	101 28 20	3489	100 7 43	3479	98 46 55	3468
19	Pollux	W.	98 38 27	3027	100 8 6	3020	101 37 54	3011	103 7 53	3003
	Regulus	W.	61 46 34	2989	63 17 1	2979	64 47 40	2969	66 18 31	2960
	Antares	E.	38 8 59	2992	36 38 36	2985	35 8 4	2977	33 37 23	2971
	Jupiter	E.	39 2 19	3007	37 32 15	2998	36 2 0	2989	34 31 34	2980
	Saturn	E.	62 48 39	2994	61 18 19	2985	59 47 48	2976	58 17 5	2967
	α Aquilæ	E.	92 0 7	3421	90 38 14	3412	89 16 11	3403	87 53 58	3395
20	Regulus	W.	73 55 57	2906	75 28 8	2894	77 0 34	2882	78 33 16	2870
	Jupiter	E.	26 56 24	2930	25 24 43	2920	23 52 49	2909	22 20 41	2898
	Saturn	E.	50 40 26	2914	49 8 25	2903	47 36 10	2891	46 3 39	2878
	α Aquilæ	E.	81 0 35	3356	79 37 28	3350	78 14 14	3343	76 50 52	3317
	Sun	E.	109 33 56	3275	108 9 15	3262	106 44 19	3249	105 19 8	3235
21	Regulus	W.	86 20 53	2802	87 55 18	2788	89 30 1	2773	91 5 4	2758
	Spica	W.	32 18 23	2793	33 53 0	2779	35 27 56	2765	37 3 12	2747
	Saturn	E.	38 16 56	2811	36 42 42	2797	35 8 10	2782	33 33 18	2766
	α Aquilæ	E.	69 52 25	3313	68 28 29	3311	67 4 30	3309	65 40 29	3307
	Sun	E.	98 9 0	3162	96 42 5	3146	95 14 51	3130	93 47 18	3114
22	Spica	W.	45 4 53	2666	46 42 19	2649	48 20 8	2631	49 58 21	2613
	Saturn	E.	25 33 48	2686	23 56 49	2669	22 19 28	2652	20 41 44	2634
	α Aquilæ	E.	58 40 32	3322	57 16 46	3330	55 53 9	3340	54 29 44	3352
	Sun	E.	86 24 25	3028	84 54 47	3009	83 24 45	2989	81 54 19	2971
23	Spica	W.	58 15 32	2522	59 56 14	2504	61 37 21	2485	63 18 55	2467
	α Aquilæ	E.	47 37 27	3468	46 16 27	3505	44 56 8	3548	43 36 37	3598
	Sun	E.	74 16 17	2876	72 43 28	2856	71 10 13	2836	69 36 32	2817
24	Spica	W.	71 53 21	2373	73 37 34	2355	75 22 14	2335	77 7 22	2317
	Jupiter	W.	24 51 20	2407	26 34 45	2387	28 18 39	2367	30 3 1	2348
	Sun	E.	61 41 44	2719	60 5 29	2699	58 28 48	2680	56 51 41	2661
25	Spica	W.	85 59 40	2228	87 47 26	2210	89 35 38	2194	91 24 15	2177
	Jupiter	W.	38 51 44	2256	40 38 49	2238	42 26 20	2220	44 14 17	2203
	Sun	E.	48 39 50	2571	47 0 15	2554	45 20 17	2538	43 39 56	2522
26	Spica	W.	100 33 22	2101	102 24 19	2087	104 15 38	2074	106 7 17	2061
	Antares	W.	55 5 57	2113	56 36 37	2098	58 47 39	2085	60 39 2	2071
	Jupiter	W.	53 20 11	2126	55 10 31	2112	57 1 12	2098	58 52 14	2085
	Saturn	W.	29 40 6	2120	31 30 35	2105	33 21 26	2092	35 12 37	2079
	Sun	E.	35 13 9	2457	33 30 55	2446	31 48 26	2438	30 5 46	2433
27	Jupiter	W.	68 12 4	2030	70 4 51	2021	71 57 52	2013	73 51 6	2005
	Saturn	W.	44 33 11	2025	46 26 7	2017	48 19 15	2009	50 12 36	2000
	Sun	E.	21 31 15	2439	19 48 36	2456	18 6 21	2481	16 24 41	2520

182

MARCH, 1900.

AT GREENWICH APPARENT NOON.

THE SUN'S

Day of the Week	Day of the Month	Apparent Right Ascension	Diff. for 1 Hour.	Apparent Declination	Diff. for 1 Hour.	Semi-diameter.	Sidereal Time of Semi-diameter Passing Meridian.	Equation of Time, to be Added to Apparent Time.	Diff. for 1 Hour.
		h m s	s	° ′ ″	″	′ ″	s	m s	s
Thur.	1	22 47 57.02	9.371	S. 7 38 24.5	+56.96	16 9.26	65.37	12 34.70	0.485
Frid.	2	22 51 41.63	9.350	7 15 34.2	57.22	16 9.02	65.30	12 22.80	0.506
Sat.	3	22 55 25.75	9.330	6 52 37.9	57.47	16 8.77	65.23	12 10.41	0.527
SUN.	4	22 59 9.38	9.310	6 29 35.9	+57.70	16 8.52	65.16	11 57.52	0.547
Mon.	5	23 2 52.53	9.291	6 6 28.7	57.91	16 8.27	65.09	11 44.16	0.566
Tues.	6	23 6 35.24	9.272	5 43 16.5	58.11	16 8.02	65.03	11 30.35	0.584
Wed.	7	23 10 17.52	9.254	5 19 59.8	+58.29	16 7.77	64.97	11 16.12	0.602
Thur.	8	23 13 59.37	9.237	4 56 39.0	58.46	16 7.52	64.91	11 1.46	0.619
Frid.	9	23 17 40.83	9.221	4 33 14.4	58.61	16 7.26	64.86	10 46.40	0.635
Sat.	10	23 21 21.91	9.206	4 9 46.5	+58.74	16 7.00	64.81	10 30.97	0.650
SUN.	11	23 25 2.64	9.192	3 46 15.5	58.86	16 6.74	64.76	10 15.19	0.665
Mon.	12	23 28 43.04	9.178	3 22 41.9	58.96	16 6.48	64.71	9 59.07	0.679
Tues.	13	23 32 23.11	9.165	2 59 6.1	+59.05	16 6.22	64.66	9 42.65	0.691
Wed.	14	23 36 2.90	9.154	2 35 28.3	59.12	16 5.96	64.62	9 25.93	0.702
Thur	15	23 39 42.42	9.144	2 11 48.9	59.18	16 5.69	64.58	9 8.95	0.712
Frid.	16	23 43 21.72	9.134	1 48 8.4	+59.23	16 5.42	64.55	8 51.73	0.721
Sat.	17	23 47 0.80	9.125	1 24 27.0	59.26	16 5.15	64.52	8 34.31	0.730
SUN.	18	23 50 39.68	9.118	1 0 45.0	59.27	16 4.88	64.49	8 16.68	0.738
Mon.	19	23 54 18.38	9.112	0 37 2.8	+59.27	16 4.61	64.47	7 58.87	0.744
Tues.	20	23 57 56.94	9.106	S. 0 13 20.8	59.26	16 4.34	64.45	7 40.93	0.750
Wed.	21	0 1 35.37	9.101	N. 0 10 20.7	59.23	16 4.07	64.43	7 22.85	0.755
Thur.	22	0 5 13.71	9.098	0 34 1.3	+59.18	16 3.79	64.41	7 4.69	0.759
Frid.	23	0 8 51.97	9.095	0 57 40.8	59.12	16 3.51	64.40	6 46.44	0.762
Sat.	24	0 12 30.17	9.093	1 21 18.6	59.05	16 3.24	64.39	6 28.14	0.764
SUN.	25	0 16 8.33	9.092	1 44 54.6	+58.97	16 2.96	64.39	6 9.80	0.765
Mon.	26	0 19 46.47	9.092	2 8 28.1	58.86	16 2.69	64.38	5 51.45	0.765
Tues.	27	0 23 24.61	9.093	2 31 59.1	58.74	16 2.41	64.38	5 33.09	0.764
Wed.	28	0 27 2.78	9.094	2 55 27.0	+58.60	16 2.13	64.38	5 14.77	0.763
Thur.	29	0 30 41.00	9.096	3 18 51.5	58.44	16 1.86	64.39	4 56.48	0.761
Frid.	30	0 34 19.27	9.098	3 42 12.2	58.27	16 1.58	64.40	4 38.23	0.759
Sat.	31	0 37 57.60	9.101	4 5 28.7	58.09	16 1.31	64.41	4 20.05	0.756
SUN.	32	0 41 36.01	9.105	N. 4 28 40.7	+57.90	16 1.03	64.42	4 1.98	0.752

NOTE.—The mean time of semidiameter passing may be found by subtracting 0ˢ.18 from the sidereal time.
The sign + prefixed to the hourly change of declination indicates that south declinations are decreasing; north declinations, increasing.

II. MARCH, 1900.

AT GREENWICH MEAN NOON.

Day of the Week	Day of the Month	THE SUN'S Apparent Right Ascension.	Diff. for 1 Hour.	THE SUN'S Apparent Declination.	Diff. for 1 Hour.	Equation of Time, to be Subtracted from Mean Time.	Diff. for 1 Hour.	Sidereal Time, or Right Ascension of Mean Sun.
		h m s	s	° ′ ″	″	m s	s	h m s
Thur.	1	22 47 55.05	9.371	S. 7 38 36.4	+56.96	12 34.80	0.485	22 35 20.24
Frid.	2	22 51 39.70	9.350	7 15 46.0	57.22	12 22.90	0.506	22 39 16.80
Sat.	3	22 55 23.86	9.330	6 52 49.6	57.47	12 10.51	0.527	22 43 13.35
SUN.	4	22 59 7.53	9.310	6 29 47.4	+57.70	11 57.63	0.547	22 47 9.90
Mon.	5	23 2 50.72	9.291	6 6 40.0	57.91	11 44.27	0.566	22 51 6.45
Tues.	6	23 6 33.47	9.272	5 43 27.6	58.11	11 30.46	0.584	22 55 3.01
Wed.	7	23 10 15.79	9.254	5 20 10.7	+58.29	11 16.23	0.602	22 58 59.56
Thur.	8	23 13 57.68	9.237	4 56 49.7	58.46	11 1.57	0.619	23 2 56.11
Frid.	9	23 17 39.18	9.221	4 33 24.9	58.61	10 46.51	0.635	23 6 52.66
Sat.	10	23 21 20.30	9.206	4 9 56.8	+58.74	10 31.08	0.650	23 10 49.22
SUN.	11	23 25 1.07	9.192	3 46 25.6	58.86	10 15.30	0.665	23 14 45.77
Mon.	12	23 28 41.51	9.178	3 22 51.8	58.96	9 59.18	0.679	23 18 42.32
Tues.	13	23 32 21.63	9.165	2 59 15.7	+59.05	9 42.76	0.691	23 22 38.88
Wed.	14	23 36 1.47	9.154	2 35 37.7	59.12	9 26.04	0.702	23 26 35.43
Thur.	15	23 39 41.04	9.144	2 11 58.0	59.18	9 9.06	0.712	23 30 31.98
Frid.	16	23 43 20.38	9.134	1 48 17.2	+59.23	8 51.84	0.721	23 34 28.54
Sat.	17	23 46 59.50	9.125	1 24 35.5	59.26	8 34.41	0.730	23 38 25.09
SUN.	18	23 50 38.42	9.118	1 0 53.2	59.27	8 16.78	0.738	23 42 21.64
Mon.	19	23 54 17.17	9.112	0 37 10.7	+59.27	7 58.98	0.744	23 46 18.19
Tues.	20	23 57 55.77	9.106	S. 0 13 28.4	59.26	7 41.03	0.750	23 50 14.74
Wed.	21	0 1 34.25	9.101	N. 0 10 13.4	59.23	7 22.95	0.755	23 54 11.30
Thur.	22	0 5 12.63	9.098	0 33 54.3	+59.18	7 4.78	0.759	23 58 7.85
Frid.	23	0 8 50.93	9.095	0 57 34.1	59.12	6 46.53	0.762	0 2 4.40
Sat.	24	0 12 29.18	9.093	1 21 12.2	59.05	6 28.22	0.764	0 6 0.96
SUN.	25	0 16 7.39	9.092	1 44 48.5	+58.97	6 9.88	0.765	0 9 57.51
Mon.	26	0 19 45.58	9.092	2 8 22.3	58.86	5 51.52	0.765	0 13 54.06
Tues.	27	0 23 23.77	9.093	2 31 53.6	58.74	5 33.16	0.764	0 17 50.61
Wed.	28	0 27 1.99	9.094	2 55 21.8	+58.60	5 14.83	0.763	0 21 47.16
Thur.	29	0 30 40.25	9.096	3 18 46.7	58.44	4 56.54	0.761	0 25 43.72
Frid.	30	0 34 18.56	9.098	3 42 7.7	58.27	4 38.29	0.759	0 29 40.27
Sat.	31	0 37 56.94	9.101	4 5 24.5	58.09	4 20.12	0.756	0 33 36.82
SUN.	32	0 41 35.40	9.105	N. 4 28 36.8	+57.90	4 2.03	0.752	0 37 33.37

NOTE.—The semidiameter for mean noon may be assumed the same as that for apparent noon. The sign + prefixed to the hourly change of declination indicates that south declinations are decreasing; north declinations, increasing.

Diff. for 1 Hour, +9ˢ.8565. (Table III.)

II. APRIL, 1900.

		AT GREENWICH MEAN NOON.						
Day of the Week	Day of the Month	THE SUN'S				Equation of Time, to be Subtracted from Added to Mean Time.	Diff. for 1 Hour.	Sidereal Time, or Right Ascension of Mean Sun.
		Apparent Right Ascension.	Diff. for 1 Hour.	Apparent Declination.	Diff. for 1 Hour.			
SUN.	1	h m s 0 41 35.40	9.105	° ′ ″ N. 4 28 36.8	″	m s +57.90	m s 4 2.03	s 0.752	h m s 0 37 33.37
Mon.	2	0 45 13.96	9.109	4 51 44.2	57.69	3 44.04	0.748	0 41 29.93
Tues.	3	0 48 52.63	9.114	5 14 46.3	57.47	3 26.15	0.743	0 45 26.48
Wed.	4	0 52 31.44	9.120	5 37 42.7	+57.23	3 8.40	0.737	0 49 23.03
Thur.	5	0 56 10.39	9.126	6 0 33.2	56.97	2 50.81	0.731	0 53 19.59
Frid.	6	0 59 49.49	9.133	6 23 17.3	56.70	2 33.36	0.724	0 57 16.14
Sat.	7	1 3 28.79	9.141	6 45 54.7	+56.41	2 16.09	0.716	1 1 12.69
SUN.	8	1 7 8.27	9.150	7 8 25.1	56.11	1 59.02	0.707	1 5 9.24
Mon.	9	1 10 47.97	9.160	7 30 48.2	55.80	1 42.17	0.697	1 9 5.80
Tues.	10	1 14 27.91	9.170	7 53 3.7	+55.48	1 25.57	0.687	1 13 2.35
Wed.	11	1 18 8.11	9.181	8 15 11.1	55.15	1 9.21	0.676	1 16 58.90
Thur.	12	1 21 48.58	9.193	8 37 10.3	54.80	0 53.13	0.664	1 20 55.46
Frid.	13	1 25 29.35	9.206	8 59 0.8	+54.43	0 37.34	0.651	1 24 52.01
Sat.	14	1 29 10.43	9.219	9 20 42.4	54.05	0 21.86	0.638	1 28 48.56
SUN.	15	1 32 51.84	9.233	9 42 14.7	53.65	0 6.72	0.624	1 32 45.12
Mon.	16	1 36 33.60	9.248	10 3 37.4	+53.24	0 8.07	0.609	1 36 41.67
Tues.	17	1 40 15.72	9.264	10 24 50.1	52.82	0 22.50	0.593	1 40 38.22
Wed.	18	1 43 58.24	9.280	10 45 52.7	52.39	0 36.54	0.576	1 44 34.78
Thur.	19	1 47 41.17	9.297	11 6 44.7	+51.94	0 50.16	0.559	1 48 31.33
Frid.	20	1 51 24.51	9.315	11 27 25.9	51.48	1 3.38	0.541	1 52 27.88
Sat.	21	1 55 8.29	9.334	11 47 55.8	51.01	1 16.15	0.523	1 56 24.44
SUN.	22	1 58 52.53	9.353	12 8 14.2	+50.52	1 28.46	0.504	2 0 20.99
Mon.	23	2 2 37.24	9.373	12 28 20.7	50.02	1 40.32	0.484	2 4 17.55
Tues.	24	2 6 22.42	9.393	12 48 15.0	49.50	1 51.68	0.464	2 8 14.10
Wed.	25	2 10 8.09	9.413	13 7 56.8	+48.97	2 2.56	0.443	2 12 10.65
Thur.	26	2 13 54.26	9.434	13 27 25.7	48.43	2 12.95	0.422	2 16 7.21
Frid.	27	2 17 40.94	9.455	13 46 41.4	47.87	2 22.82	0.401	2 20 3.76
Sat.	28	2 21 28.13	9.476	14 5 43.5	+47.29	2 32.18	0.380	2 24 0.32
SUN.	29	2 25 15.84	9.498	14 24 31.7	46.70	2 41.03	0.358	2 27 56.87
Mon.	30	2 29 4.07	9.520	14 43 5.6	46.10	2 49.36	0.336	2 31 53.42
Tues.	31	2 32 52.82	9.542	N.15 1 25.0	+45.49	2 57.16	0.314	2 35 49.98

NOTE.—The semidiameter for mean noon may be assumed the same as that for apparent noon. The sign + prefixed to the hourly change of declination indicates that north declinations are increasing.

Diff. for 1 Hour,
+9^s.8565.
(Table III.)

AUGUST, 1900.

AT GREENWICH MEAN NOON.

Day of the Week	Day of the Month	THE SUN'S				Equation of Time, to be Subtracted from Added to Mean Time.	Diff. for 1 Hour.	Sidereal Time, or Right Ascension of Mean Sun.
		Apparent Right Ascension.	Diff. for 1 Hour.	Apparent Declination.	Diff. for 1 Hour.			
		h m s	s	° ′ ″	″	m s	s	h m s
Wed.	1	8 44 41.30	9.716	N.18 5 1.1	−37.61	6 8.12	0.140	8 38 33.18
Thur.	2	8 48 34.18	9.690	17 49 49.7	38.34	6 4.44	0.166	8 42 29.74
Frid.	3	8 52 26.44	9.664	17 34 21.0	39.05	6 0.15	0.191	8 46 26.29
Sat.	4	8 56 18.08	9.638	17 18 35.2	−39.75	5 55.23	0.217	8 50 22.85
SUN.	5	9 0 9.10	9.613	17 2 32.6	40.45	5 49.70	0.243	8 54 19.40
Mon.	6	9 3 59.51	9.587	16 46 13.5	41.13	5 43.55	0.269	8 58 15.96
Tues.	7	9 7 49.31	9.562	16 29 38.3	−41.80	5 36.80	0.294	9 2 12.51
Wed.	8	9 11 38.51	9.537	16 12 47.1	42.46	5 29.44	0.319	9 6 9.07
Thur.	9	9 15 27.11	9.513	15 55 40.3	43.10	5 21.48	0.344	9 10 5.62
Frid.	10	9 19 15.12	9.489	15 38 18.2	−43.73	5 12.94	0.368	9 14 2.18
Sat.	11	9 23 2.56	9.465	15 20 41.0	44.35	5 3.82	0.392	9 17 58.73
SUN.	12	9 26 49.43	9.442	15 2 49.1	44.96	4 54.14	0.415	9 21 55.29
Mon.	13	9 30 35.75	9.419	14 44 42.7	−45.56	4 43.91	0.438	9 25 51.84
Tues.	14	9 34 21.54	9.397	14 26 22.0	46.15	4 33.14	0.460	9 29 48.40
Wed.	15	9 38 6.80	9.375	14 7 47.5	46.72	4 21.85	0.482	9 33 44.95
Thur.	16	9 41 51.55	9.354	13 48 59.4	−47.28	4 10.04	0.503	9 37 41.50
Frid.	17	9 45 35.79	9.333	13 29 57.9	47.83	3 57.73	0.524	9 41 38.06
Sat.	18	9 49 19.54	9.313	13 10 43.5	48.36	3 44.93	0.544	9 45 34.61
SUN.	19	9 53 2.81	9.293	12 51 16.4	−48.88	3 31.64	0.564	9 49 31.17
Mon.	20	9 56 45.61	9.273	12 31 37.0	49.38	3 17.89	0.583	9 53 27.72
Tues.	21	10 0 27.94	9.254	12 11 45.6	49.88	3 3.67	0.602	9 57 24.28
Wed.	22	10 4 9.83	9.235	11 51 42.6	−50.36	2 49.00	0.620	10 1 20.83
Thur.	23	10 7 51.27	9.217	11 31 28.2	50.83	2 33.89	0.638	10 5 17.38
Frid.	24	10 11 32.28	9.199	11 11 2.8	51.28	2 18.34	0.656	10 9 13.94
Sat.	25	10 15 12.86	9.182	10 50 26.8	−51.72	2 2.37	0.674	10 13 10.49
SUN.	26	10 18 53.03	9.165	10 29 40.5	52.14	1 45.99	0.691	10 17 7.04
Mon.	27	10 22 32.81	9.149	10 8 44.3	52.55	1 29.21	0.707	10 21 3.60
Tues.	28	10 26 12.19	9.133	9 47 38.4	−52.94	1 12.04	0.723	10 25 0.15
Wed.	29	10 29 51.20	9.118	9 26 23.2	53.32	0 54.50	0.738	10 28 56.70
Thur.	30	10 33 29.86	9.104	9 4 59.1	53.69	0 36.60	0.753	10 32 53.26
Frid.	31	10 37 8.16	9.090	8 43 26.3	54.04	0 18.35	0.767	10 36 49.81
Sat.	32	10 40 46.13	9.076	N. 8 21 45.2	−54.38	0 0.23	0.781	10 40 46.36

NOTE.—The semidiameter for mean noon may be assumed the same as that for apparent noon. The sign — prefixed to the hourly change of declination indicates that north declinations are decreasing.

Diff. for 1 Hour,
+9s.8565.
(Table III.)

SEPTEMBER, 1900   III.

AT GREENWICH MEAN NOON.

Day of the Month	Day of the Year	THE SUN'S				Logarithm of the Radius Vector of the Earth	Diff. for 1 Hour	Mean Time of Sidereal Noon
		TRUE LONGITUDE		Diff. for 1 Hour	LATITUDE			
		λ	λ'					
		° ′ ″	° ′ ″	″	″		″	h m s
1	244	158 34 13.5	33 23.3	145.25	− 0.18	0.0038298	−44.4	13 17 2.70
2	245	159 32 20.2	31 29.9	145.31	− 0.06	0.0037217	44.9	13 13 6.80
3	246	160 30 28.3	29 37.9	145.37	+ 0.07	0.0036125	45.4	13 9 10.89
4	247	161 28 37.8	27 47.3	145.43	+ 0.20	0.0035024	−45.9	13 5 14.98
5	248	162 26 48.7	25 58.2	145.49	0.28	0.0033914	46.3	13 1 19.07
6	249	163 25 1.2	24 10.6	145.55	0.35	0.0032799	46.6	12 57 23.17
7	250	164 23 15.3	22 24.5	145.62	+ 0.40	0.0031678	−46.8	12 53 27.26
8	251	165 21 31.0	20 40.1	145.69	0.42	0.0030554	46.9	12 49 31.35
9	252	166 19 48.4	18 57.5	145.77	0.40	0.0029427	47.0	12 45 35.45
10	253	167 18 7.8	17 16.7	145.85	+ 0.35	0.0028297	−47.1	12 41 39.54
11	254	168 16 29.0	15 37.9	145.93	0.27	0.0027166	47.2	12 37 43.64
12	255	169 14 52.3	14 1.0	146.01	0.15	0.0026032	47.3	12 33 47.73
13	256	170 13 17.6	12 26.3	146.10	+ 0.03	0.0024894	−47.5	12 29 51.82
14	257	171 11 45.2	10 53.7	146.19	− 0.10	0.0023751	47.7	12 25 55.91
15	258	172 10 14.9	9 23.4	146.28	0.23	0.0022603	48.0	12 22 0.01
16	259	173 8 46.8	7 55.2	146.37	− 0.35	0.0021448	−48.3	12 18 4.10
17	260	174 7 20.9	6 29.2	146.46	0.47	0.0020285	48.6	12 14 8.19
18	261	175 5 57.2	5 5.5	146.56	0.56	0.0019114	49.0	12 10 12.29
19	262	176 4 35.8	3 43.9	146.65	− 0.63	0.0017934	−49.4	12 6 16.38
20	263	177 3 16.5	2 24.6	146.74	0.68	0.0016743	49.8	12 2 20.48
21	264	178 1 59.3	1 7.3	146.83	0.70	0.0015543	50.2	11 58 24.57
22	265	178 60 44.2	59 52.0	146.91	− 0.70	0.0014332	−50.6	11 54 28.66
23	266	179 59 31.0	58 38.8	147.00	0.66	0.0013112	51.0	11 50 32.76
24	267	180 58 20.0	57 27.7	147.08	0.61	0.0011882	51.4	11 46 36.85
25	268	181 57 10.9	56 18.5	147.16	− 0.53	0.0010642	−51.8	11 42 40.94
26	269	182 56 3.7	55 11.3	147.24	0.44	0.0009394	52.1	11 38 45.04
27	270	183 54 58.4	54 5.9	147.32	0.33	0.0008137	52.4	11 34 49.13
28	271	184 53 55.0	53 2.4	147.39	− 0.21	0.0006872	−52.7	11 30 53.22
29	272	185 52 53.4	52 0.7	147.47	− 0.08	0.0005603	53.0	11 26 57.32
30	273	186 51 53.6	51 0.8	147.54	+ 0.04	0.0004328	53.2	11 23 1.41
31	274	187 50 55.5	50 2.6	147.62	+ 0.17	0.0003049	−53.3	11 19 5.50

NOTE.—The numbers in column λ correspond to the true equinox of the date; in column λ' to the mean equinox of January 0°.0.

Diff. for 1 Hour.
−9ˢ.8296.
(Table II.)

II. NOVEMBER, 1900.

		AT GREENWICH MEAN NOON.						
		THE SUN'S						Sidereal Time, or Right Ascension of Mean Sun.
Day of the Week.	Day of the Month.	Apparent Right Ascension.	Diff. for 1 Hour.	Apparent Declination.	Diff. for 1 Hour.	Equation of Time, to be Added to Mean Time.	Diff. for 1 Hour.	
		h m s	s	° ′ ″	″	m s	s	h m s
Thur.	1	14 24 57.33	9.790	S. 14 22 58.5	−48.20	16 18.76	0.067	14 41 16.09
Frid.	2	14 28 52.67	9.823	14 42 8.4	47.61	16 19.97	0.034	14 45 12.64
Sat.	3	14 32 48.80	9.856	15 1 4.0	47.01	16 20.40	0.001	14 49 9.20
SUN.	4	14 36 45.72	9.889	15 19 44.8	−46.39	16 20.03	0.032	14 53 5.75
Mon.	5	14 40 43.46	9.923	15 38 10.5	45.75	16 18.85	0.065	14 57 2.31
Tues.	6	14 44 42.01	9.958	15 56 20.6	45.09	16 16.85	0.100	15 0 58.86
Wed.	7	14 48 41.40	9.993	16 14 14.9	−44.42	16 14.02	0.135	15 4 55.42
Thur.	8	14 52 41.62	10.028	16 31 52.8	43.73	16 10.35	0.170	15 8 51.97
Frid.	9	14 56 42.70	10.063	16 49 14.1	43.03	16 5.83	0.206	15 12 48.53
Sat.	10	15 0 44.63	10.099	17 6 18.2	−42.31	16 0.45	0.242	15 16 45.08
SUN.	11	15 4 47.43	10.135	17 23 4.8	41.57	15 54.21	0.278	15 20 41.64
Mon.	12	15 8 51.09	10.171	17 39 33.6	40.82	15 47.10	0.314	15 24 38.19
Tues.	13	15 12 55.63	10.207	17 55 44.1	−40.05	15 39.12	0.351	15 28 34.75
Wed.	14	15 17 1.03	10.243	18 11 35.9	39.26	15 30.27	0.387	15 32 31.30
Thur.	15	15 21 7.30	10.279	18 27 8.6	38.46	15 20.56	0.423	15 36 27.86
Frid.	16	15 25 14.43	10.315	18 42 21.8	−37.64	15 9.98	0.459	15 40 24.42
Sat.	17	15 29 22.42	10.351	18 57 15.1	36.80	14 58.55	0.494	15 44 20.97
SUN.	18	15 33 31.26	10.386	19 11 48.1	35.95	14 46.27	0.529	15 48 17.53
Mon.	19	15 37 40.94	10.421	19 26 0.5	−35.08	14 33.14	0.564	15 52 14.08
Tues.	20	15 41 51.46	10.455	19 39 51.8	34.19	14 19.18	0.598	15 56 10.64
Wed.	21	15 46 2.79	10.489	19 53 21.8	33.29	14 4.40	0.632	16 0 7.20
Thur.	22	15 50 14.94	10.522	20 6 29.8	−32.37	13 48.82	0.666	16 4 3.75
Frid.	23	15 54 27.87	10.555	20 19 15.8	31.44	13 32.43	0.699	16 8 0.31
Sat.	24	15 58 41.59	10.587	20 31 39.2	30.50	13 15.27	0.731	16 11 56.86
SUN.	25	16 2 56.07	10.619	20 43 39.8	−29.54	12 57.35	0.762	16 15 53.42
Mon.	26	16 7 11.29	10.649	20 55 17.2	28.57	12 38.69	0.792	16 19 49.98
Tues.	27	16 11 27.24	10.678	21 6 31.0	27.58	12 19.30	0.822	16 23 46.53
Wed.	28	16 15 43.88	10.707	21 17 20.9	−26.58	11 59.21	0.851	16 27 43.09
Thur.	29	16 20 1.22	10.736	21 27 46.7	25.56	11 38.43	0.880	16 31 39.65
Frid.	30	16 24 19.22	10.764	21 37 48.0	24.54	11 16.99	0.907	16 35 36.20
Sat.	31	16 28 37.86	10.790	S. 21 47 24.5	−23.51	10 54.90	0.933	16 39 32.76

NOTE.—The semidiameter for mean noon may be assumed the same as that for apparent noon. The sign − prefixed to the hourly change of declination indicates that south declinations are increasing.

Diff. for 1 Hour,
+9s.8565.
(Table III.)

VENUS, 1900.

GREENWICH MEAN TIME.

Day of Month	JANUARY. Apparent Right Ascension. Noon.	Var. of R.A. for 1 Hour. Noon.	Apparent Declination. Noon.	Var. of Decl. for 1 Hour. Noon.	Meridian Passage.	Day of Month	FEBRUARY. Apparent Right Ascension. Noon.	Var. of R.A. for 1 Hour. Noon.	Apparent Declination. Noon.	Var. of Decl. for 1 Hour. Noon.	Meridian Passage.
	h m s	s	° ′ ″	″	h m		h m s	s	° ′ ″	″	h m
1	20 39 23.93	+12.819	−20 9 51.7	+46.64	1 56.8	1	23 7 46.06	+11.233	−6 58 20.4	+75.82	2 22.9
2	20 44 30.90	12.763	19 50 55.3	48.06	1 57.9	2	23 12 15.26	11.199	6 27 55.5	76.24	2 23.4
3	20 49 36.51	12.706	19 31 25.0	49.45	1 59.1	3	23 16 43.65	11.167	5 57 20.9	76.63	2 23.9
4	20 54 40.75	12.648	19 11 21.7	50.82	2 0.2	4	23 21 11.26	11.136	5 26 37.4	76.99	2 24.5
5	20 59 43.61	12.590	18 50 46.0	52.15	2 1.3	5	23 25 38.14	11.106	4 55 45.8	77.31	2 25.0
6	21 4 45.08	+12.532	−18 29 38.8	+53.45	2 2.4	6	23 30 4.31	+11.077	−4 24 46.7	+77.60	2 25.5
7	21 9 45.15	12.474	18 8 0.8	54.72	2 3.5	7	23 34 29.81	11.050	3 53 41.1	77.86	2 25.9
8	21 14 43.82	12.415	17 45 52.7	55.95	2 4.5	8	23 38 54.68	11.024	3 22 29.6	78.09	2 26.4
9	21 19 41.08	12.357	17 23 15.4	57.15	2 5.5	9	23 43 18.95	11.000	2 51 13.0	78.29	2 26.9
10	21 24 36.95	12.299	17 0 9.6	58.32	2 6.5	10	23 47 42.66	10.978	2 19 52.0	78.46	2 27.3
11	21 29 31.42	+12.241	−16 36 36.1	+59.46	2 7.5	11	23 52 5.85	+10.957	−1 48 27.4	+78.60	2 27.7
12	21 34 24.51	12.183	16 12 35.7	60.57	2 8.4	12	23 56 28.56	10.937	1 16 59.8	78.70	2 28.2
13	21 39 16.22	12.126	15 48 9.2	61.64	2 9.3	13	0 0 50.82	10.919	0 45 29.9	78.77	2 28.6
14	21 44 6.57	12.070	15 23 17.3	62.68	2 10.2	14	0 5 12.68	10.903	−0 13 58.6	78.82	2 29.1
15	21 48 55.57	12.014	14 58 0.9	63.69	2 11.1	15	0 9 34.17	10.889	+0 17 33.5	78.84	2 29.5
16	21 53 43.25	+11.959	−14 32 20.7	+64.66	2 11.9	16	0 13 55.34	+10.876	+0 49 5.7	+78.83	2 29.9
17	21 58 29.61	11.905	14 6 17.5	65.60	2 12.8	17	0 18 16.22	10.865	1 20 37.2	78.79	2 30.3
18	22 3 14.70	11.852	13 39 52.0	66.51	2 13.6	18	0 22 36.86	10.856	1 52 7.4	78.72	2 30.7
19	22 7 58.53	11.800	13 13 5.1	67.39	2 14.3	19	0 26 57.29	10.848	2 23 35.6	78.62	2 31.1
20	22 12 41.11	11.749	12 45 57.6	68.23	2 15.1	20	0 31 17.54	10.841	2 55 1.0	78.49	2 31.5
21	22 17 22.48	+11.699	−12 18 30.2	+69.04	2 15.8	21	0 35 37.66	+10.836	+3 26 23.0	+78.33	2 31.9
22	22 22 2.67	11.650	11 50 43.7	69.83	2 16.6	22	0 39 57.69	10.833	3 57 40.8	78.14	2 32.3
23	22 26 41.70	11.602	11 22 38.8	70.58	2 17.3	23	0 44 17.66	10.831	4 28 53.8	77.92	2 32.6
24	22 31 19.60	11.556	10 54 16.6	71.29	2 18.0	24	0 48 37.62	10.832	5 0 1.2	77.68	2 33.0
25	22 35 56.41	11.511	10 25 37.6	71.97	2 18.7	25	0 52 57.59	10.833	5 31 2.4	77.41	2 33.4
26	22 40 32.15	+11.468	− 9 56 42.5	+72.62	2 19.3	26	0 57 17.62	+10.836	+6 1 56.6	+77.11	2 33.8
27	22 45 6.86	11.426	9 27 32.1	73.24	2 19.9	27	1 1 37.74	10.840	6 32 43.2	76.77	2 34.2
28	22 49 40.57	11.385	8 58 7.4	73.82	2 20.5	28	1 5 57.96	10.846	7 3 21.4	76.40	2 34.6
29	22 54 13.32	11.345	8 28 29.0	74.37	2 21.1	29	1 10 18.33	10.853	7 33 50.5	76.01	2 35.0
30	22 58 45.13	11.306	7 58 38.0	74.89	2 21.7	30	1 14 38.88	10.860	8 4 9.8	75.59	2 35.4
31	23 3 16.03	+11.269	− 7 28 34.9	+75.37	2 22.3	31	1 18 59.62	+10.869	+8 34 18.5	+75.14	2 35.8
32	23 7 46.06	+11.233	− 6 58 20.4	+75.82	2 22.9	32	1 23 20.59	+10.879	+9 4 16.0	+74.65	2 36.2

Day of the Month.	1st.	6th.	11th.	16th.	21st.	26th.	31st.	Day of the Month.	5th.	10th.	15th.	20th.	25th.
Semidiameter	5.84	5.93	6.03	6.14	6.24	6.37	6.50	Semidiameter .	6.63	6.78	6.95	7.12	7.31
Hor. Parallax	6.05	6.15	6.25	6.35	6.47	6.59	6.73	Hor. Parallax	6.87	7.03	7.19	7.37	7.57

NOTE.—The sign + indicates north declinations; the sign − indicates south declinations.

To avoid fine, this book should be returned on
or before the date last stamped below

CPSIA information can be obtained
at www.ICGtesting.com
Printed in the USA
LVOW07s1029090617
537544LV00016B/184/P

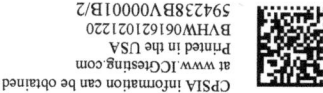

CPSIA information can be obtained
at www.ICGtesting.com
Printed in the USA
BVHW061621012220
594238BV00001B/2

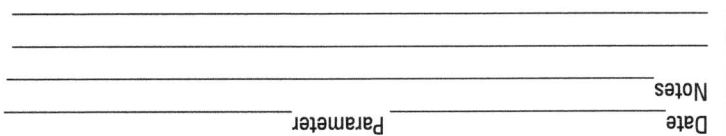

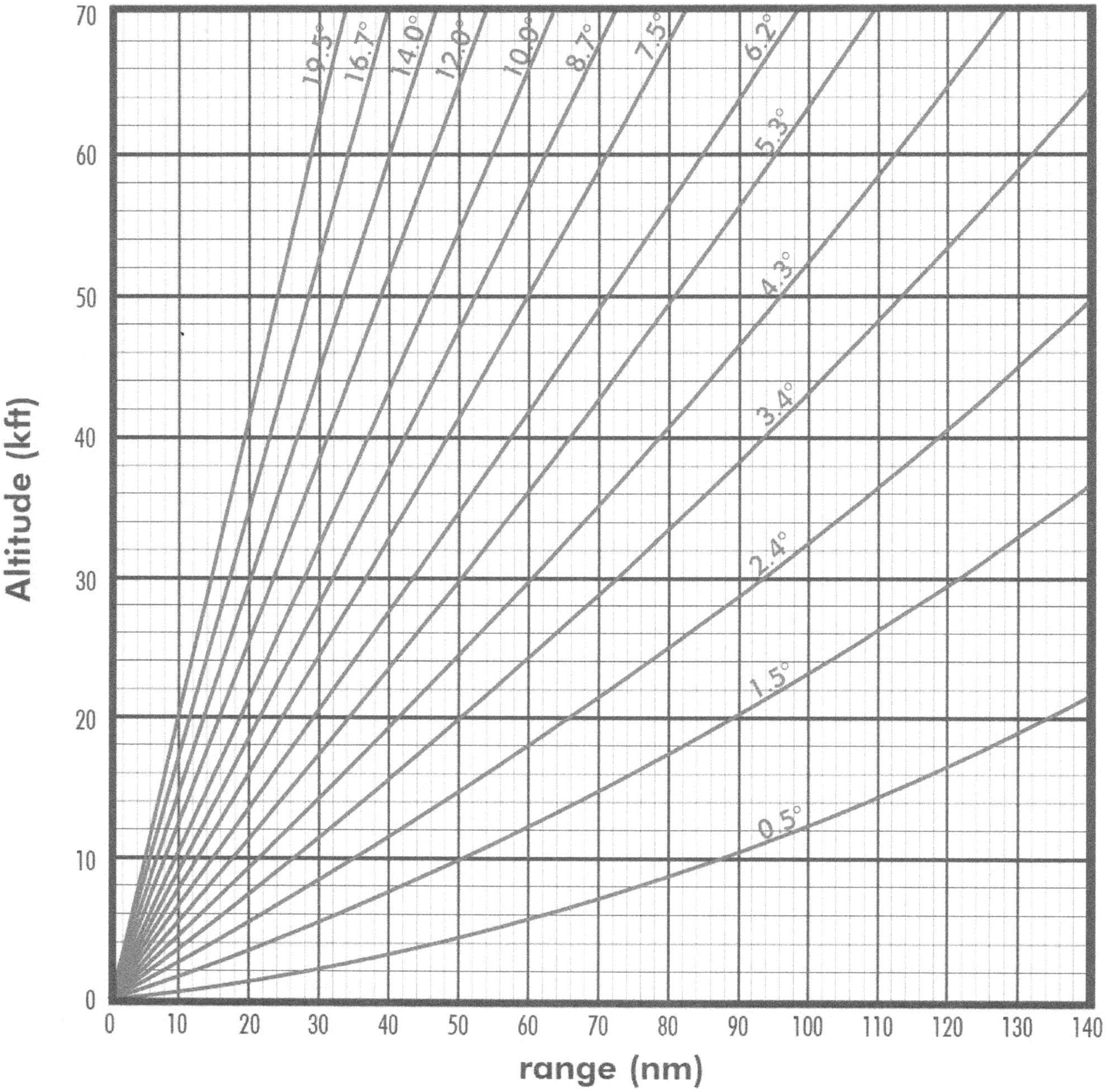

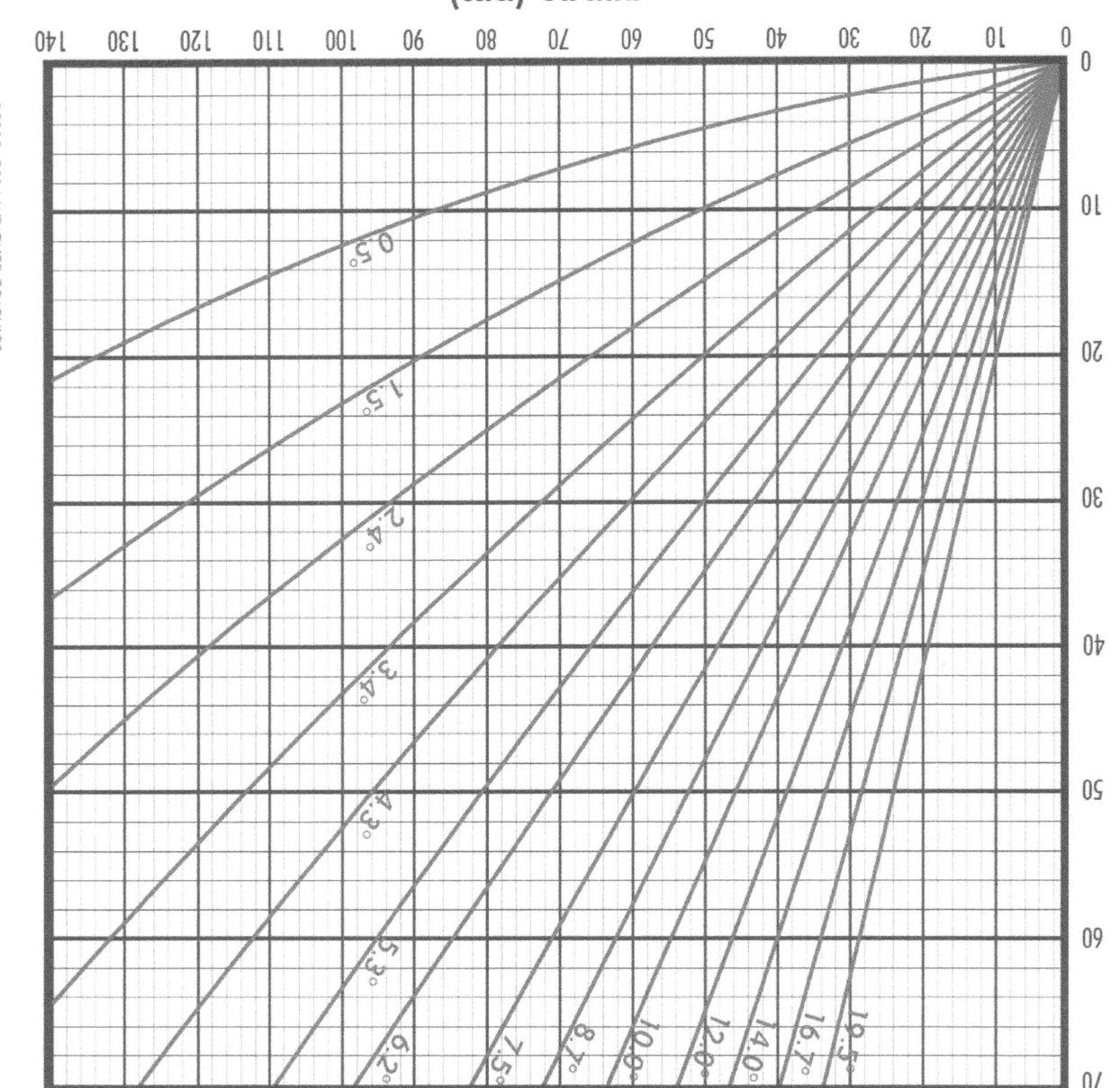

RADAR RANGE/HEIGHT DIAGRAM

WSR-88D RANGE/HEIGHT DIAGRAM

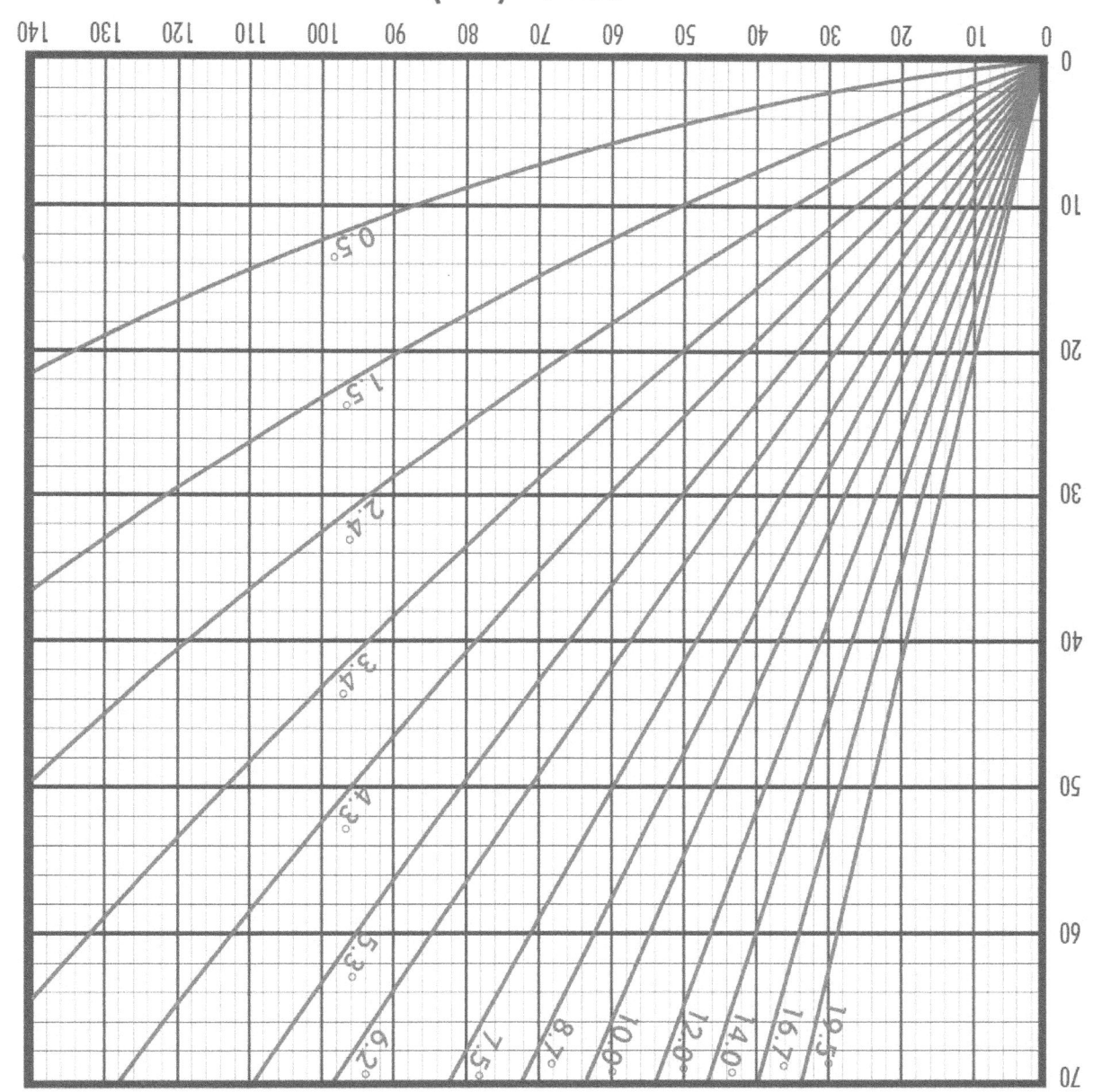

RADAR RANGE/HEIGHT DIAGRAM

WSR-88D RANGE/HEIGHT DIAGRAM

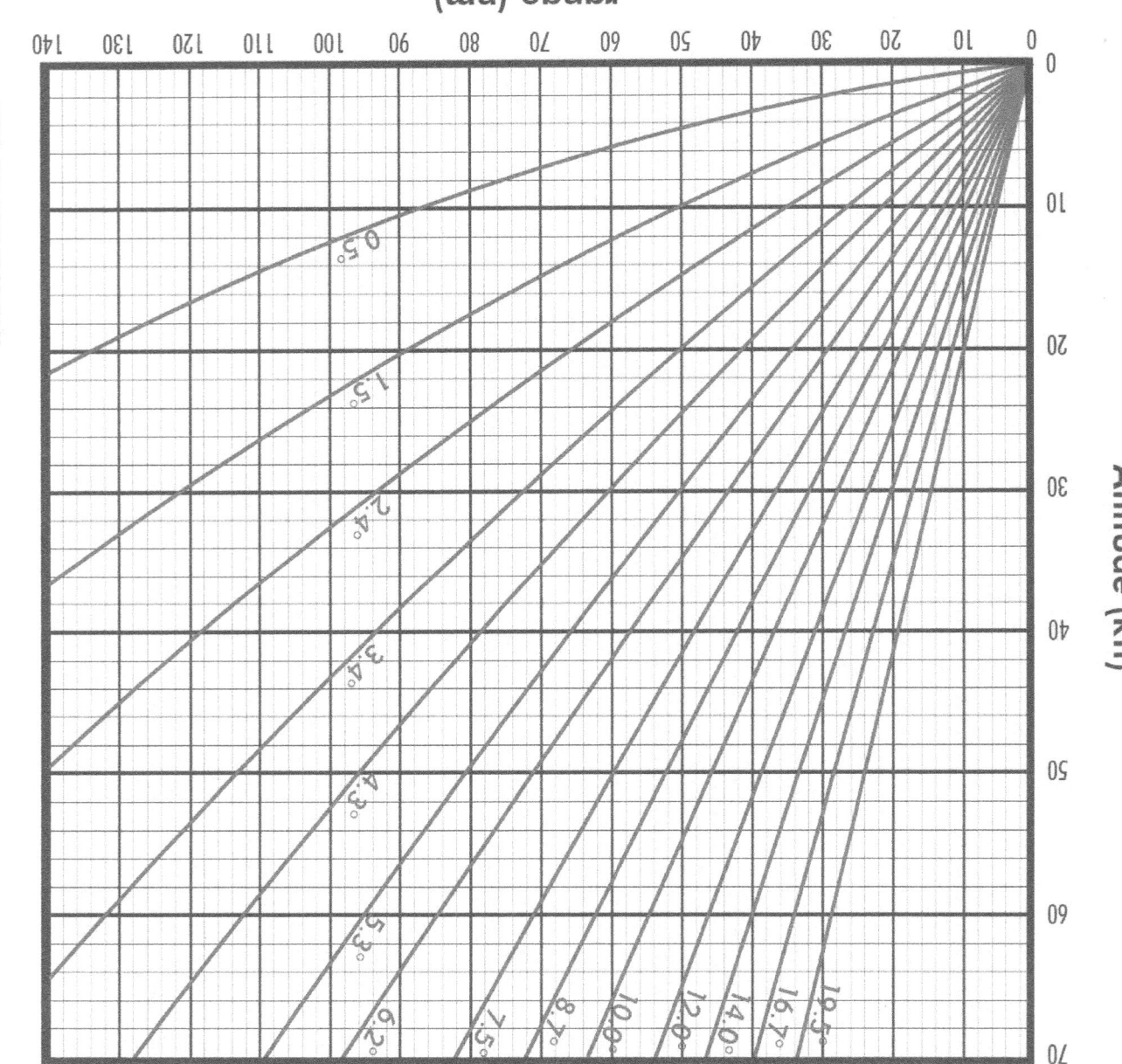

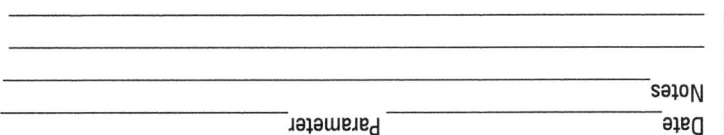

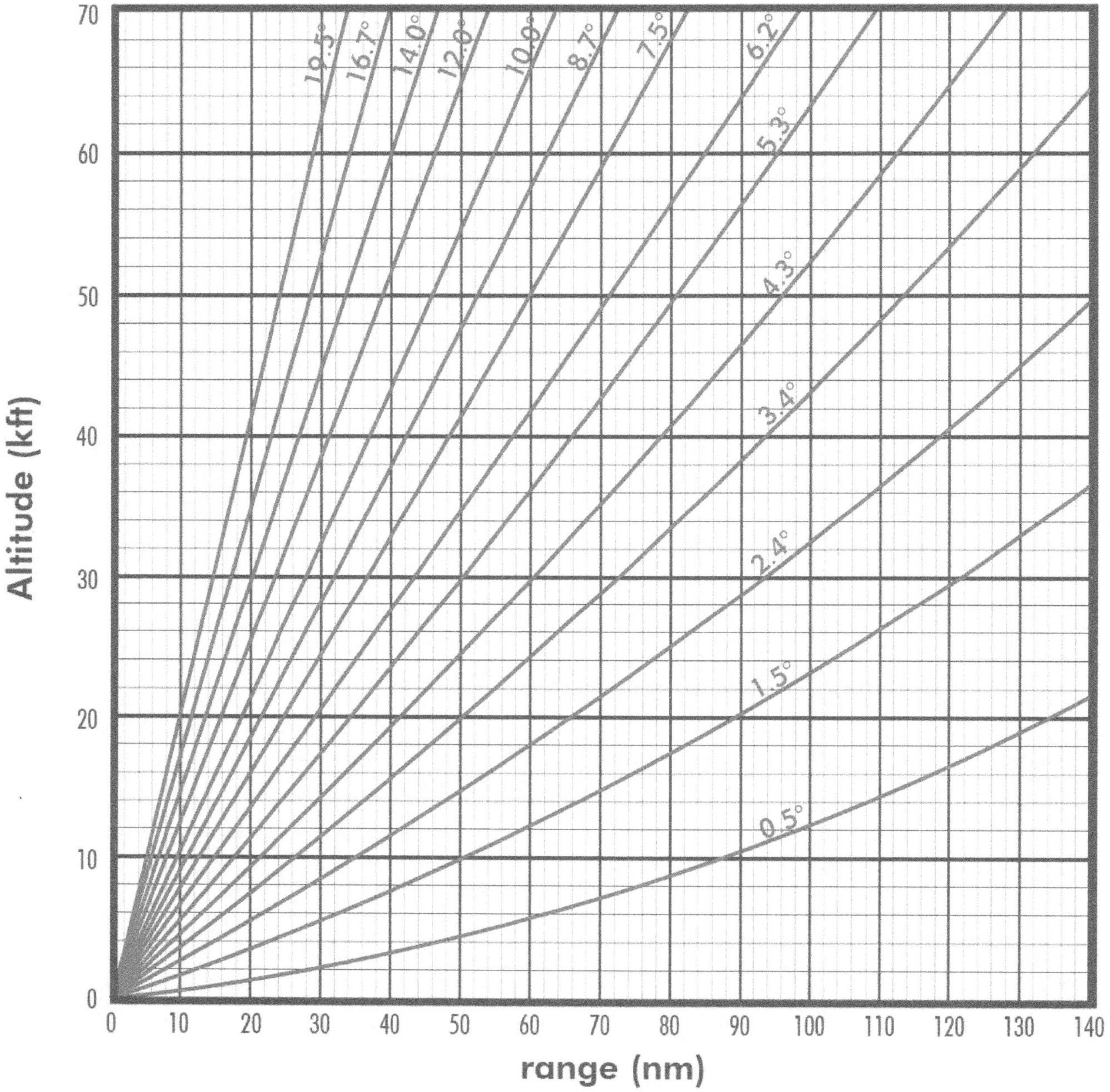

RADAR

SURFACE/UPPER AIR CHART

Vol ___ Sheet 17/17
Notes ___
Date ___ Parameter ___

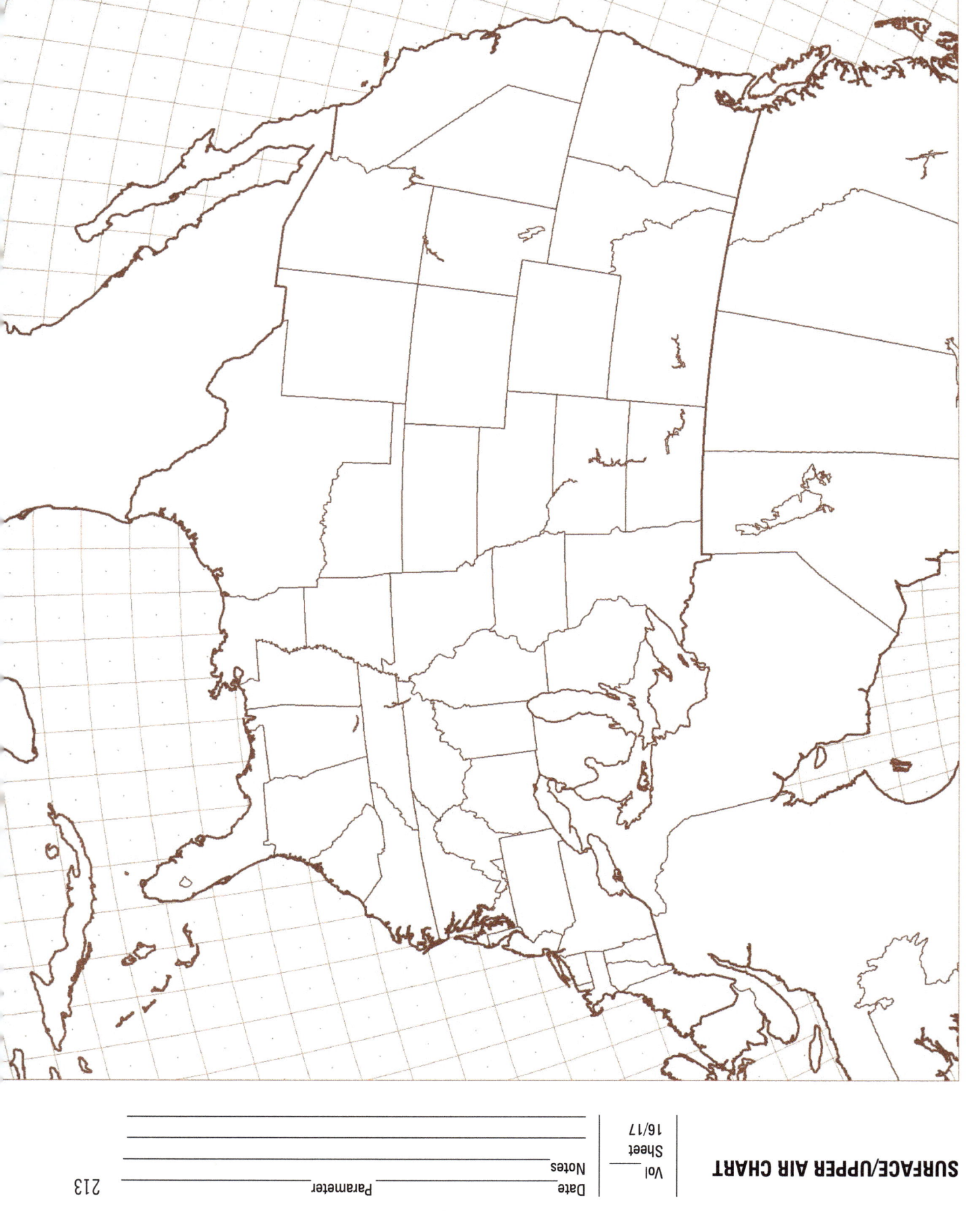

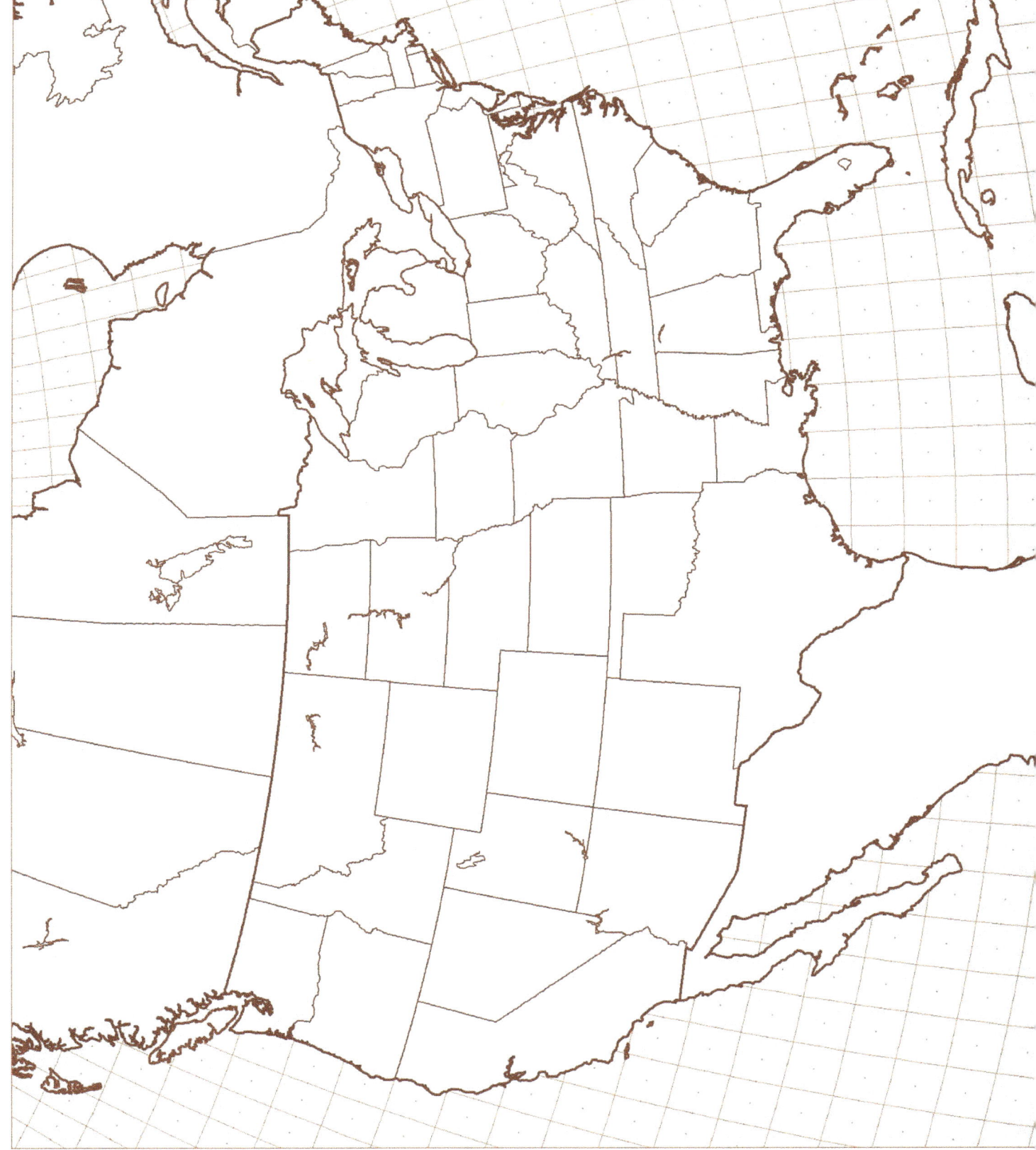

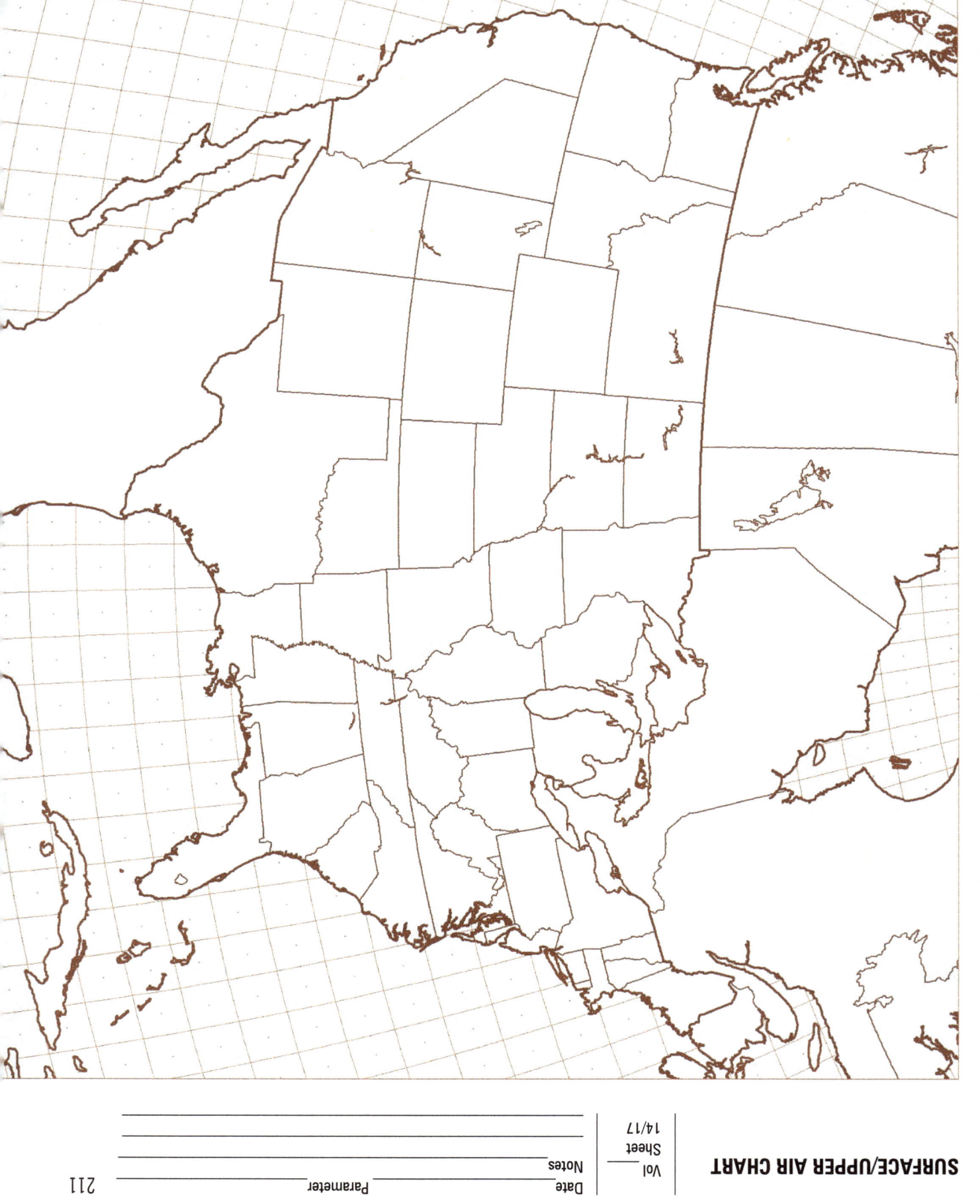

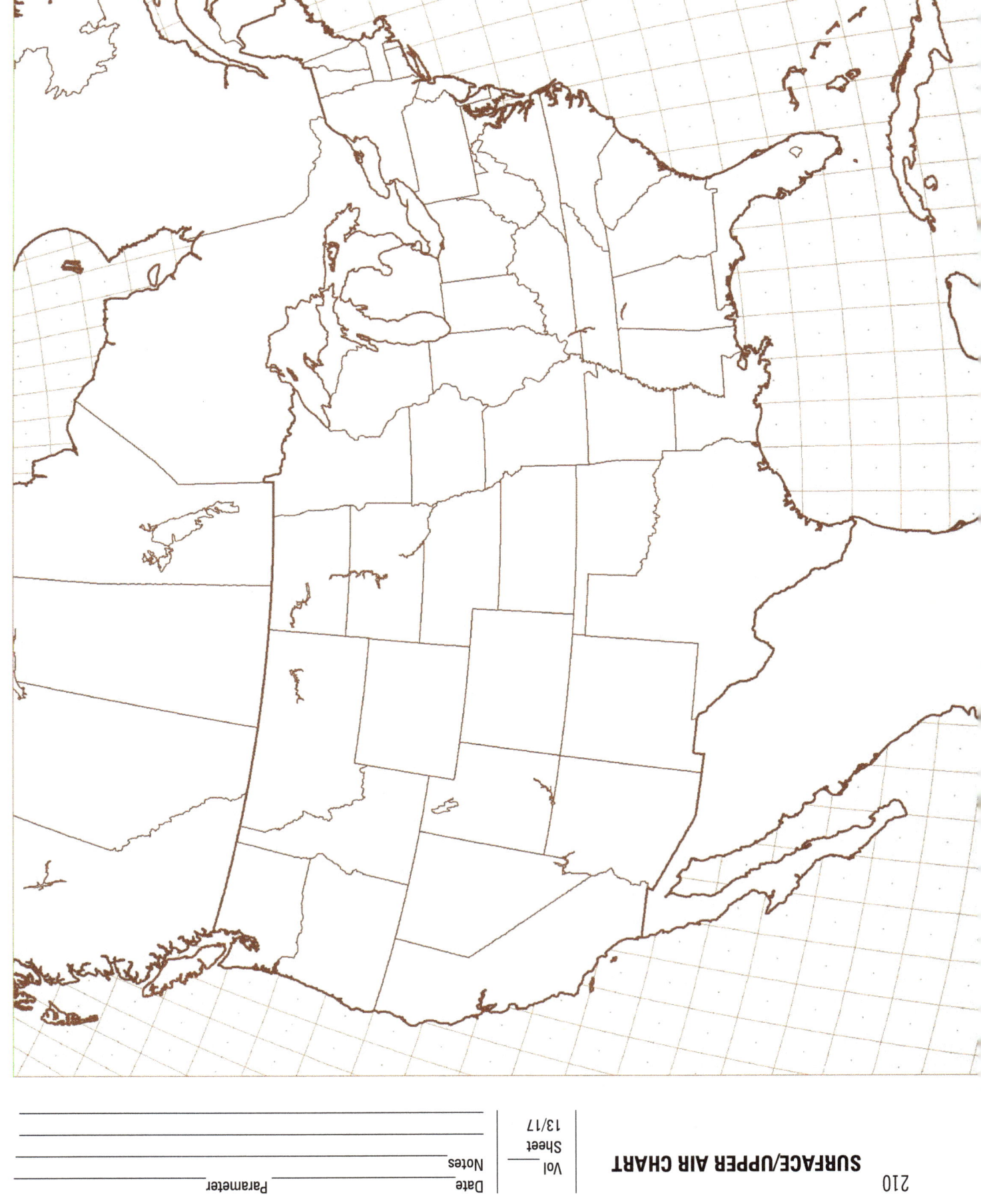

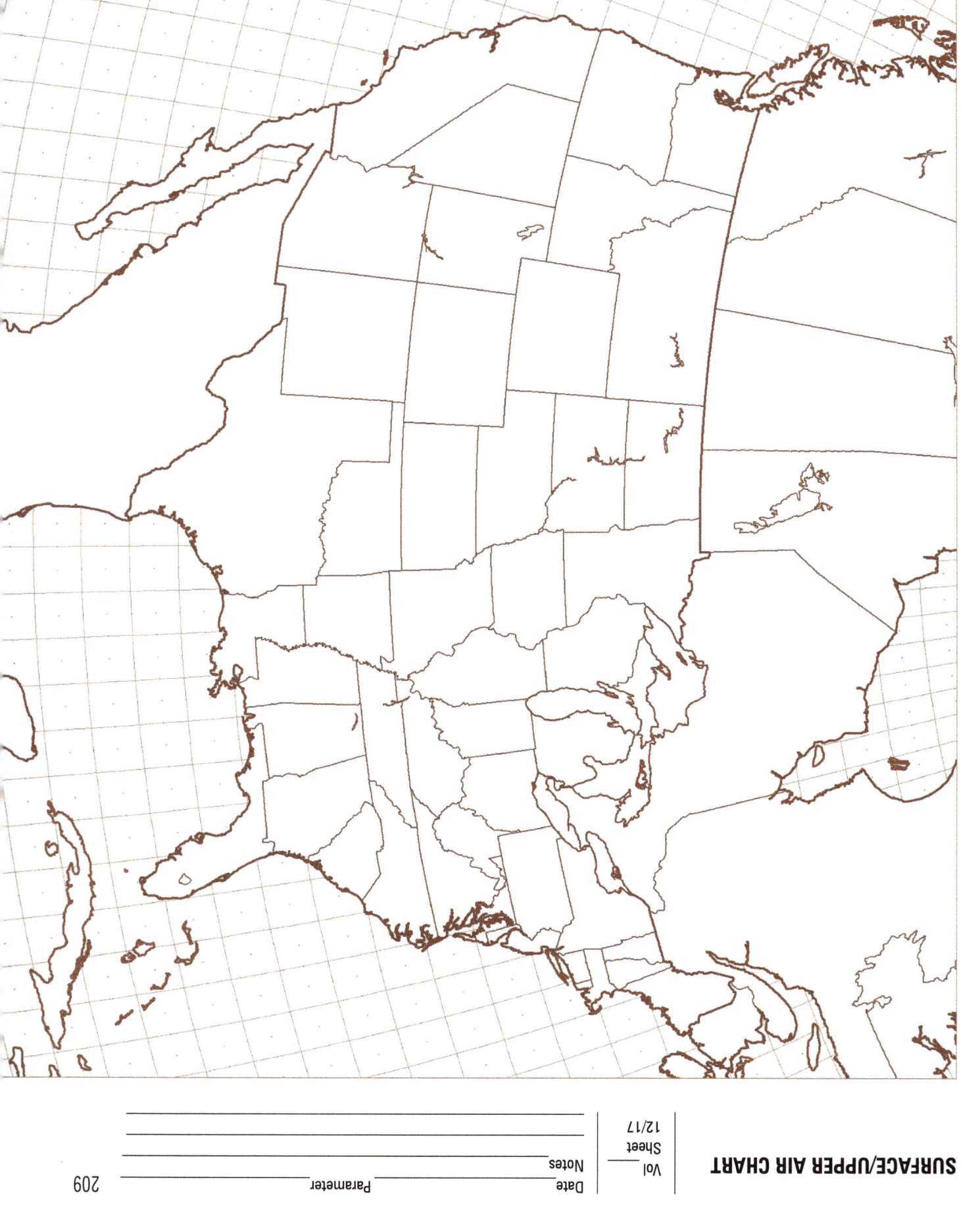

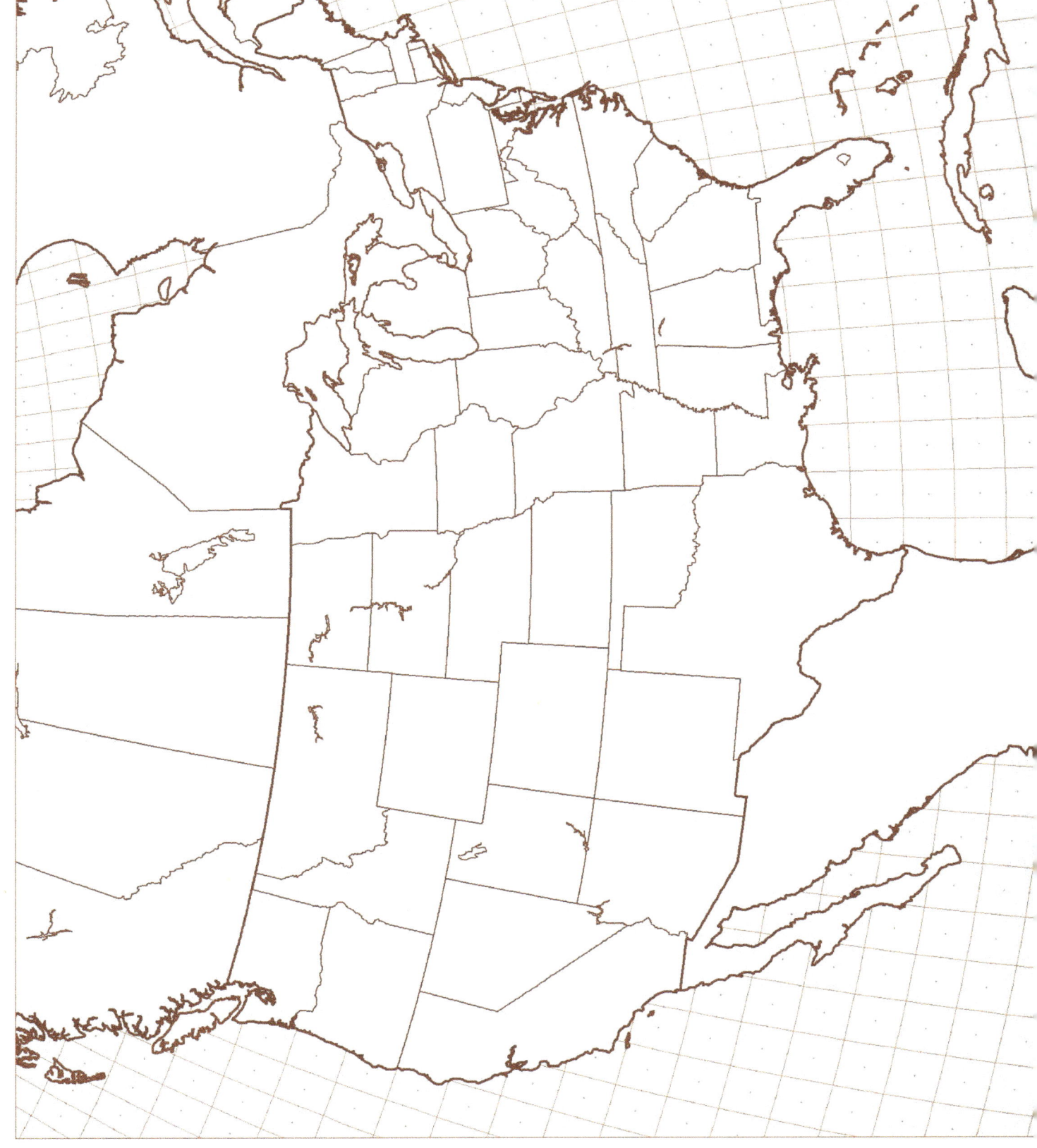

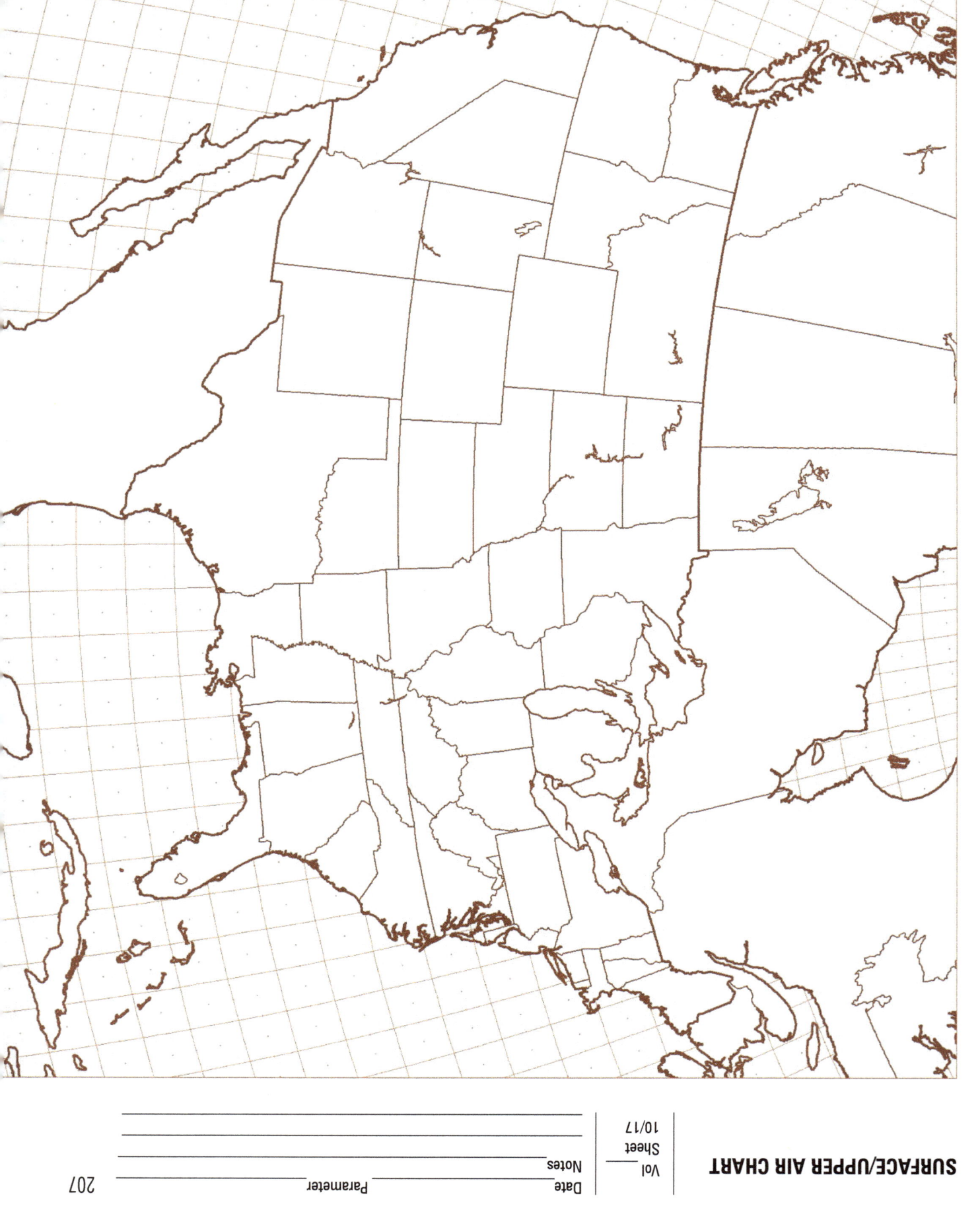

SURFACE/UPPER AIR CHART

Vol
Sheet 9/17
Date
Parameter
Notes

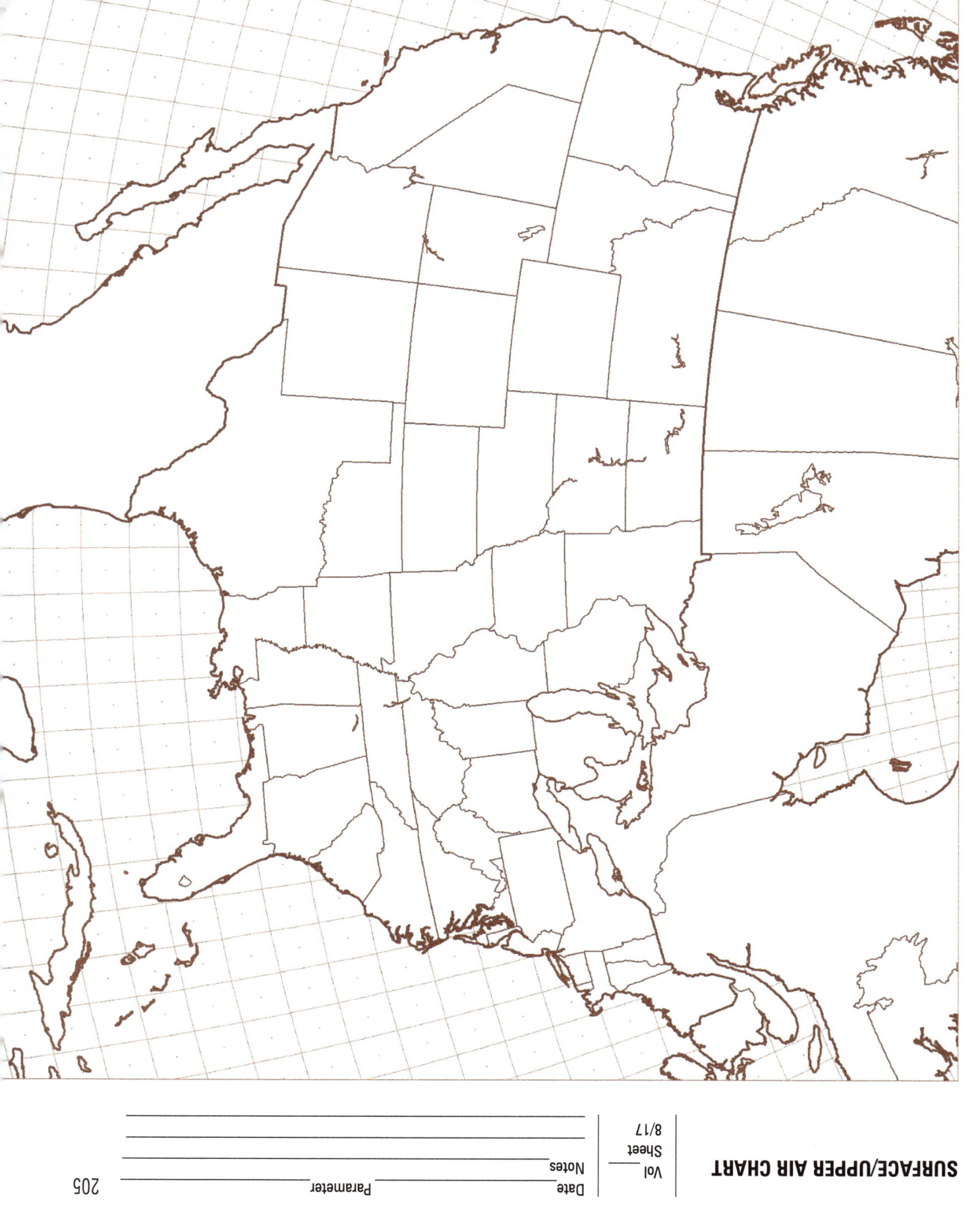

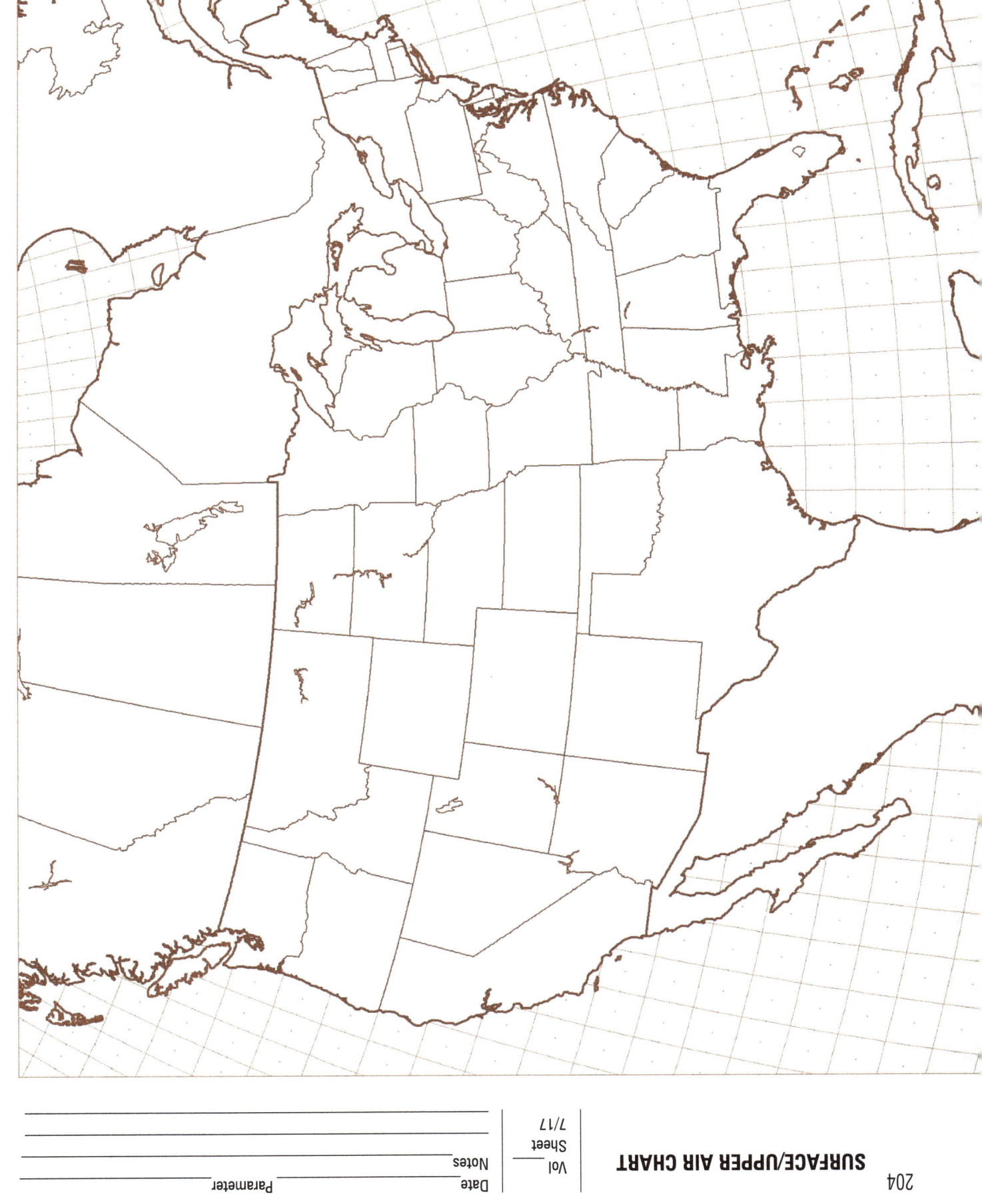

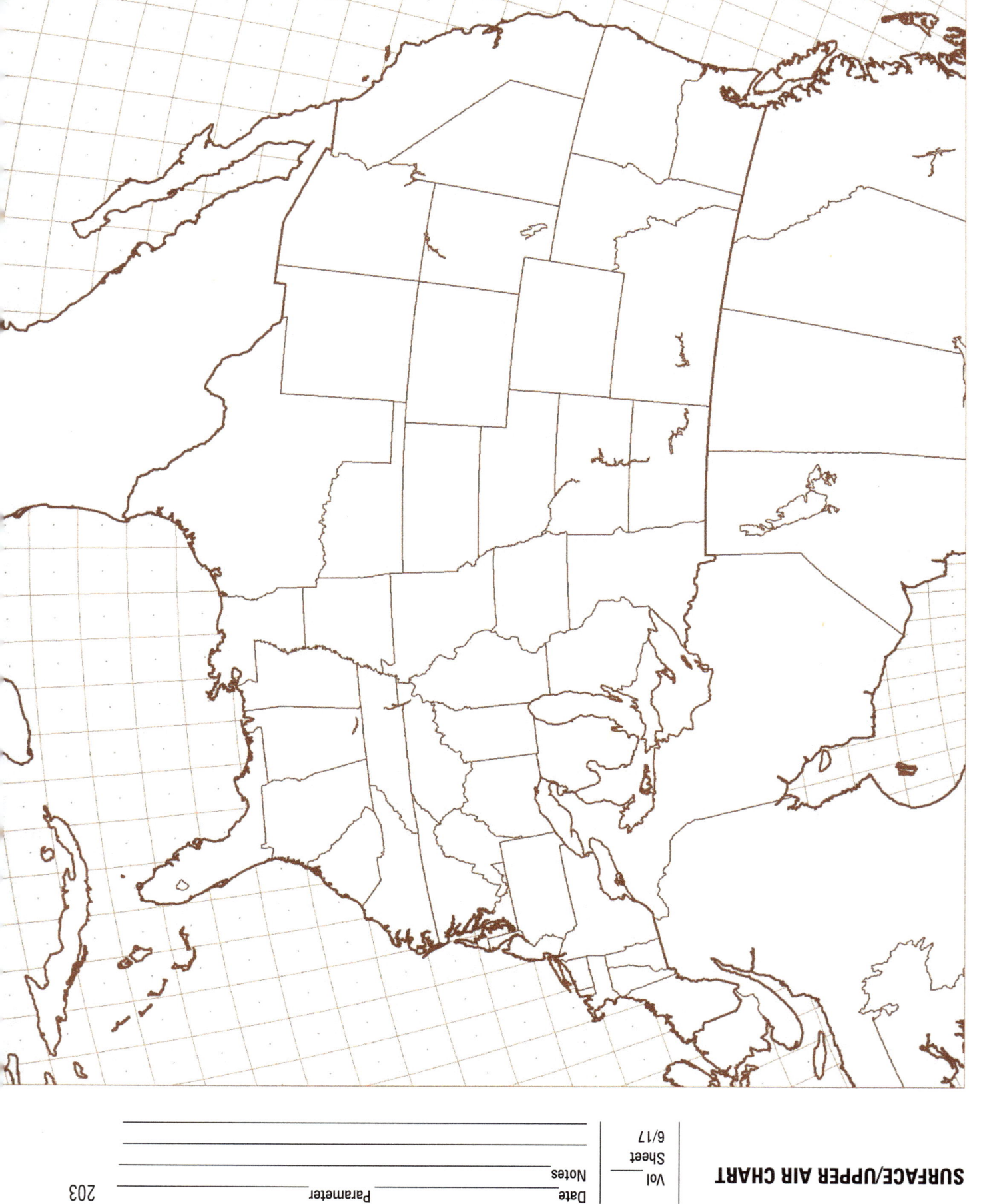

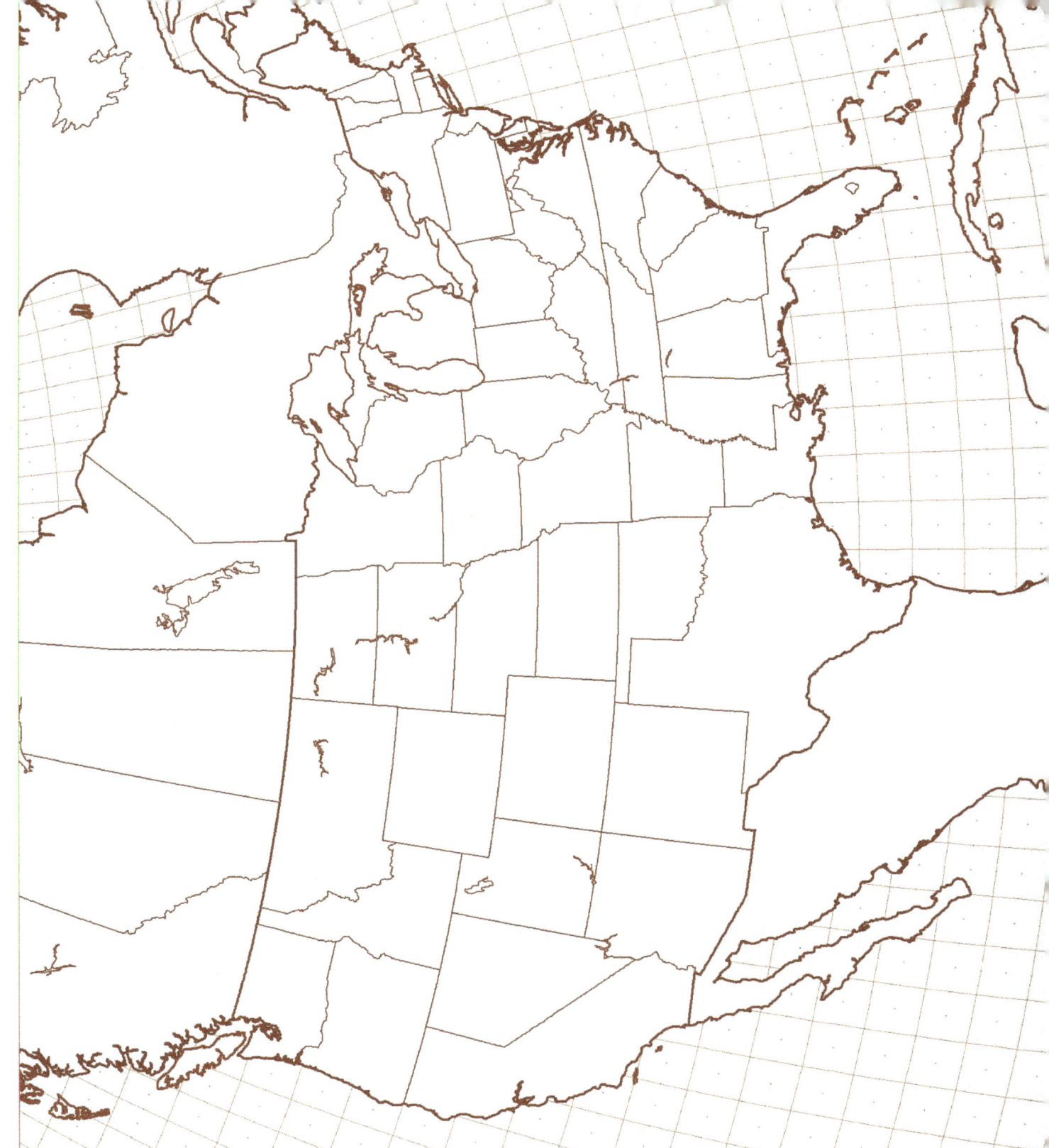

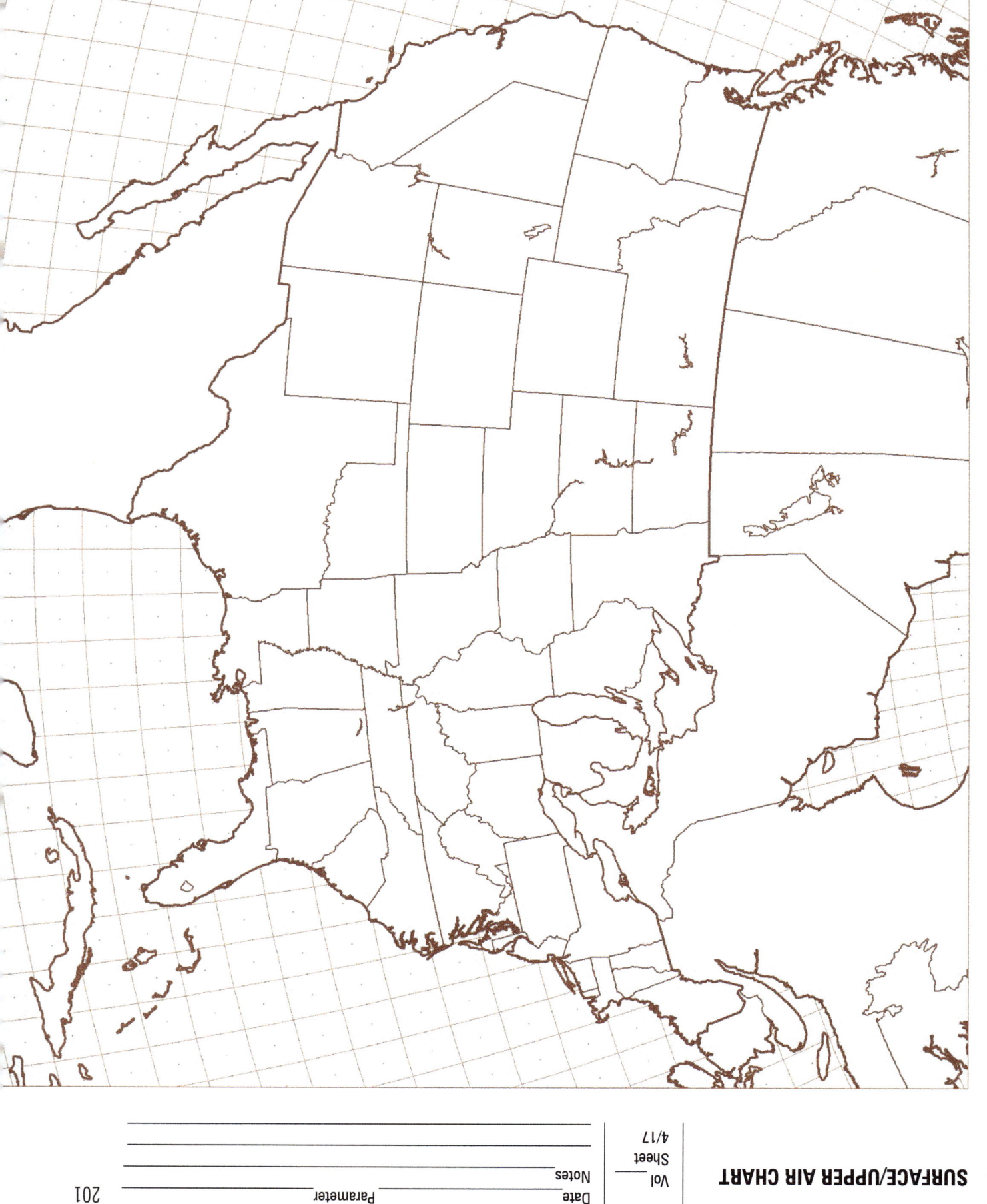

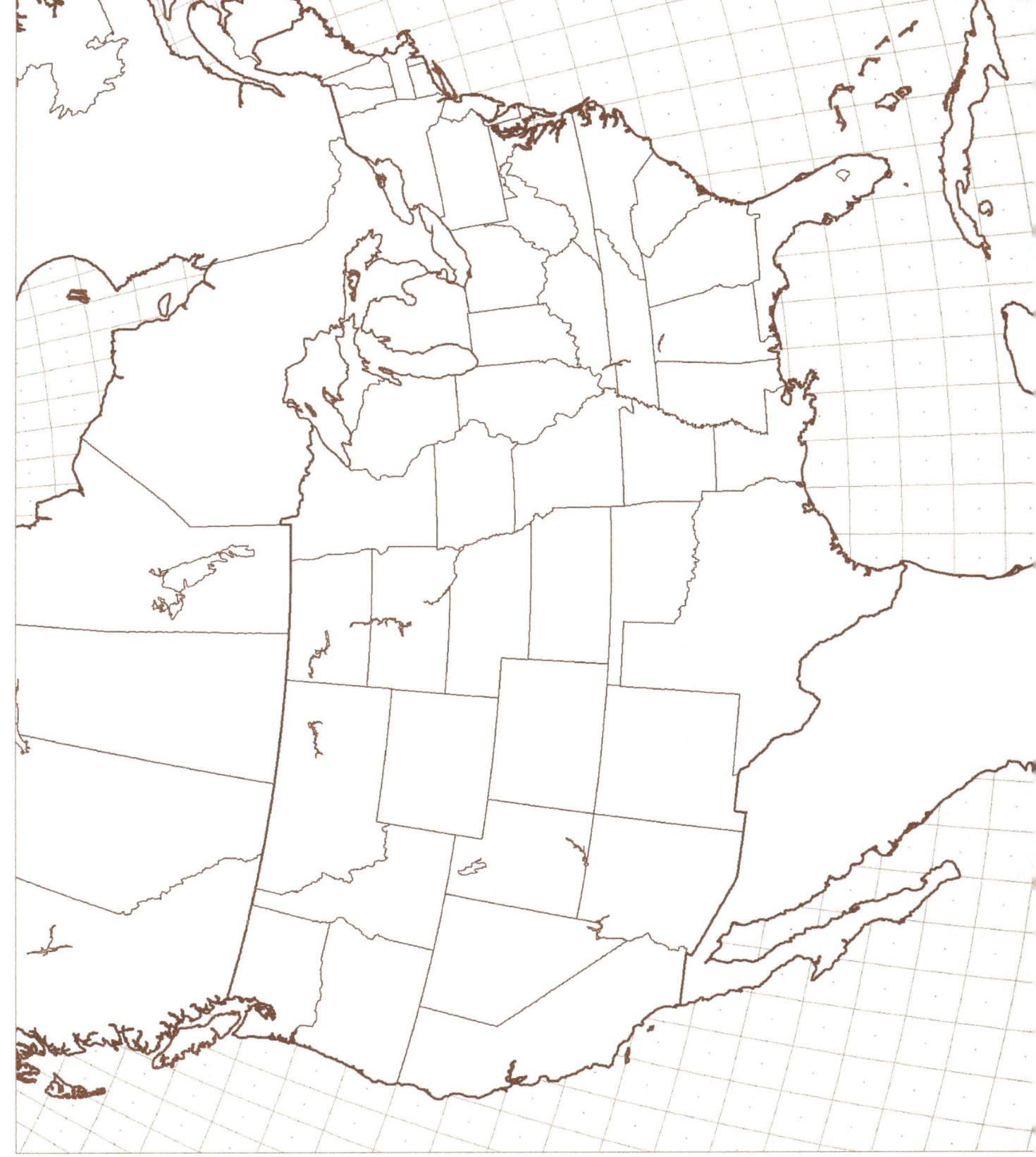

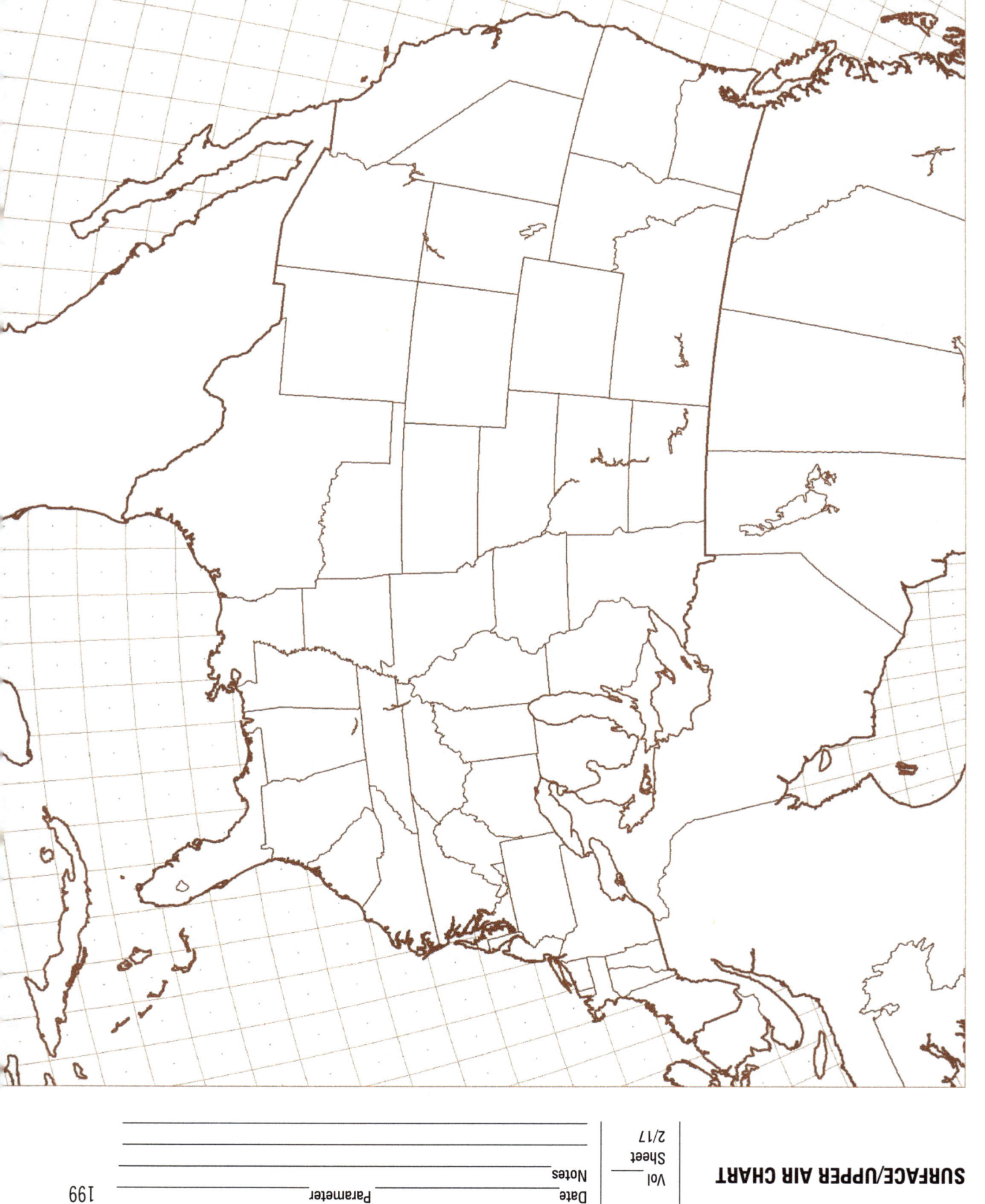

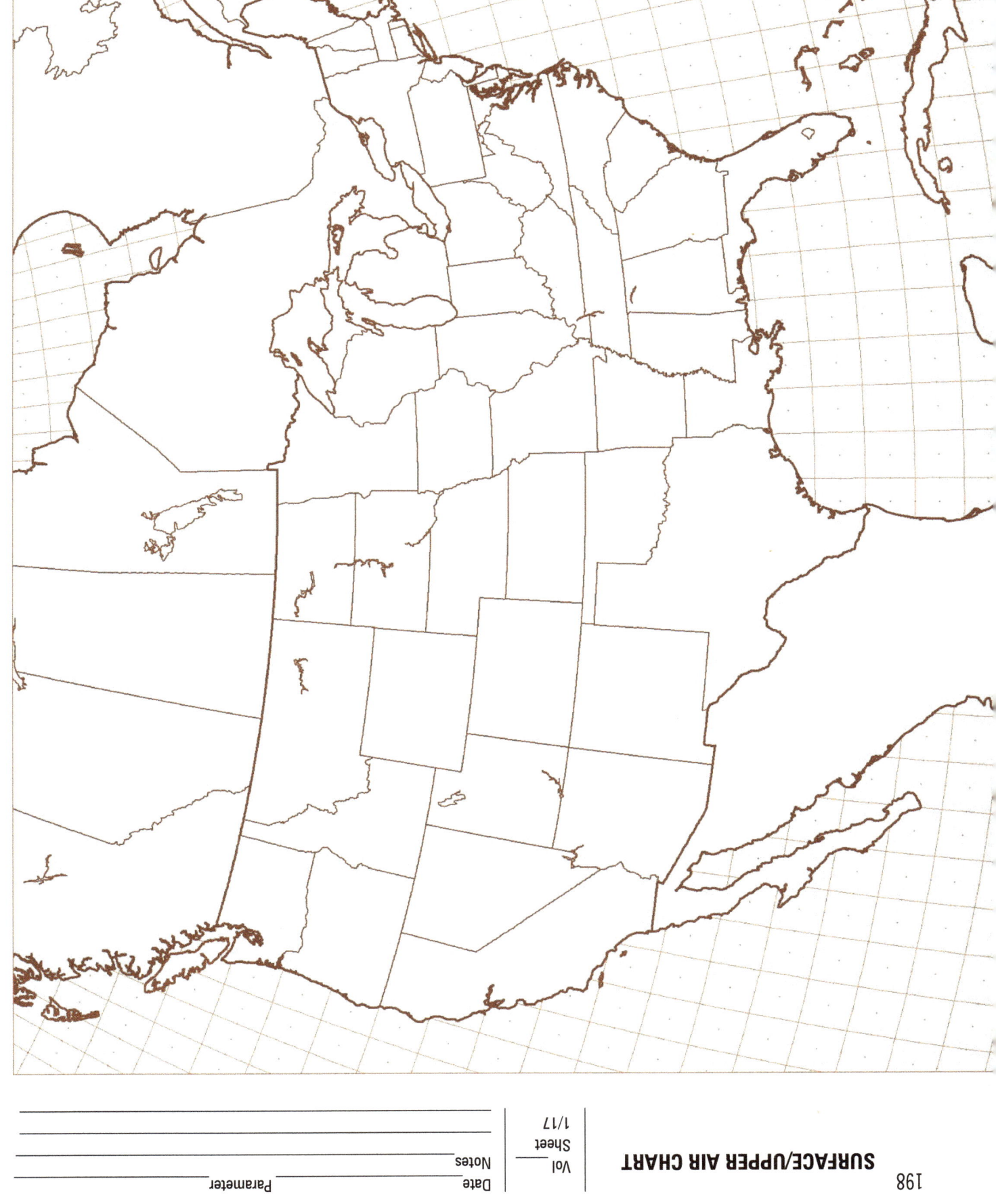

HORIZONTAL CHARTS

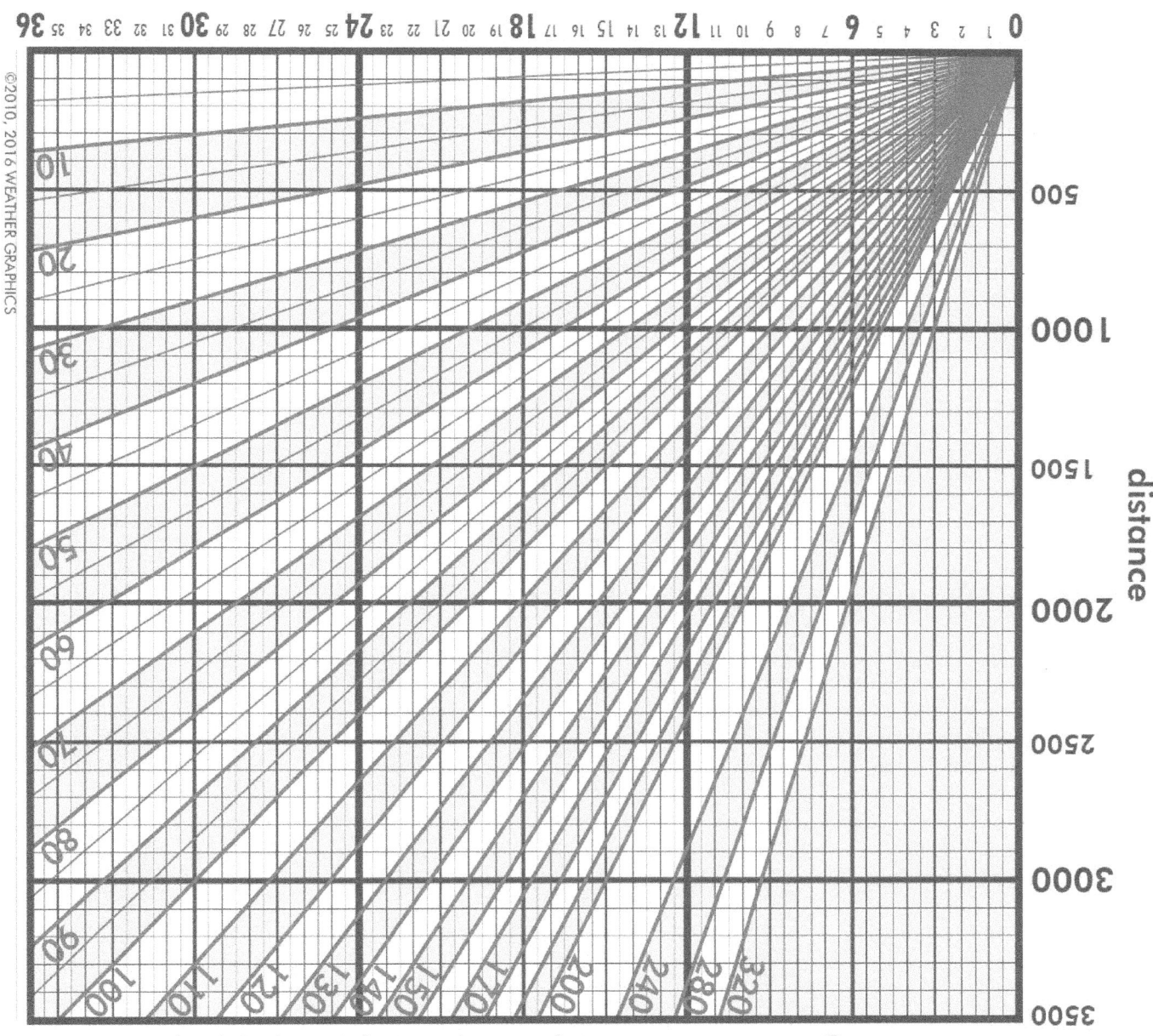

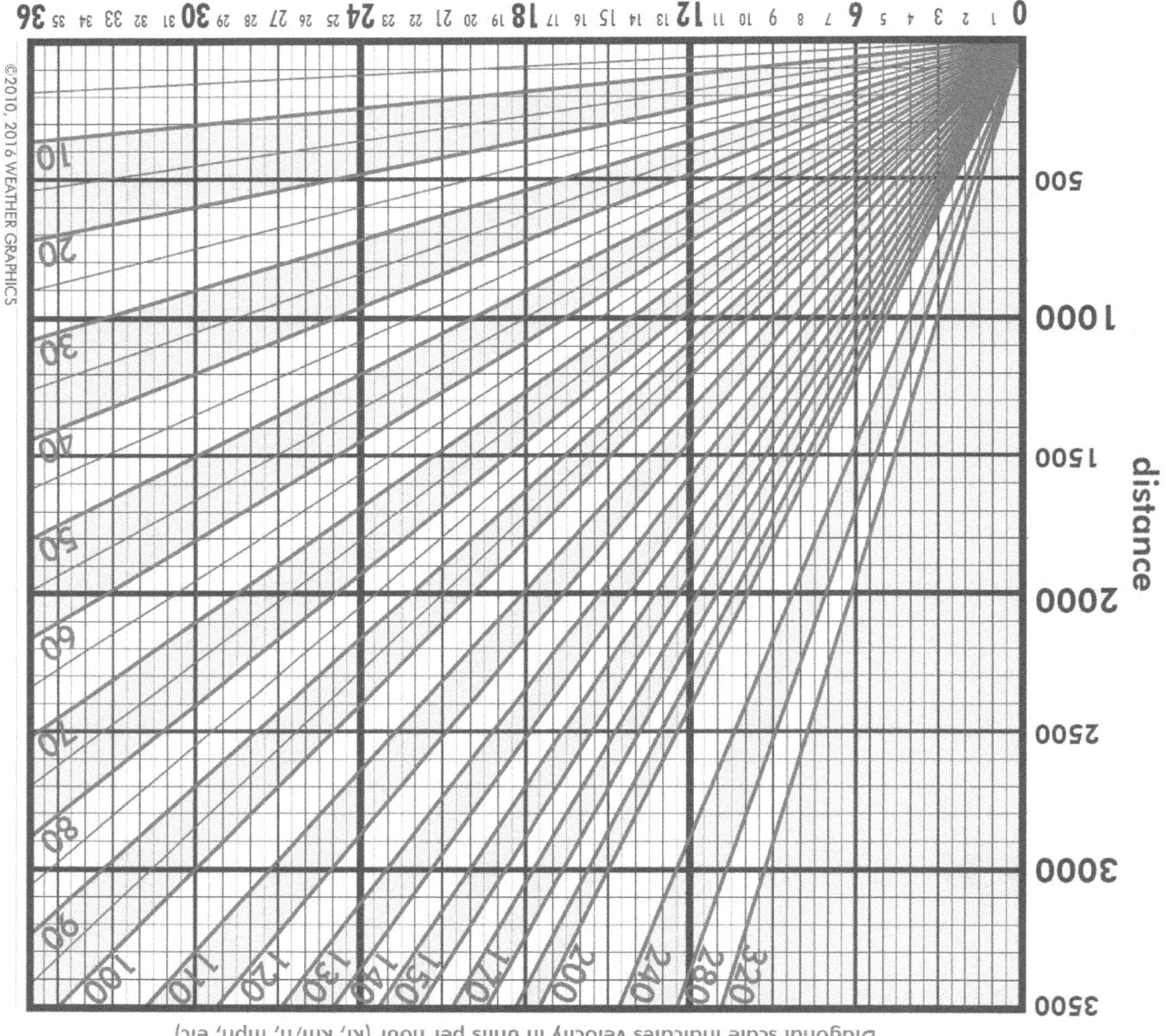

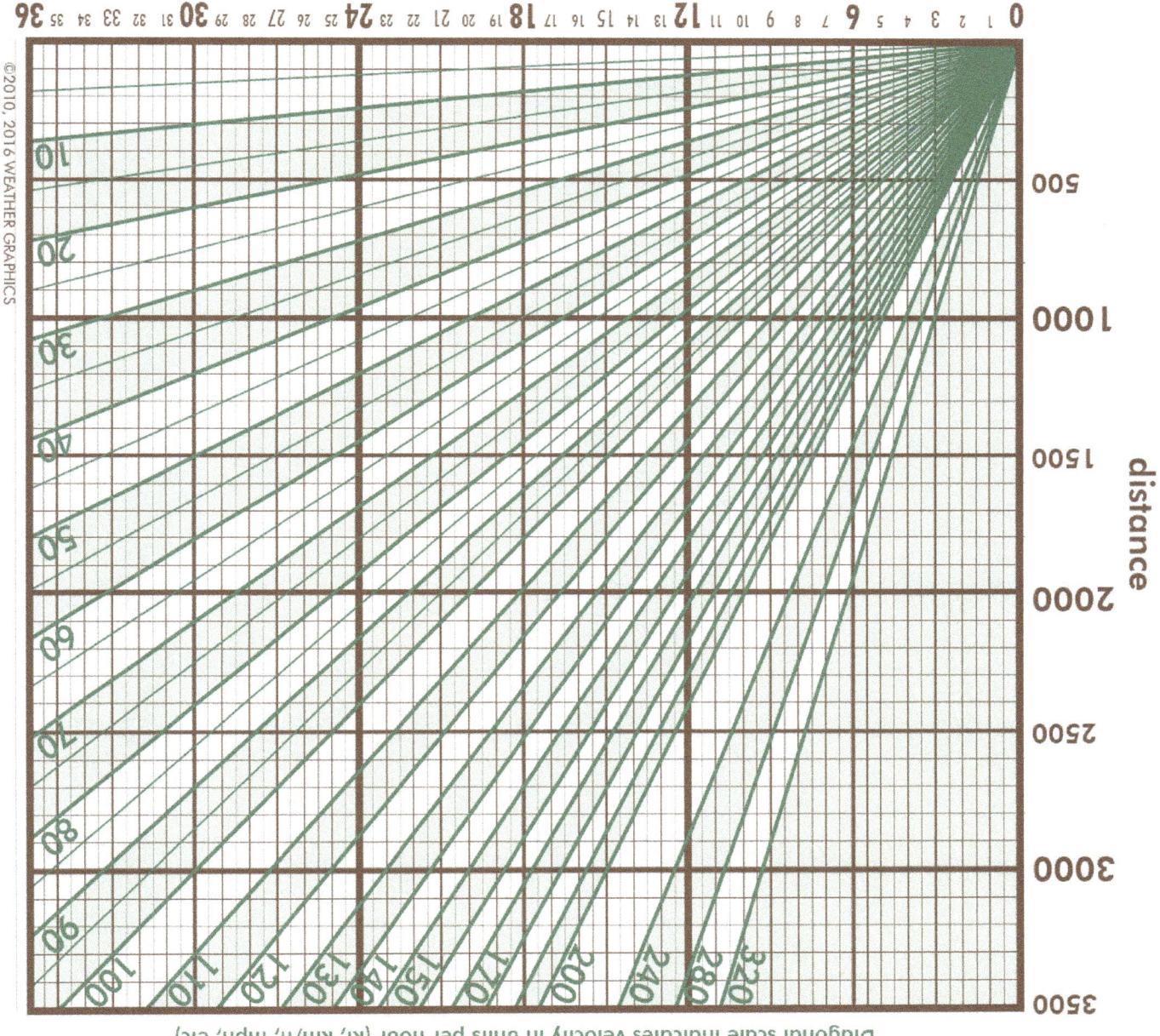

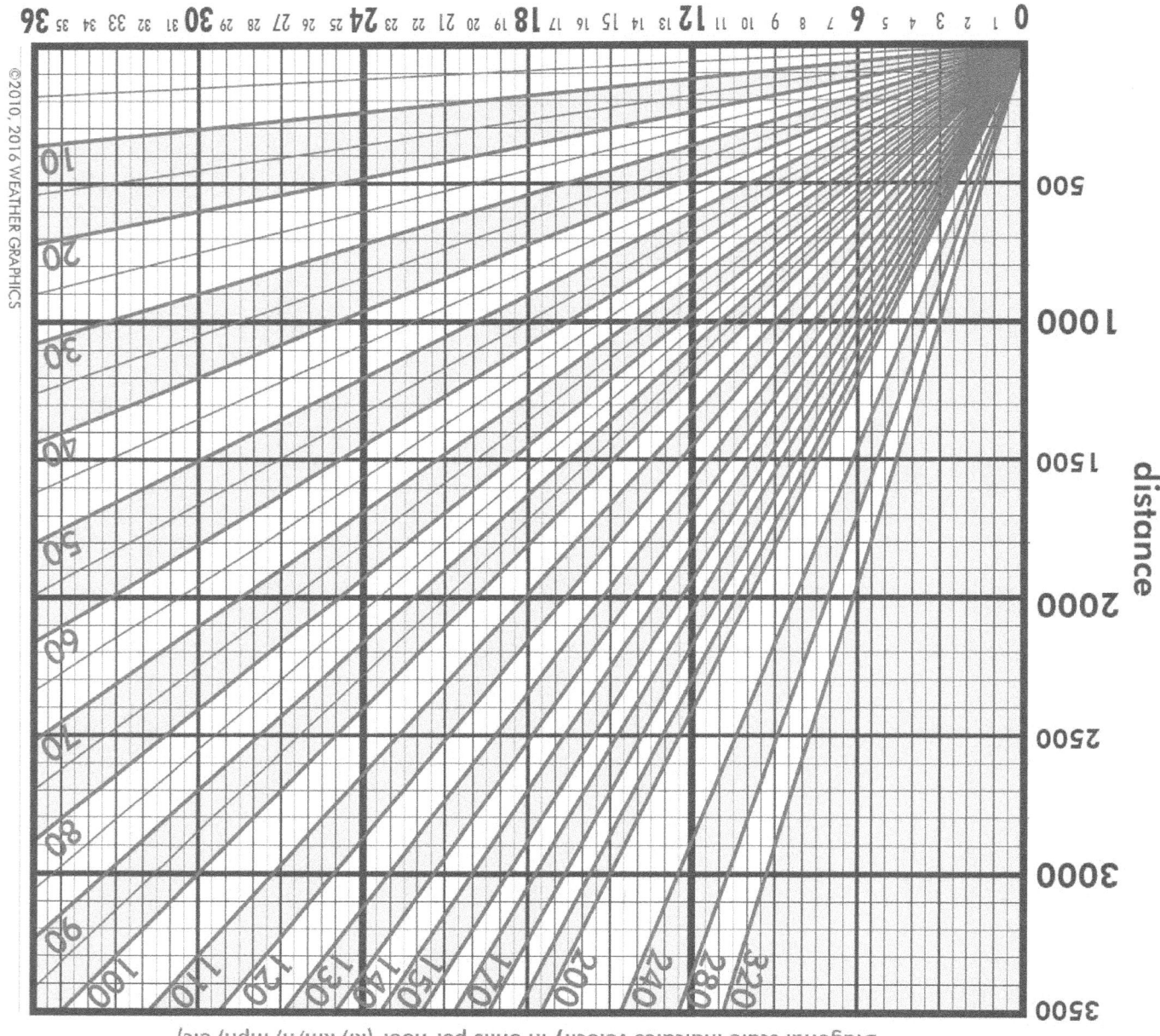

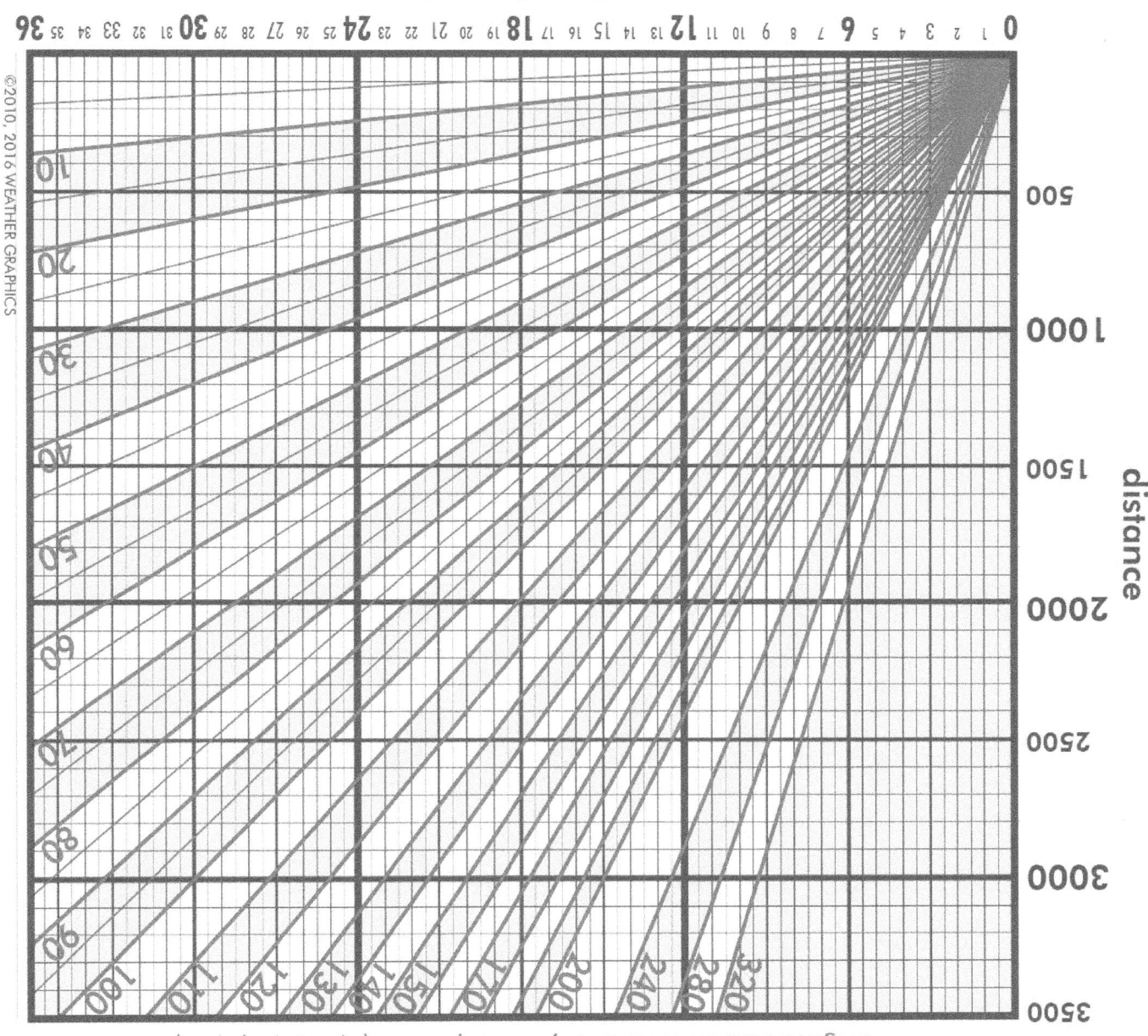

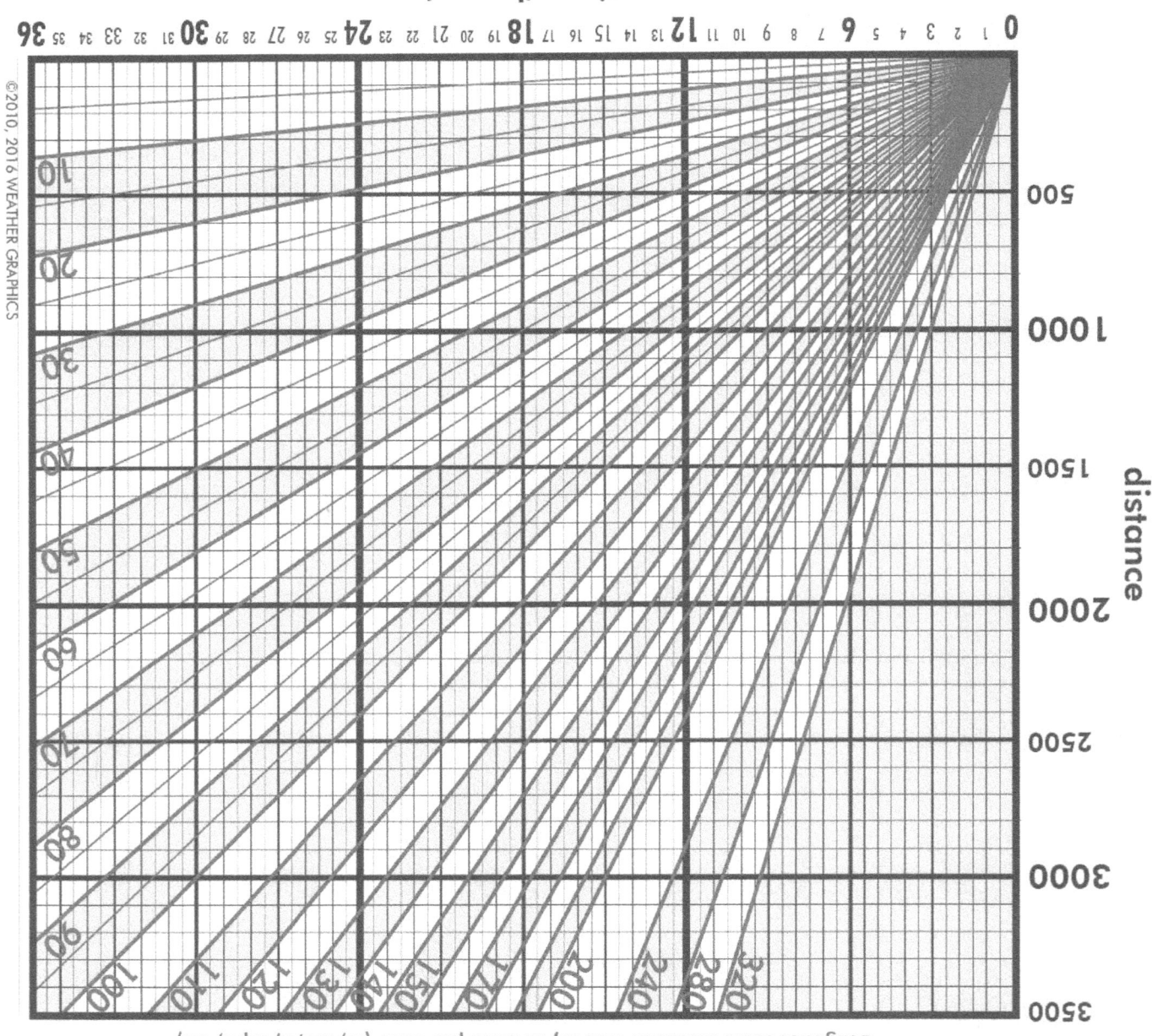

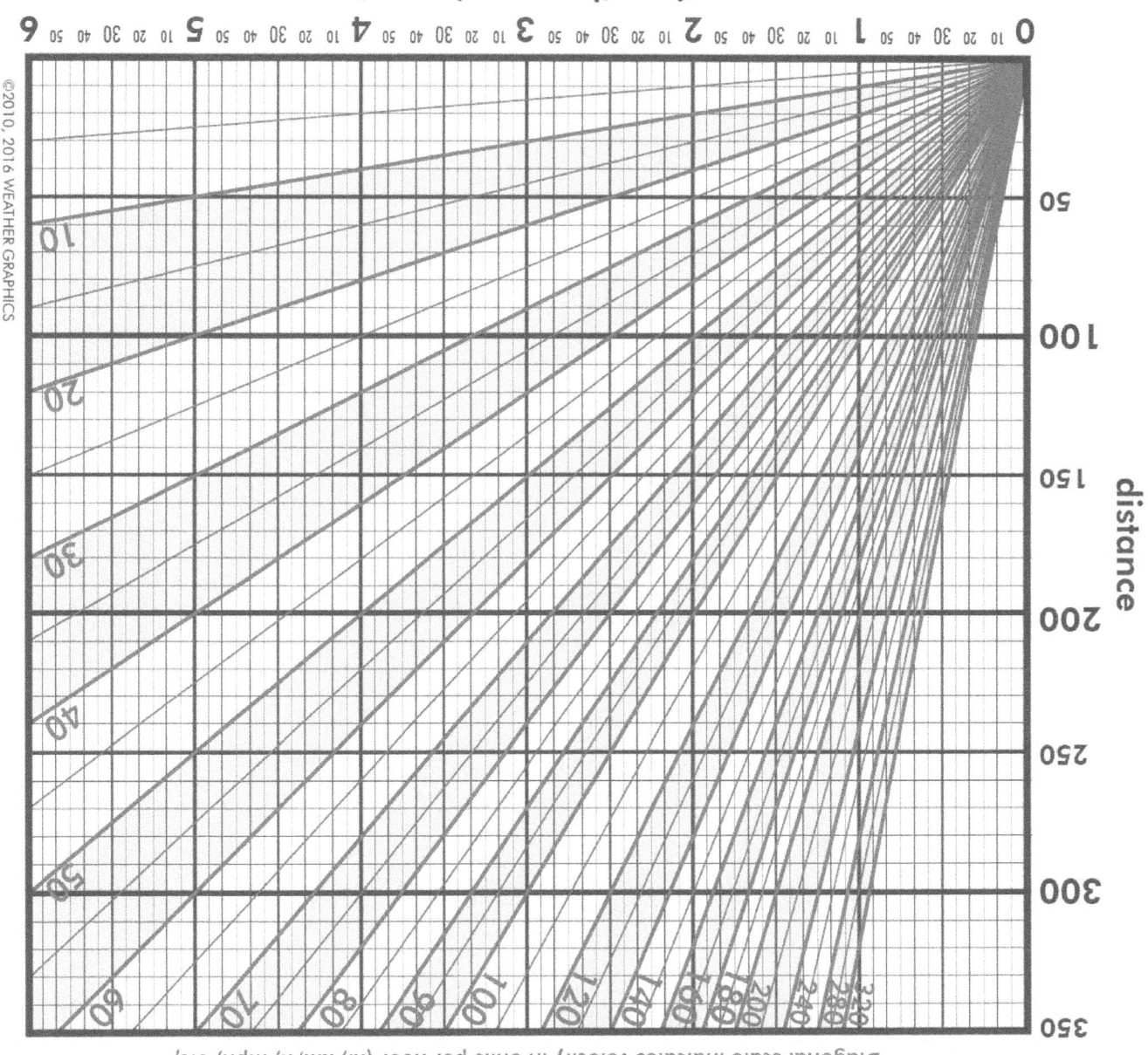

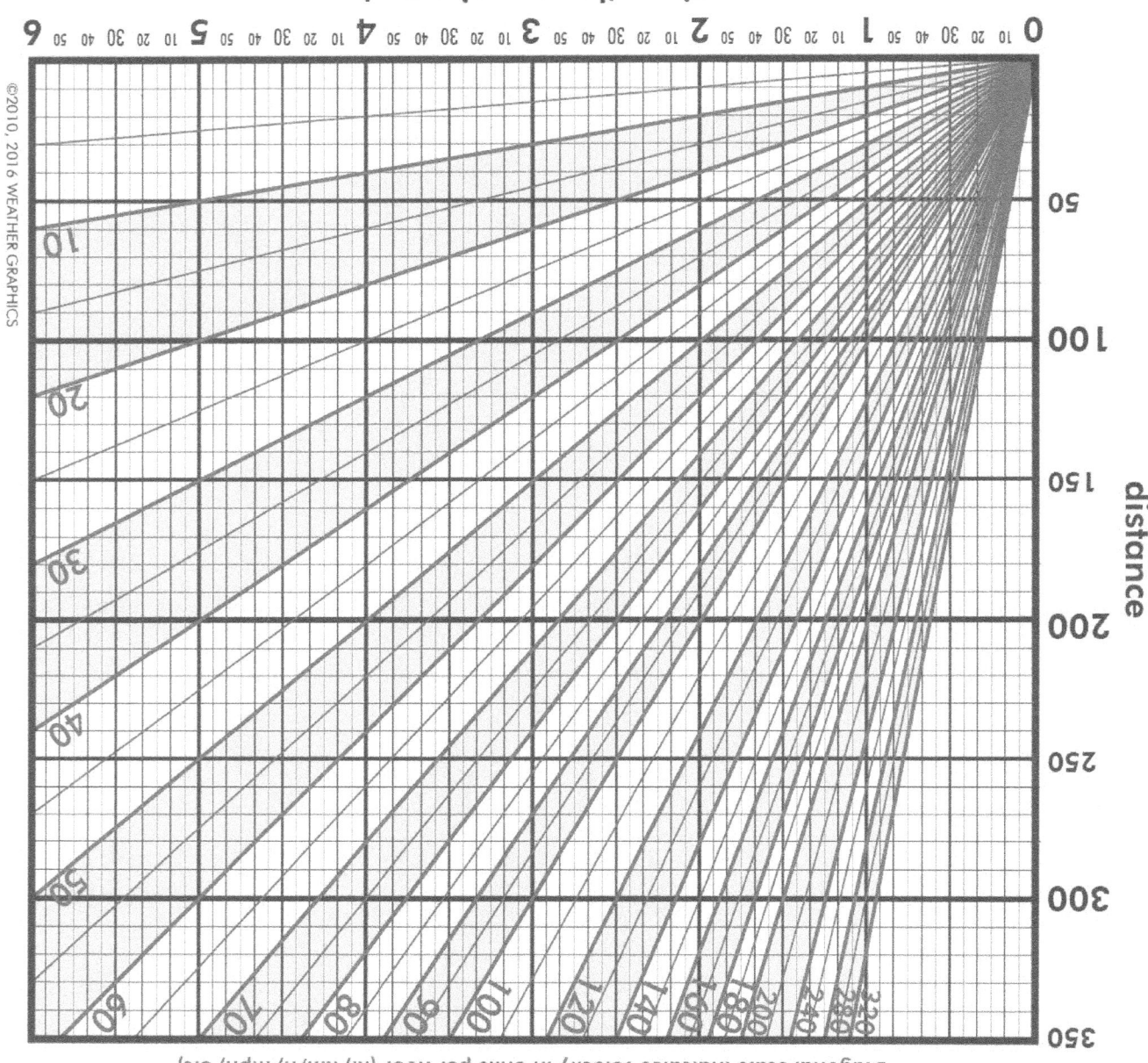

MOTION CALCULATION

Meso-β scale

Vol ____
Sheet
4/6

Date ____________ Parameter ____________
Notes ____________

FORECAST MOTION NOMOGRAM

BETA MESOSCALE

Diagonal scale indicates velocity in units per hour (kt, km/h, mph, etc)

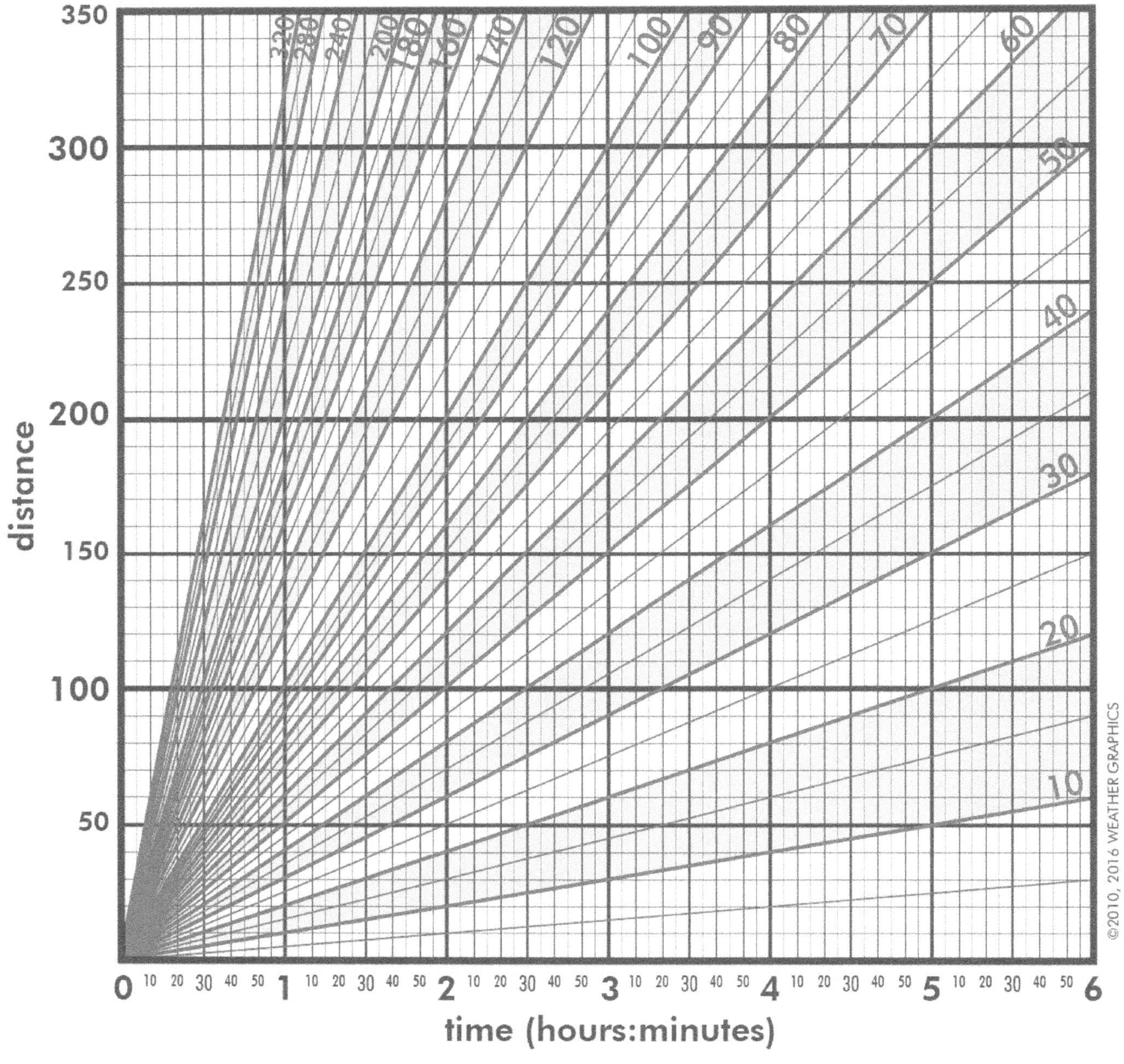

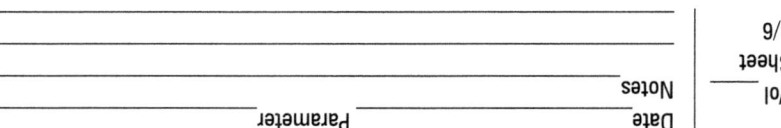

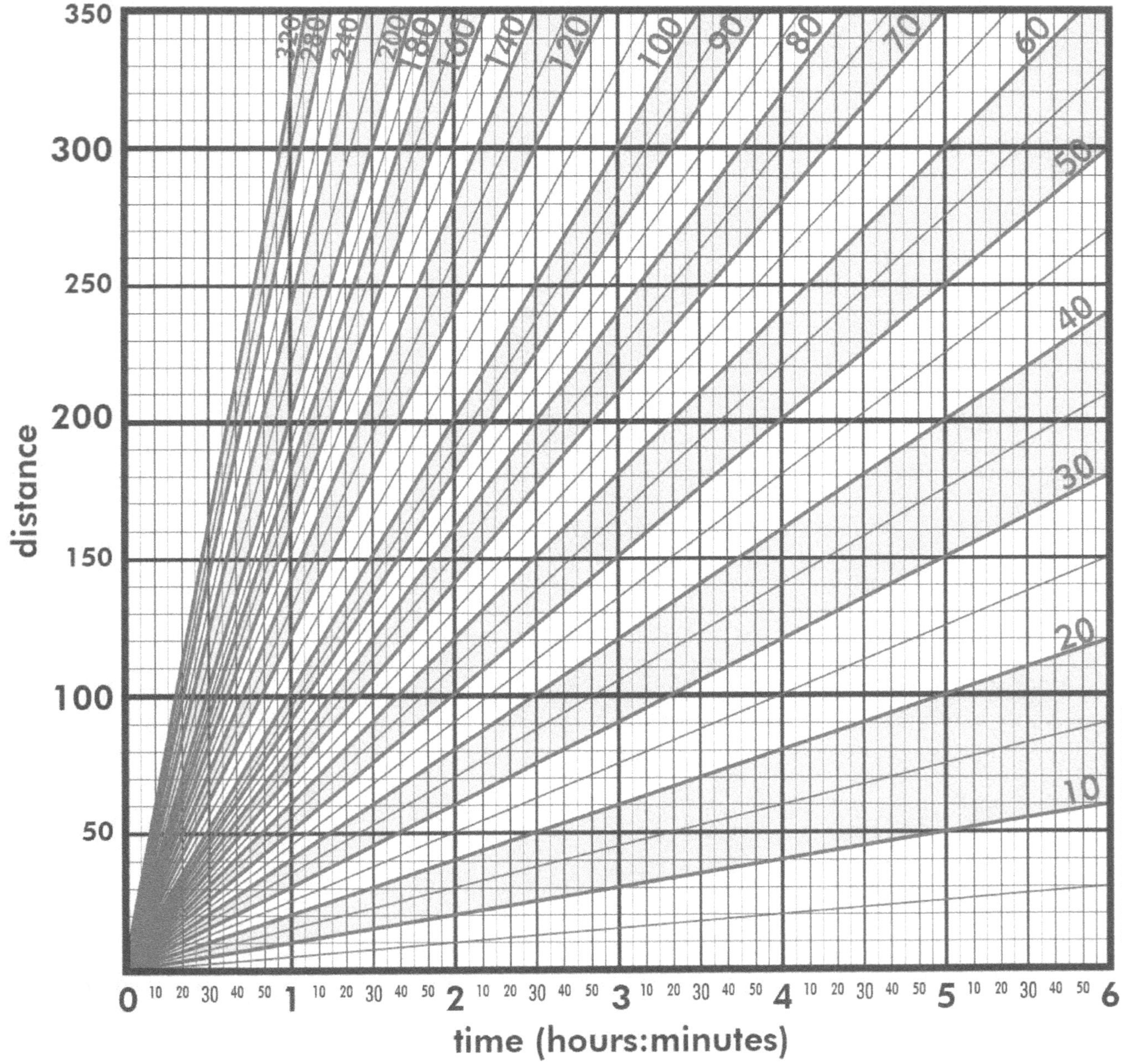

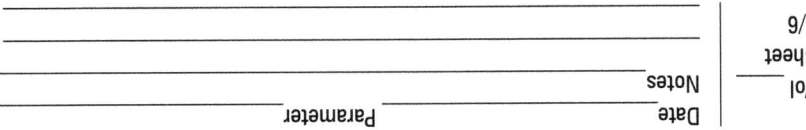

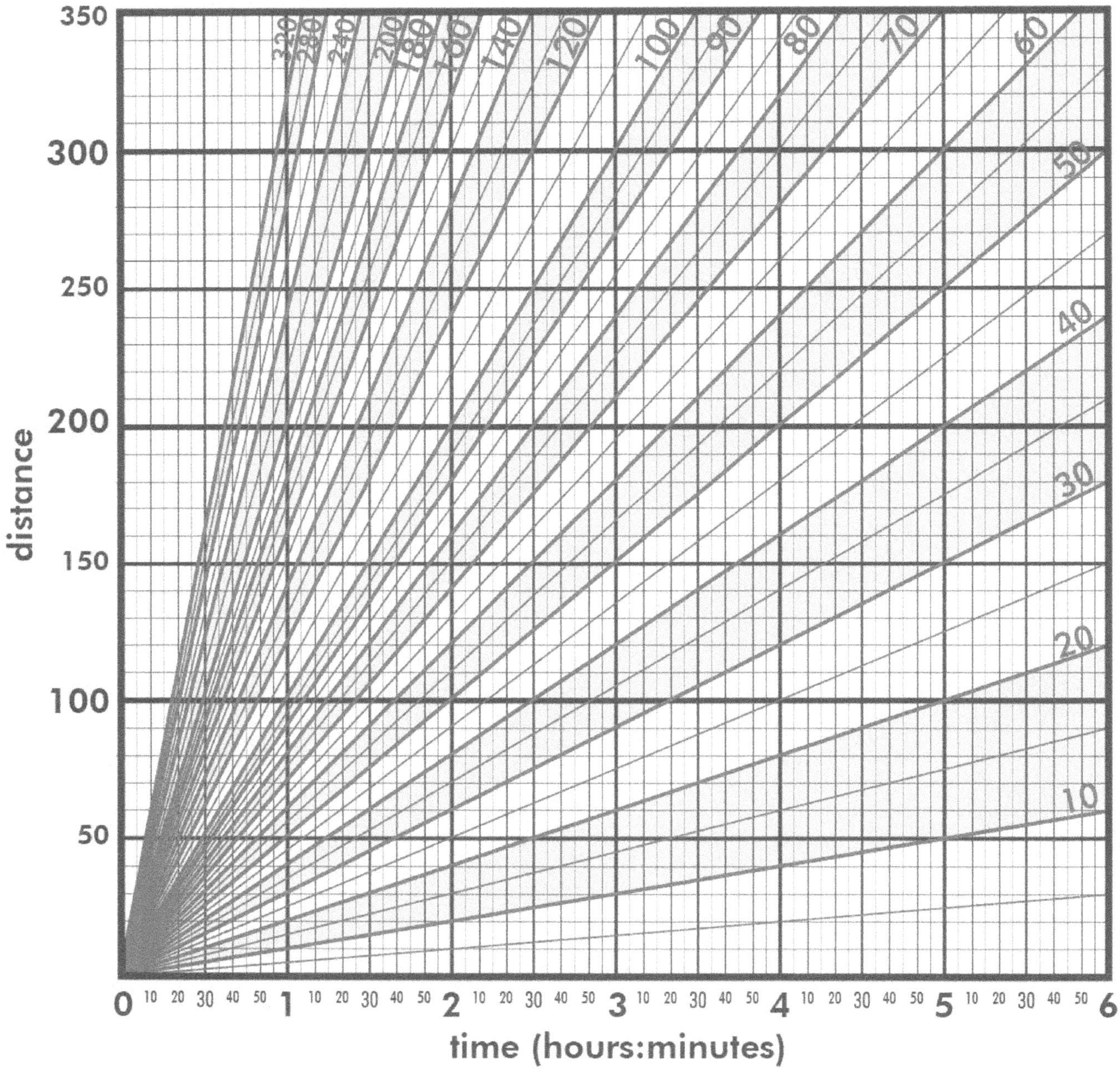

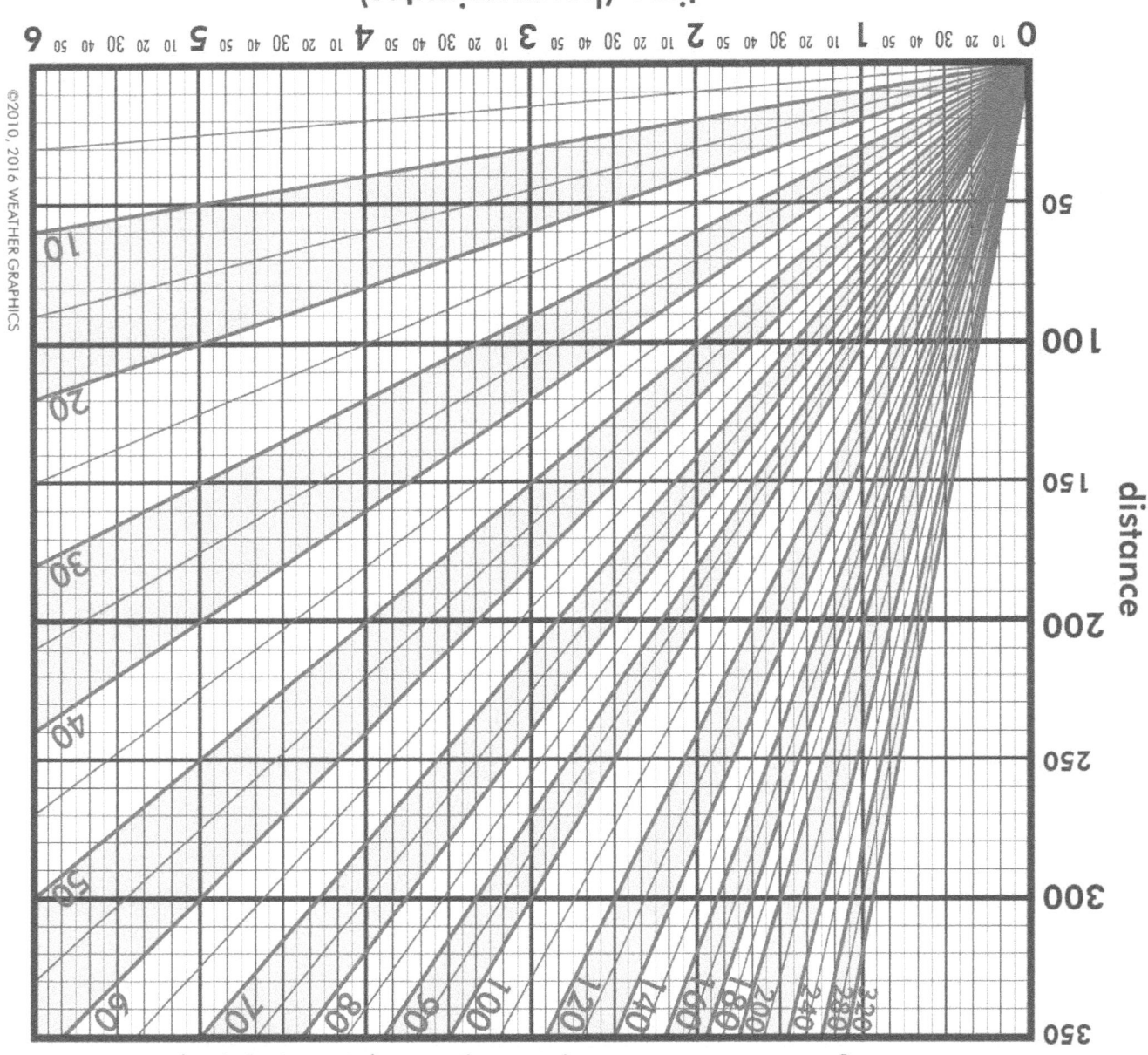

MOTION CALCULATION
Meso-γ scale

Vol ____
Sheet
8/8

Date ______________ Parameter ______________
Notes ______________

FORECAST MOTION NOMOGRAM

GAMMA MESOSCALE

Diagonal scale indicates velocity in units per hour (kt, km/h, mph, etc)

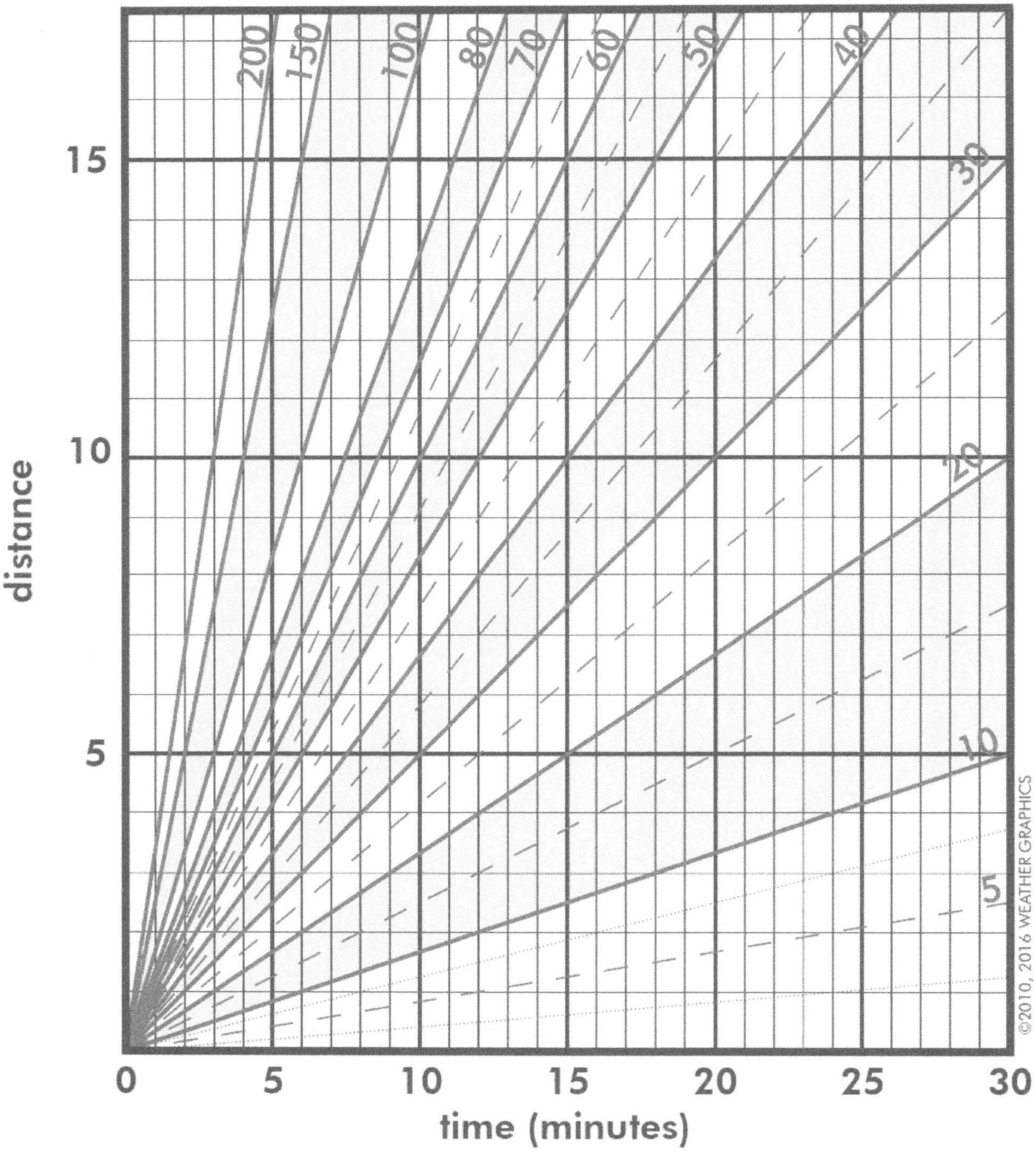

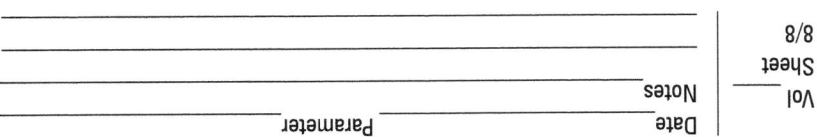

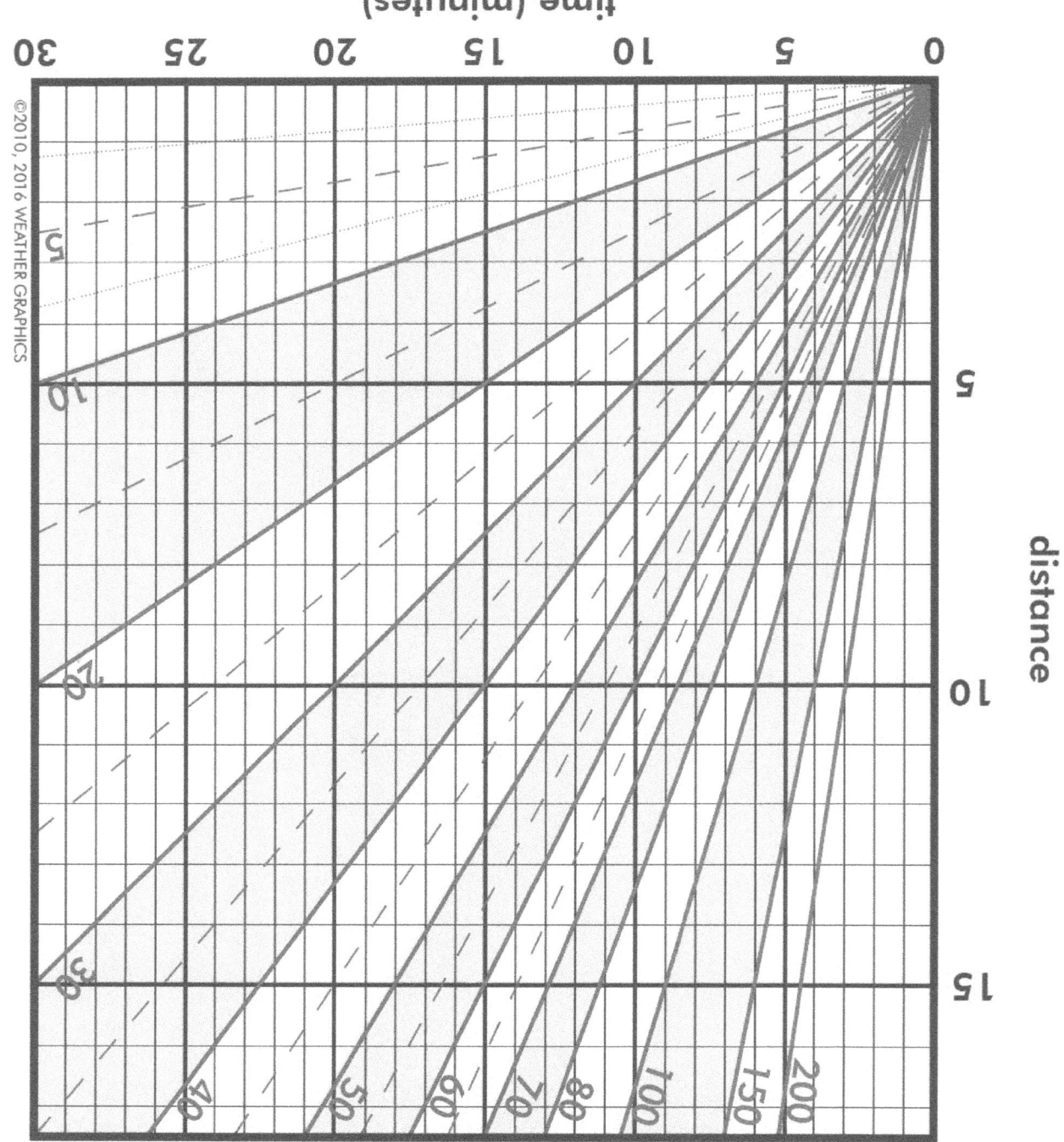

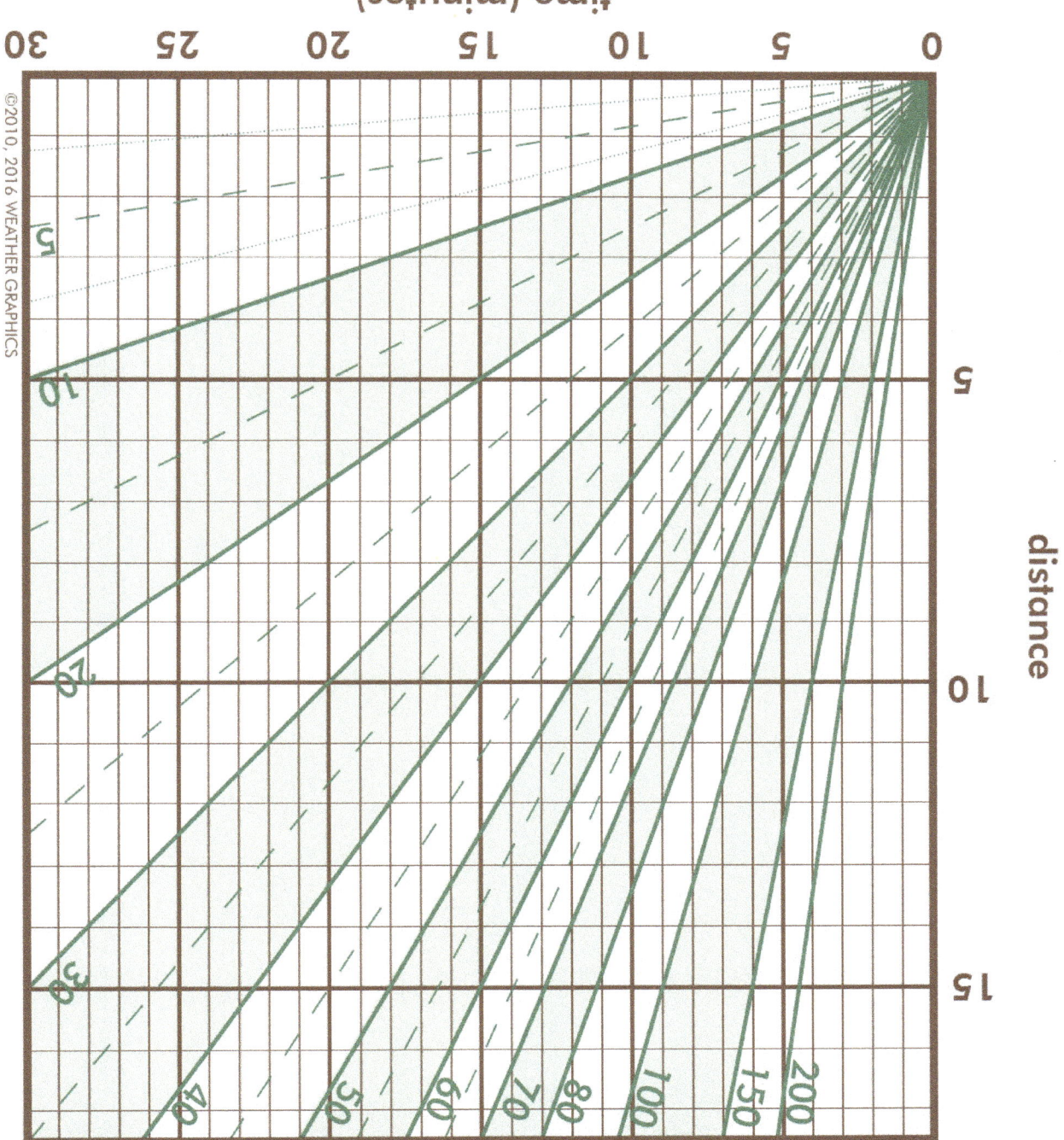

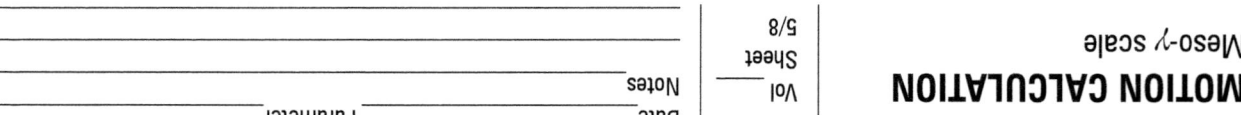

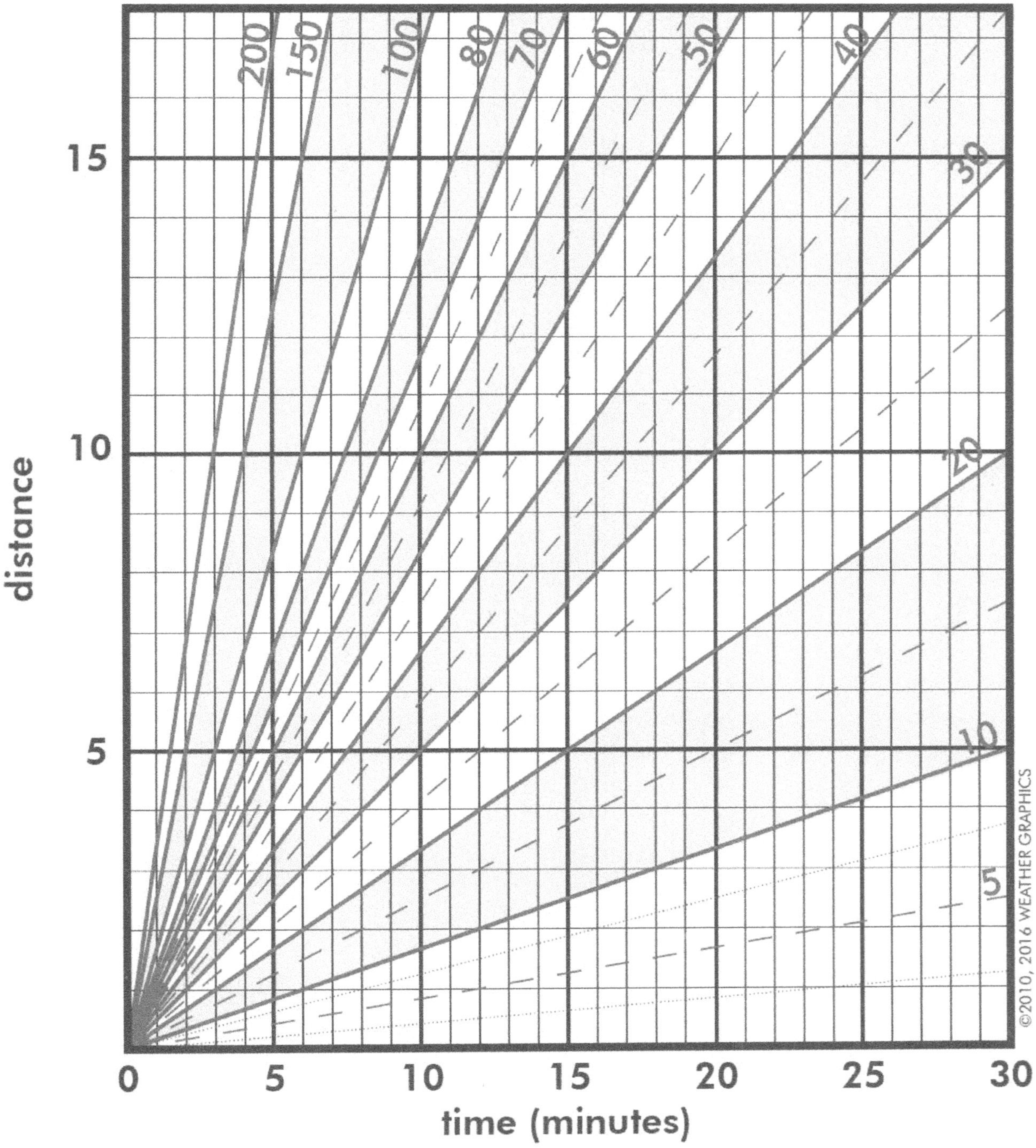

MOTION CALCULATION

Meso-γ scale

Vol ____
Sheet
4/8

Date ______________ Parameter ______________
Notes ______________

FORECAST MOTION NOMOGRAM

GAMMA MESOSCALE

Diagonal scale indicates velocity in units per hour (kt, km/h, mph, etc)

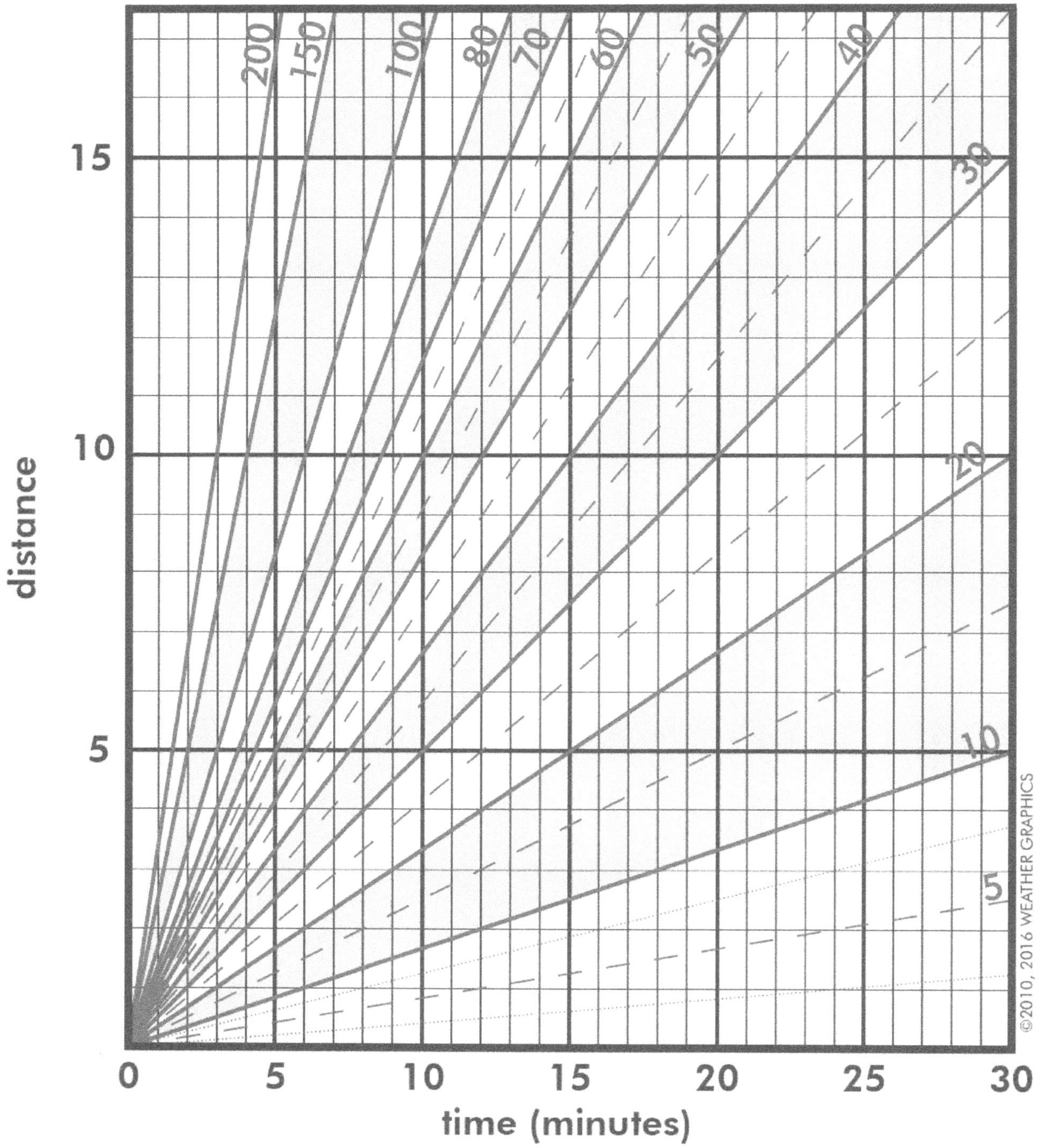

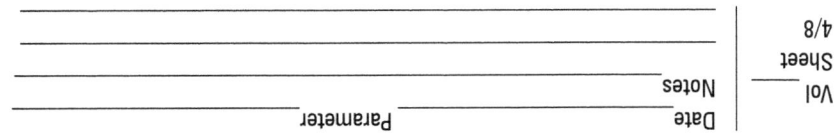

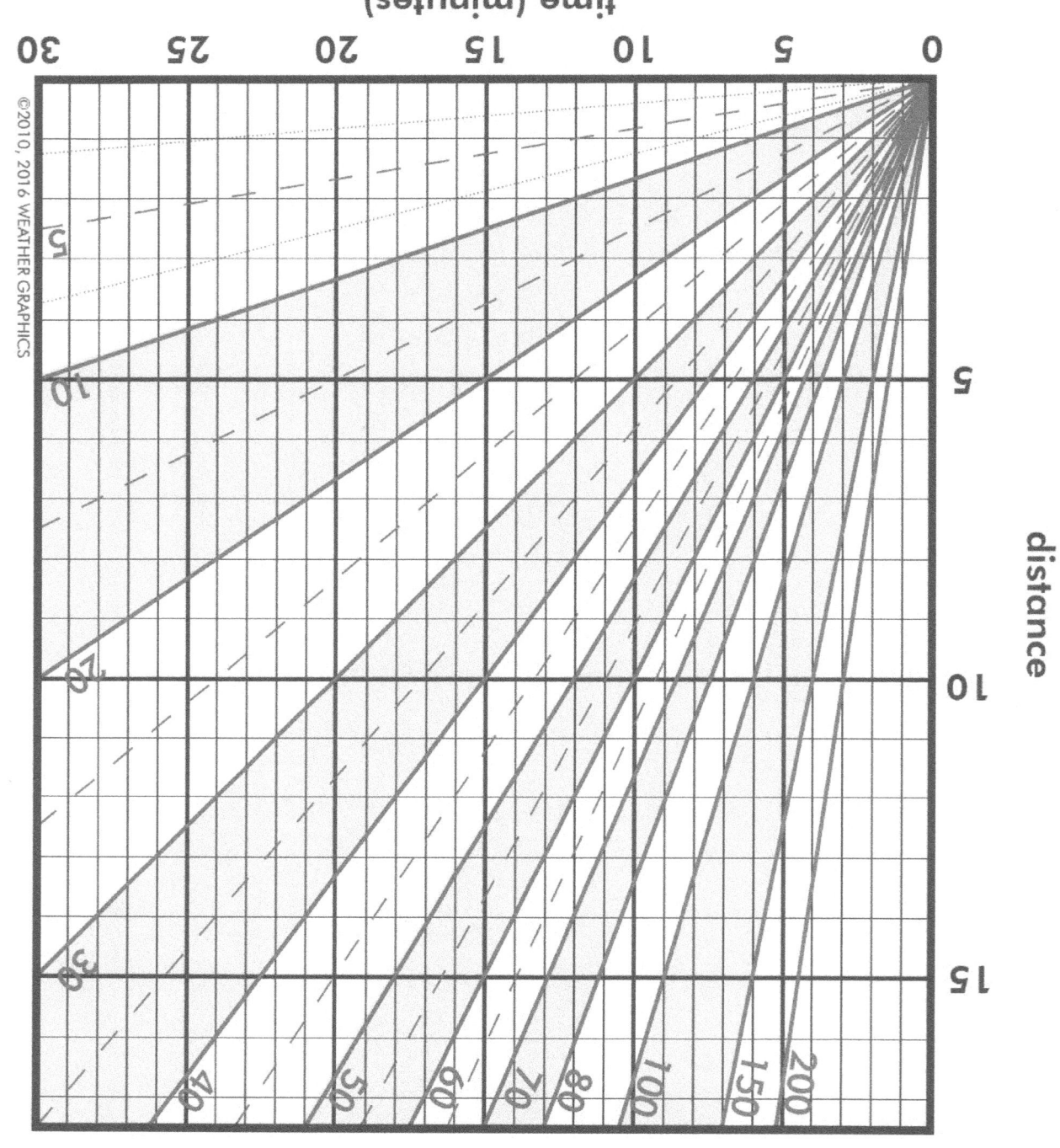

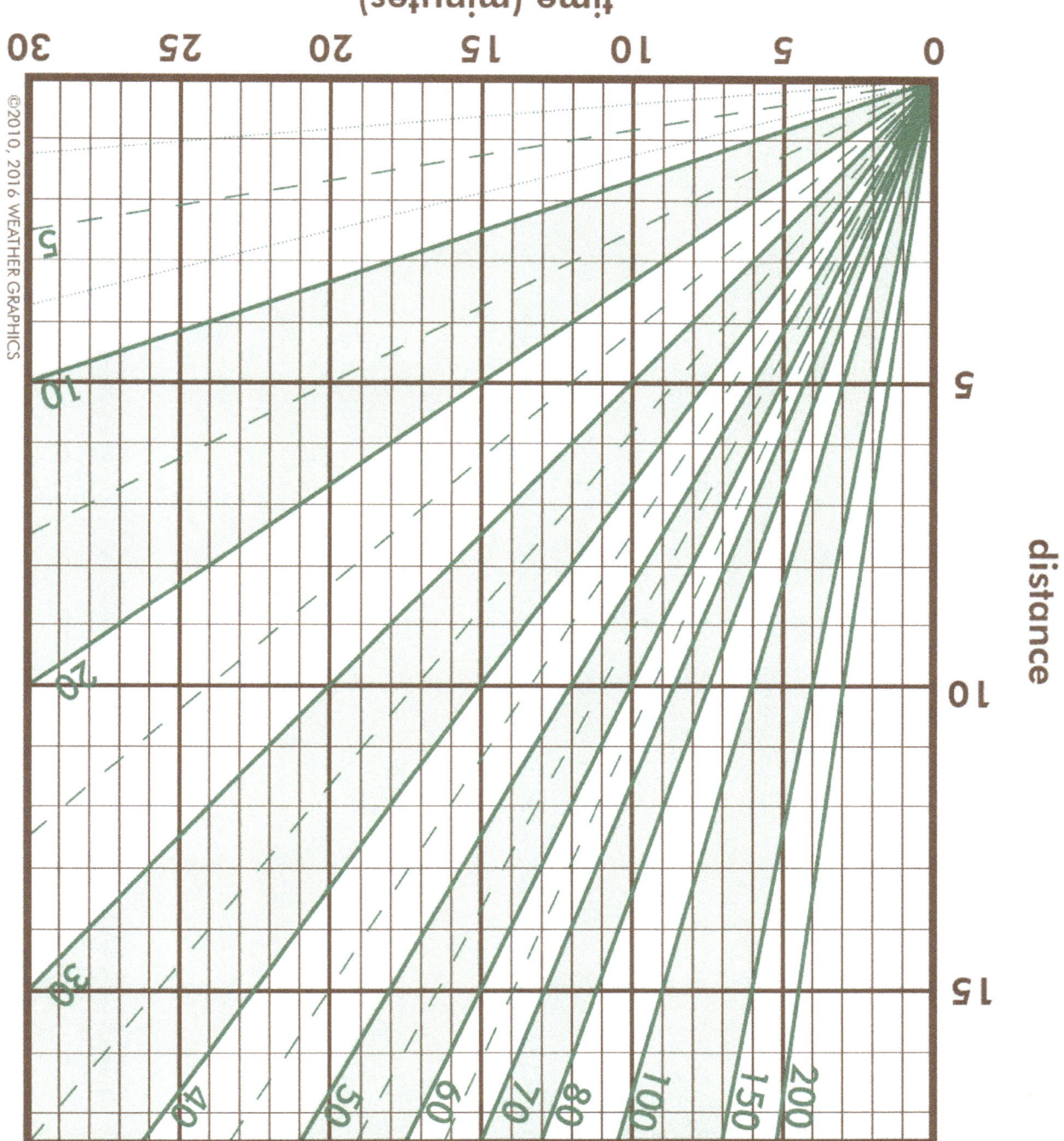

MOTION CALCULATION
Meso-γ scale

Vol ____
Sheet
1/8

Date________________ Parameter________________
Notes________________

FORECAST MOTION NOMOGRAM

GAMMA MESOSCALE

Diagonal scale indicates velocity in units per hour (kt, km/h, mph, etc)

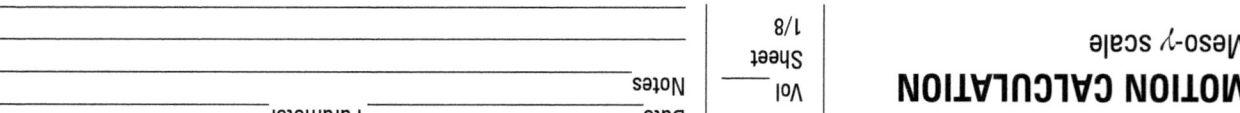

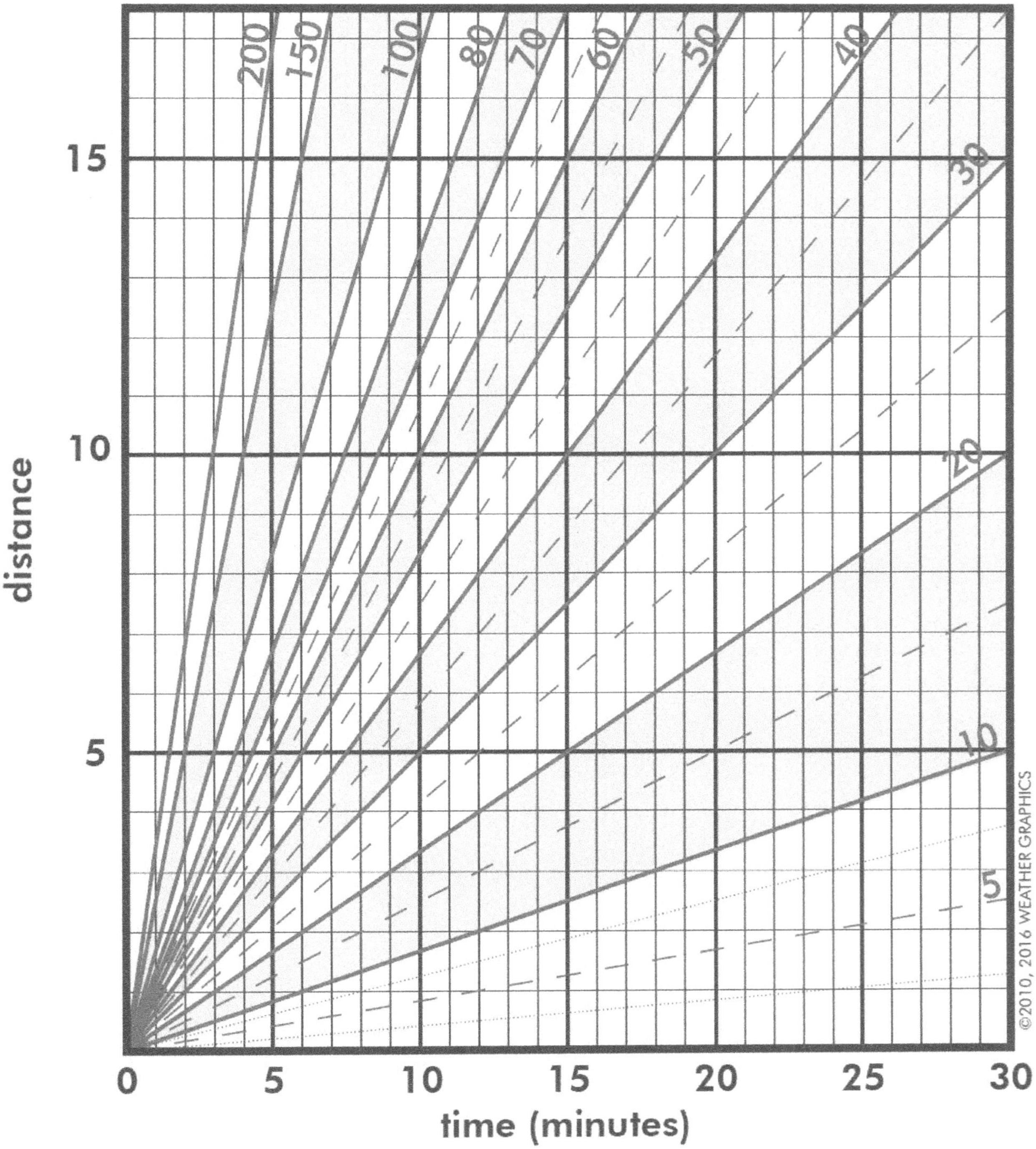

These motion calculation charts are intended for quickly assessing velocity and arrival times of weather systems. Some common uses of this diagram:

* Establishing the forward velocity of a tornado, outflow boundary, or front based on elapsed time between radar volume scans and measured distance (in GRLevelX and other programs)

* Forecasting arrival time of these features based on a known velocity (estimated from this chart or from storm data tables) and measured distance to the forecast target.

Although some software programs will calculate some of these values for you, others don't, particularly when doing satellite estimates with tools such as McIDAS, so having manual tables available will expedite the forecast process.

To determine velocity
Using the distance covered (left scale) and the elapsed time (bottom scale), read the diagonal lines to get velocity. If the distance was measured in nm (1 nm = 6076 ft), the result is kt. If it was measured in statute miles (1 sm = 5280 ft), the result is mph. If it was measured in km, the result is kph.

To determine time to a target or leadtime
Find the known velocity the diagonal scale and determine where it intersects the distance to the forecast target (left scale). From that point, read the elapsed time on the bottom scale.

Chart type	Suggested uses
Meso-α (alpha) scale	Regional fronts, hurricanes
Meso-β (beta) scale	Squall lines, local fronts, hurricanes
Meso-γ (gamma) scale	Outflow boundaries, tornadoes, velocity couplets

MOTION CALCULATION

HODOGRAPH

Vol ____
Sheet
32/32

Date/time ______________ Station ______________
Notes ______________

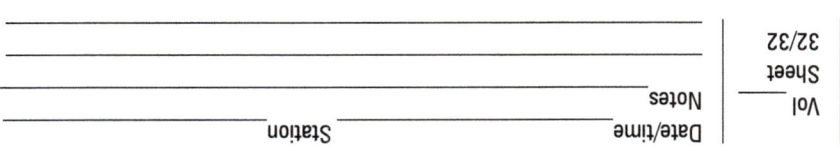

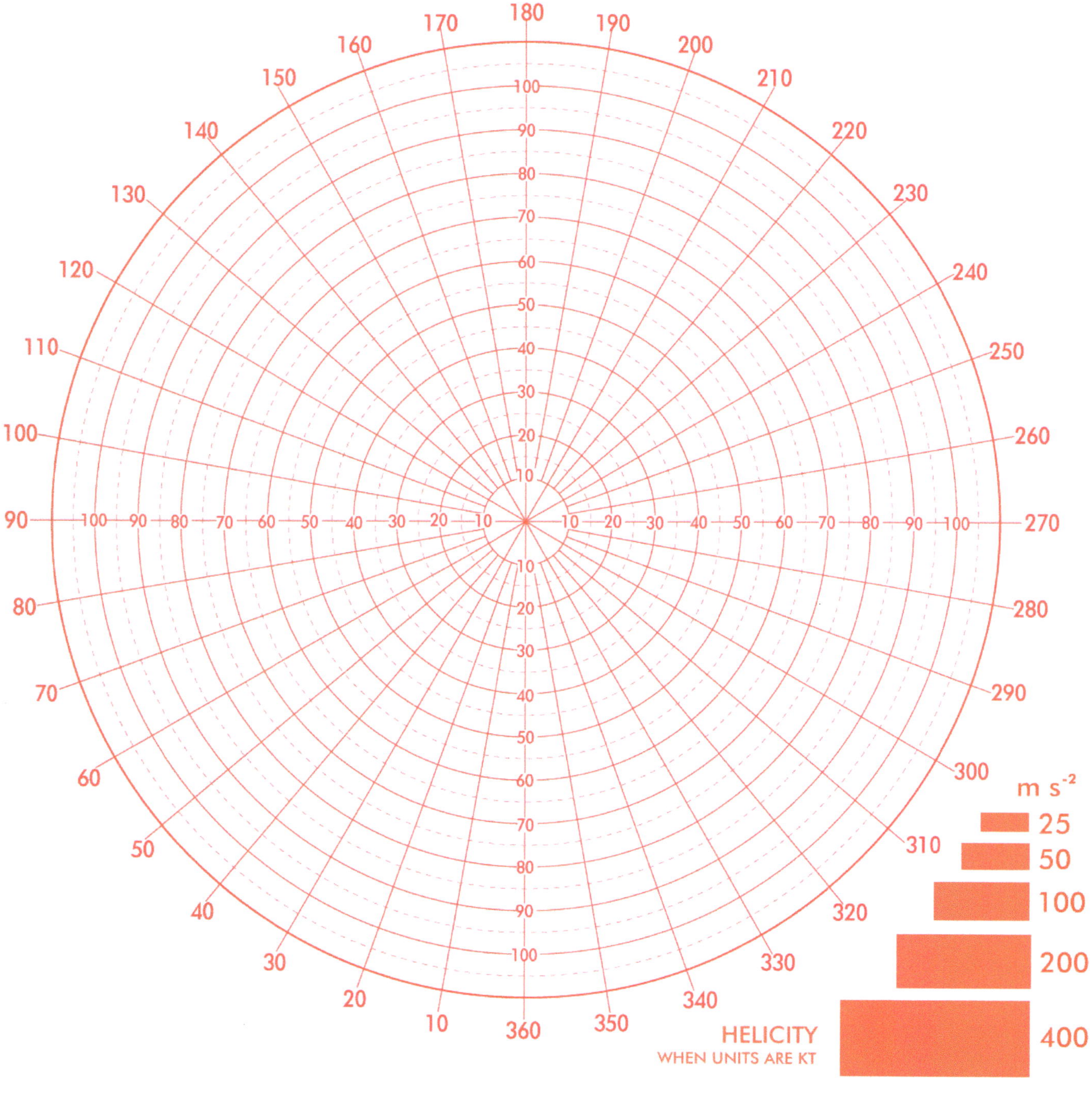

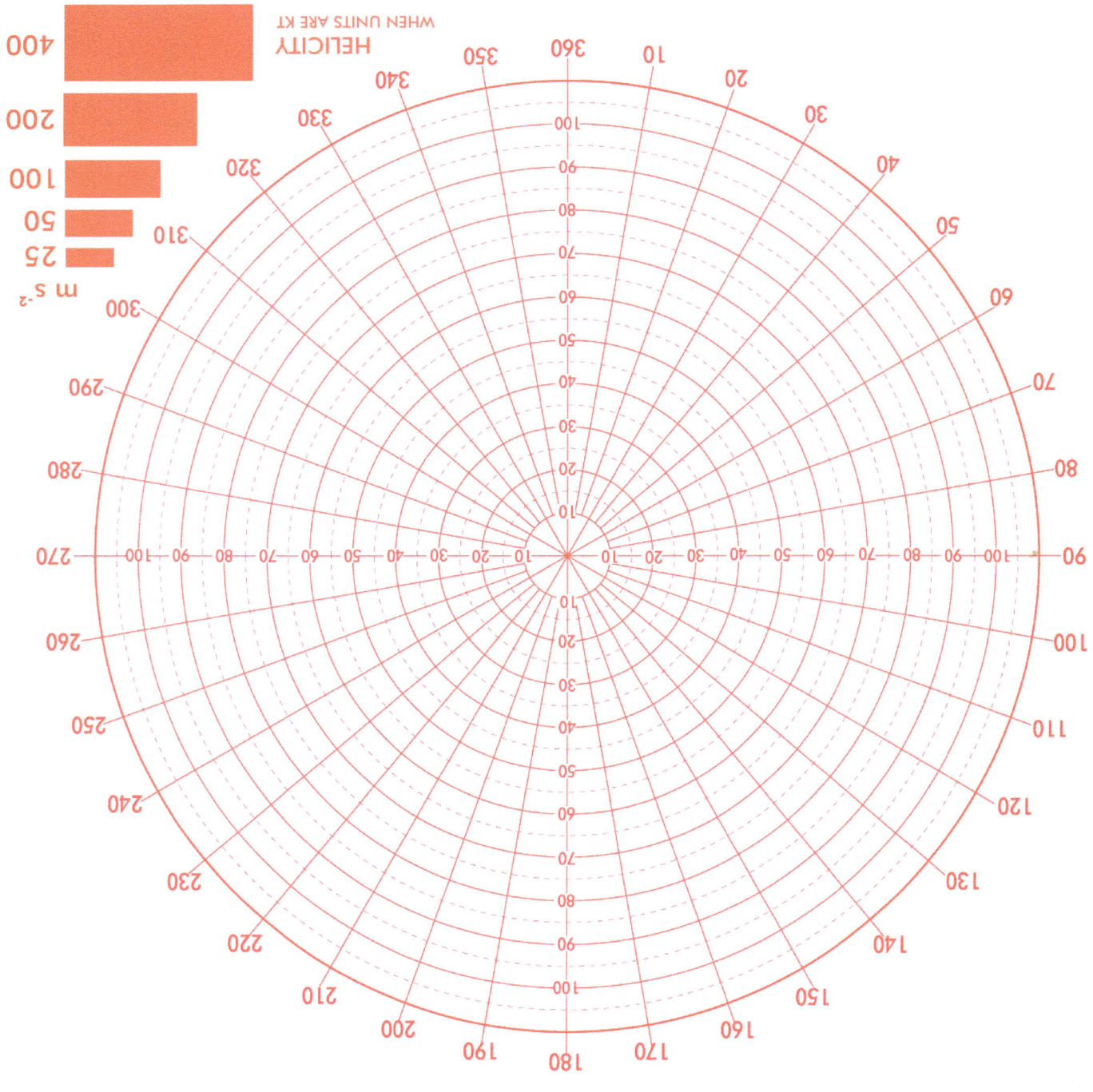

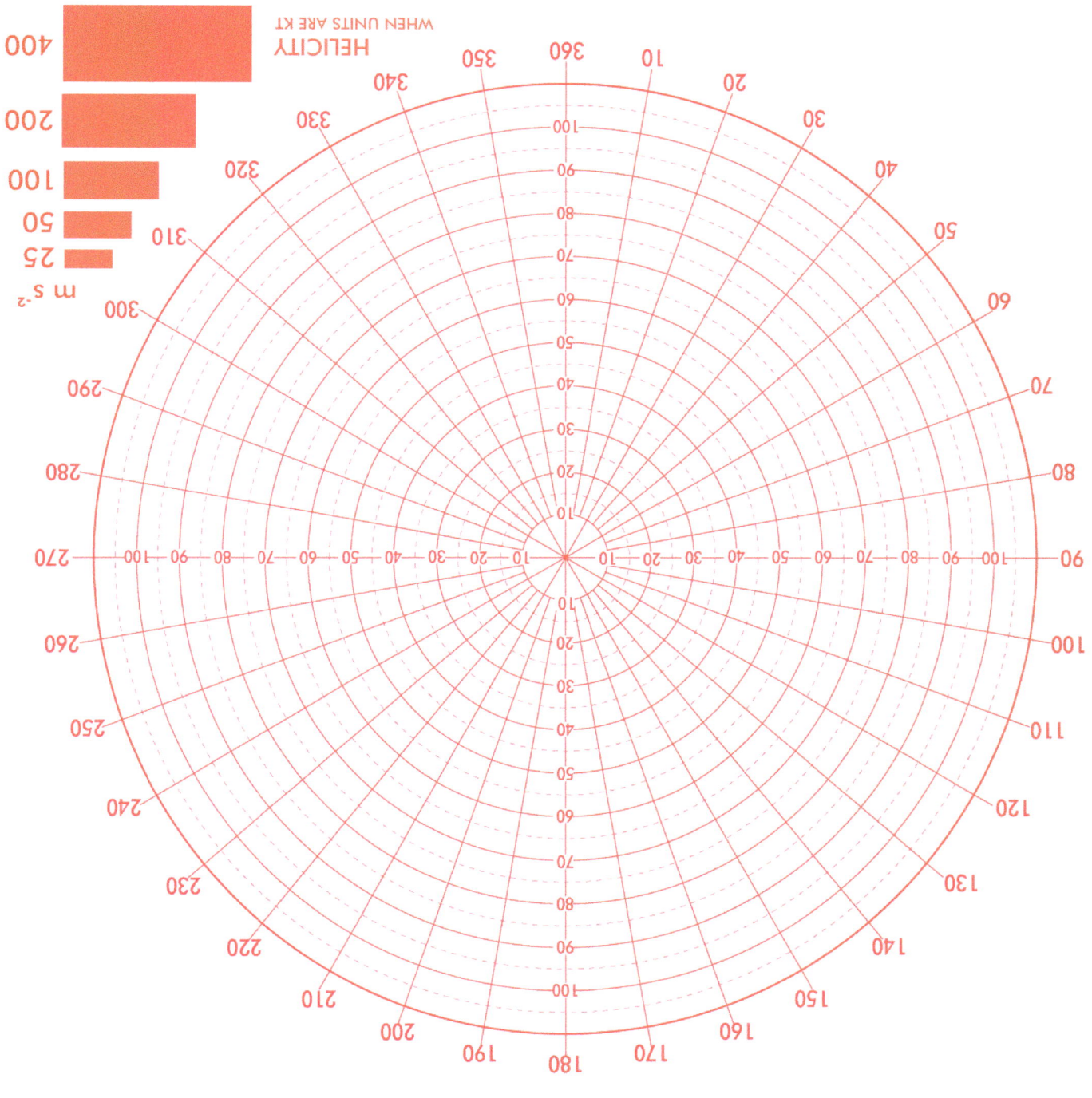

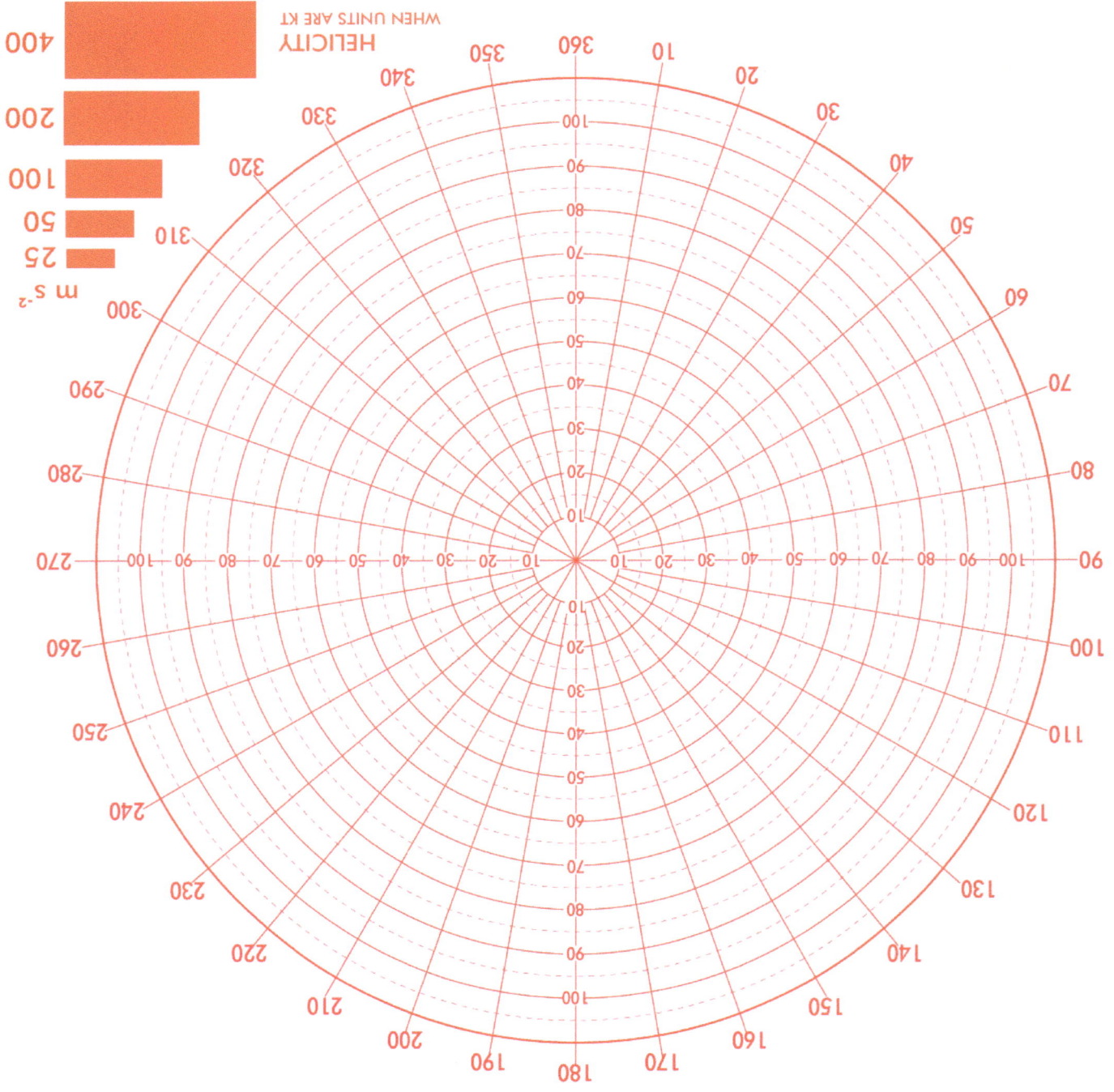

HODOGRAPH

Vol ____
Sheet
28/32

Date/time ____________ Station ____________
Notes ____________

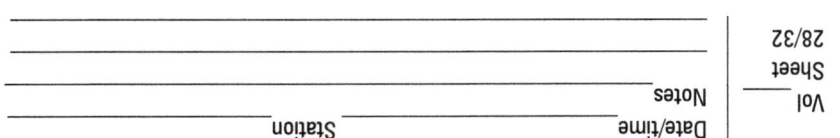

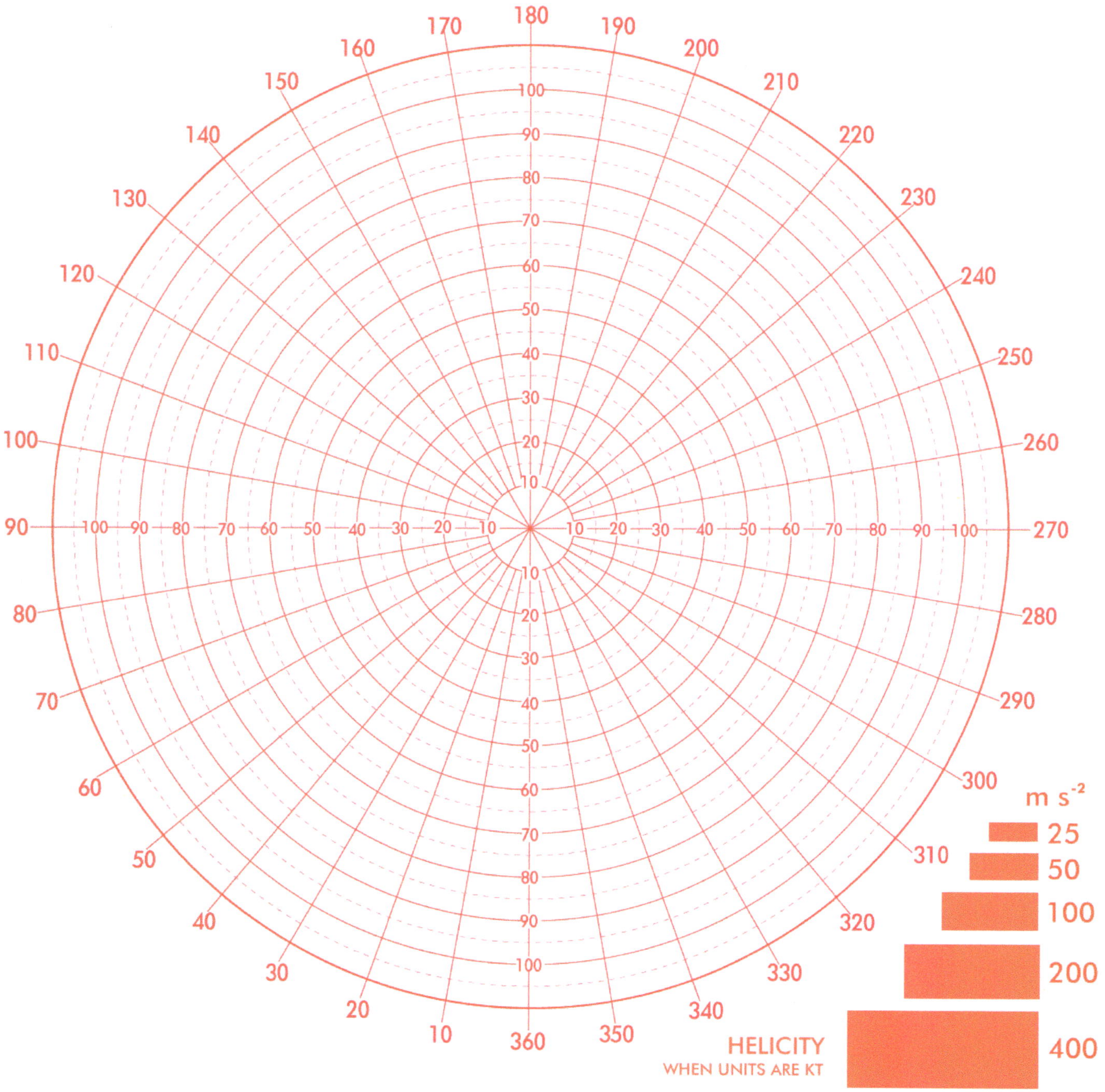

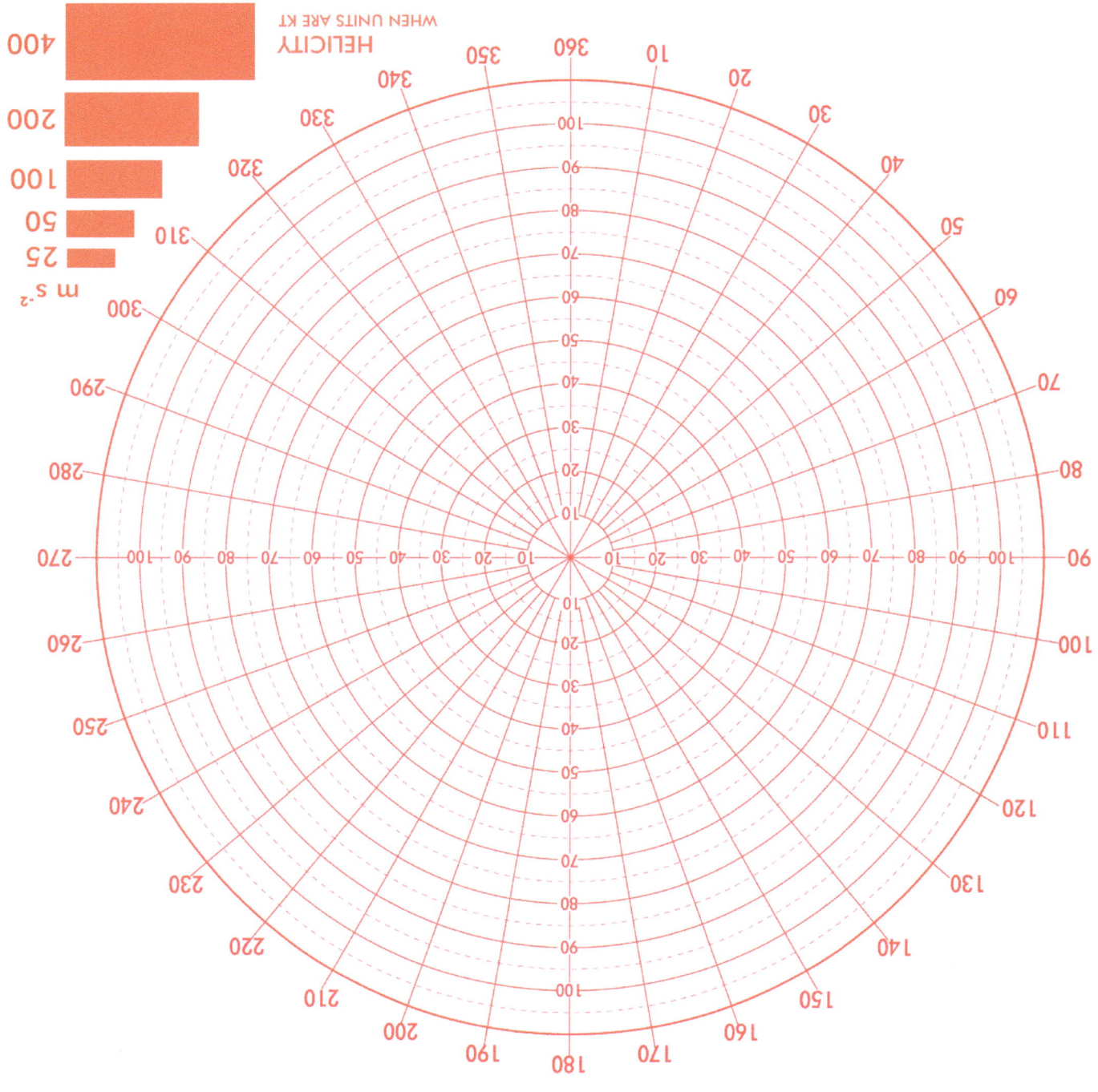

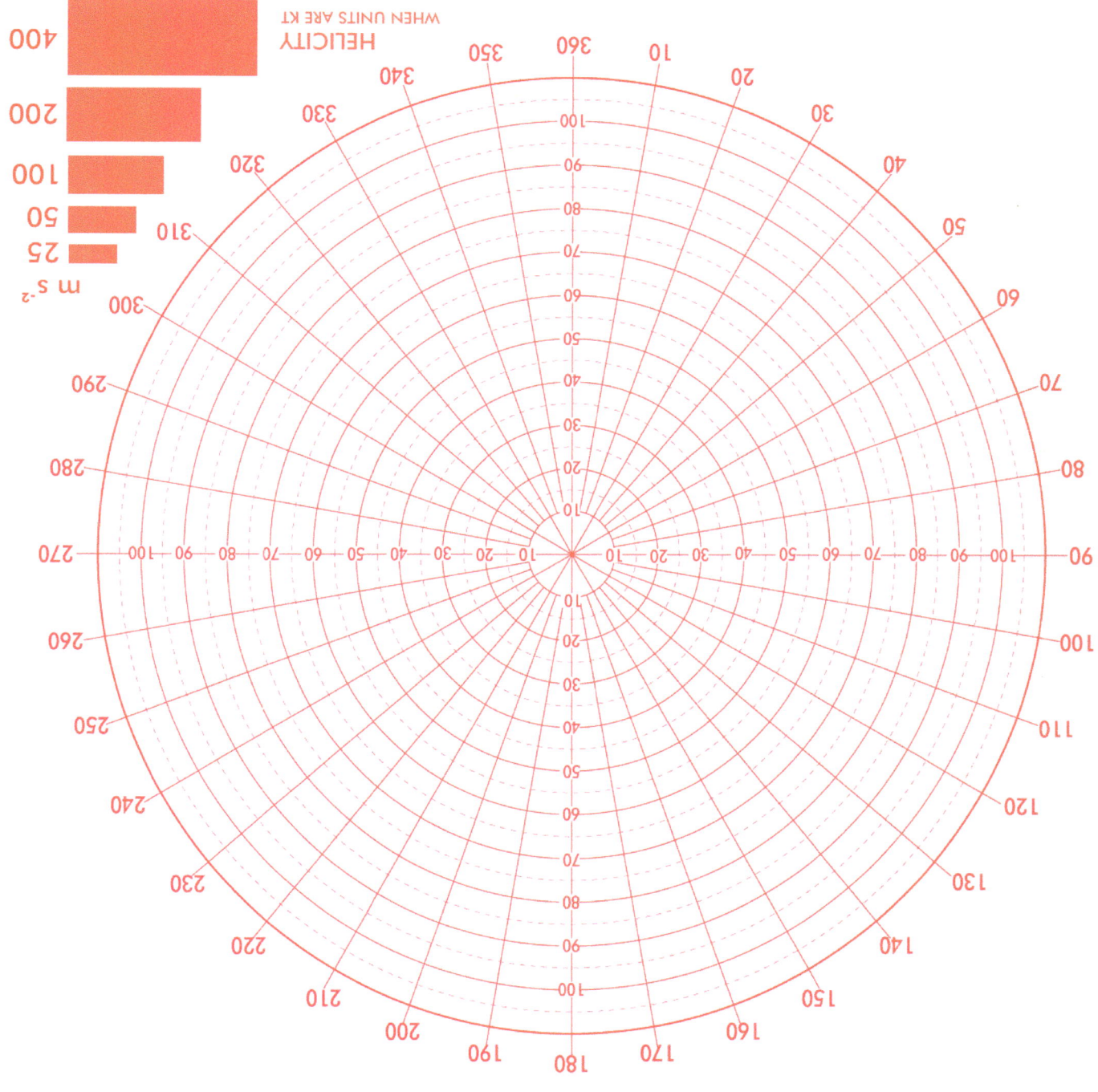

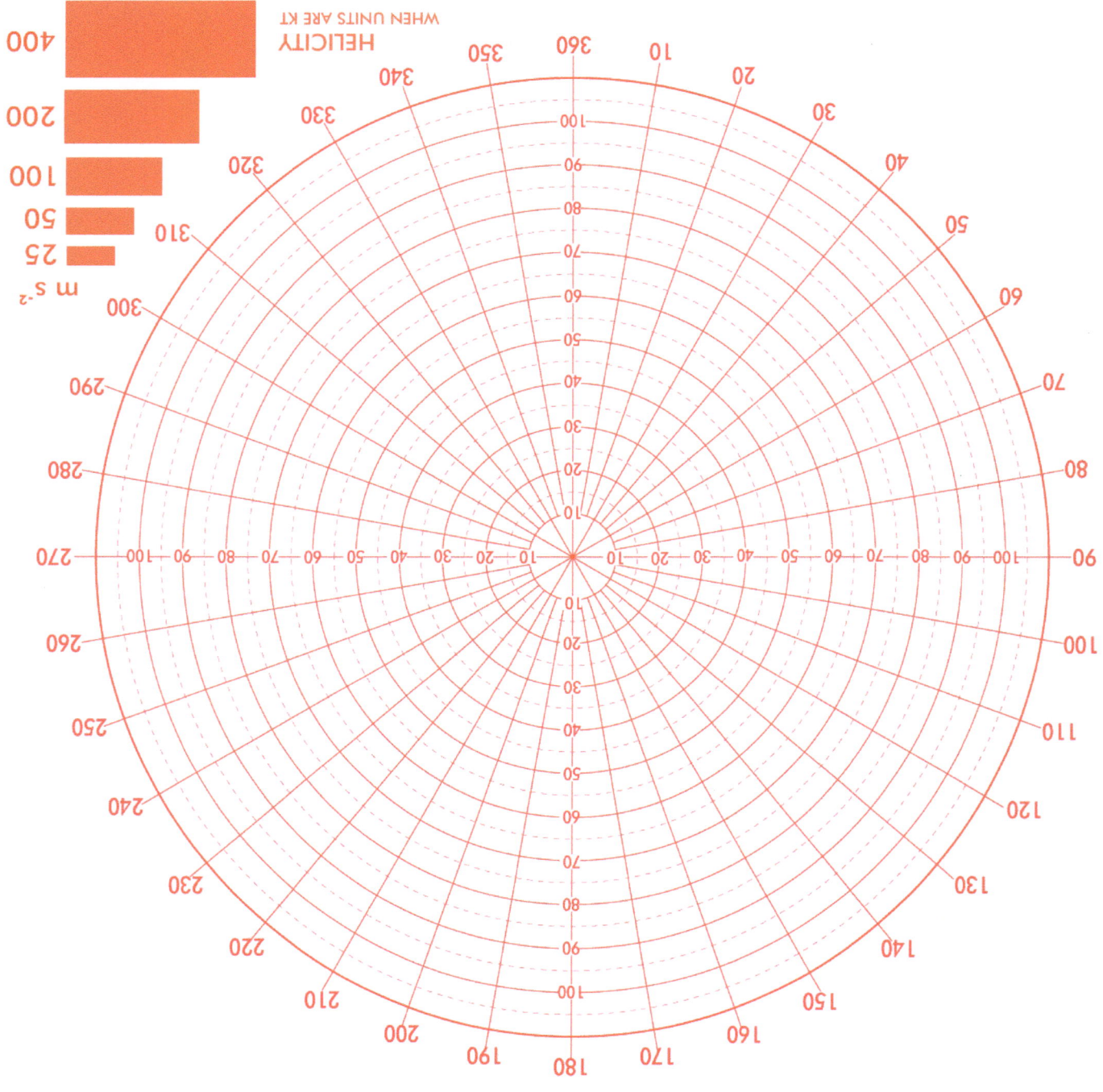

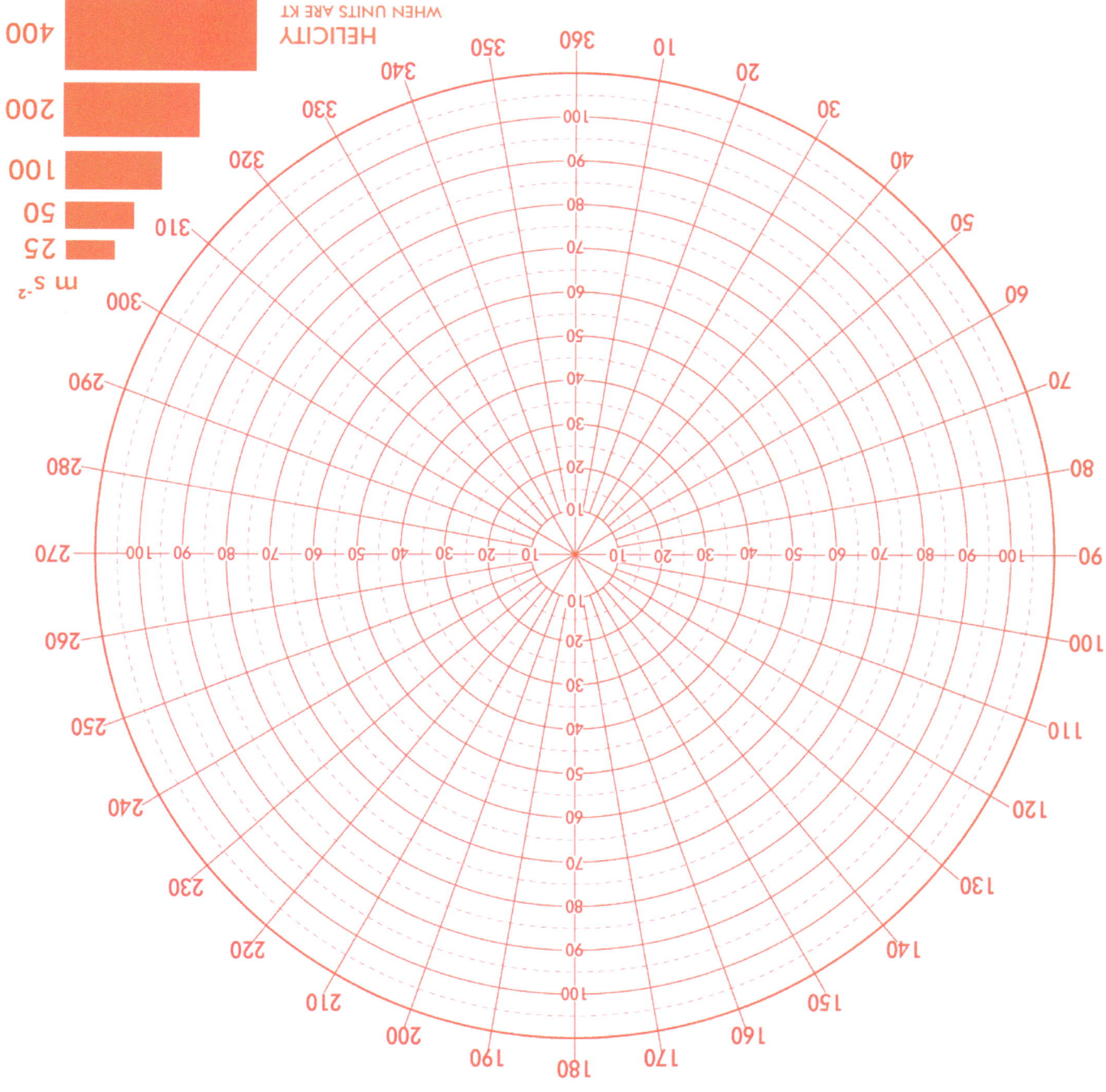

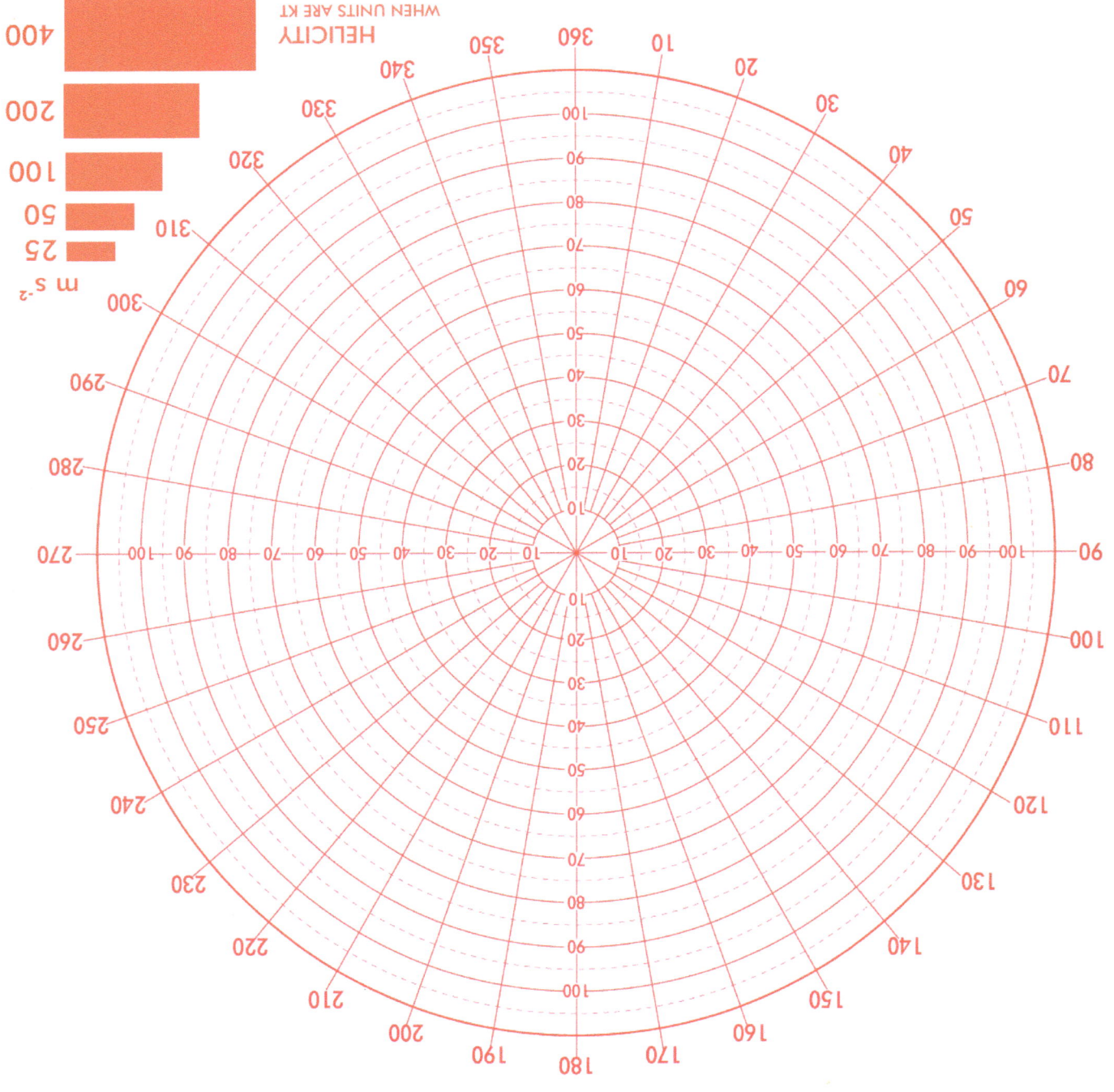

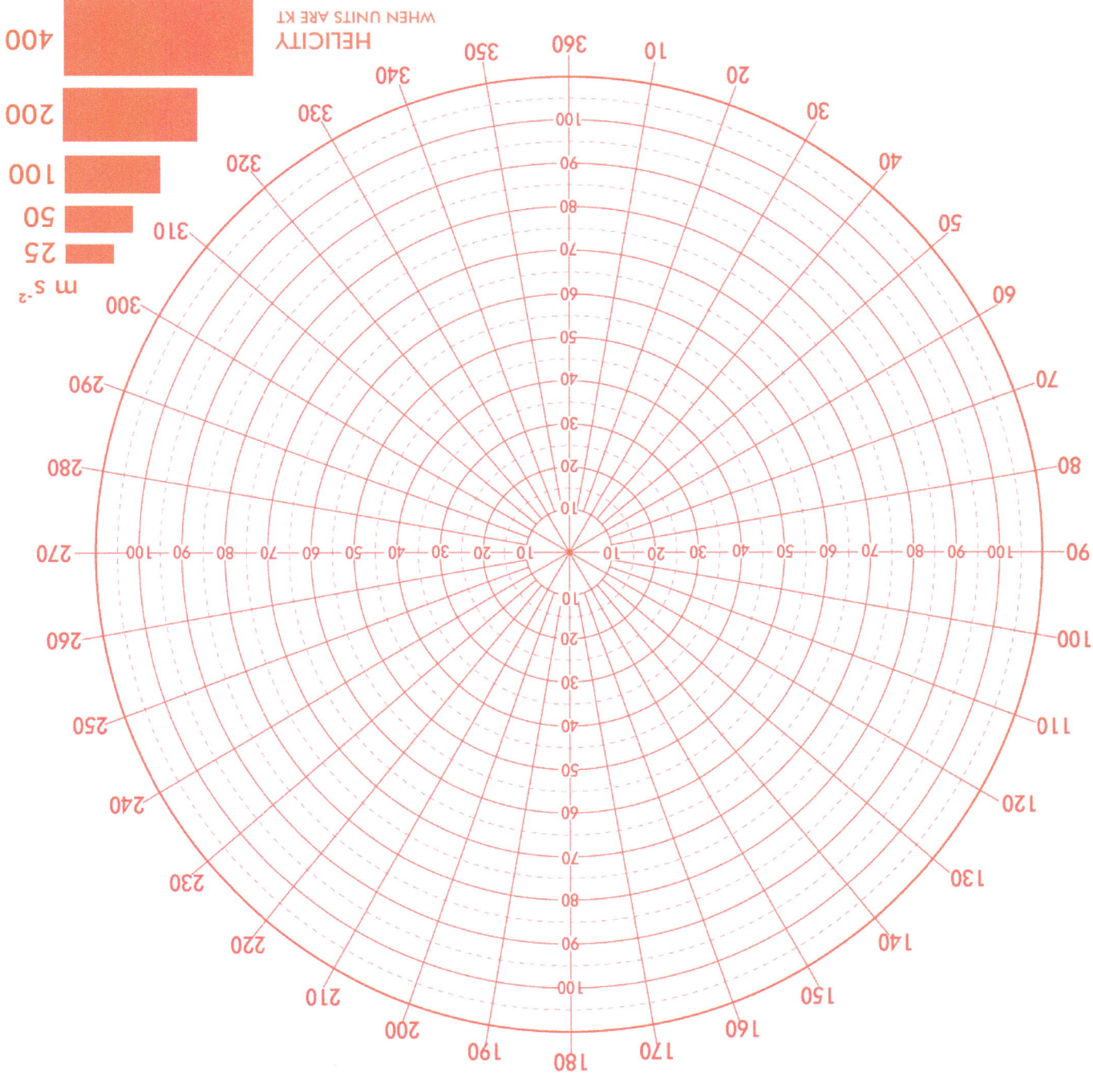

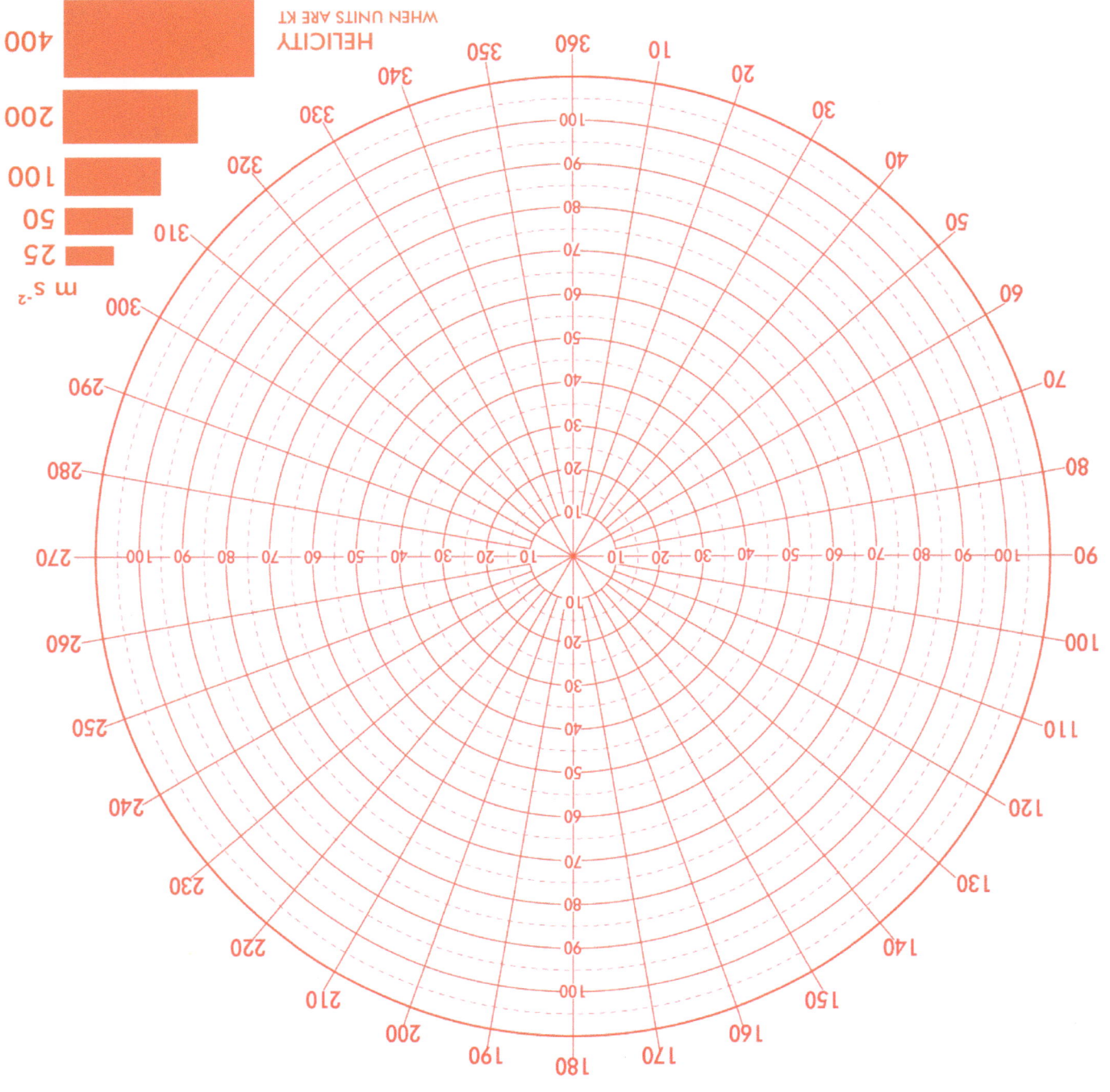

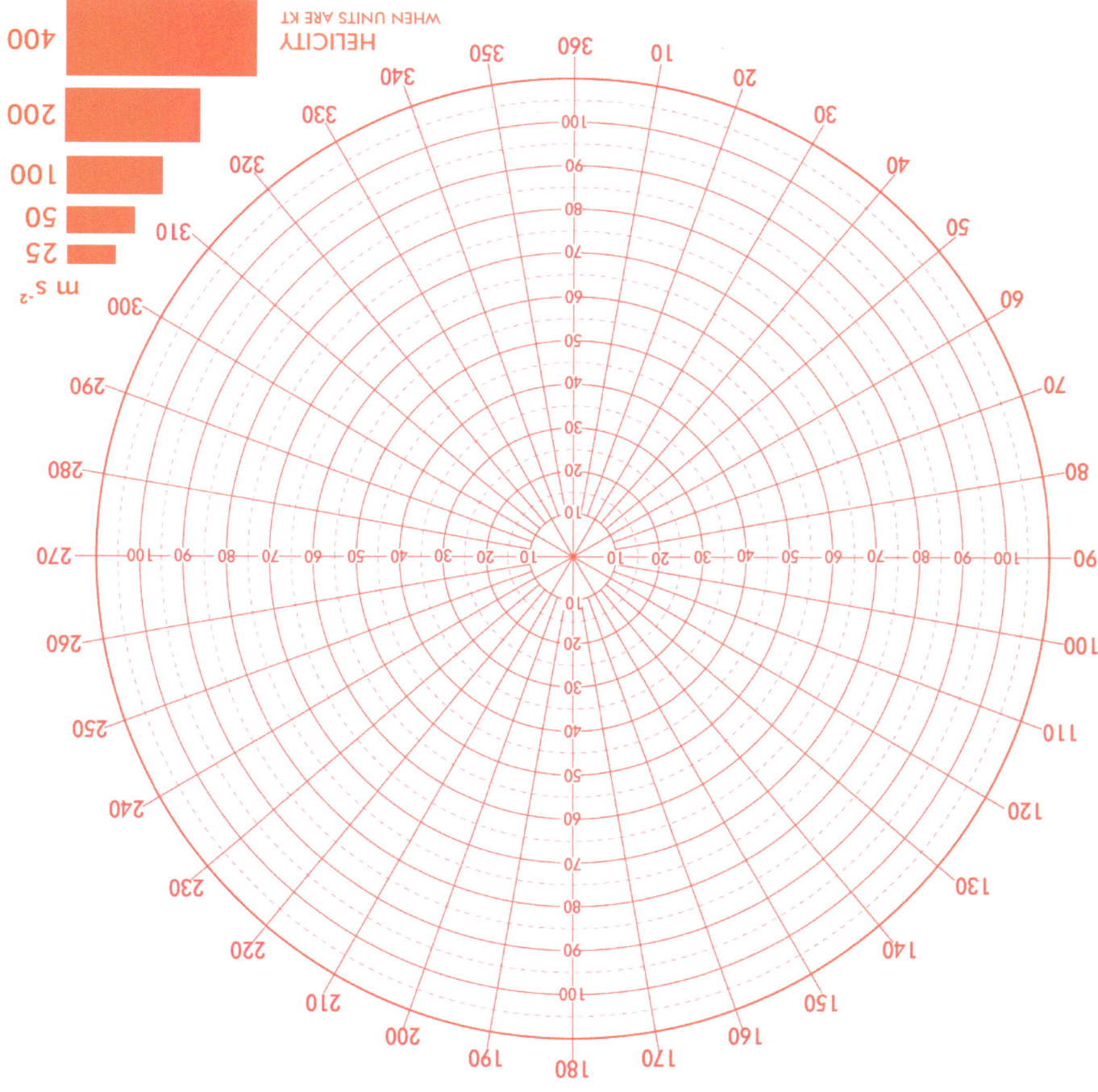

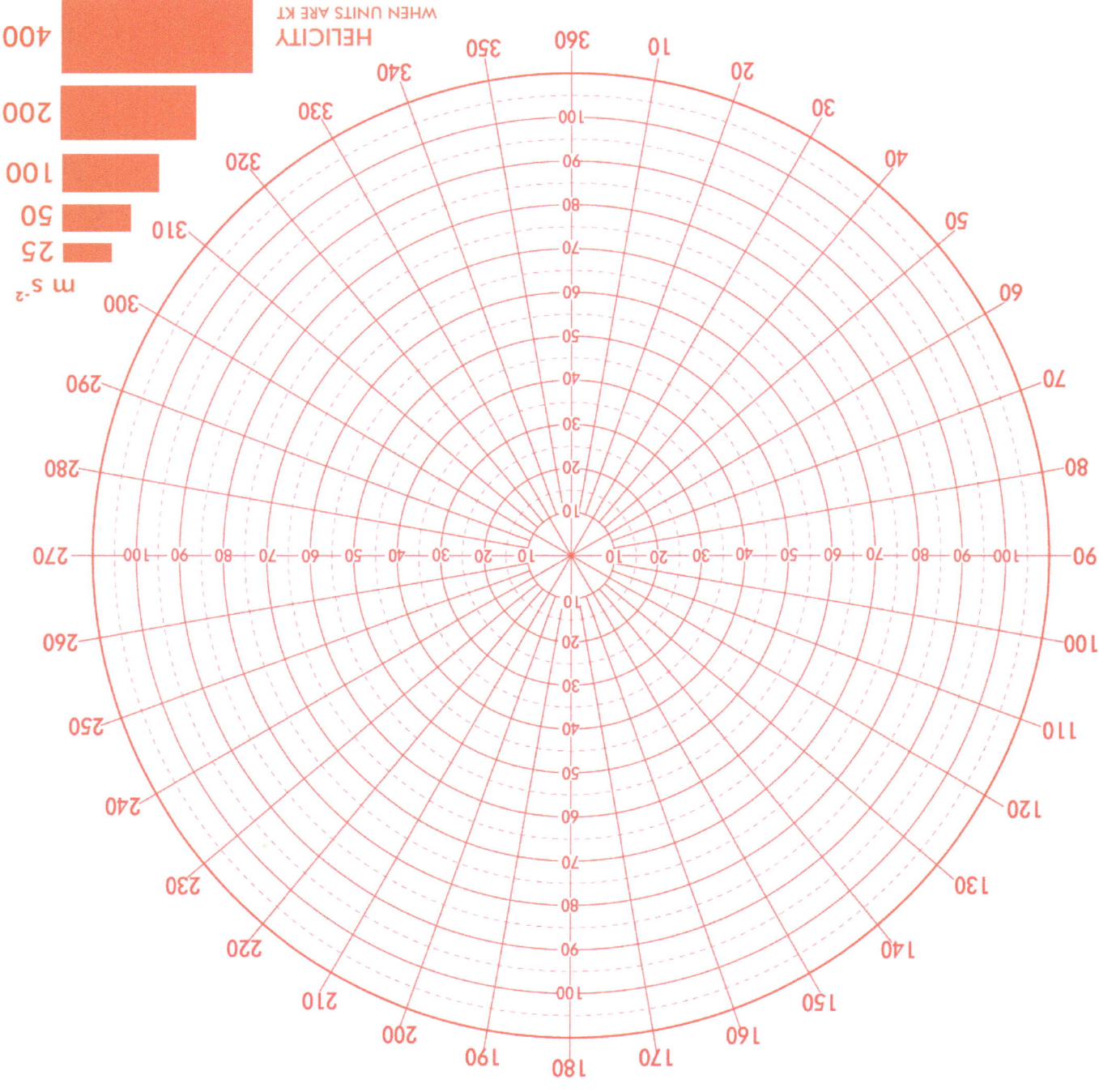

HODOGRAPH

Station
Date/time
Vol
Sheet 18/32
Notes

HELICITY
WHEN UNITS ARE KT

m s⁻²
25
50
100
200
400

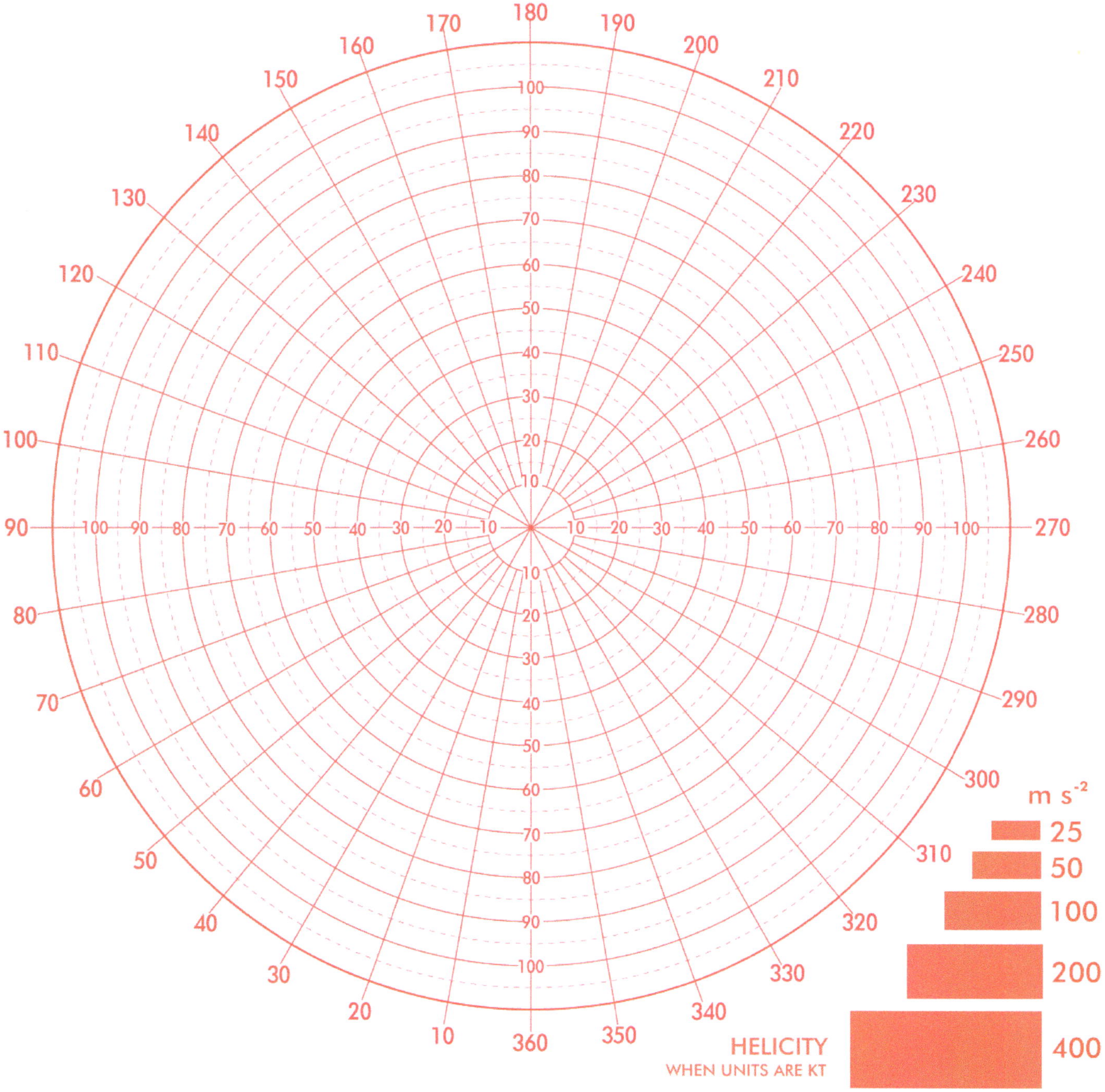

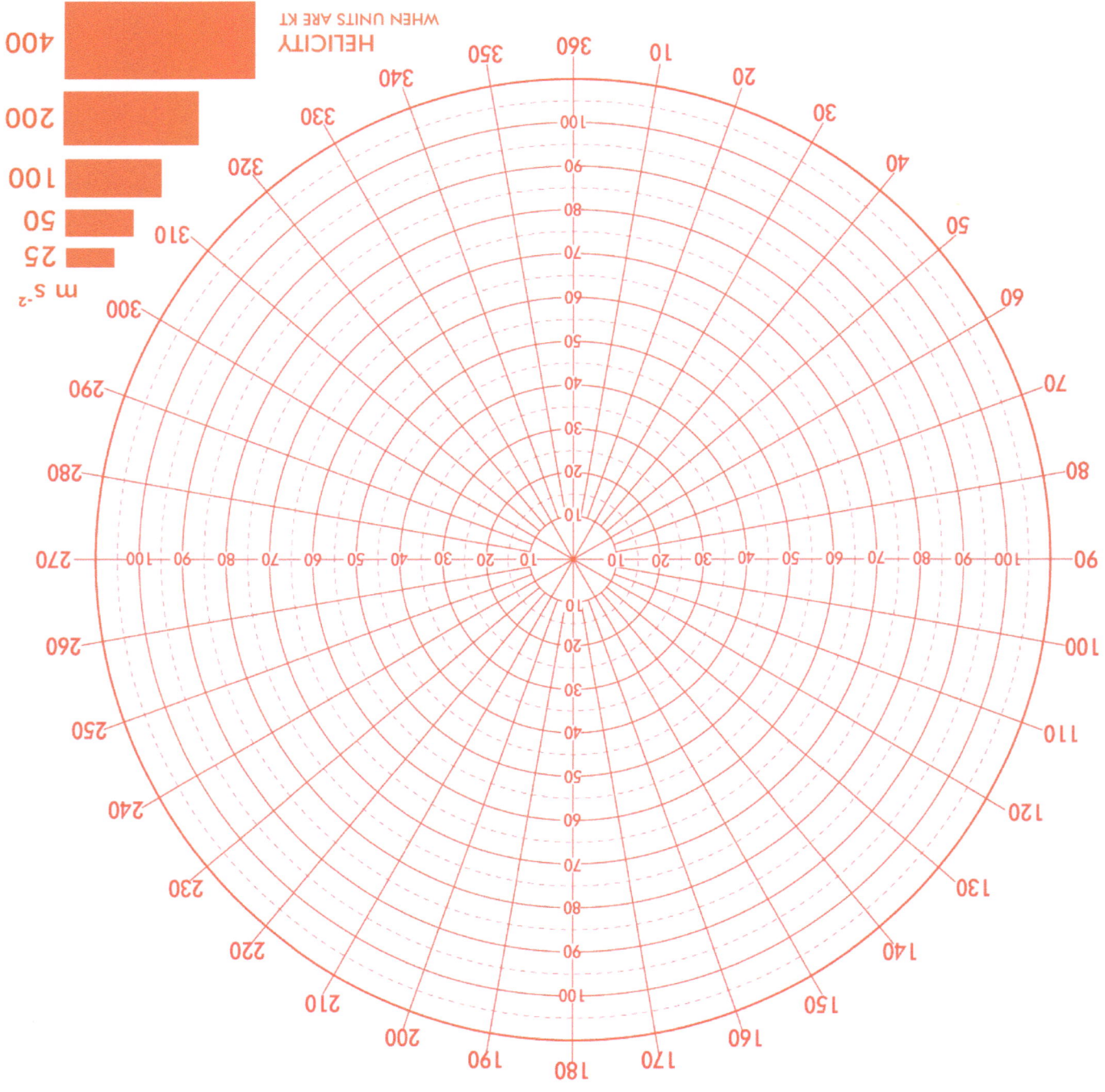

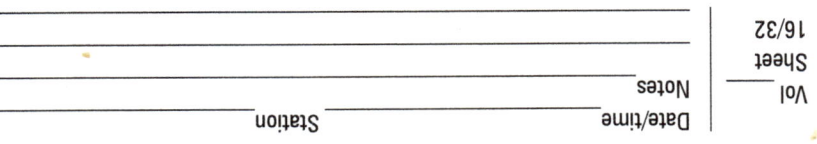

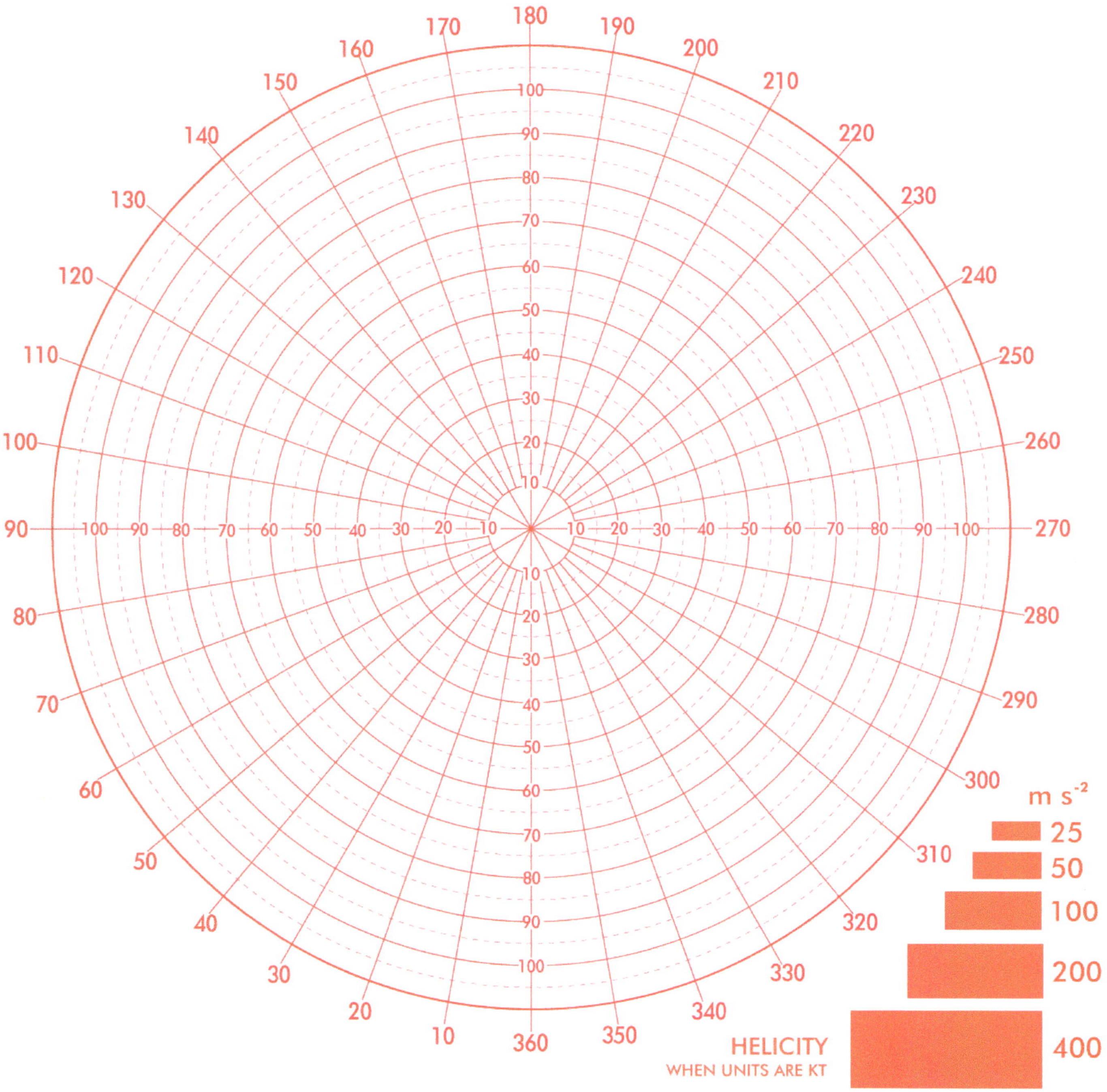

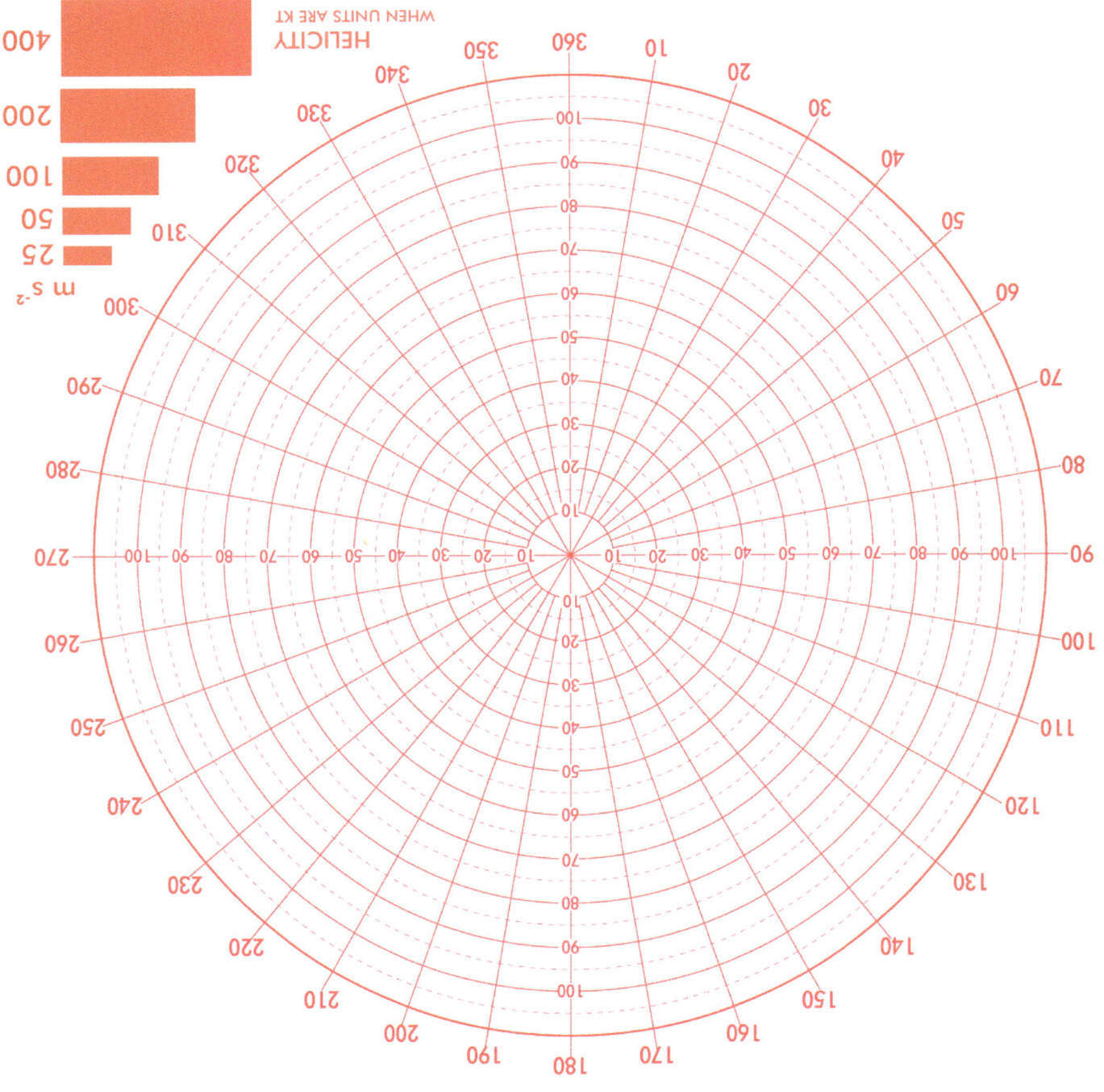

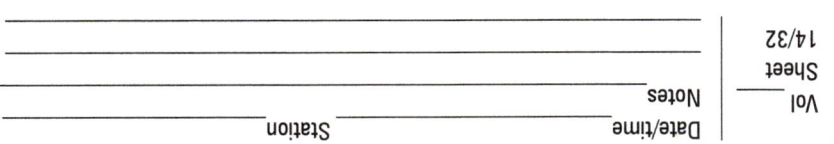

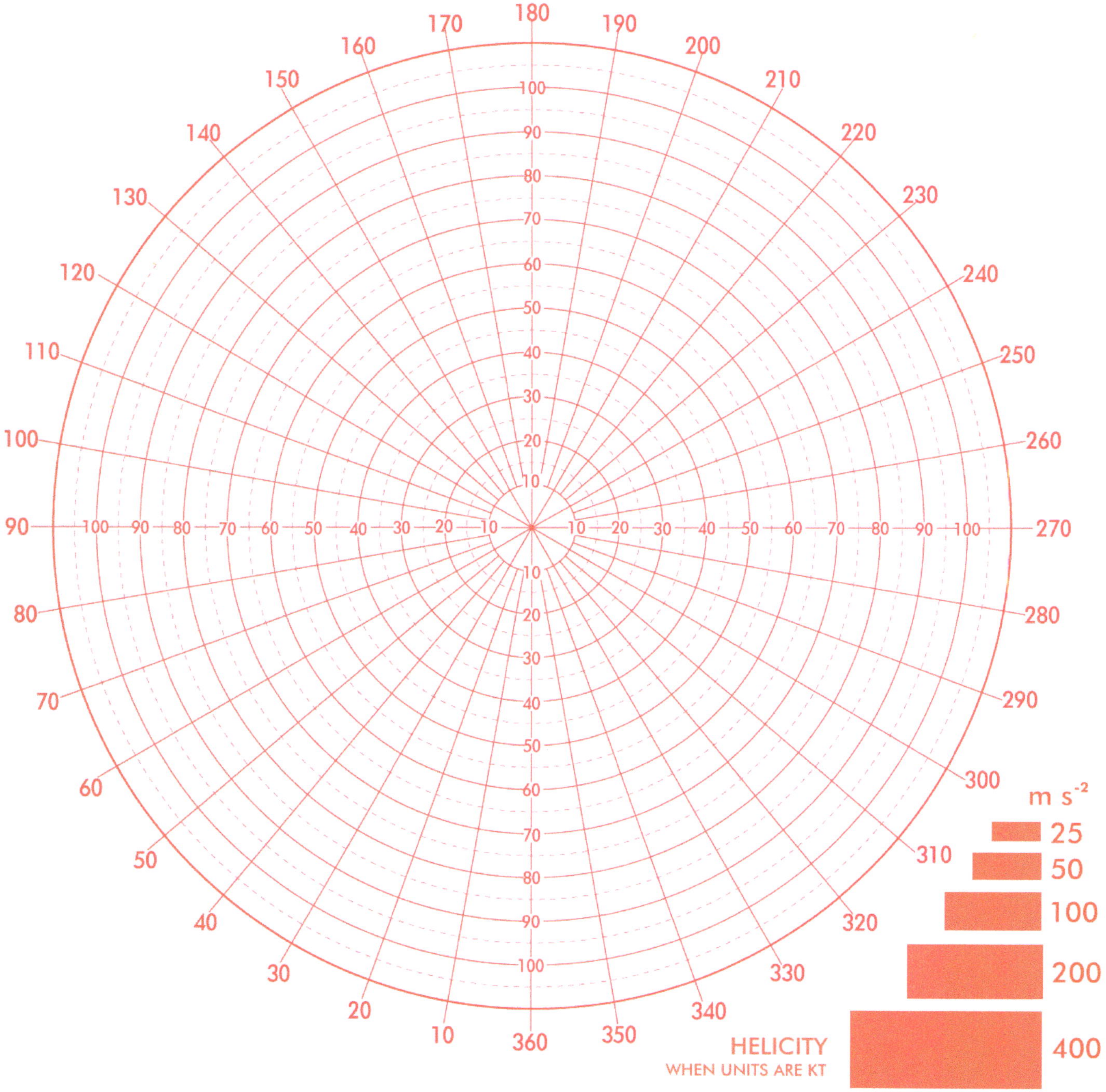

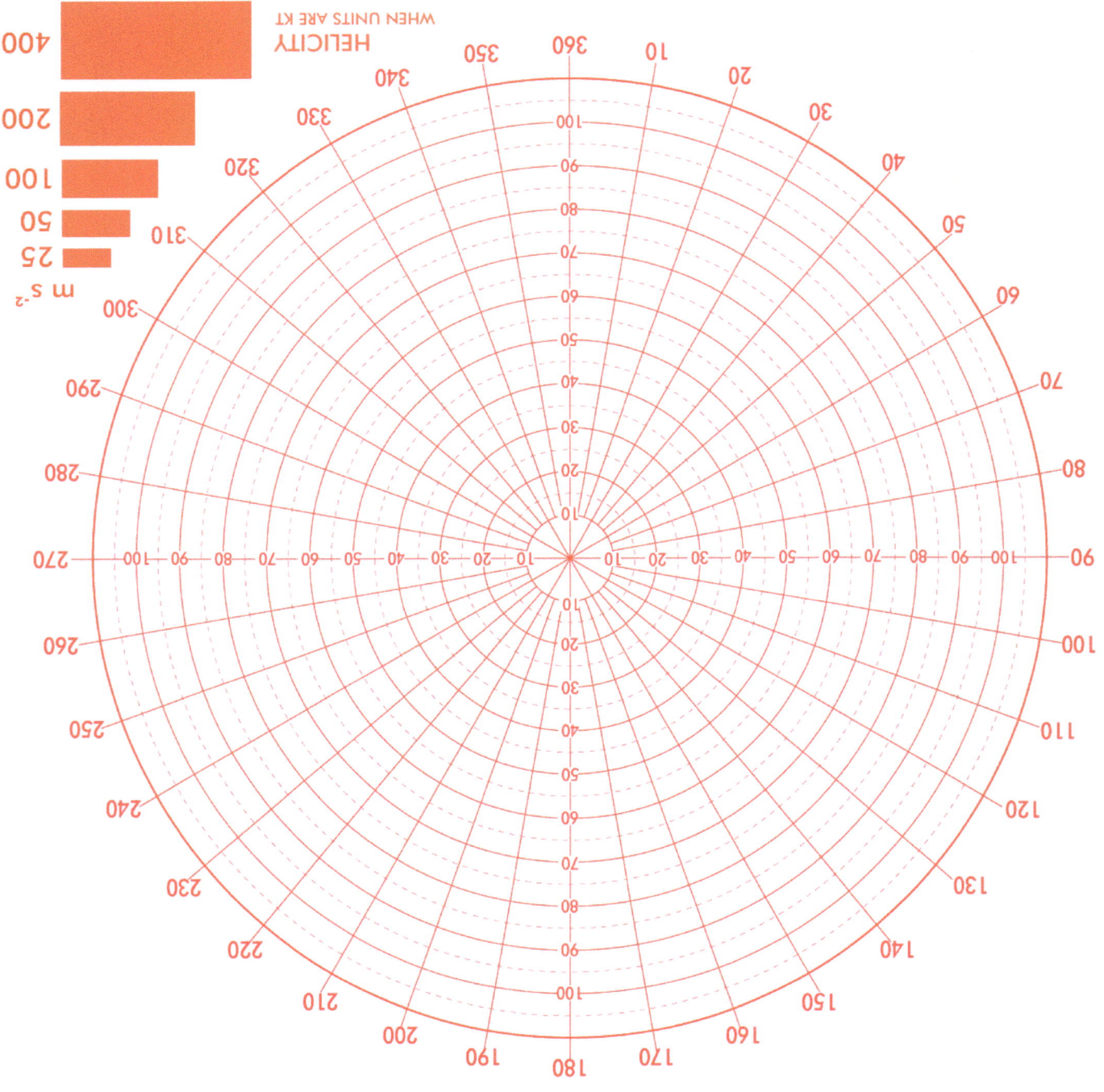

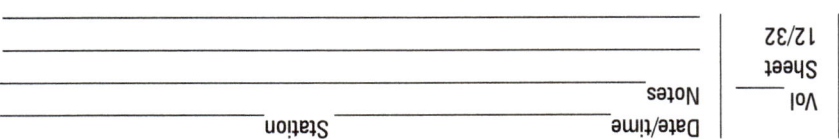

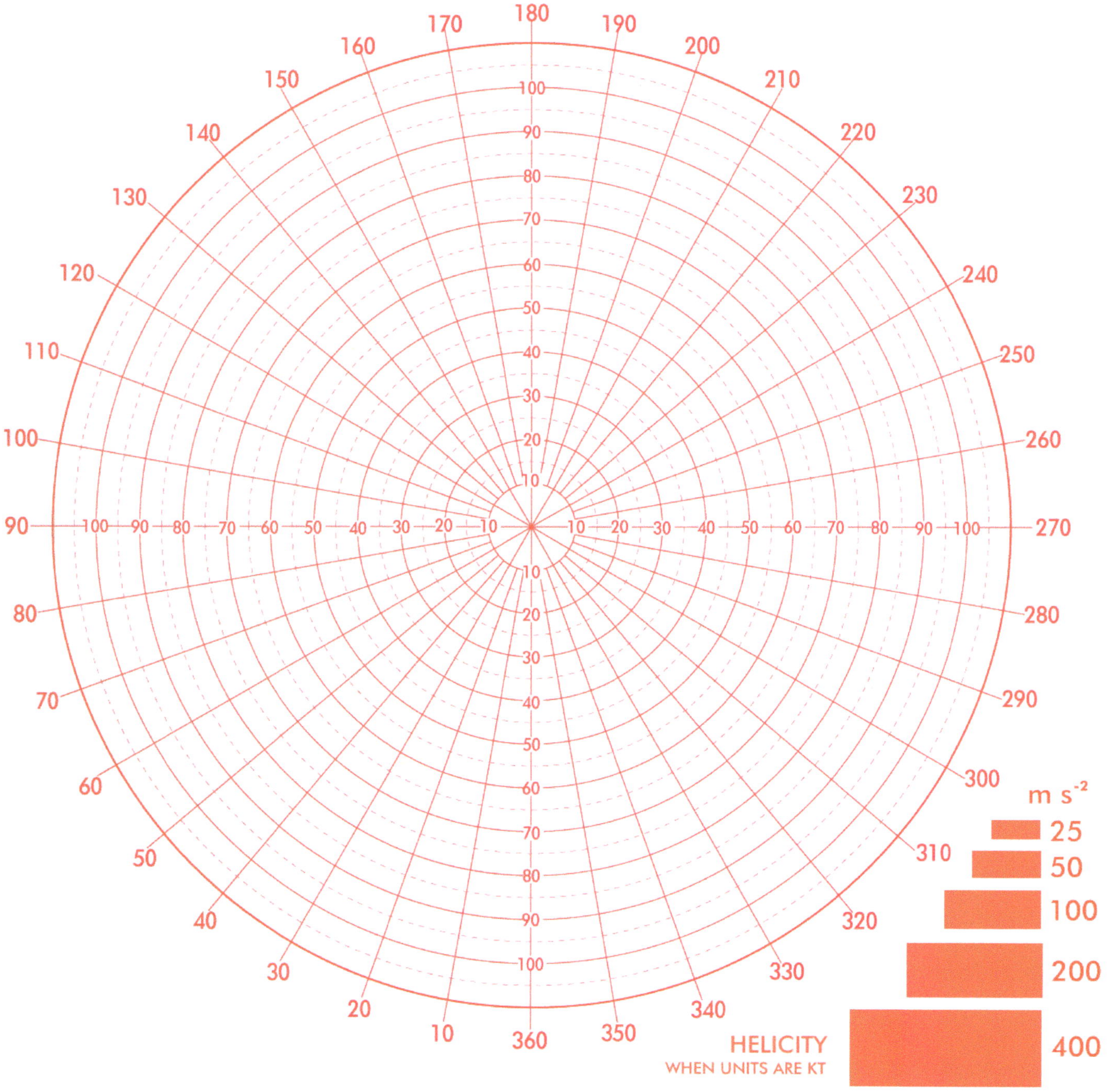

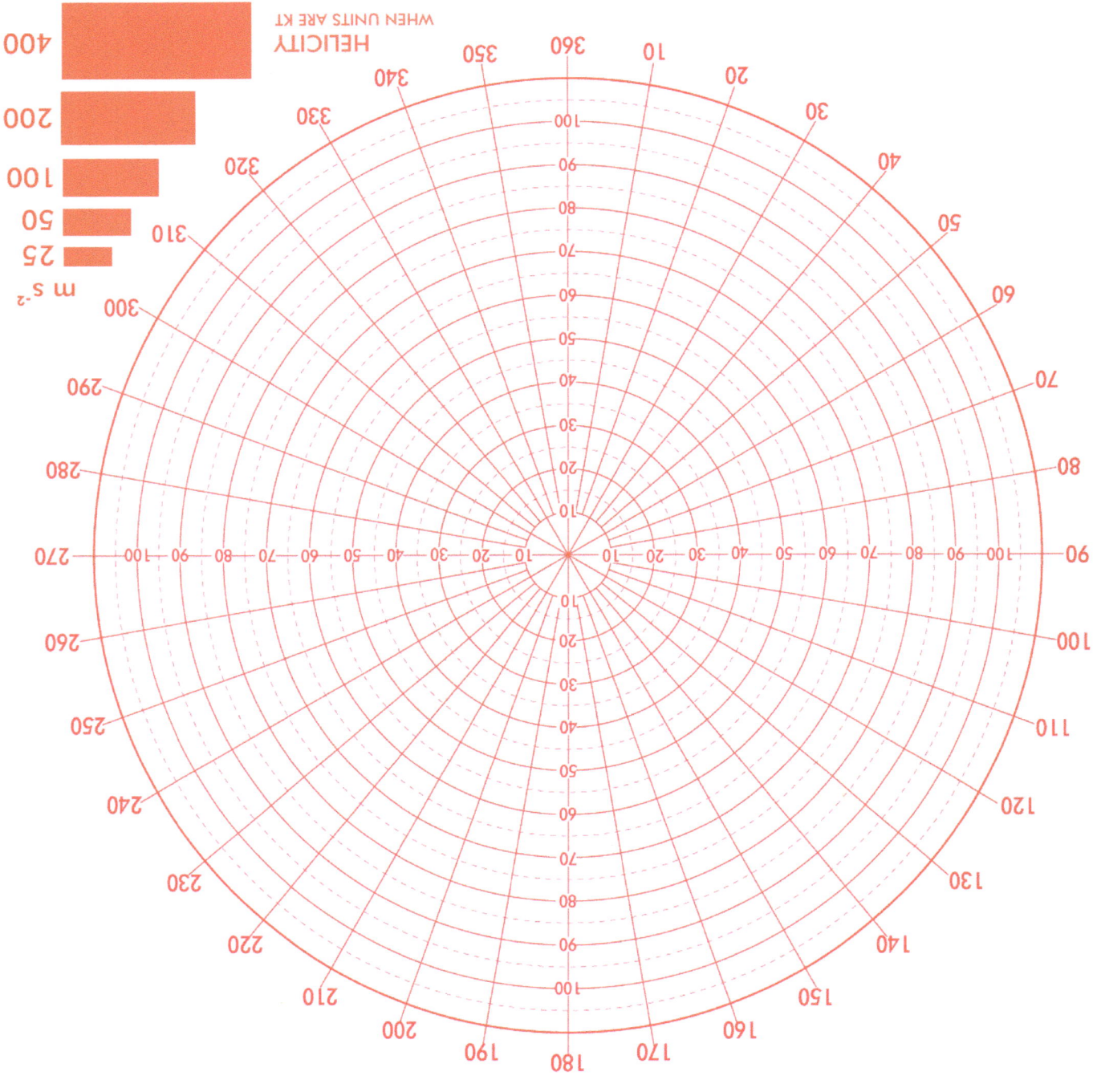

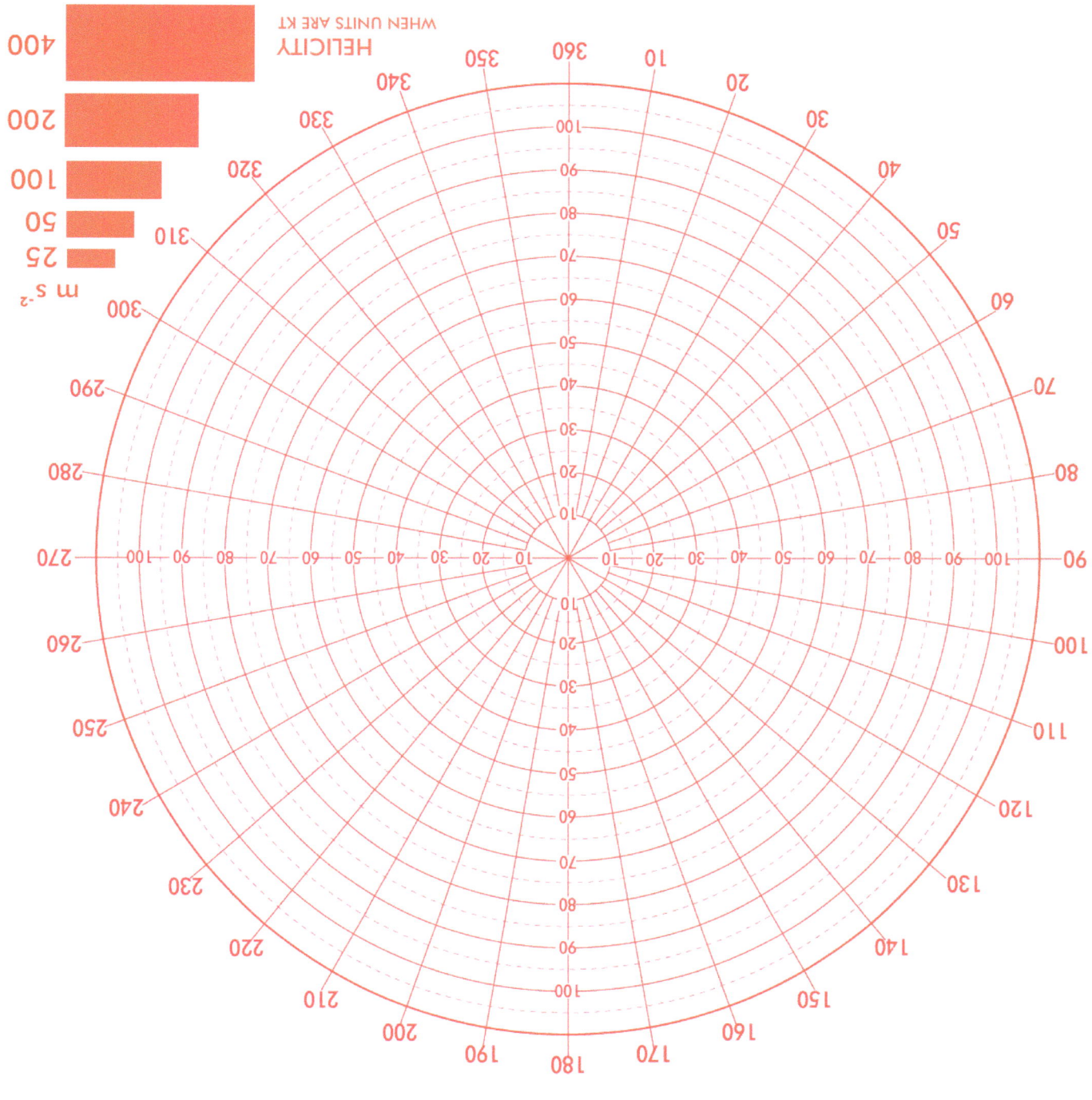

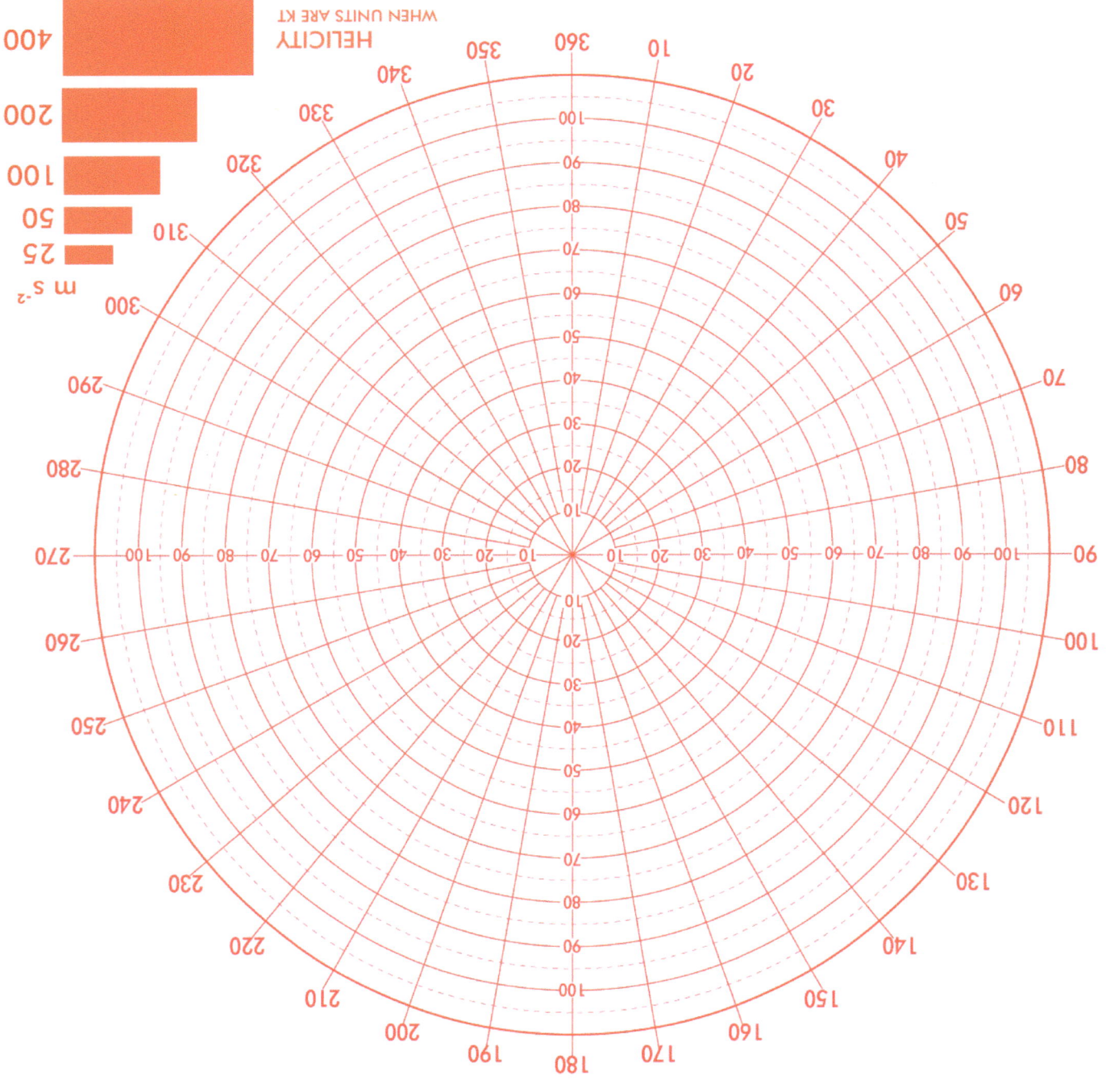

HODOGRAPH

HELICITY WHEN UNITS ARE KT

m s⁻²: 25, 50, 100, 200, 400

Date/time Station
Vol
Sheet 8/32
Notes

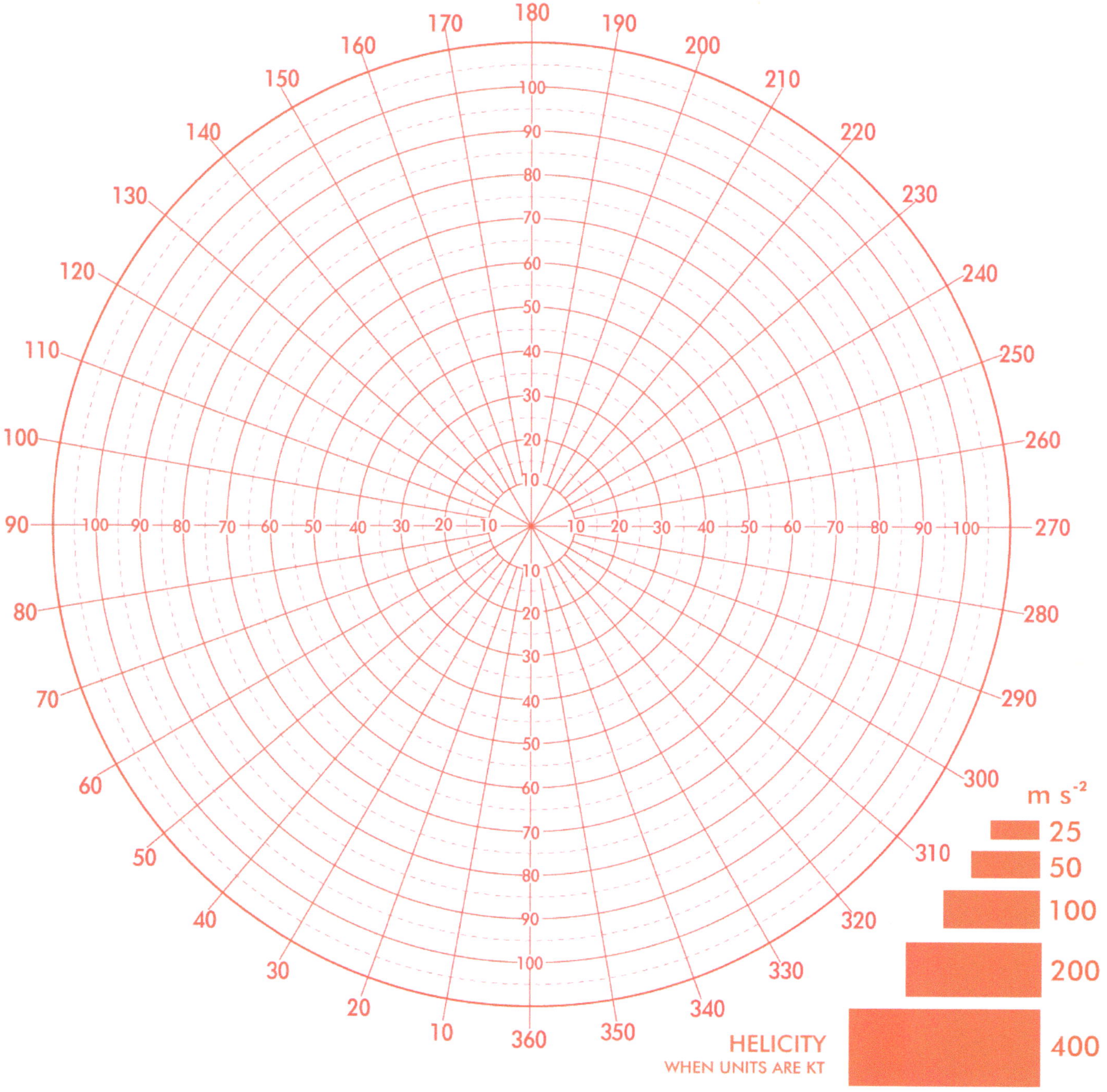

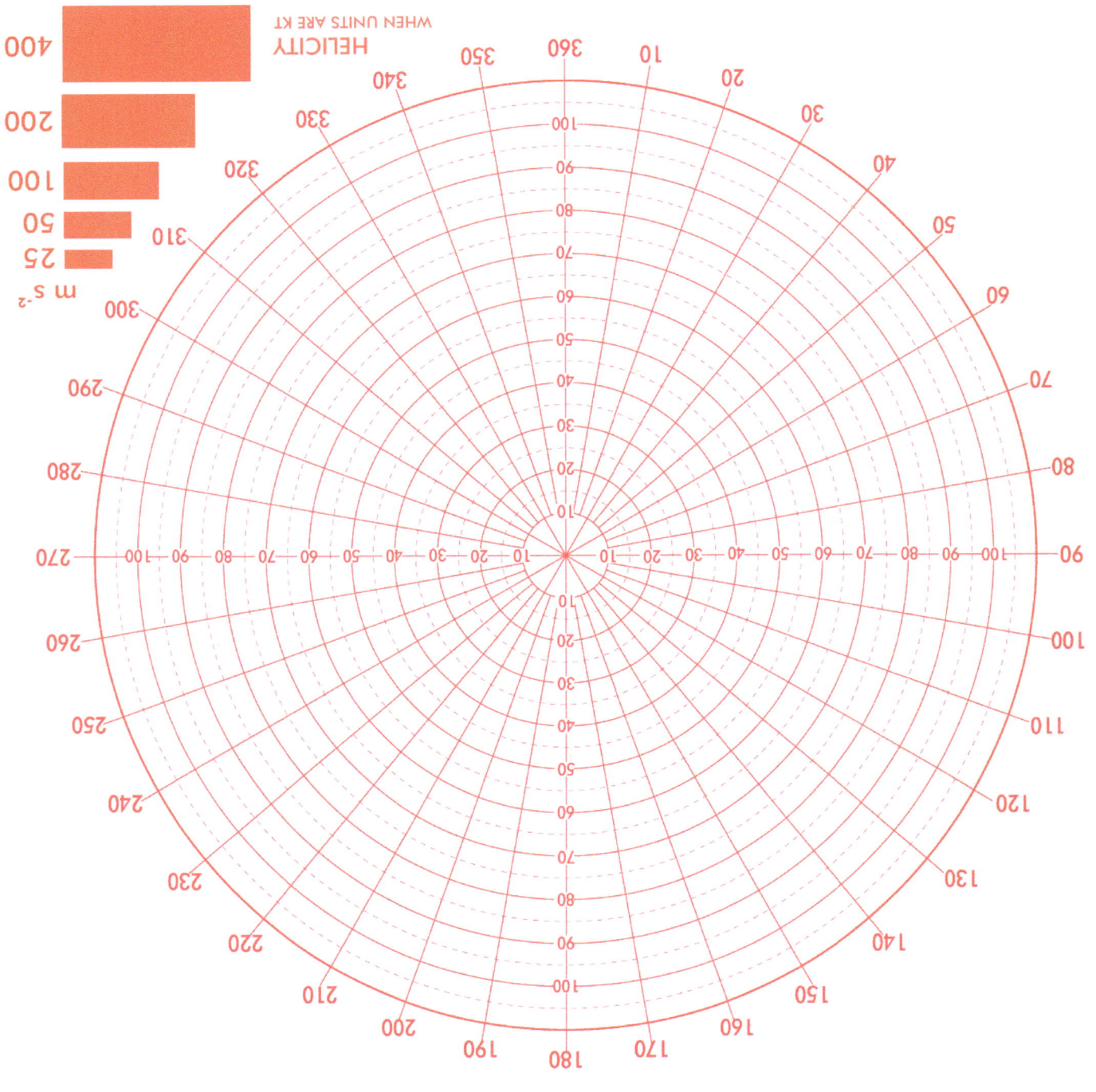

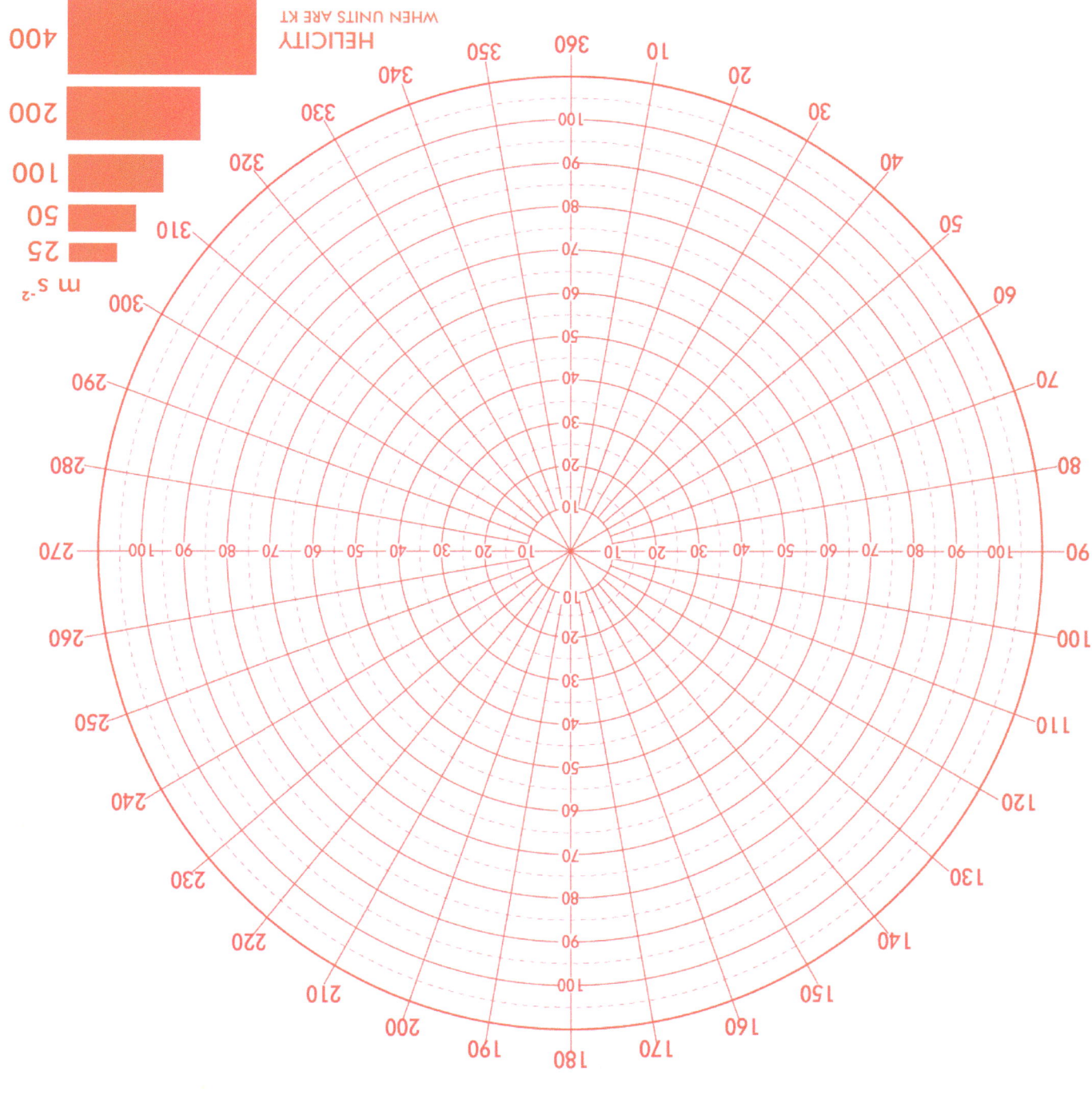

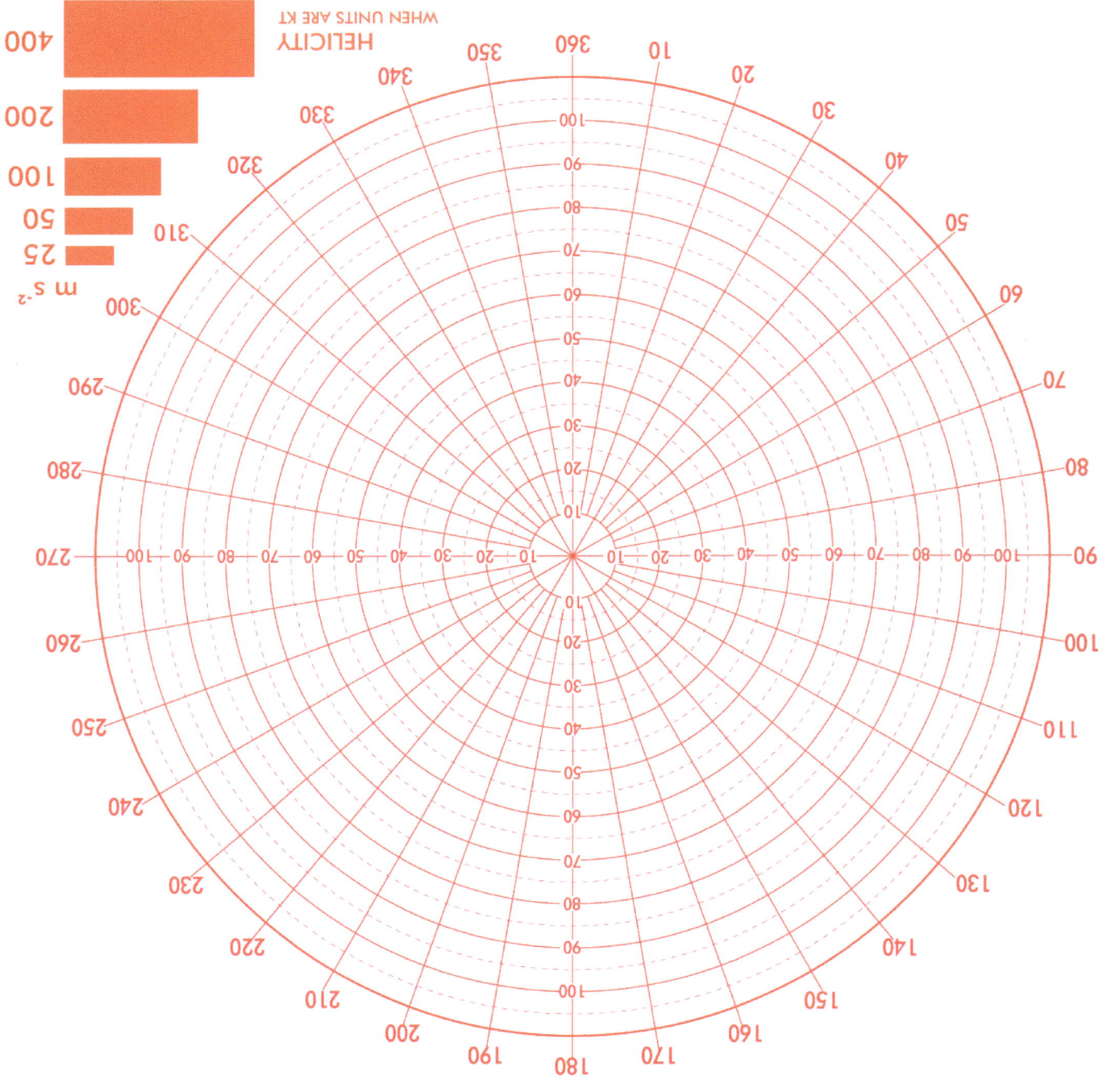

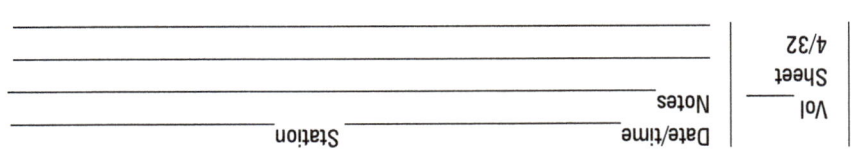

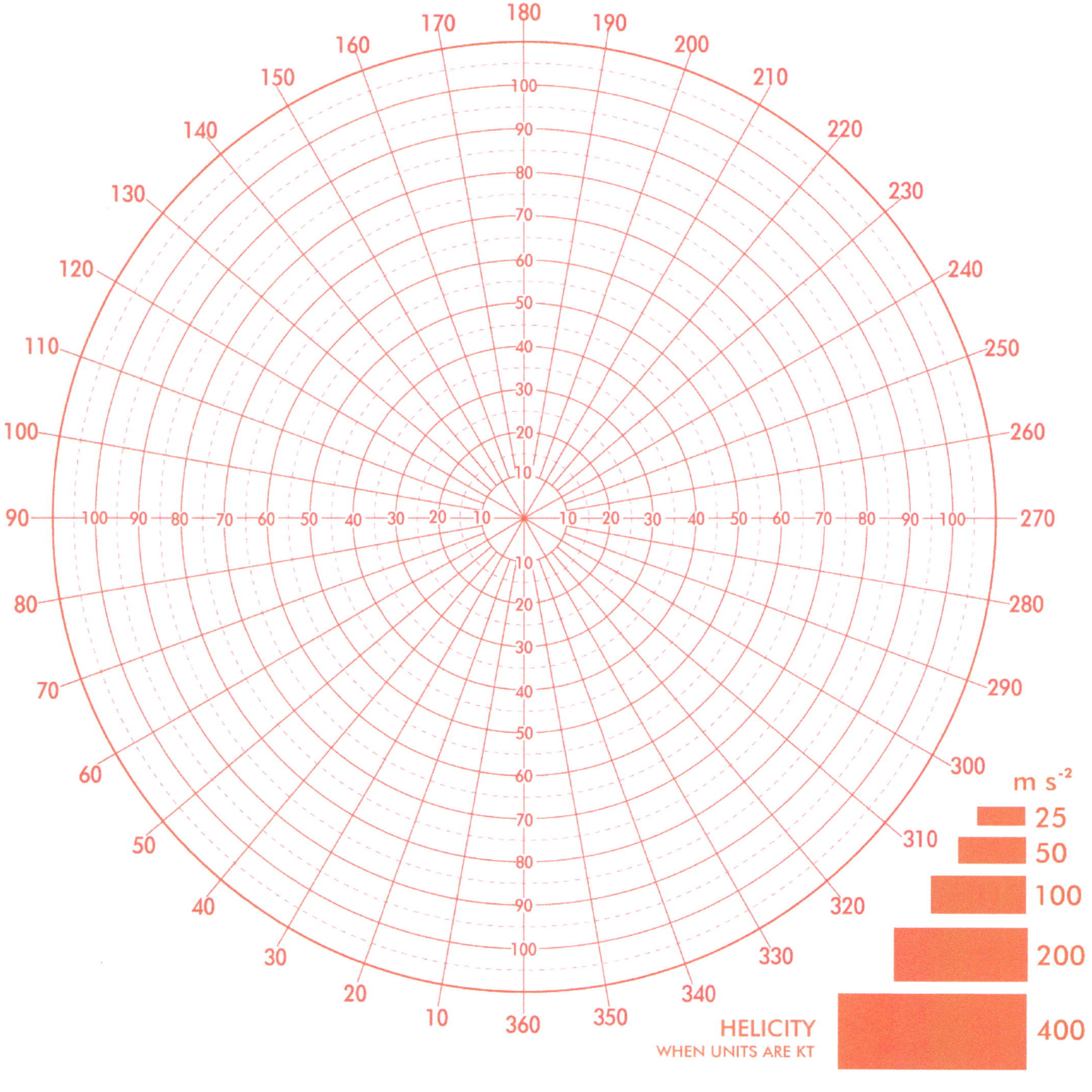

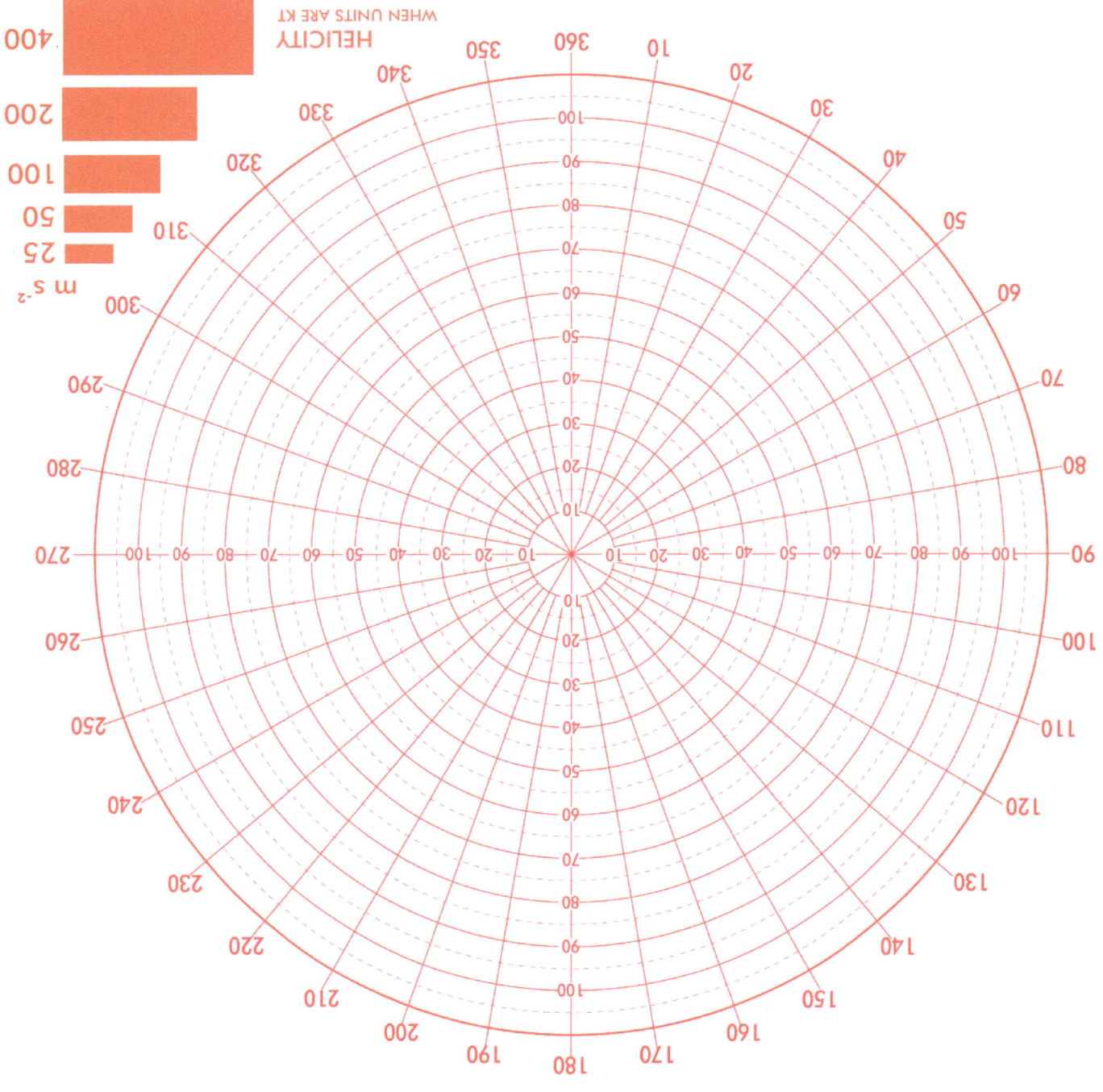

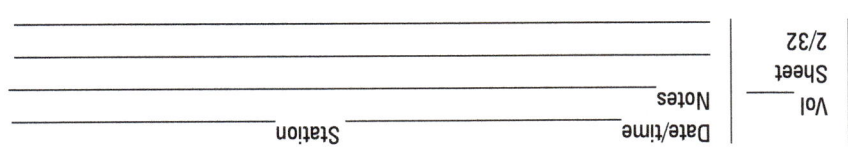

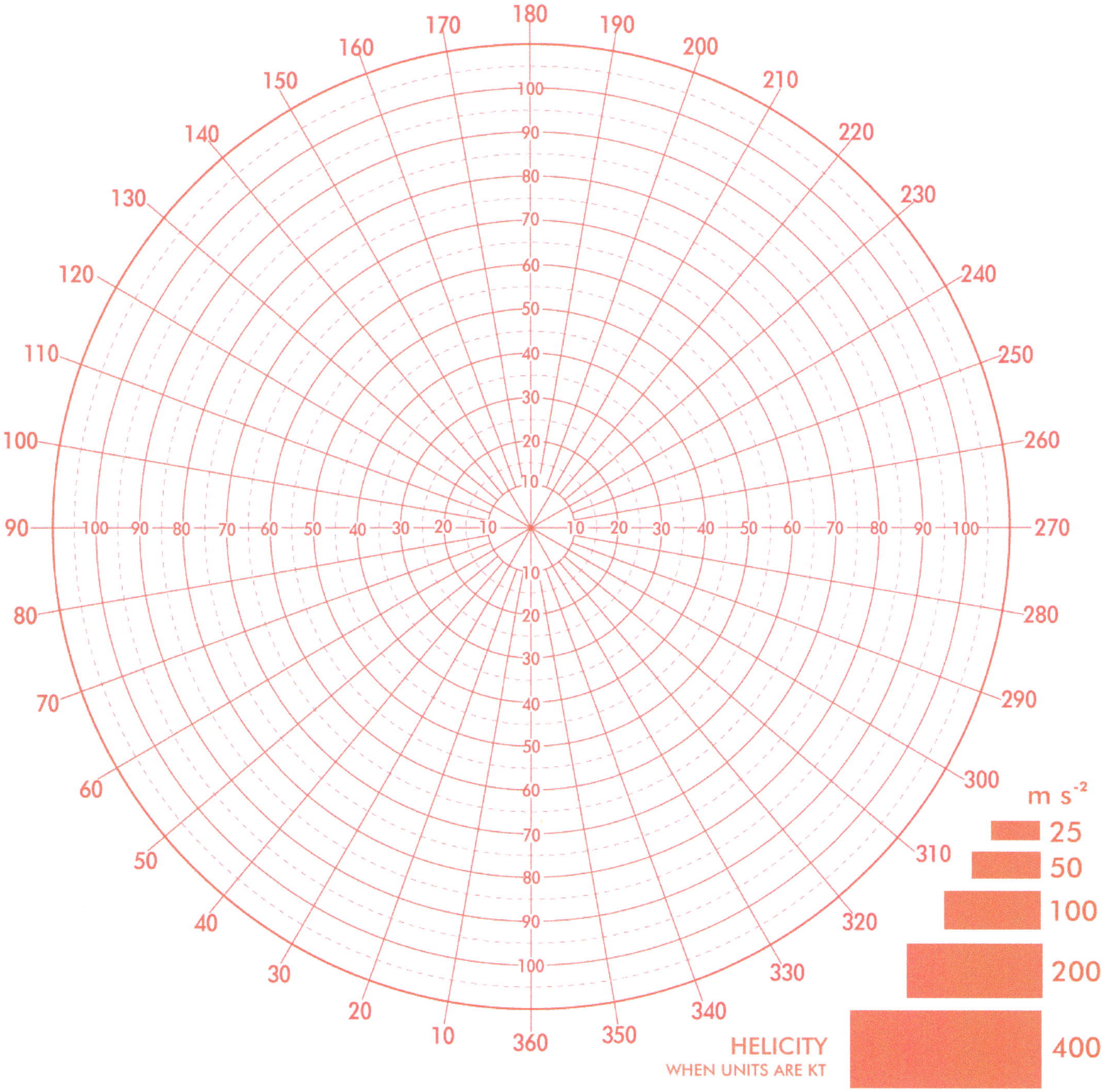

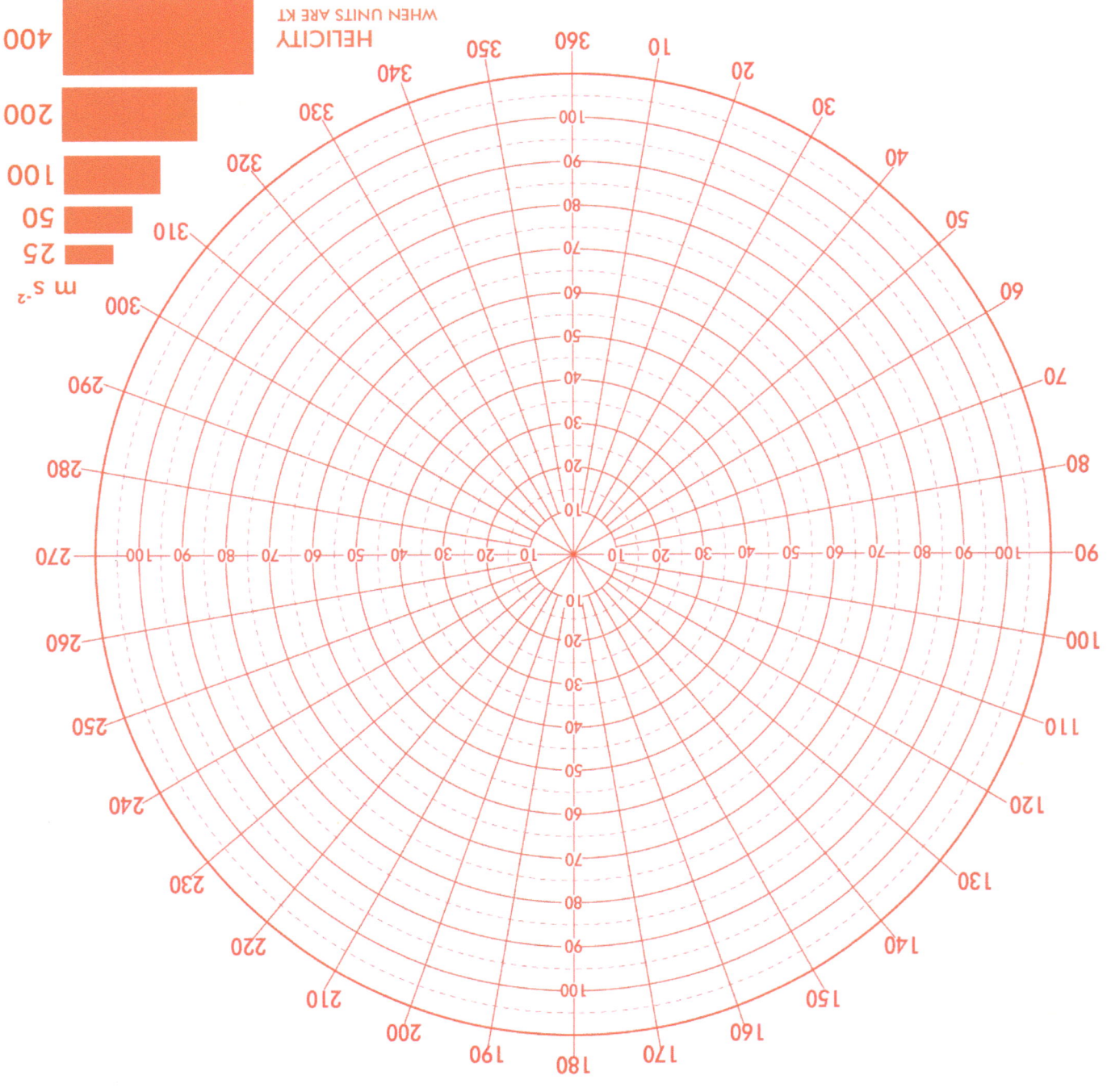

HODOGRAPH QUICK REFERENCE

This example shown here will be used in the illustrations below. Points are labeled in km AGL. We will assume data is expressed in knots (nautical miles per hour). For a station at sea level, the 6 km level that we refer to below usually corresponds to about 20,000 ft MSL or 470 mb.

Bulk shear

Bulk shear between 0 and 6 km is a common measure of shear which corresponds to the general risk of severe weather. Simply draw a vector from 0 to 6 km and measure the length. This yields a bulk shear of 73 kt.

Storm motion

A standard method of determining general storm motion is finding the mass-weighted average of winds between 0 and 6 km. To compute this manually, find a visual point that seems to represent a center of balance along a uniform distribution of points between 0 and 6 km.

Bunkers Method - Step 1

For severe storm events, the Bunkers Method should always be used to estimate motion of right- and left-moving supercells. Using a ruler, draw a bulk shear line (BS) on the sounding connecting the average of the 0-0.5 km winds and the 5.5-6.0 km winds.

Bunkers Method - Step 2

Next, plot storm motion (SM, see above). This represents the motion of nonsevere cells. Using a ruler, draw a motion axis line (MA) perpendicular to the bulk shear line (BS) which passes through the storm motion point (SM).

Bunkers Method - Step 3

Along the motion axis (MA) drawn in Step 2, draw a point 15 knots either side of the calculated base storm motion (SM). The resulting points are the right motion vector (RM), representing the motion of right-moving supercells, and left motion vector (LM), resulting left-moving supercells.

Storm relative helicity

Using any measure of storm motion, including observed motion on radar, measure the geometric area swept out between this point and the environmental winds within a given layer (usually 0-1, 0-2, or 0-3 km). In this example we've used 0-3 km SRH based on right-moving supercells. The larger the swept-out area, the greater the SRH.

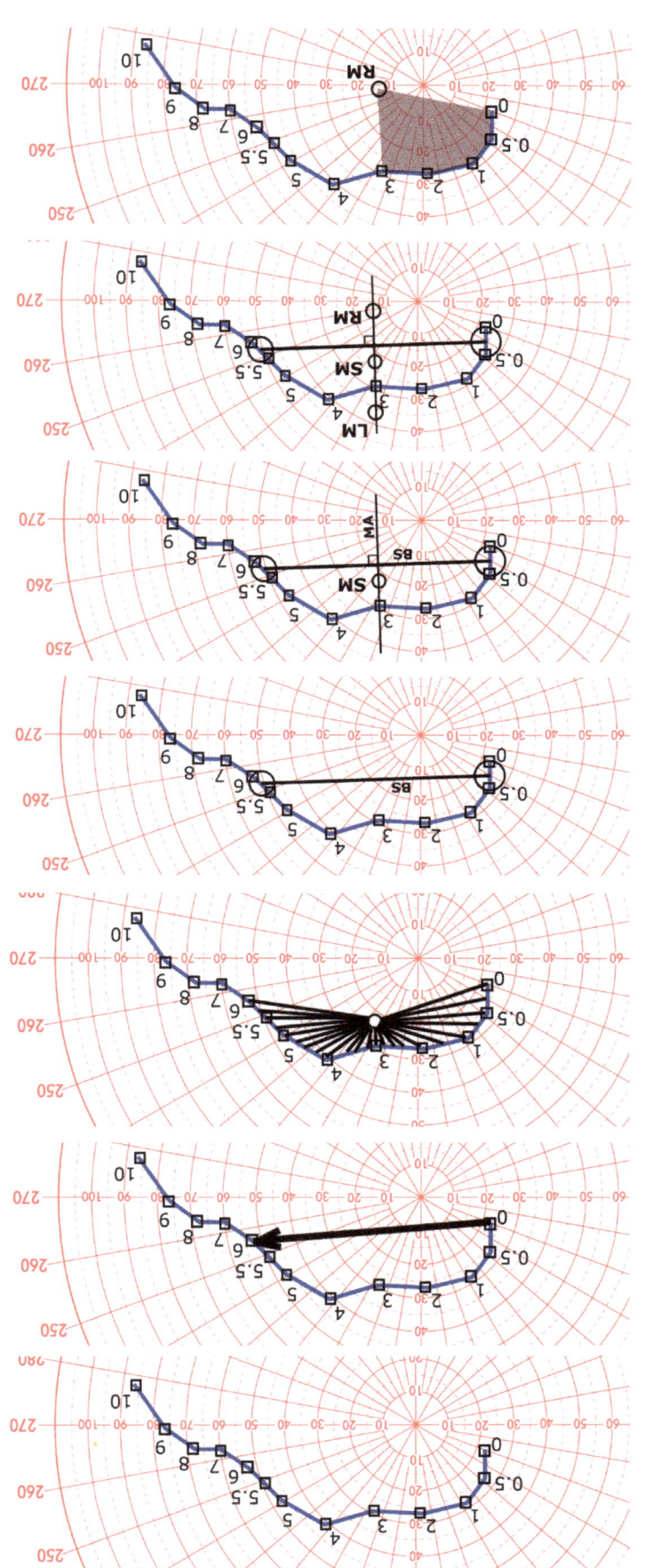

HODOGRAPHS

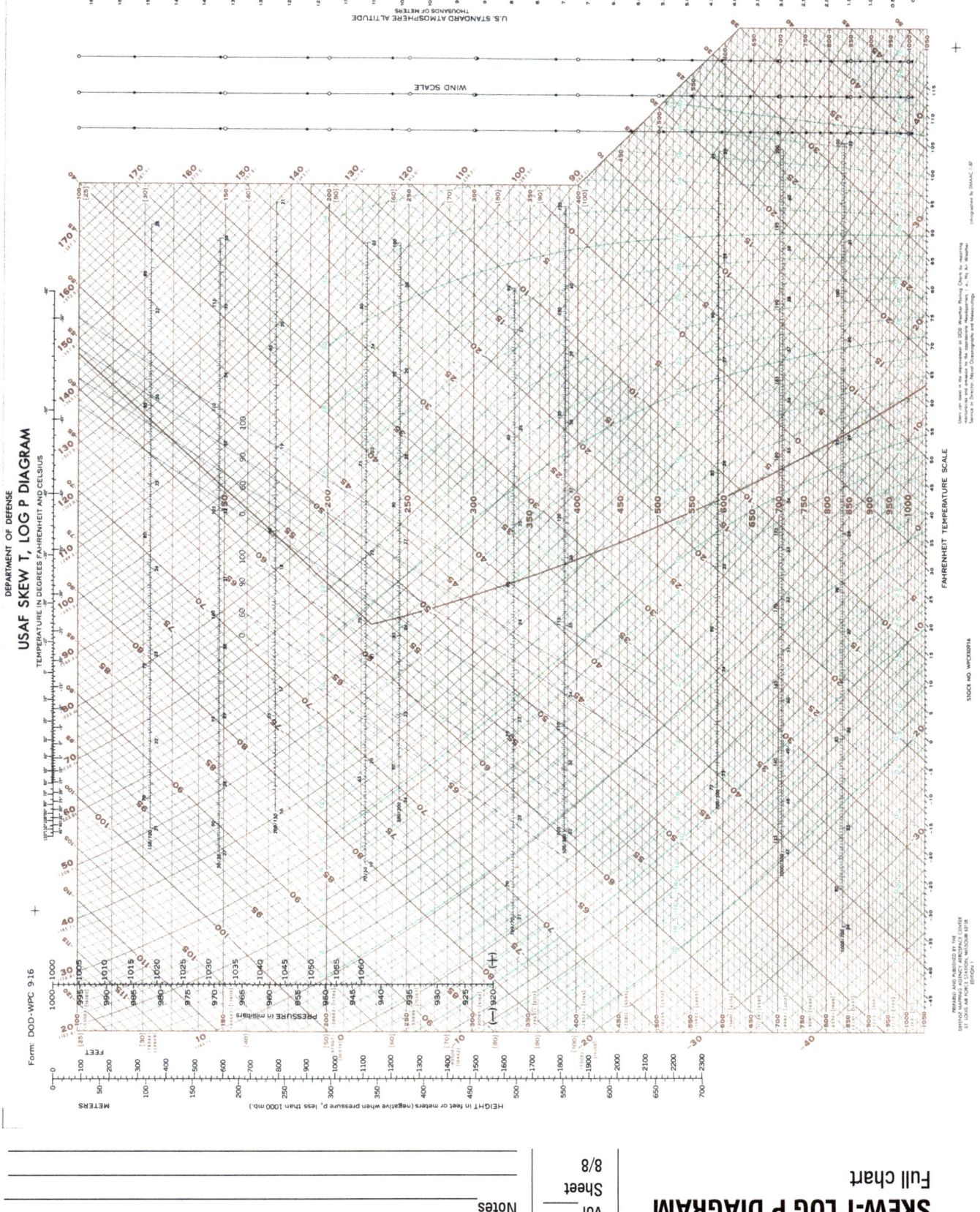

SKEW-T LOG P DIAGRAM

Full chart

Vol ____
Sheet
6/8

Date/time ____________________ Station ____________________

Notes ____________________

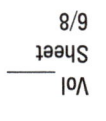

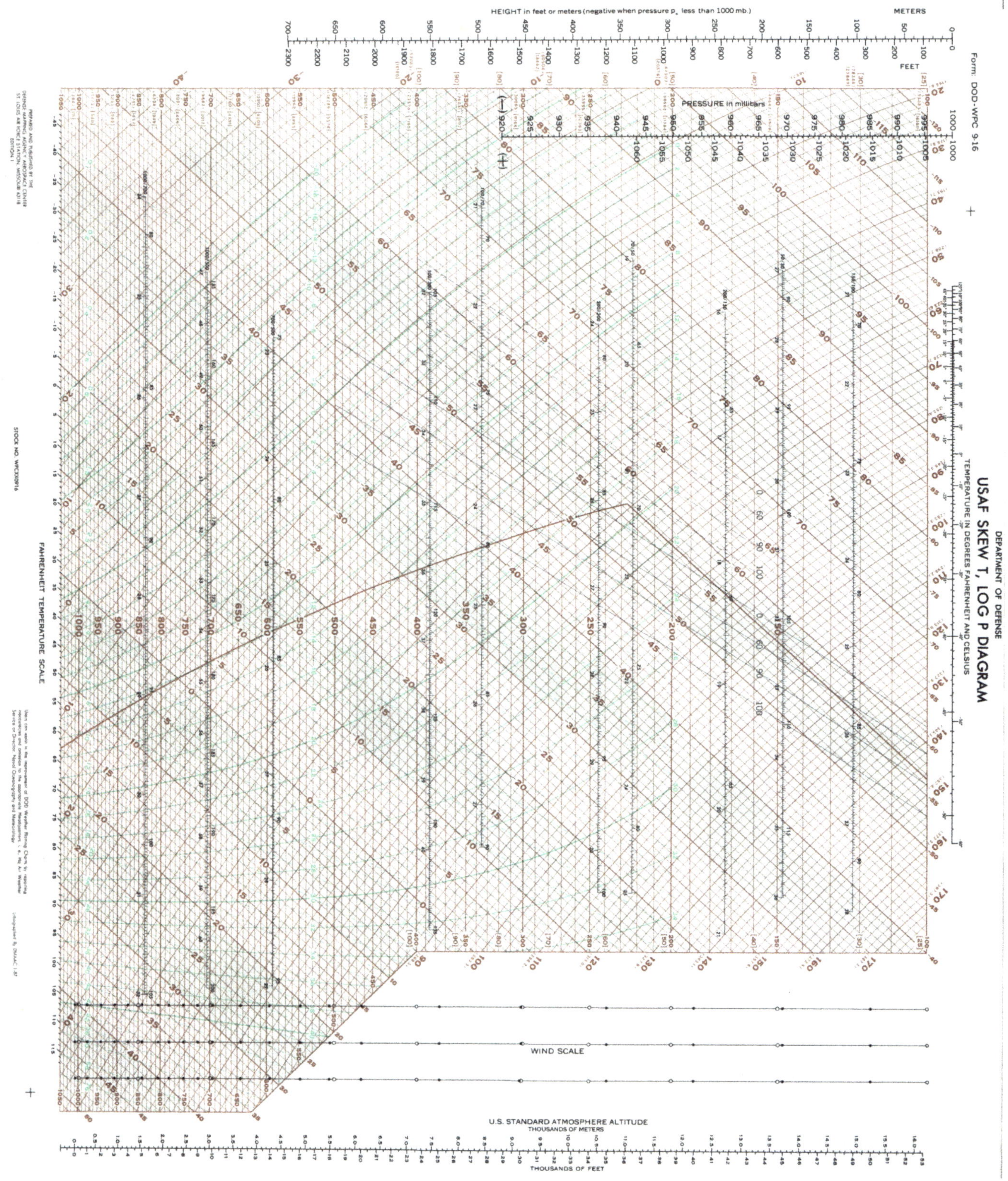

SKEW-T LOG P DIAGRAM
Full chart

USAF SKEW T, LOG P DIAGRAM
TEMPERATURE IN DEGREES FAHRENHEIT AND CELSIUS
DEPARTMENT OF DEFENSE

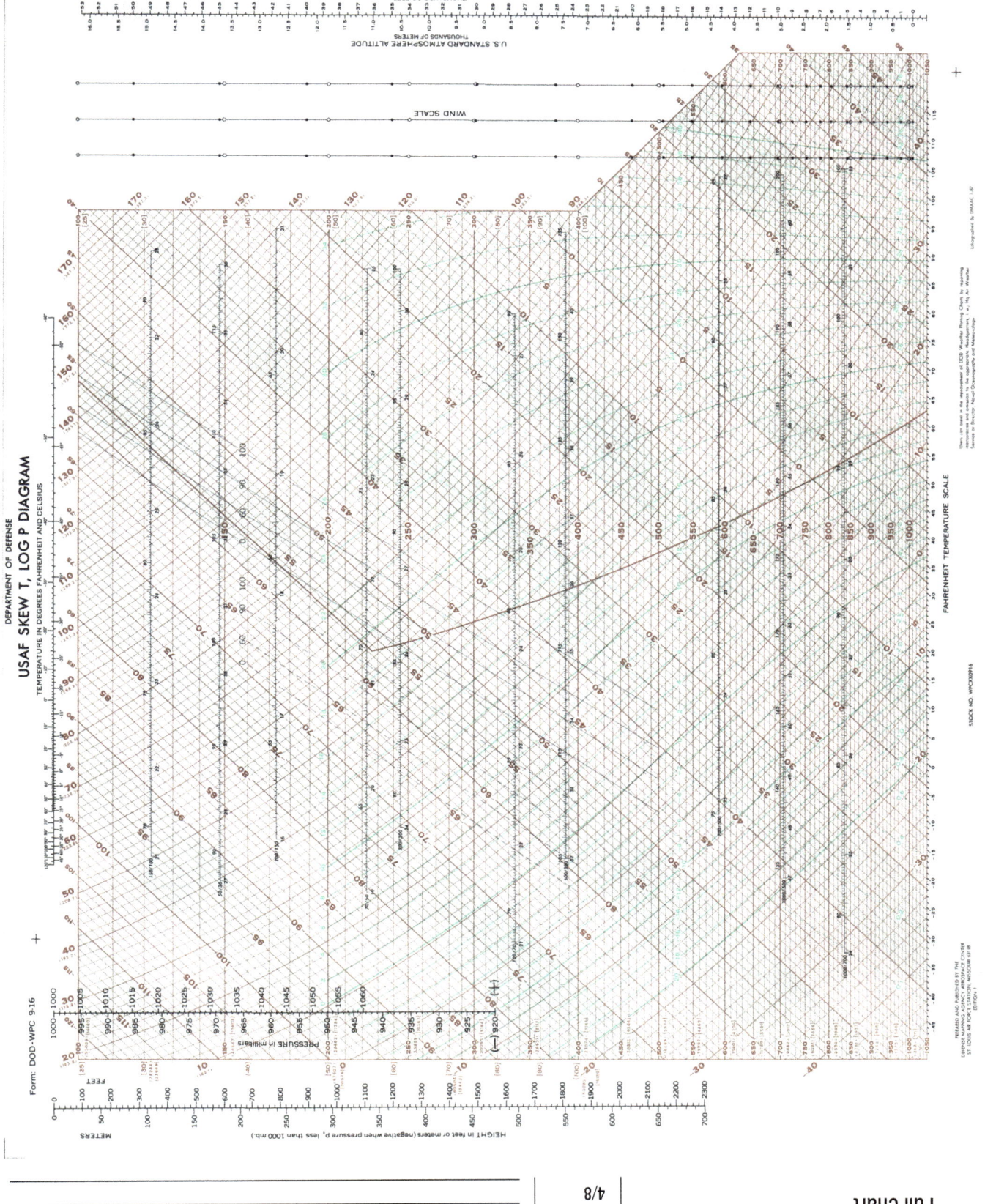

SKEW-T LOG P DIAGRAM
Full chart

USAF SKEW T, LOG P DIAGRAM
DEPARTMENT OF DEFENSE
TEMPERATURE IN DEGREES FAHRENHEIT AND CELSIUS

Form: DOD-WPC 9-16

Vol / Sheet 3/8

Date/time — Station — Notes

135

SKEW-T LOG P DIAGRAM

Full chart

Vol ____
Sheet
2/8

Date/time______________________ Station______________________
Notes____

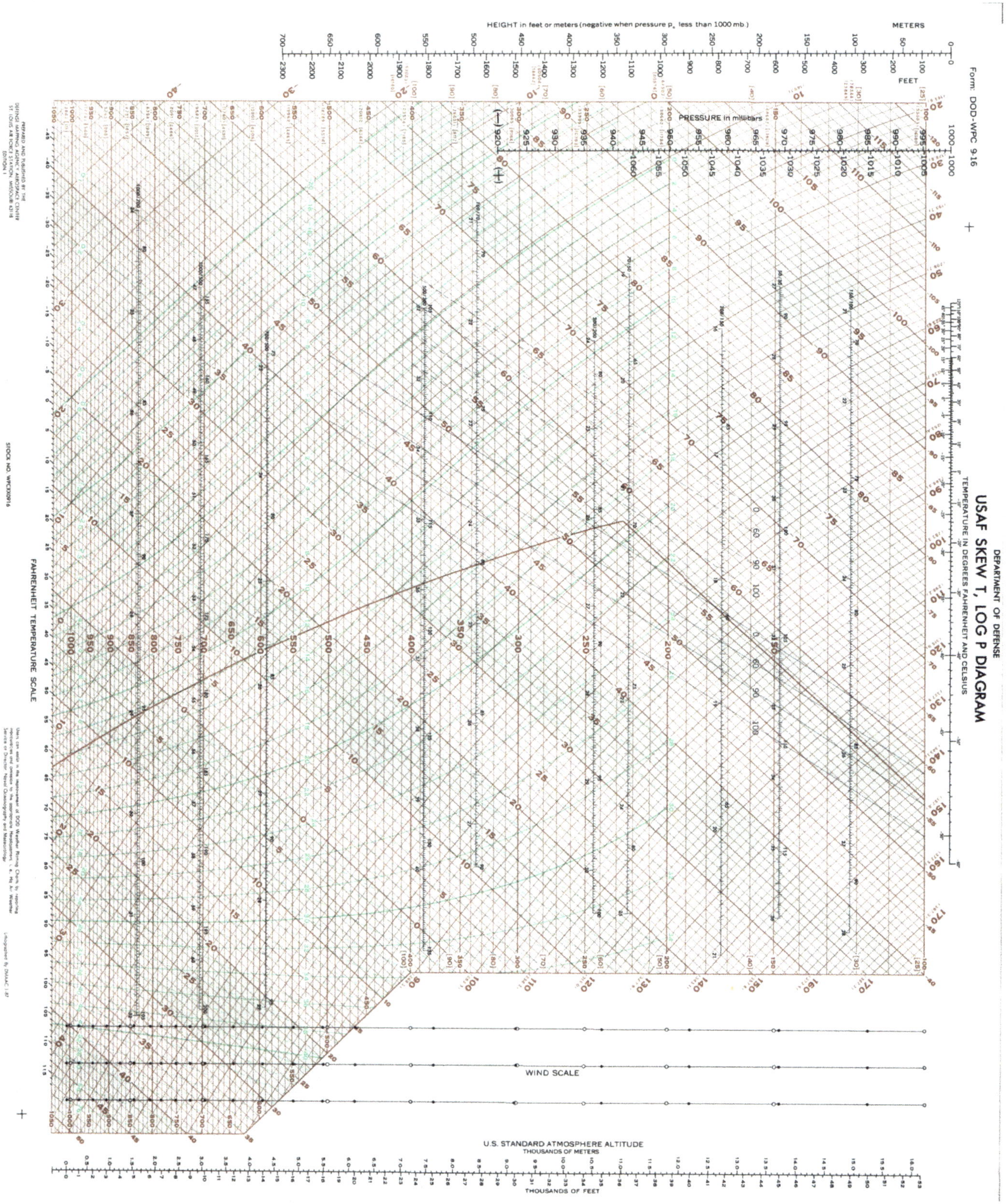

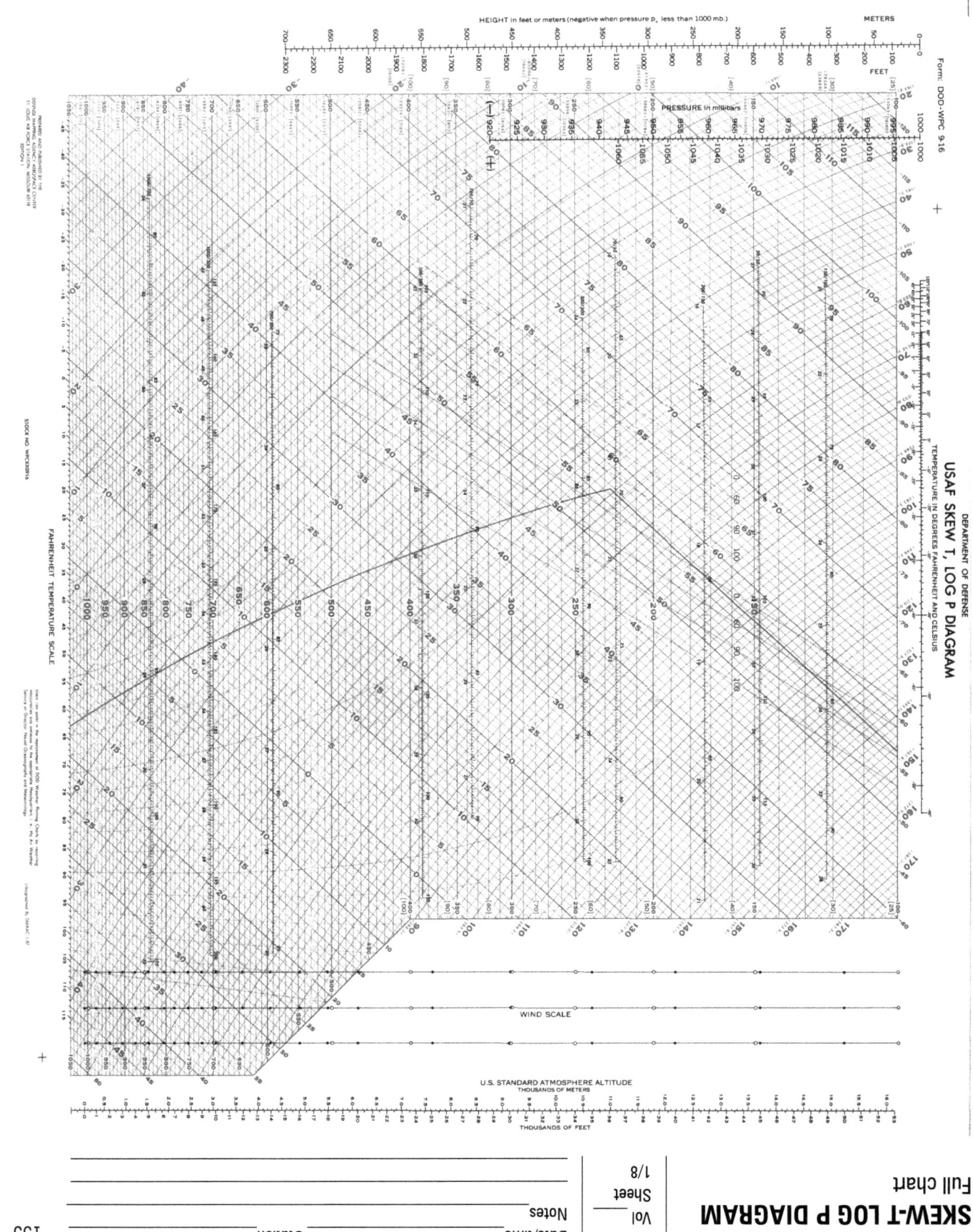

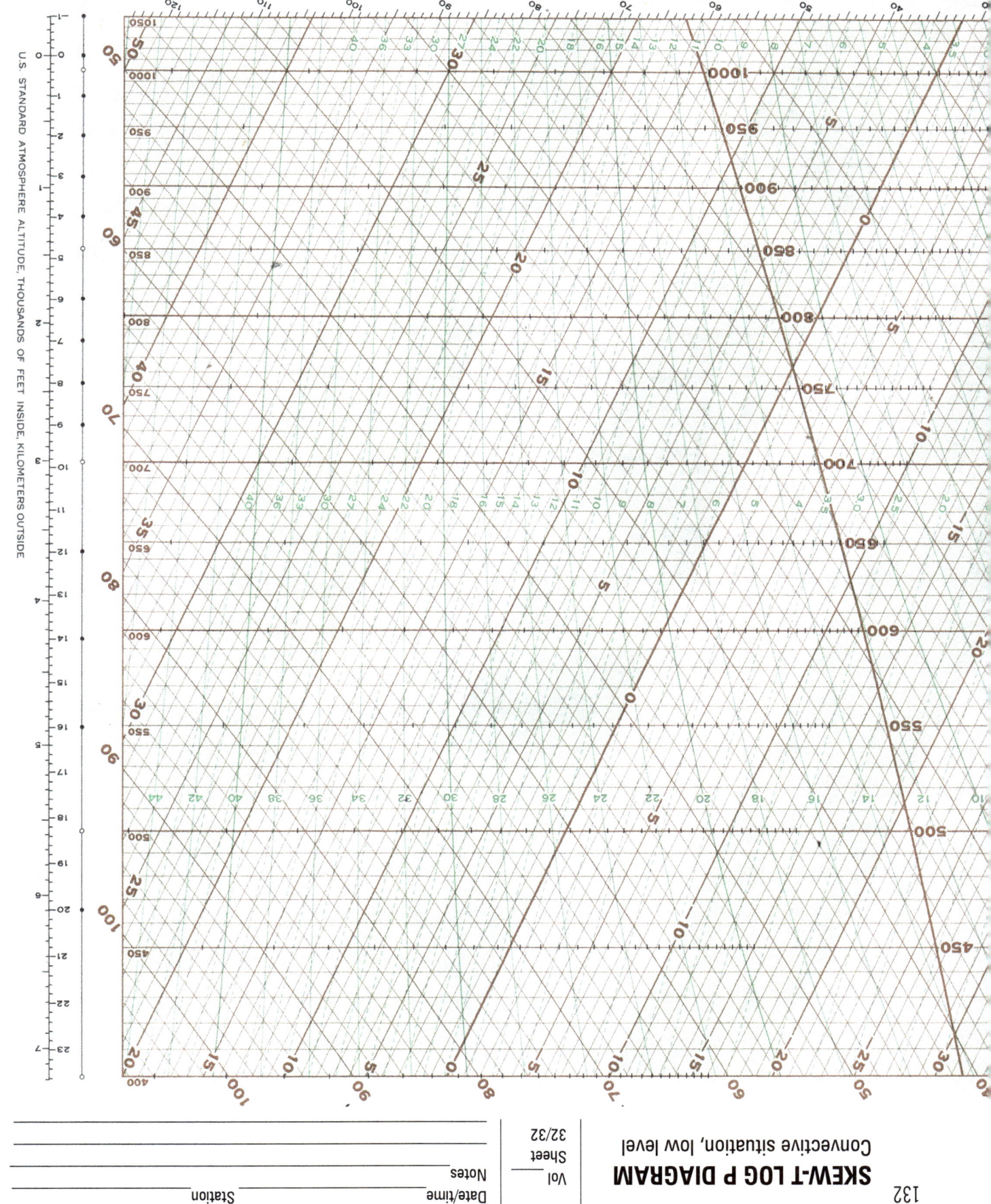

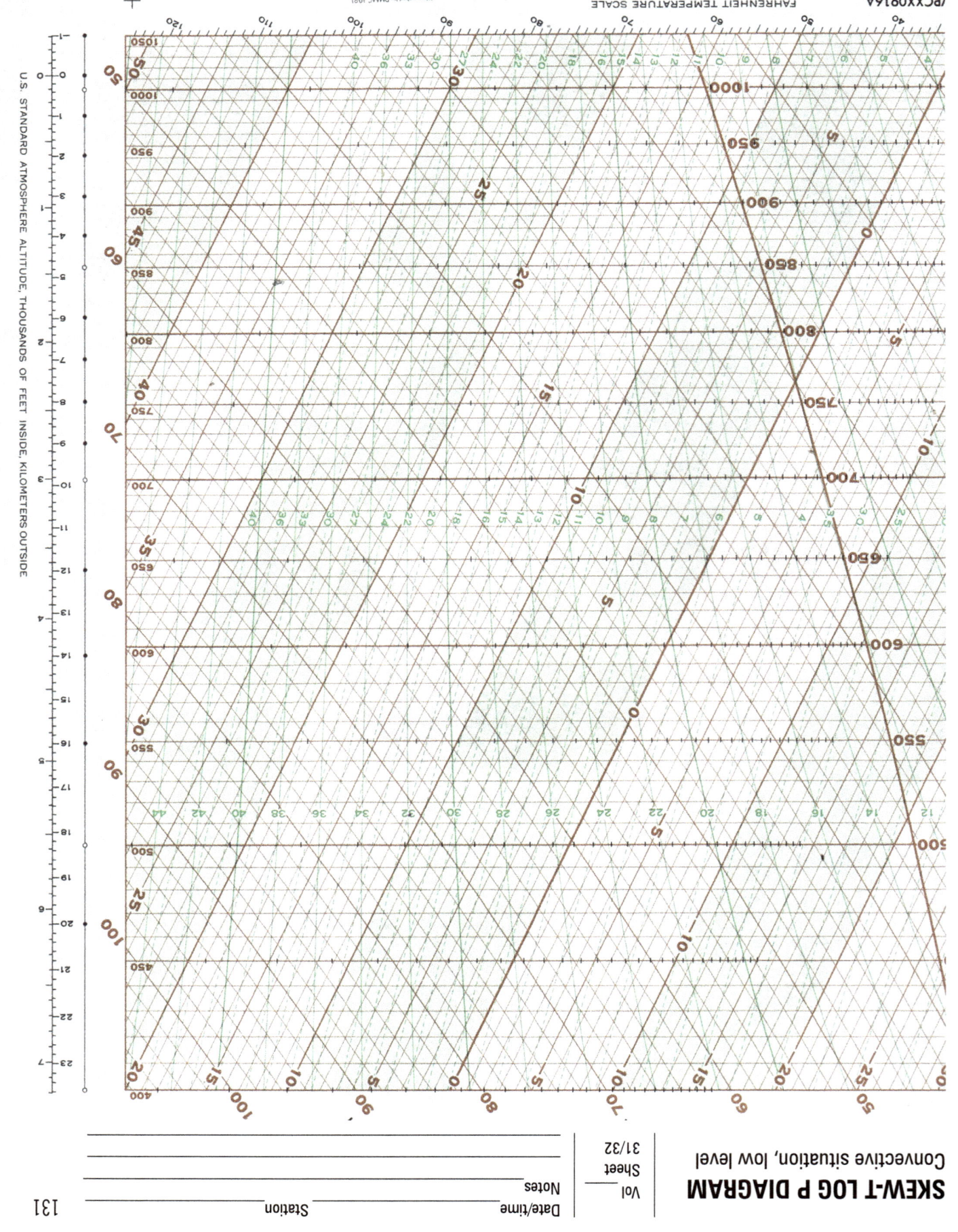

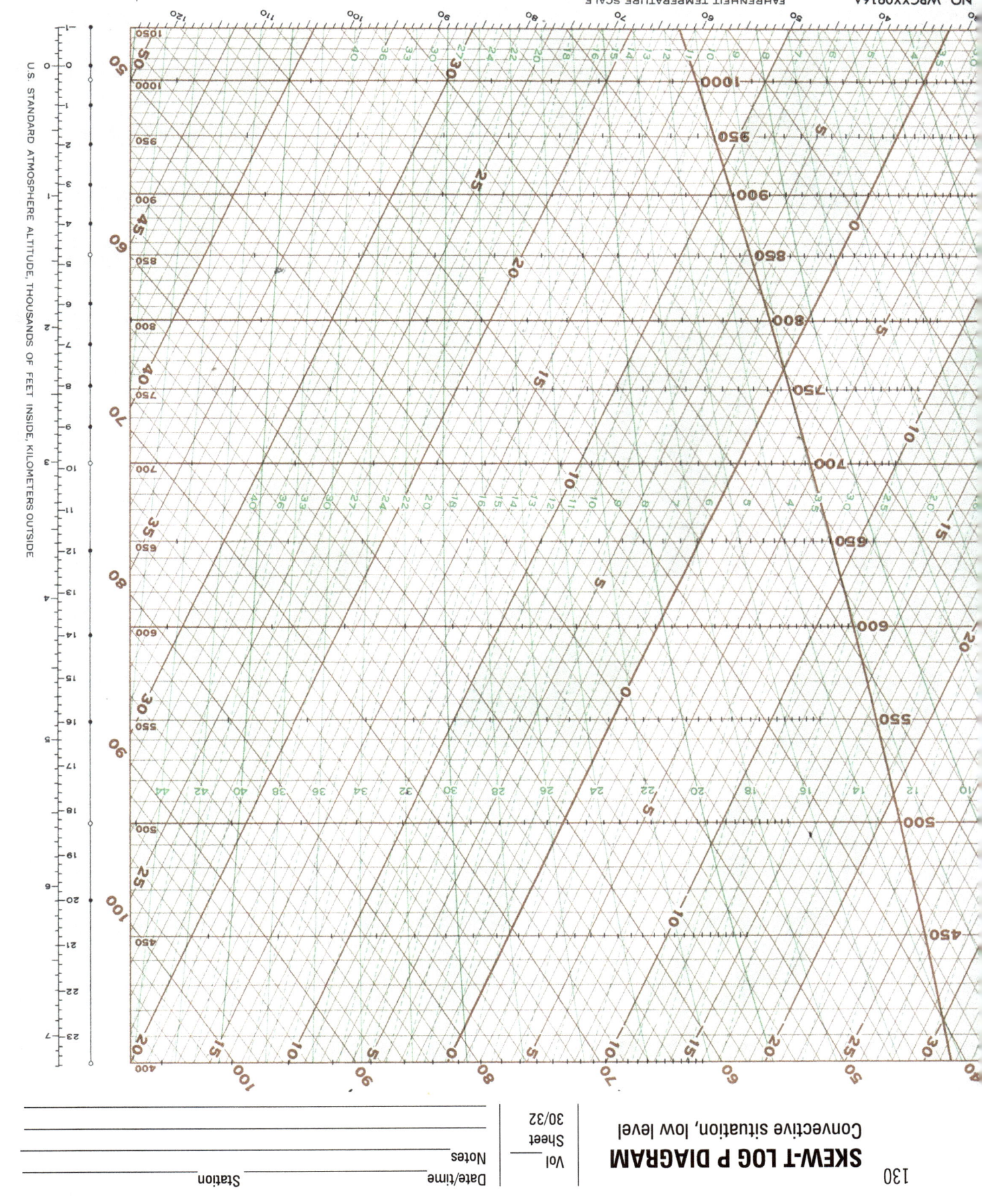

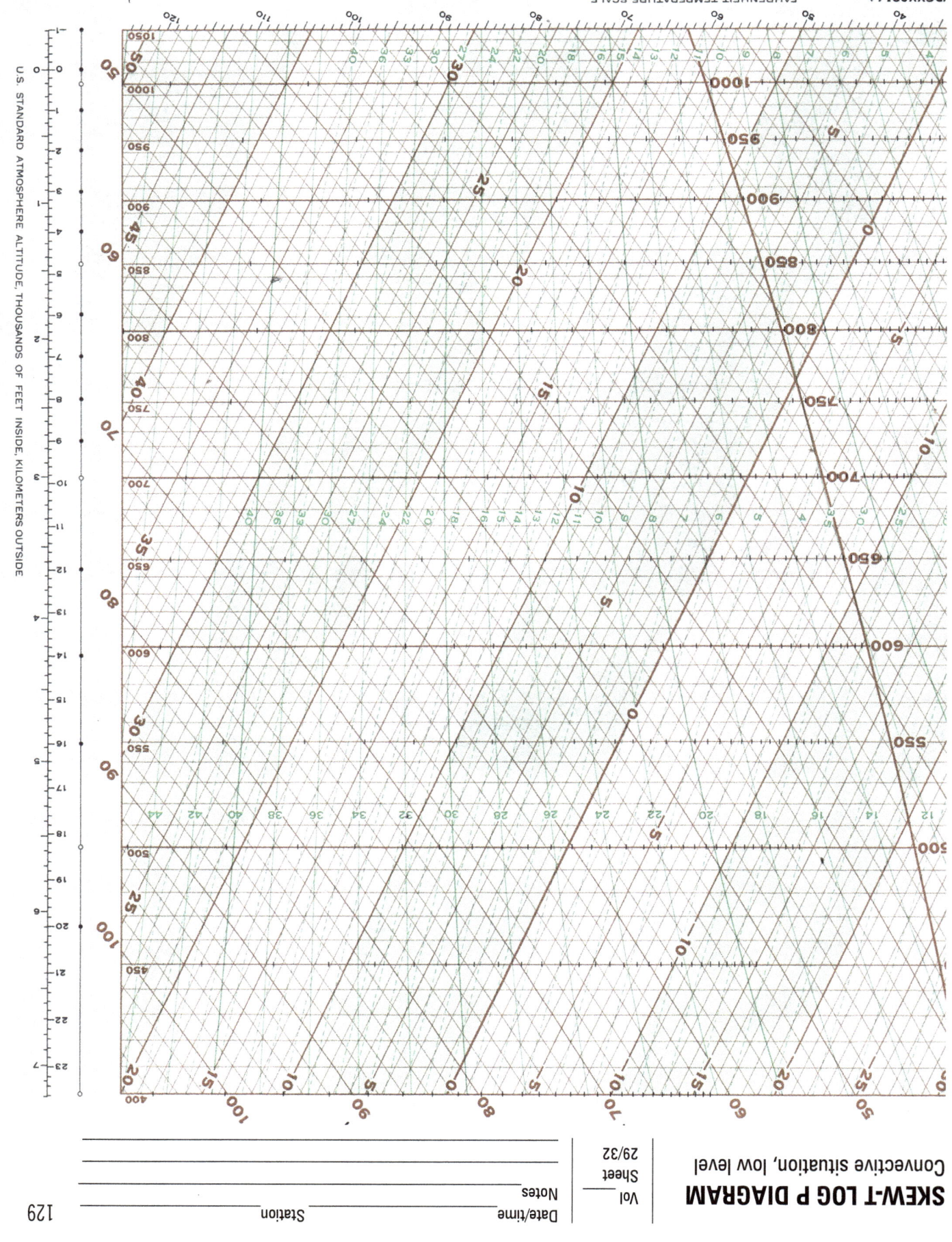

SKEW-T LOG P DIAGRAM
Convective situation, low level

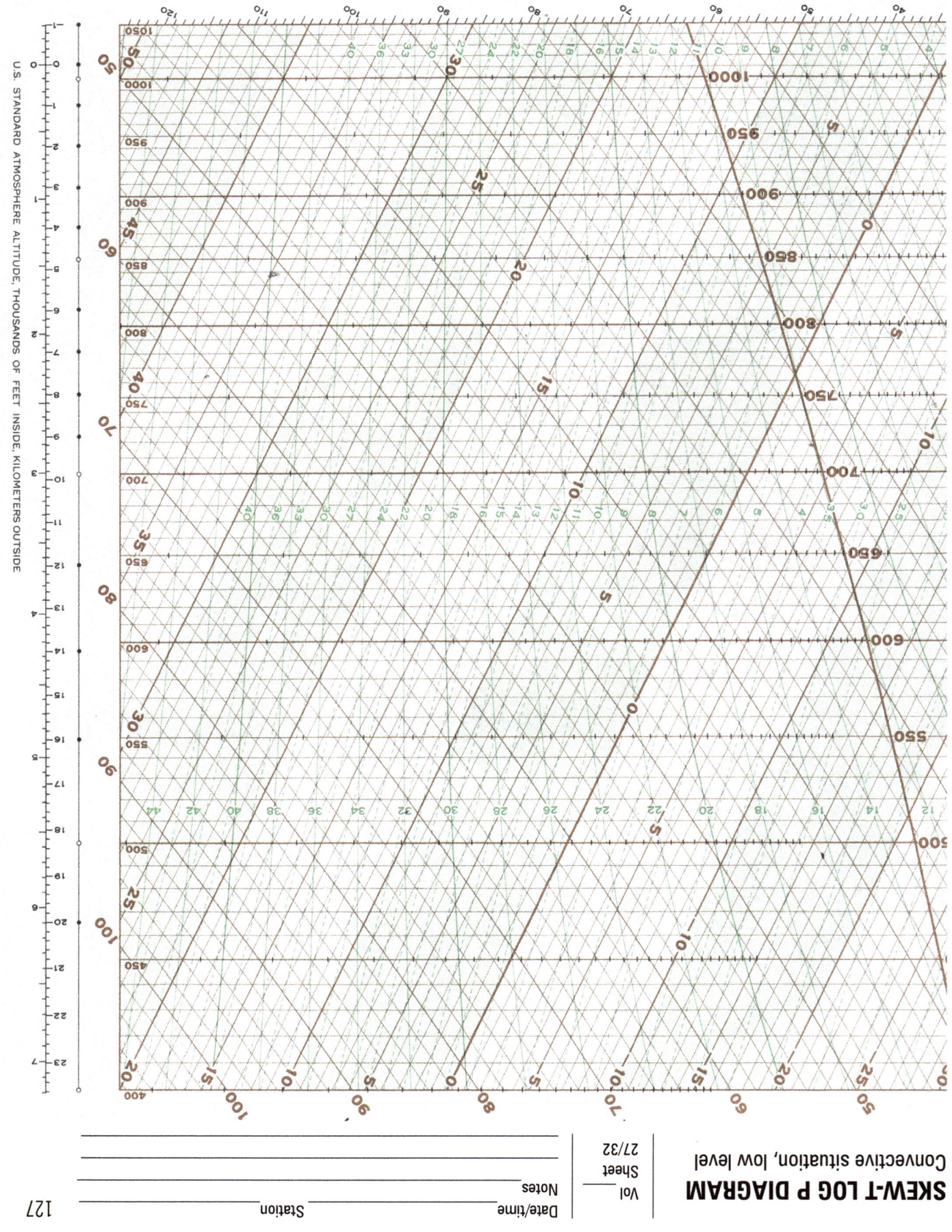

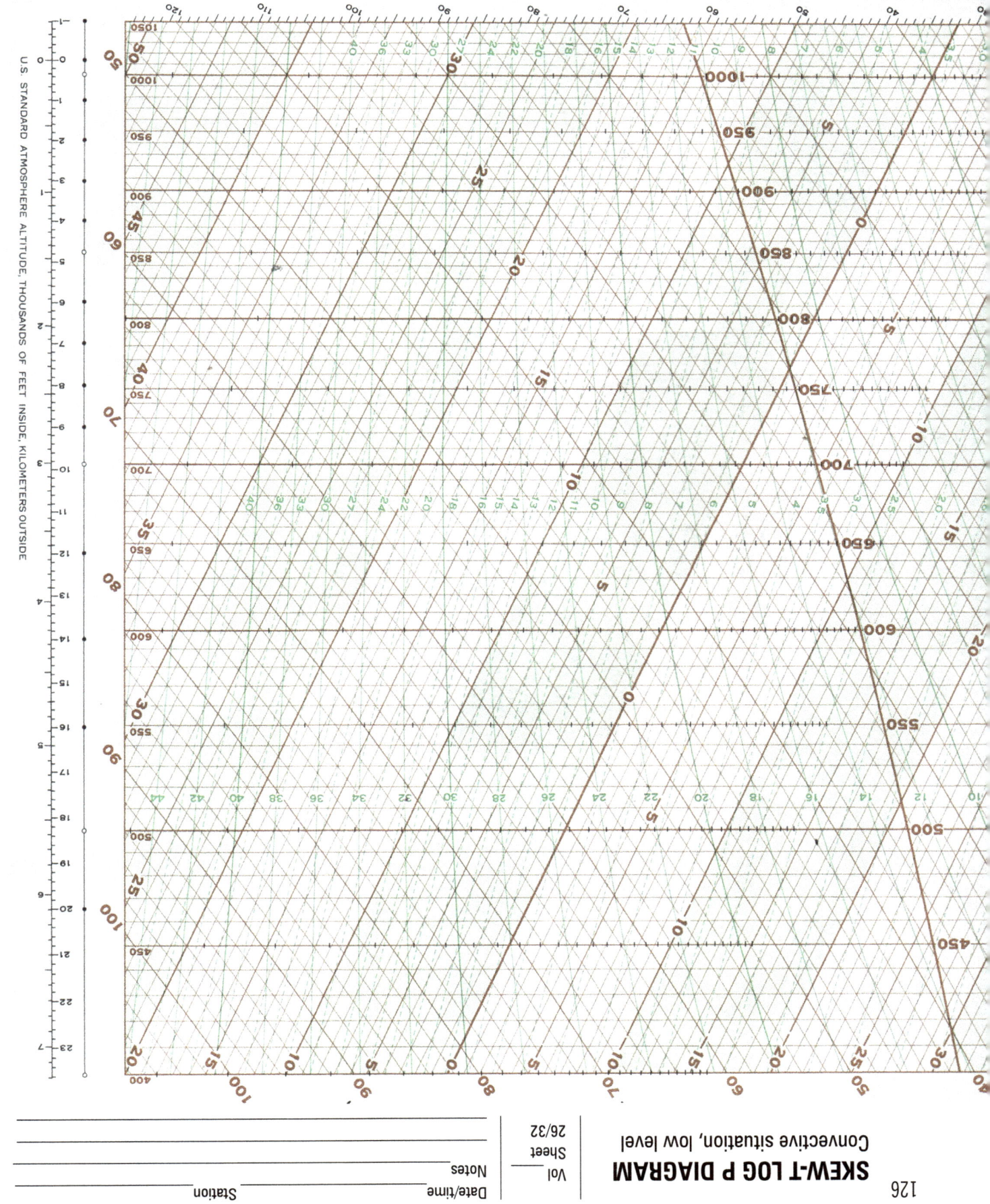

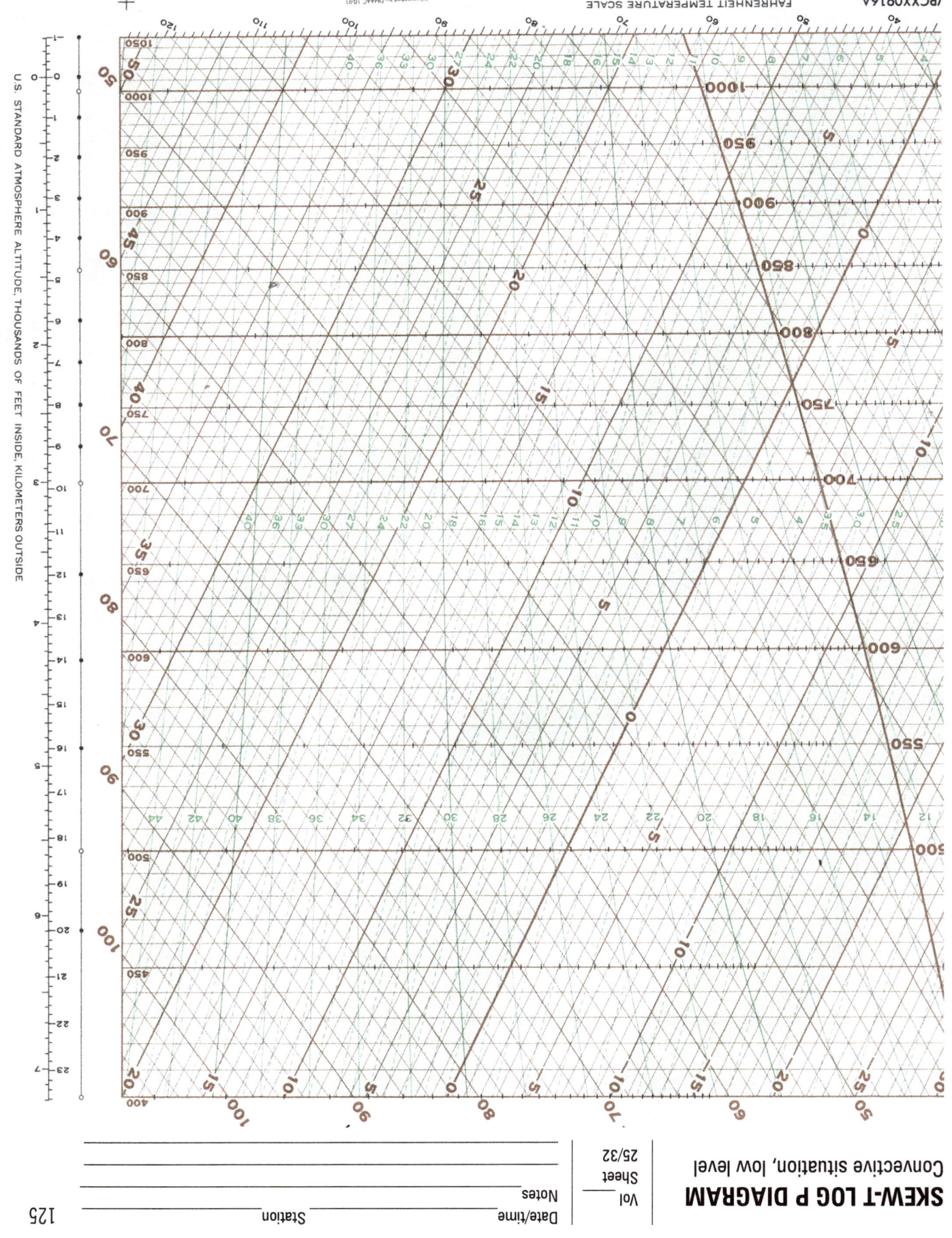

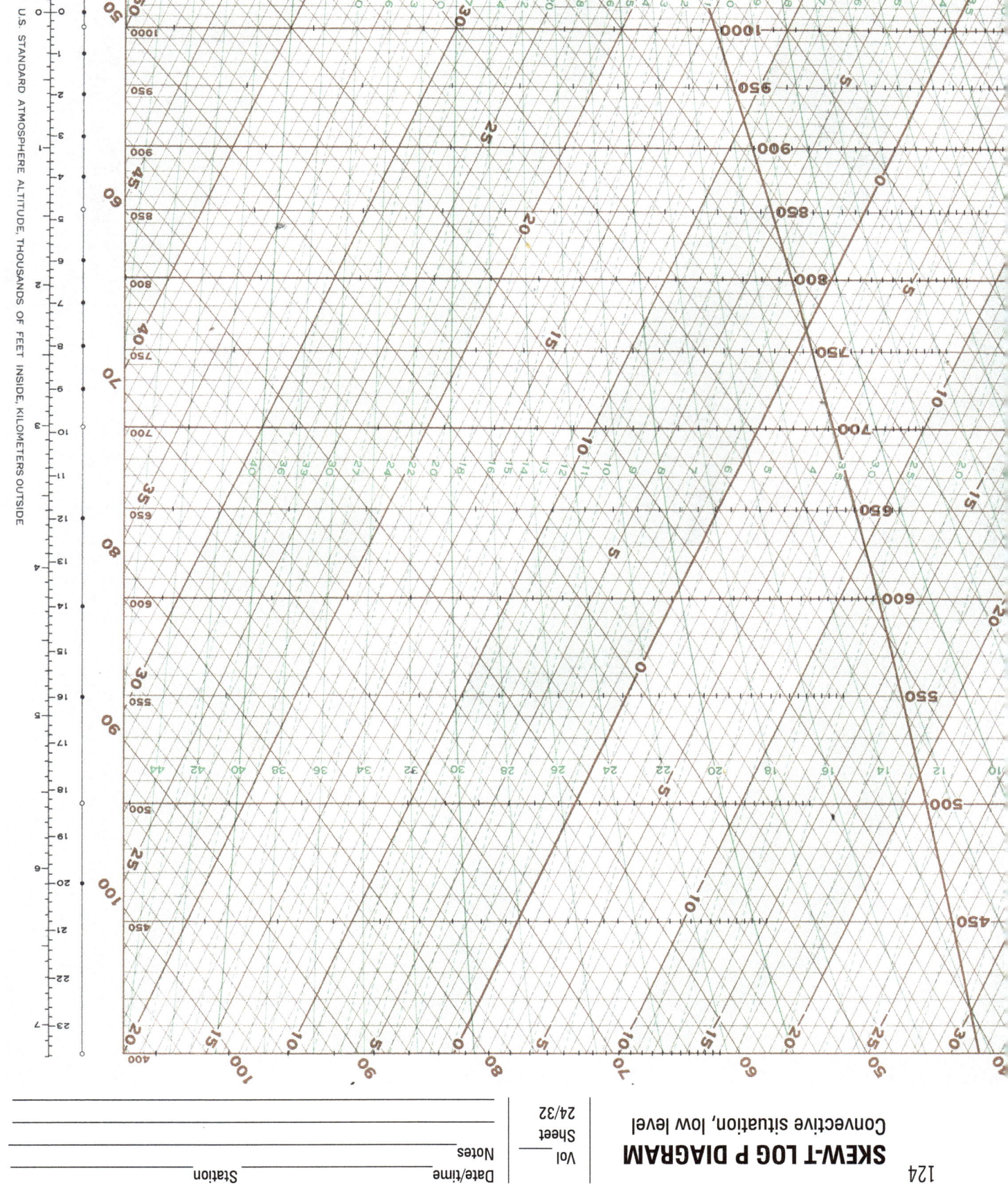

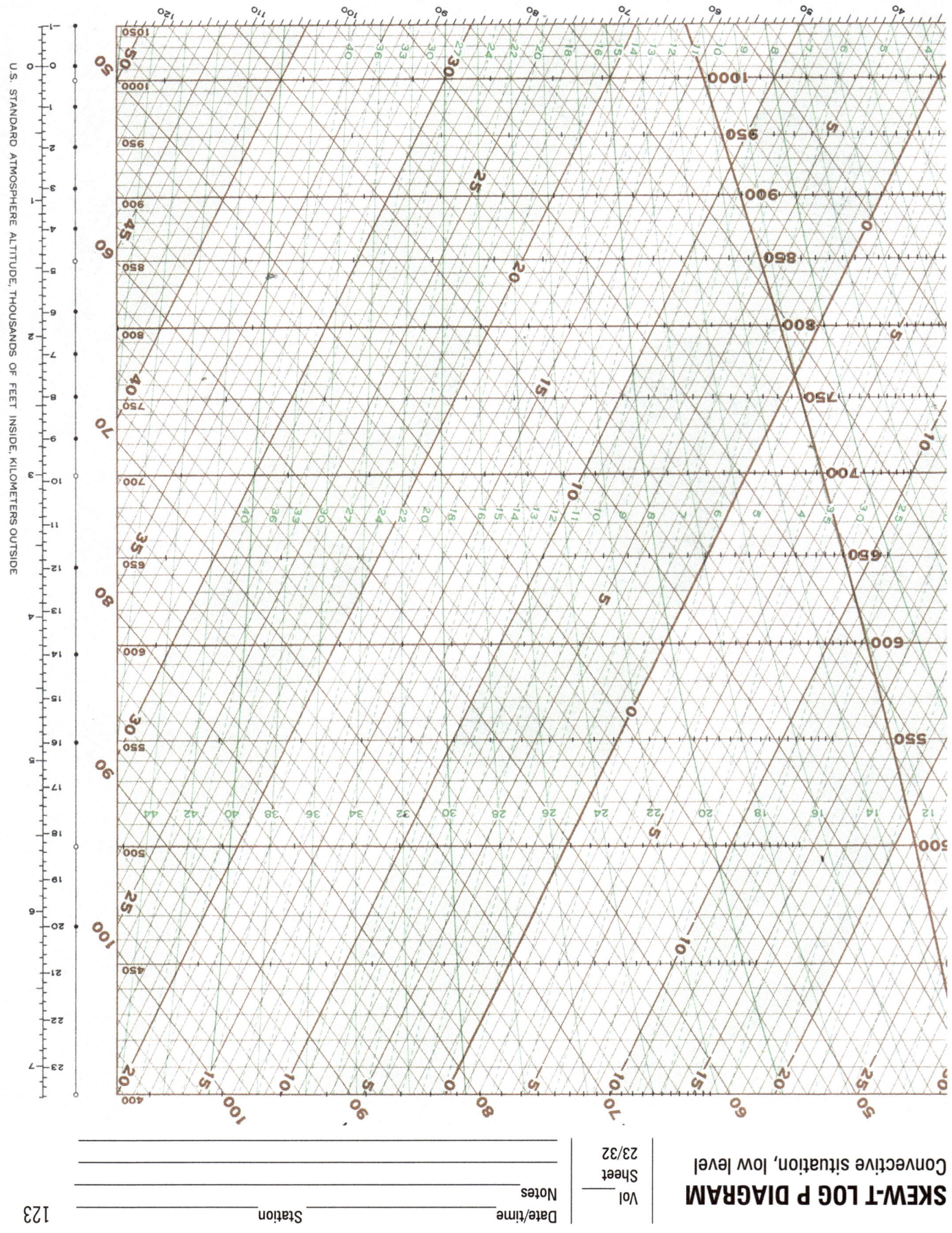

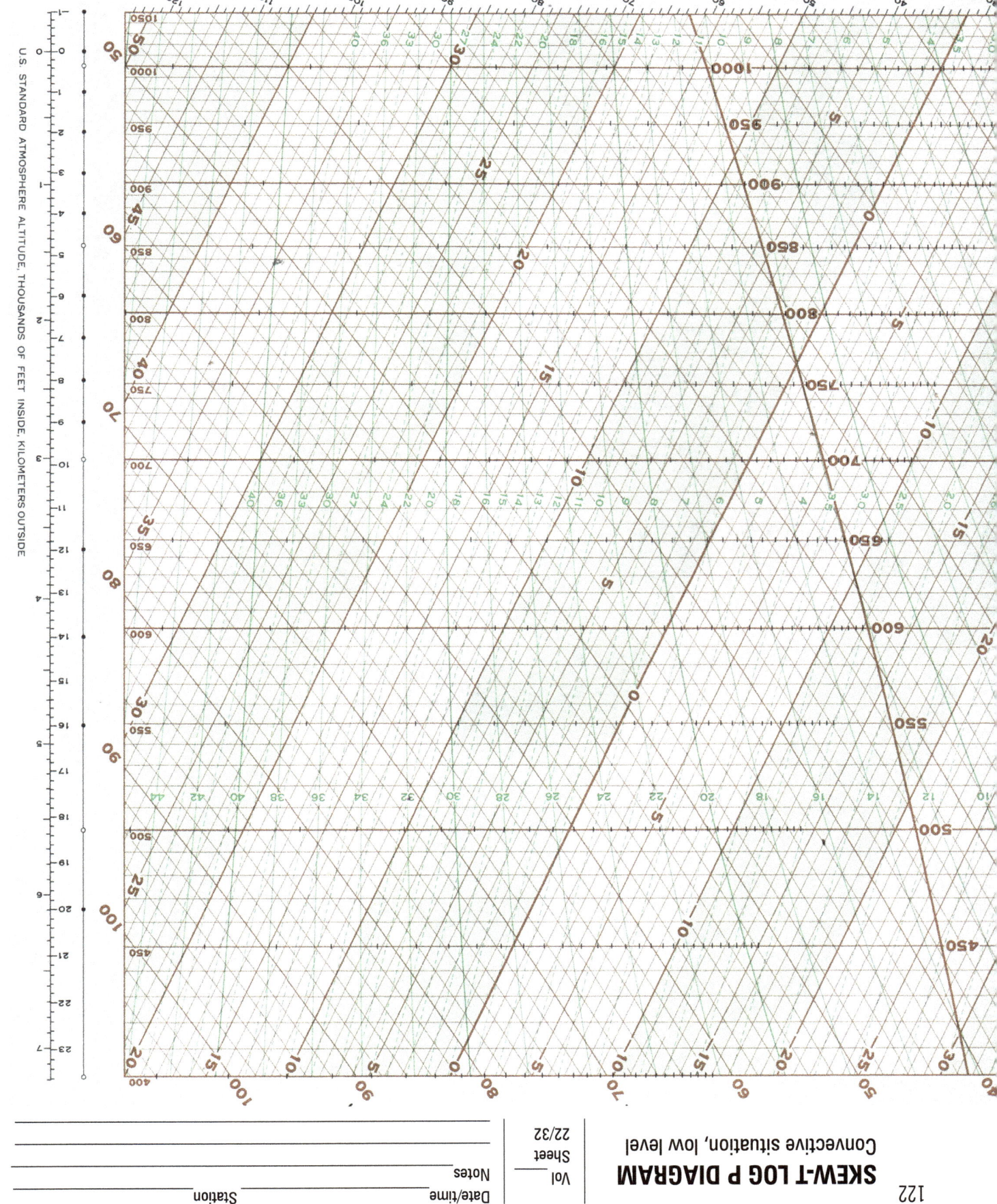

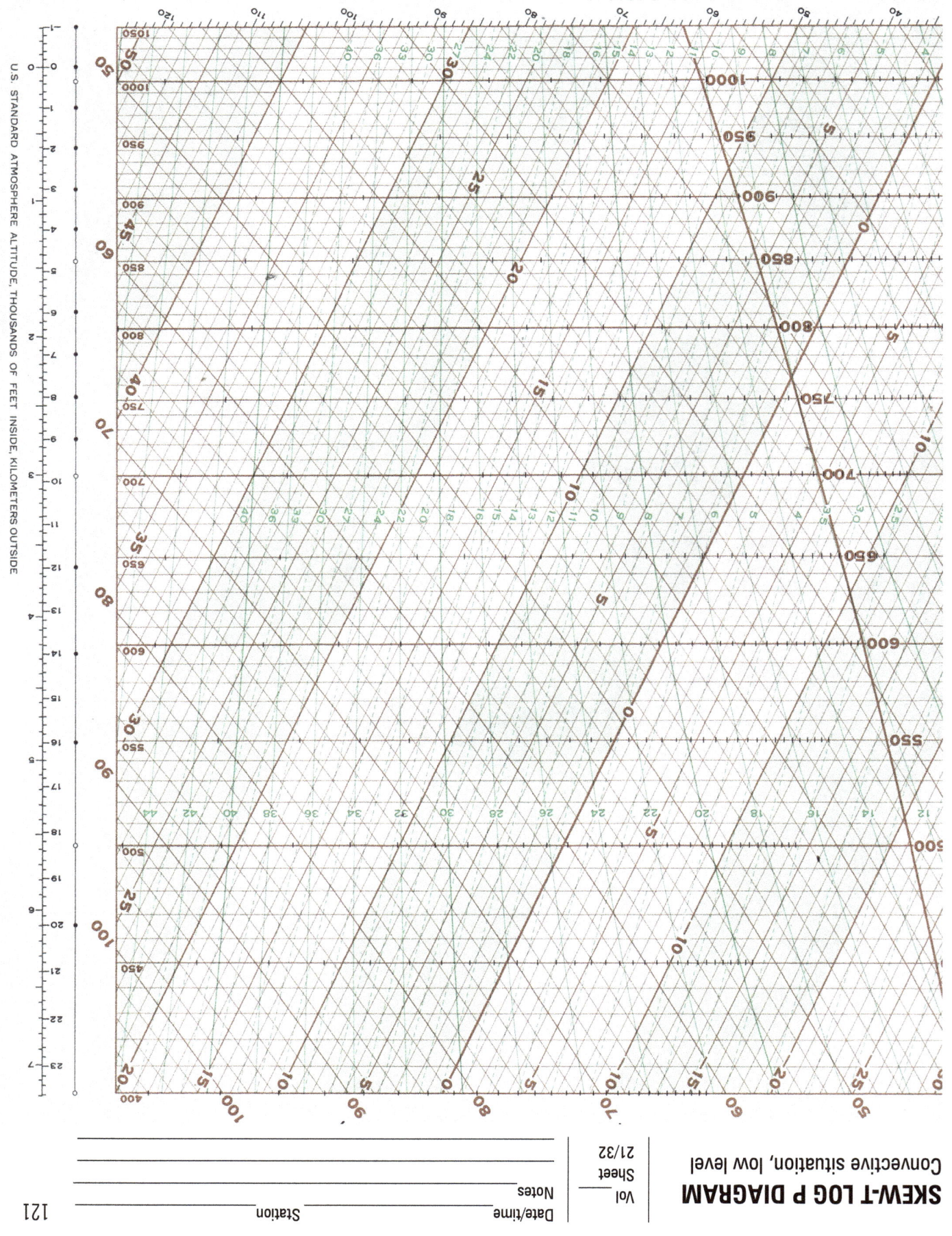

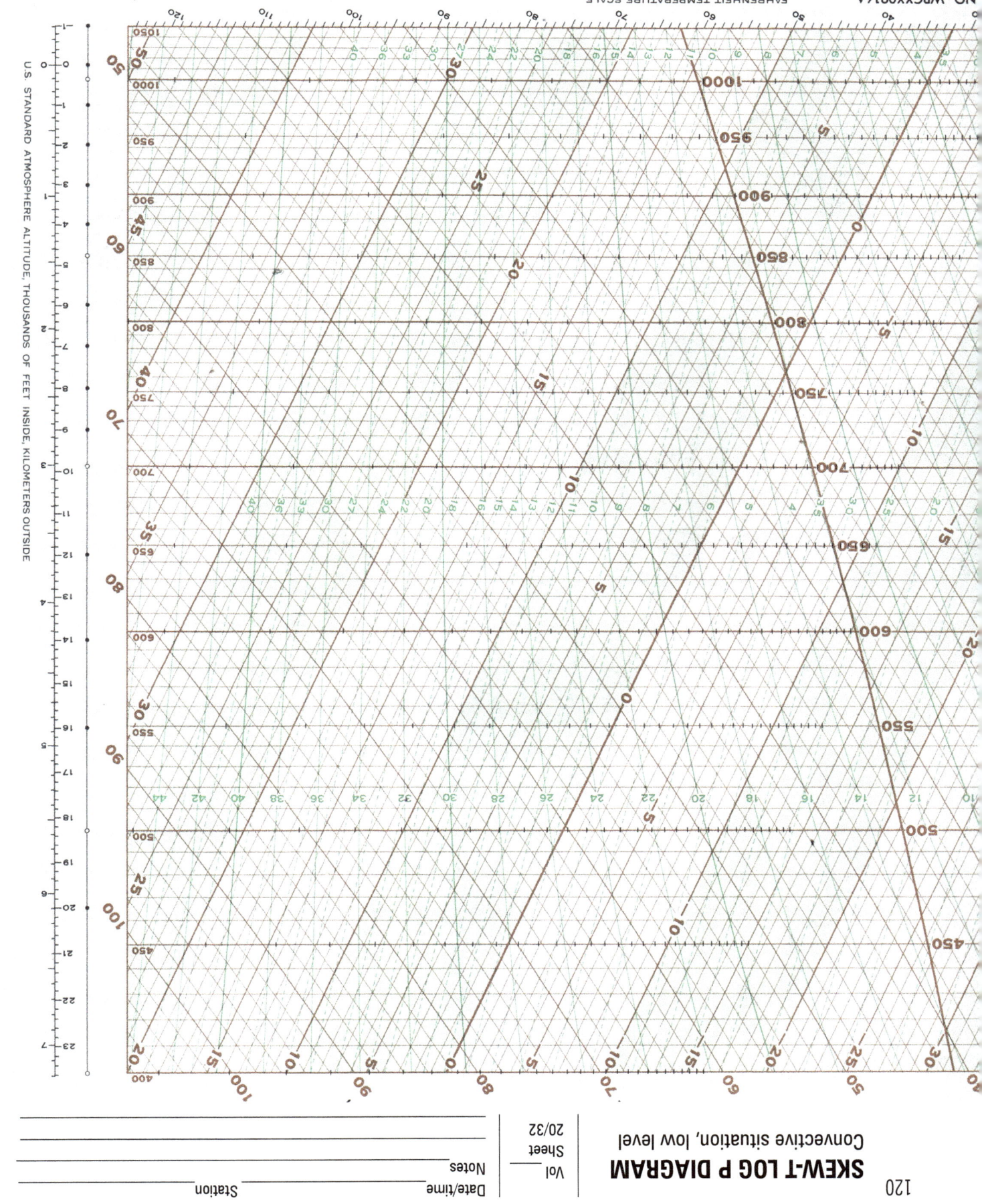

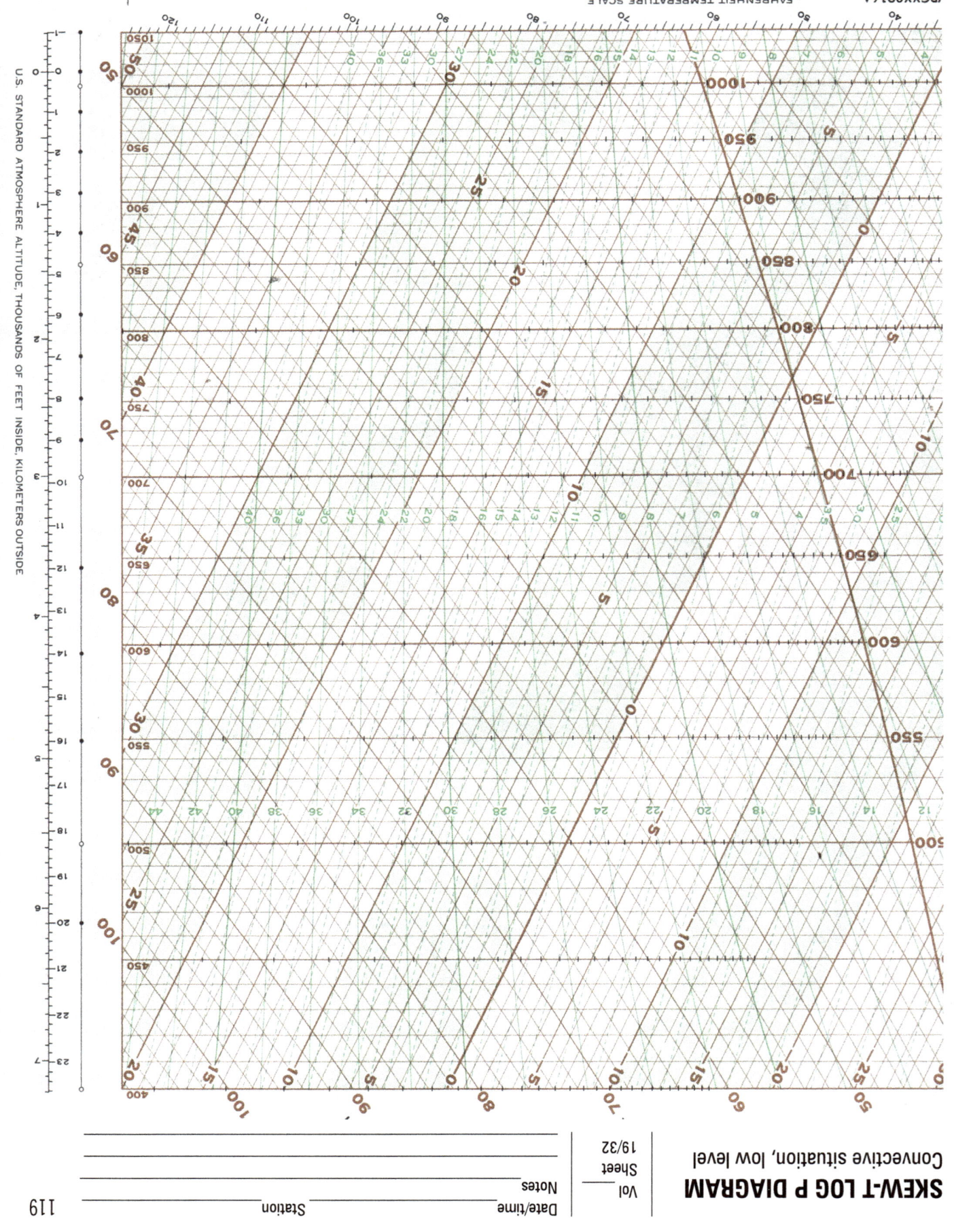

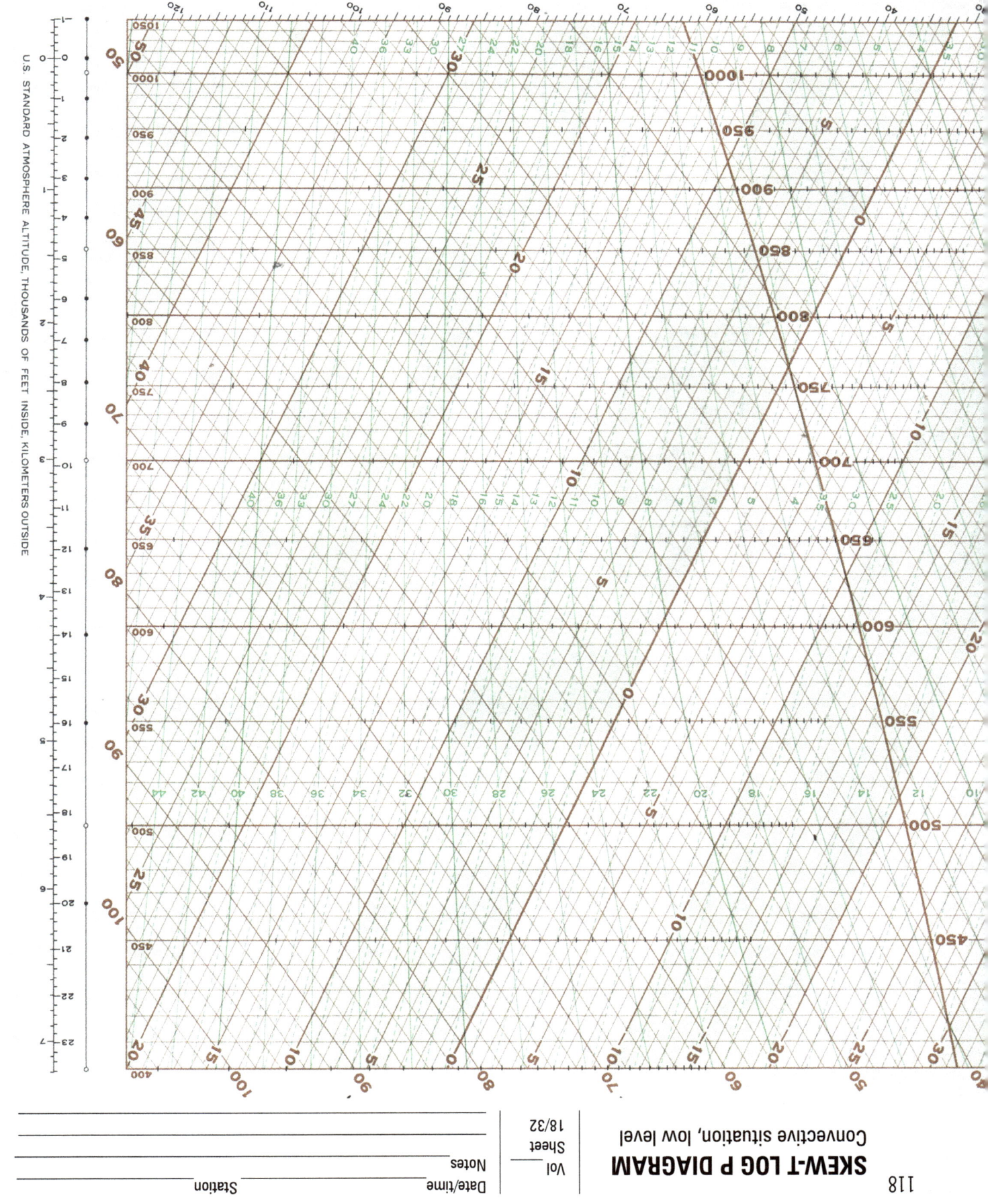

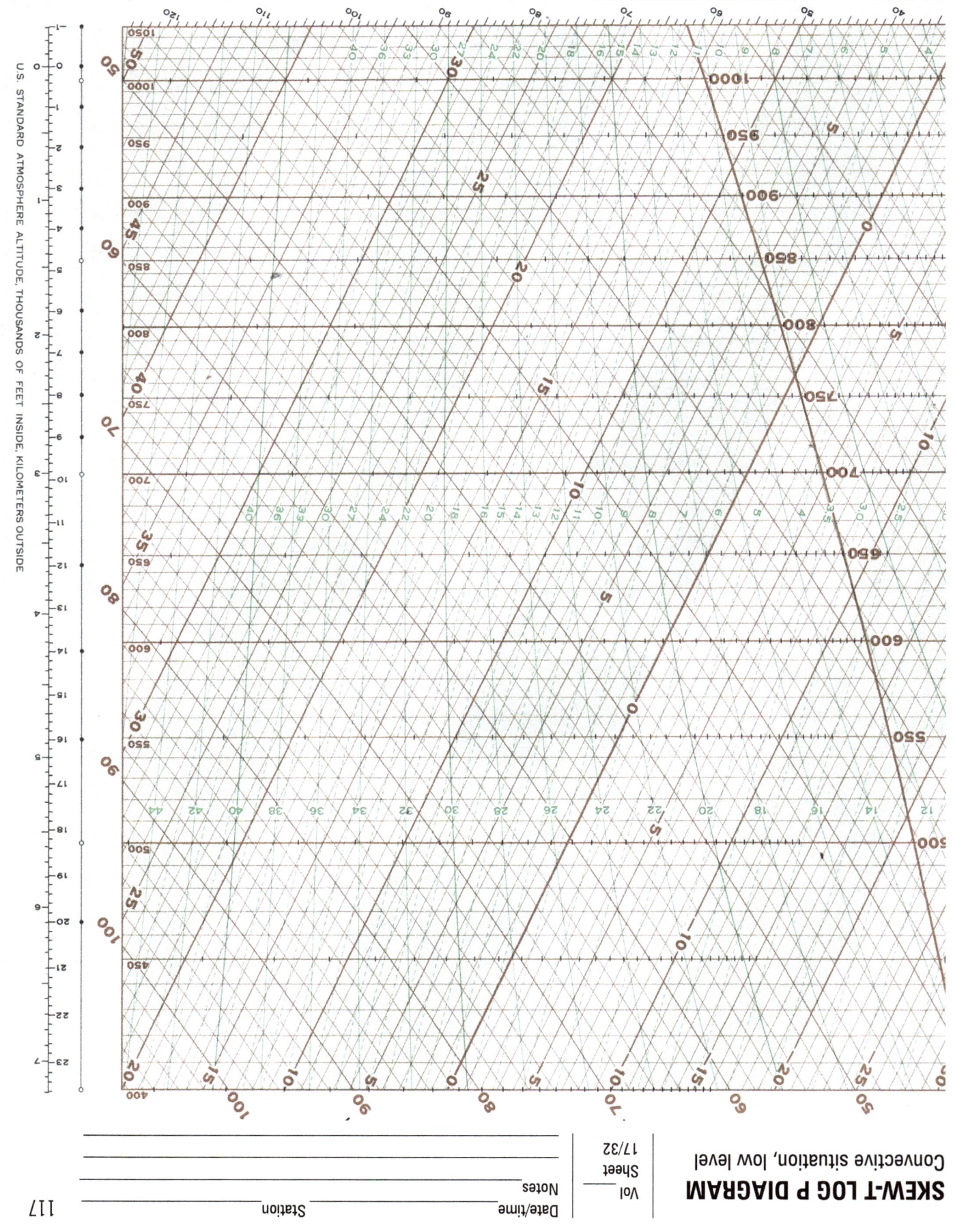

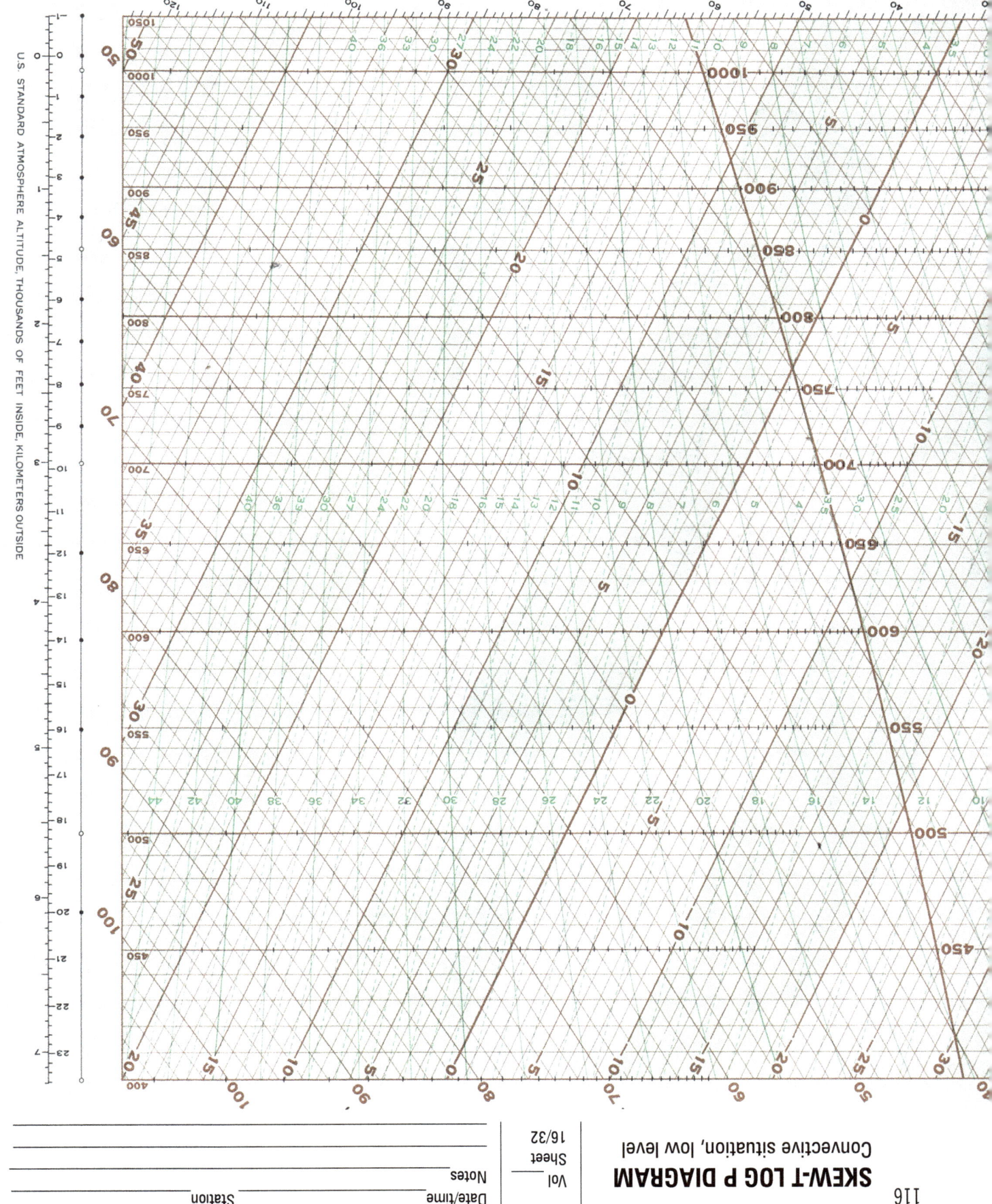

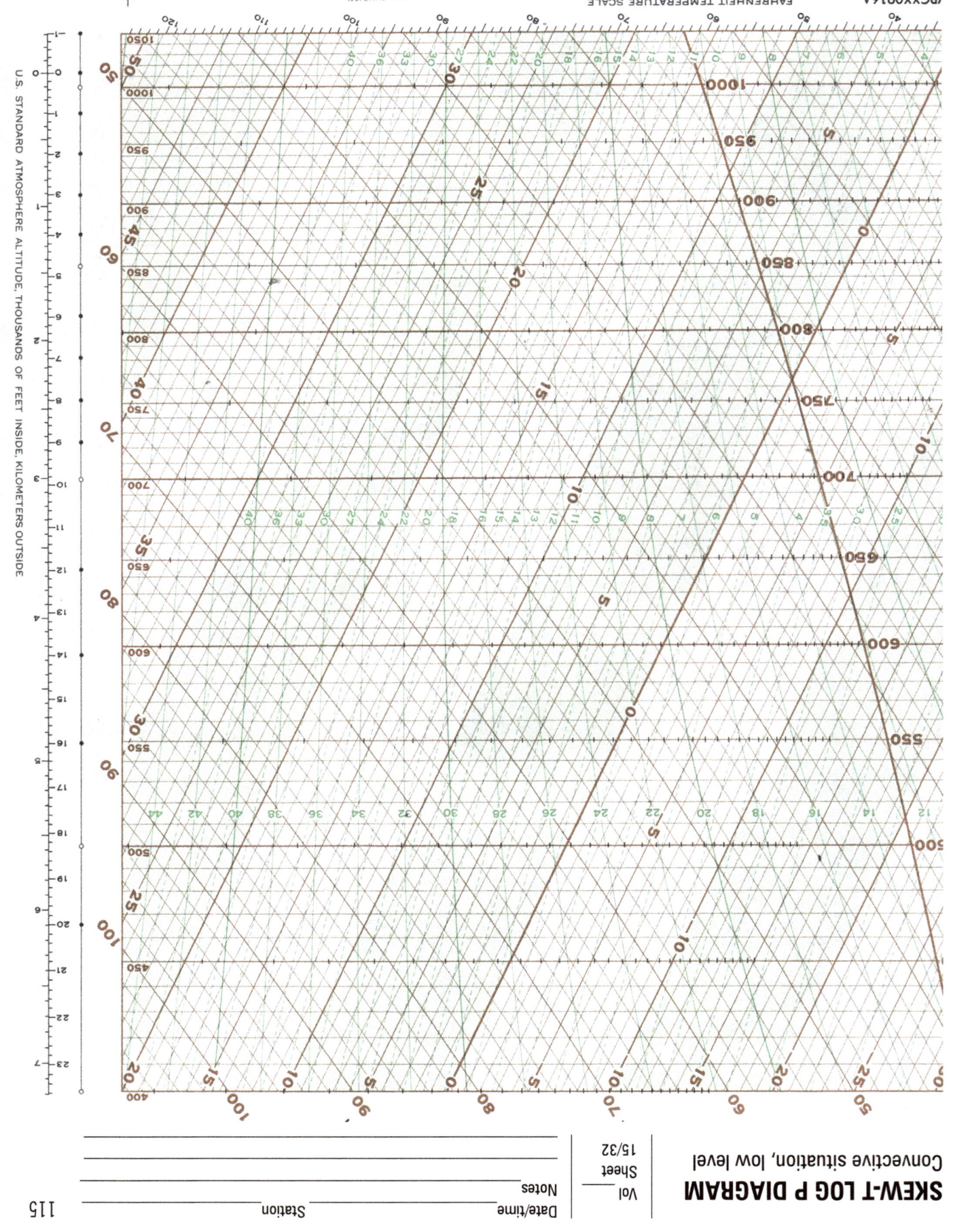

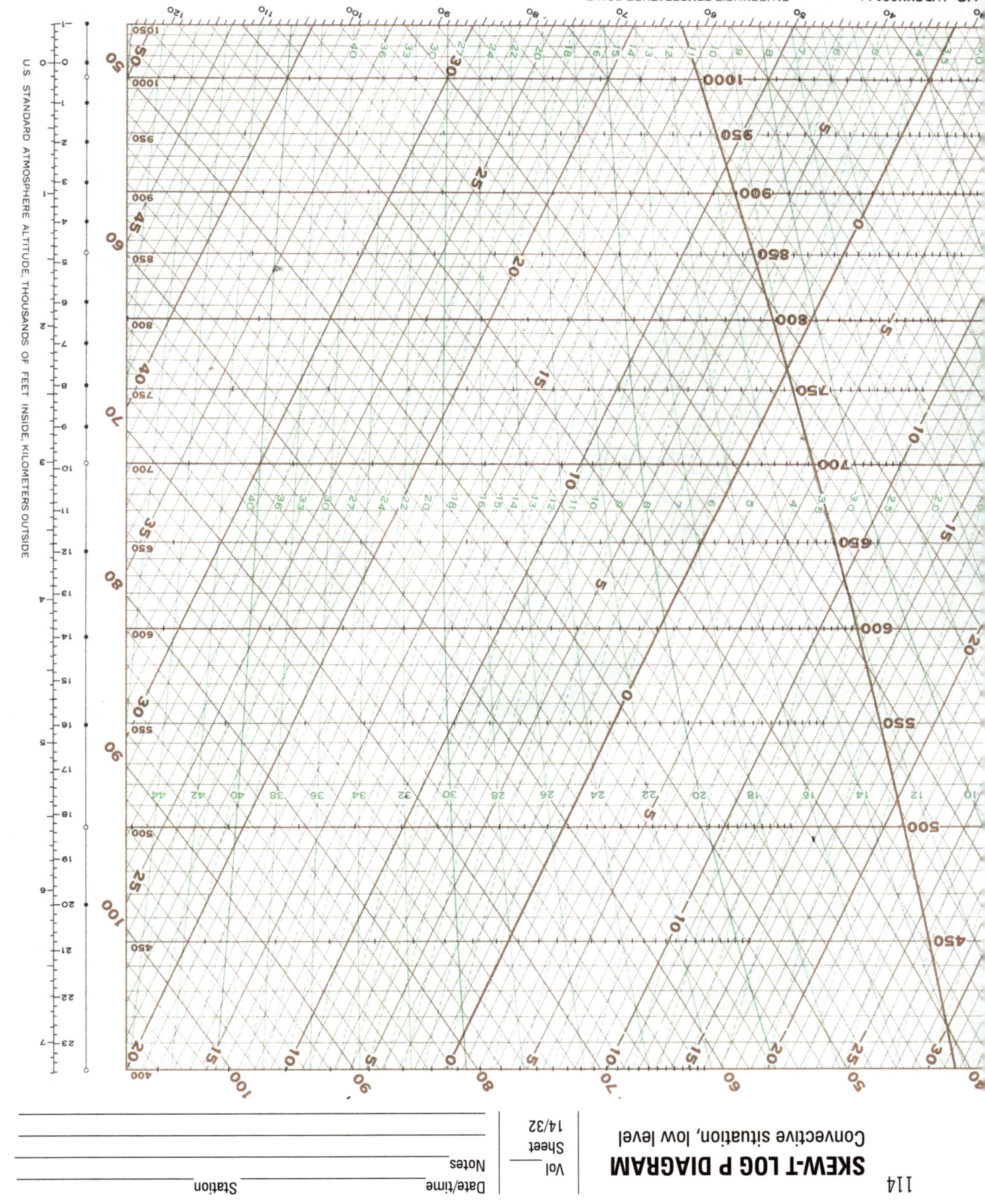

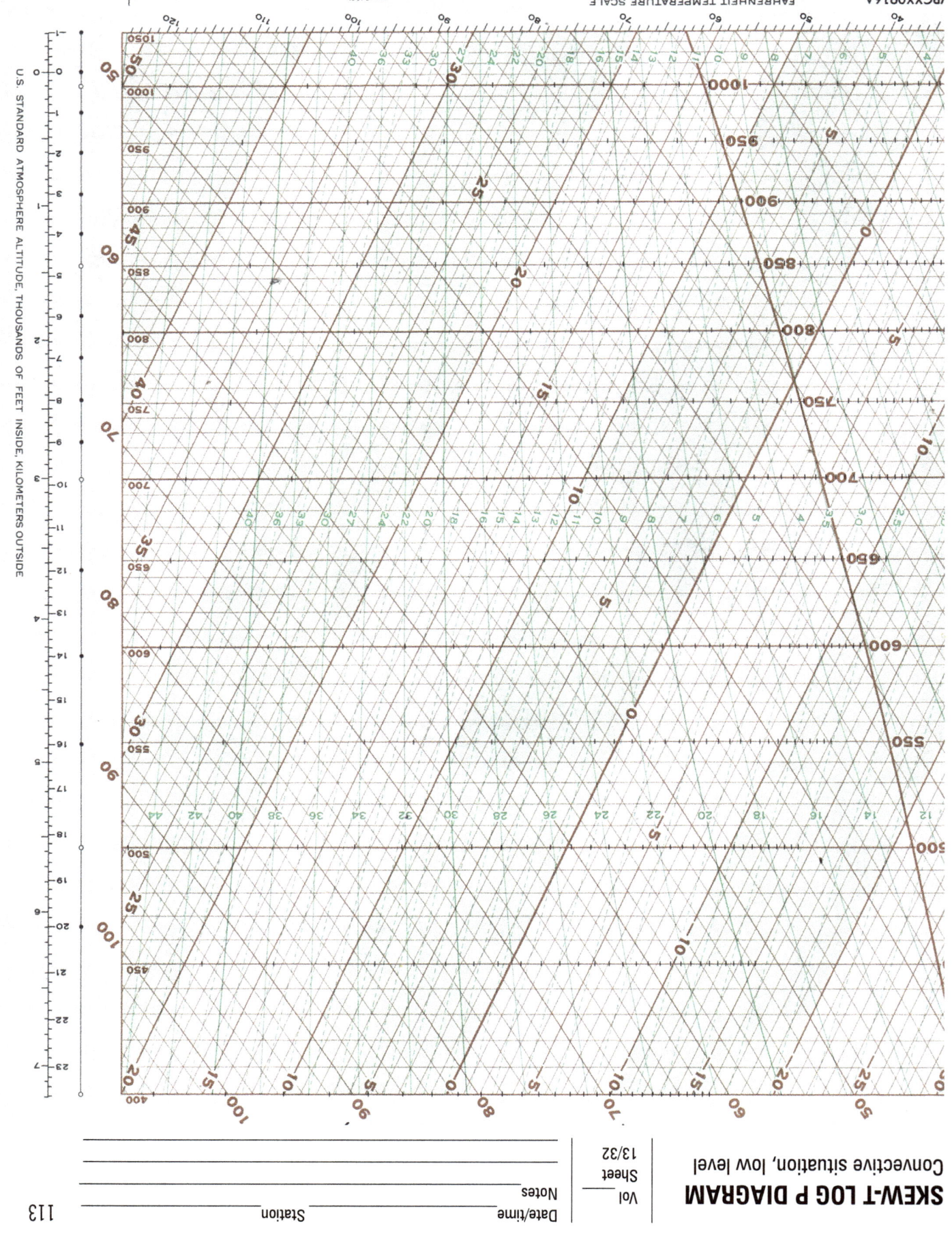

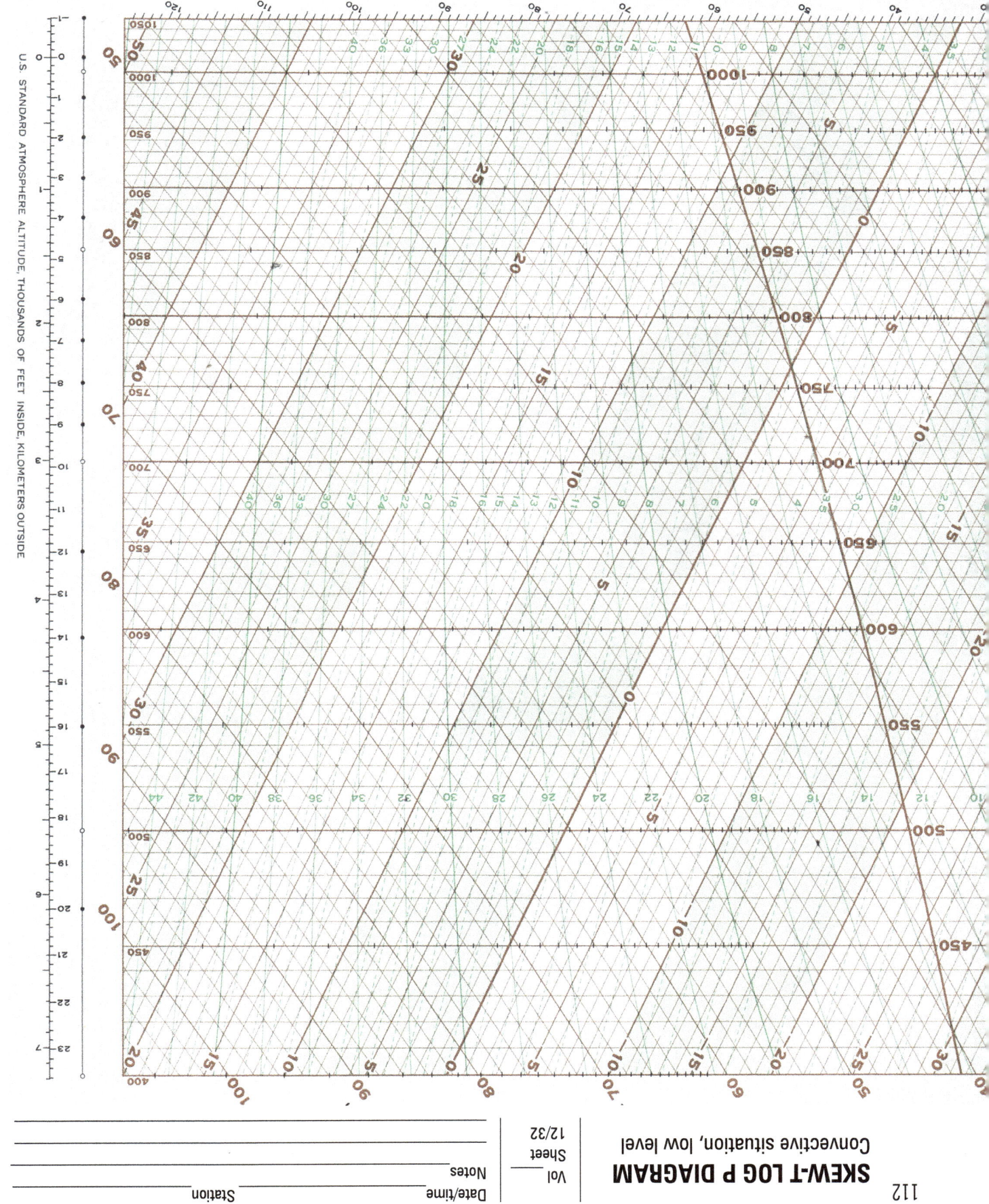

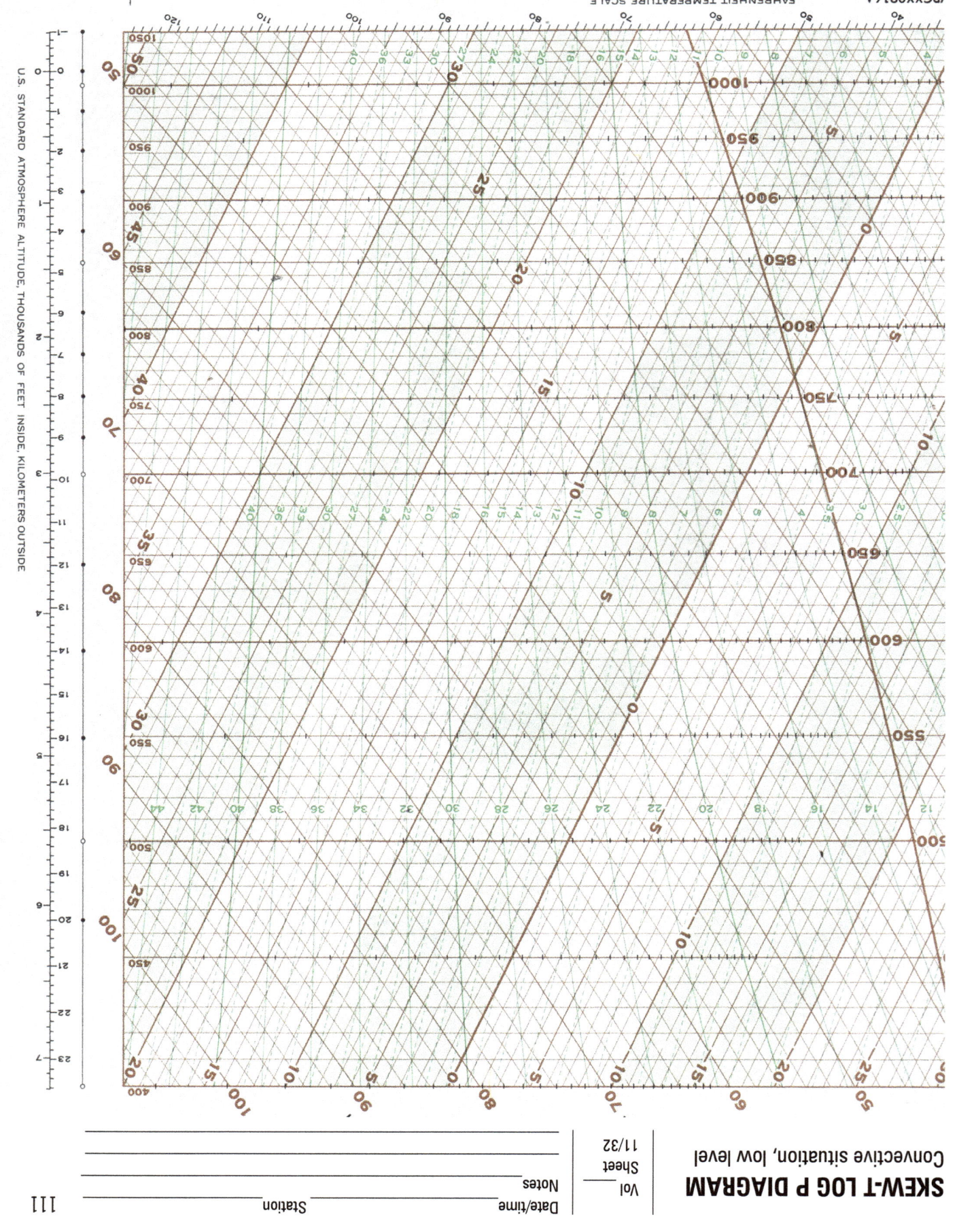

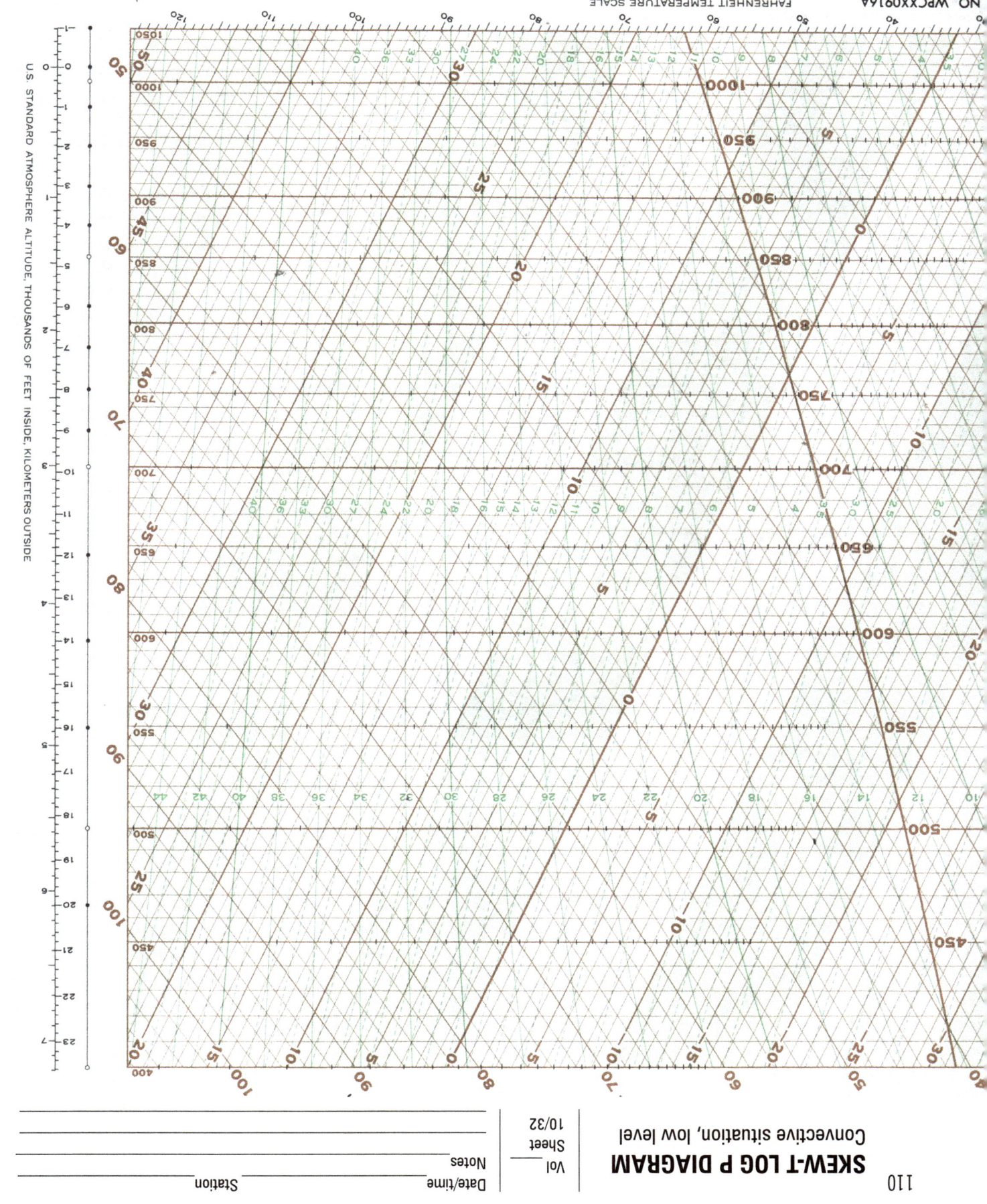

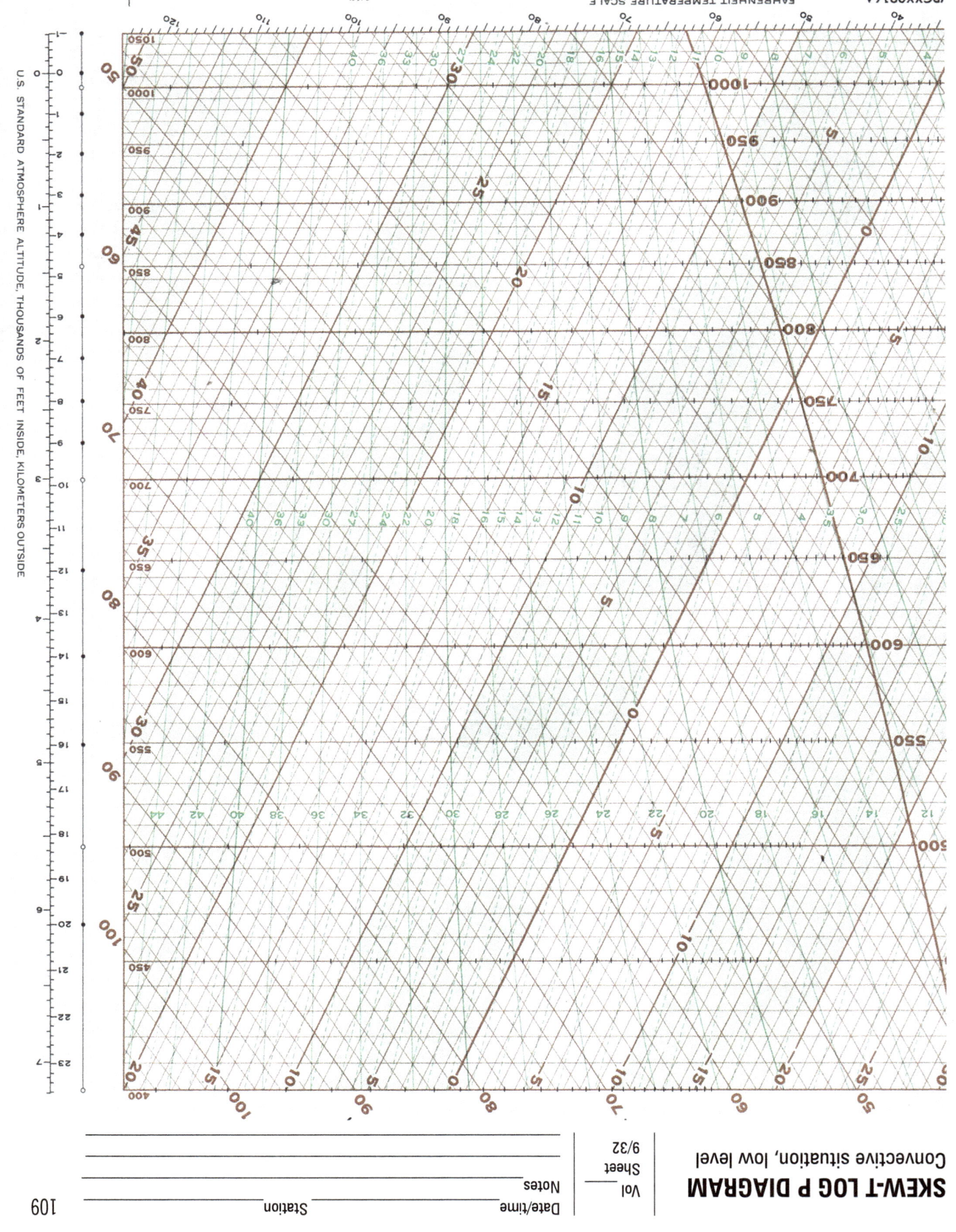

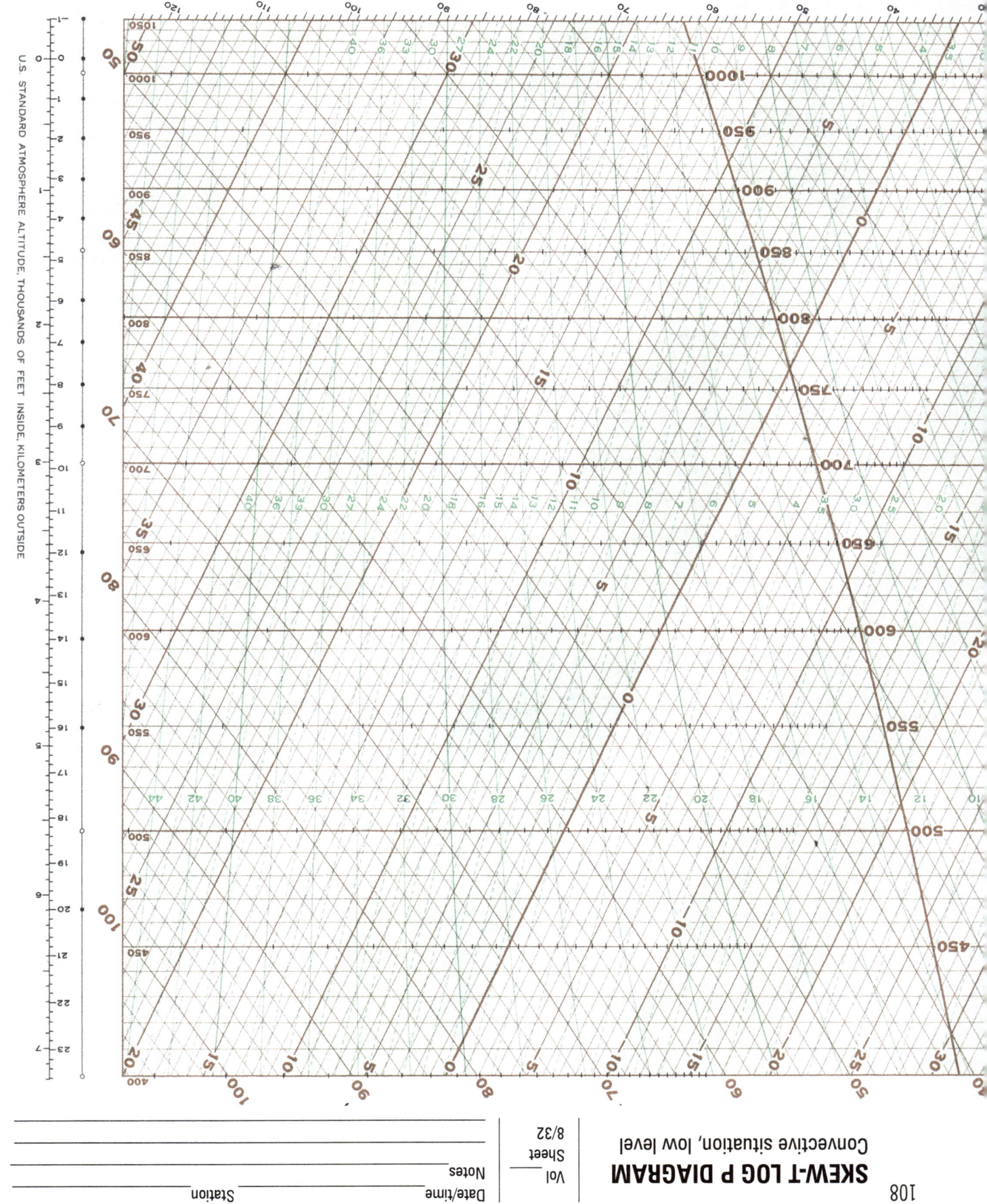

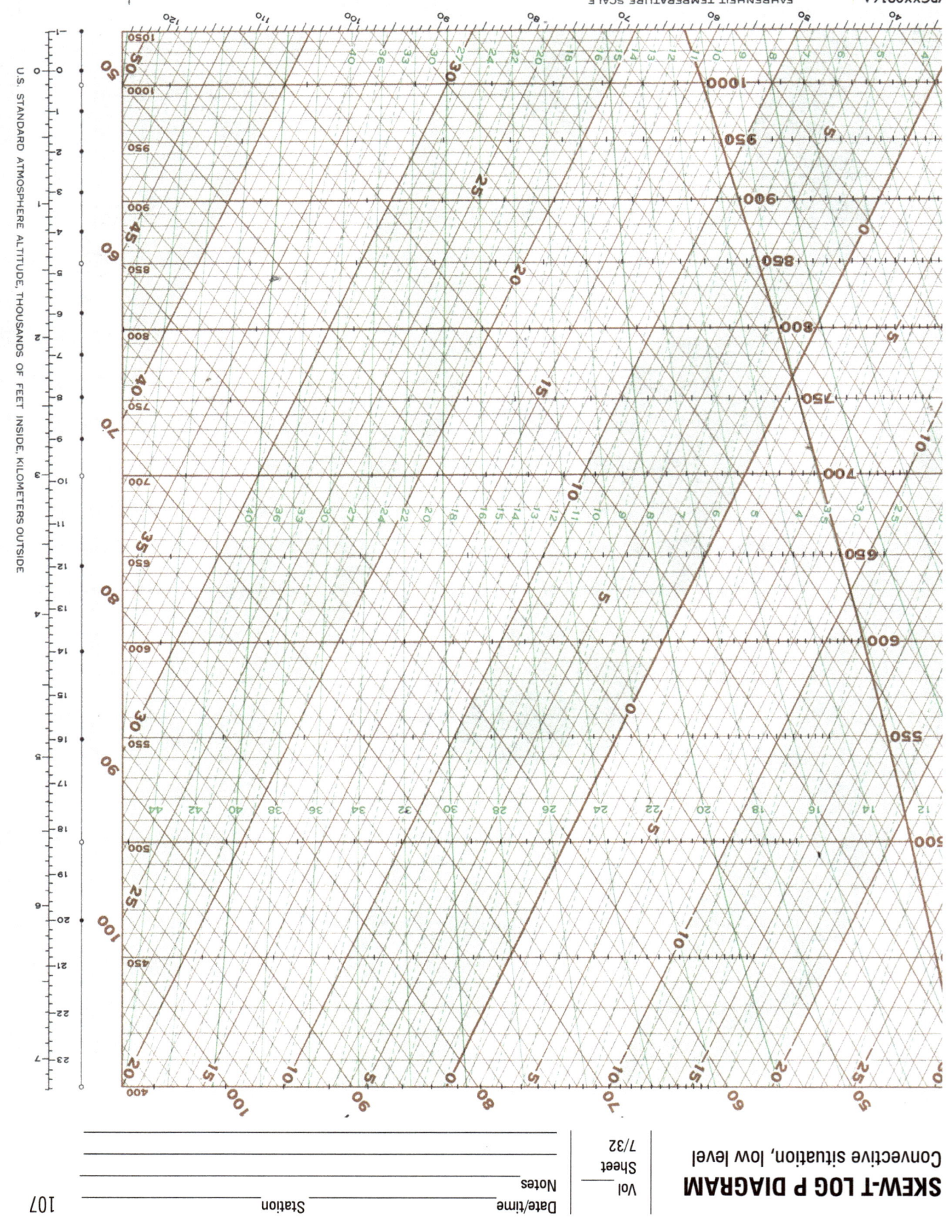

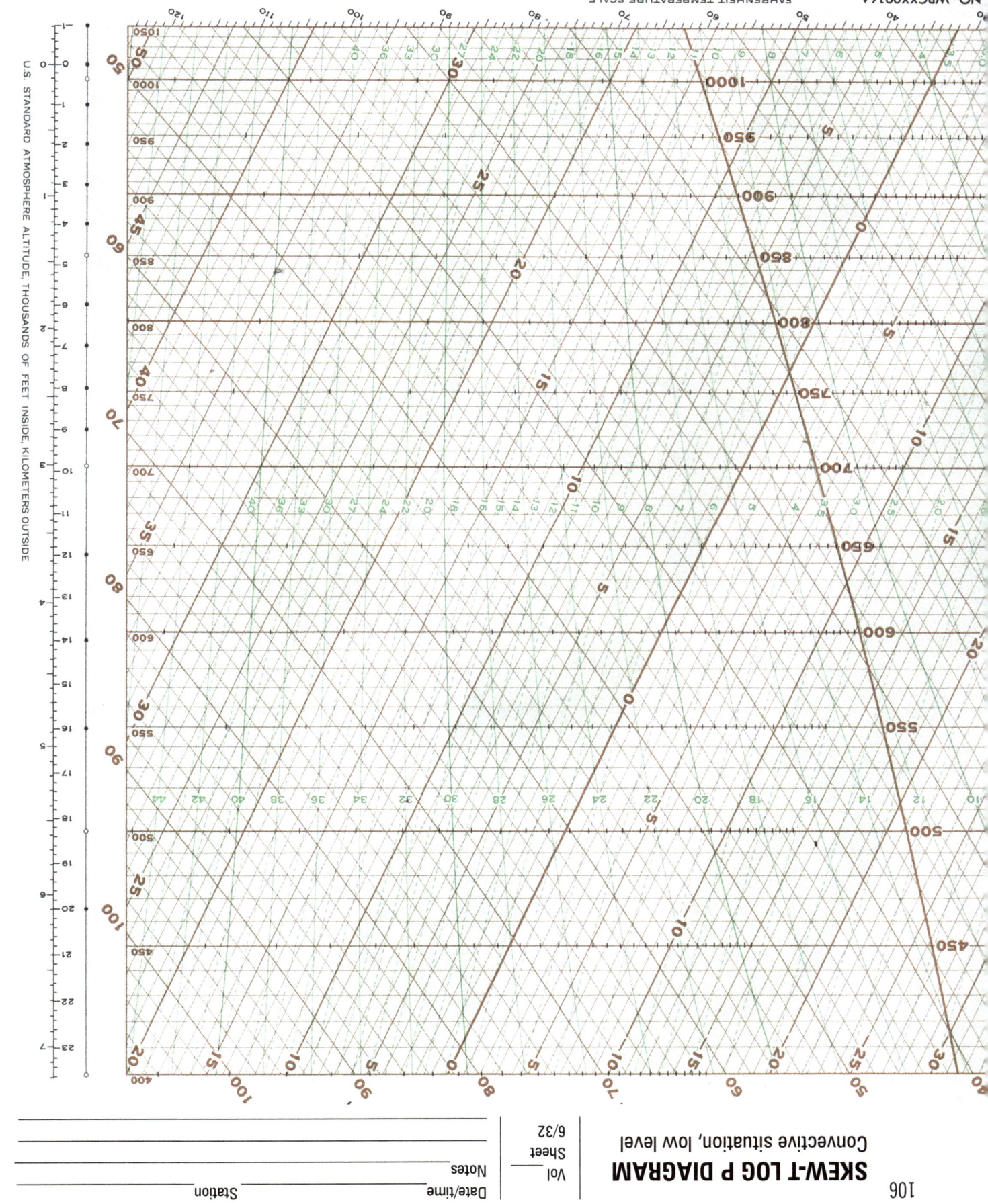

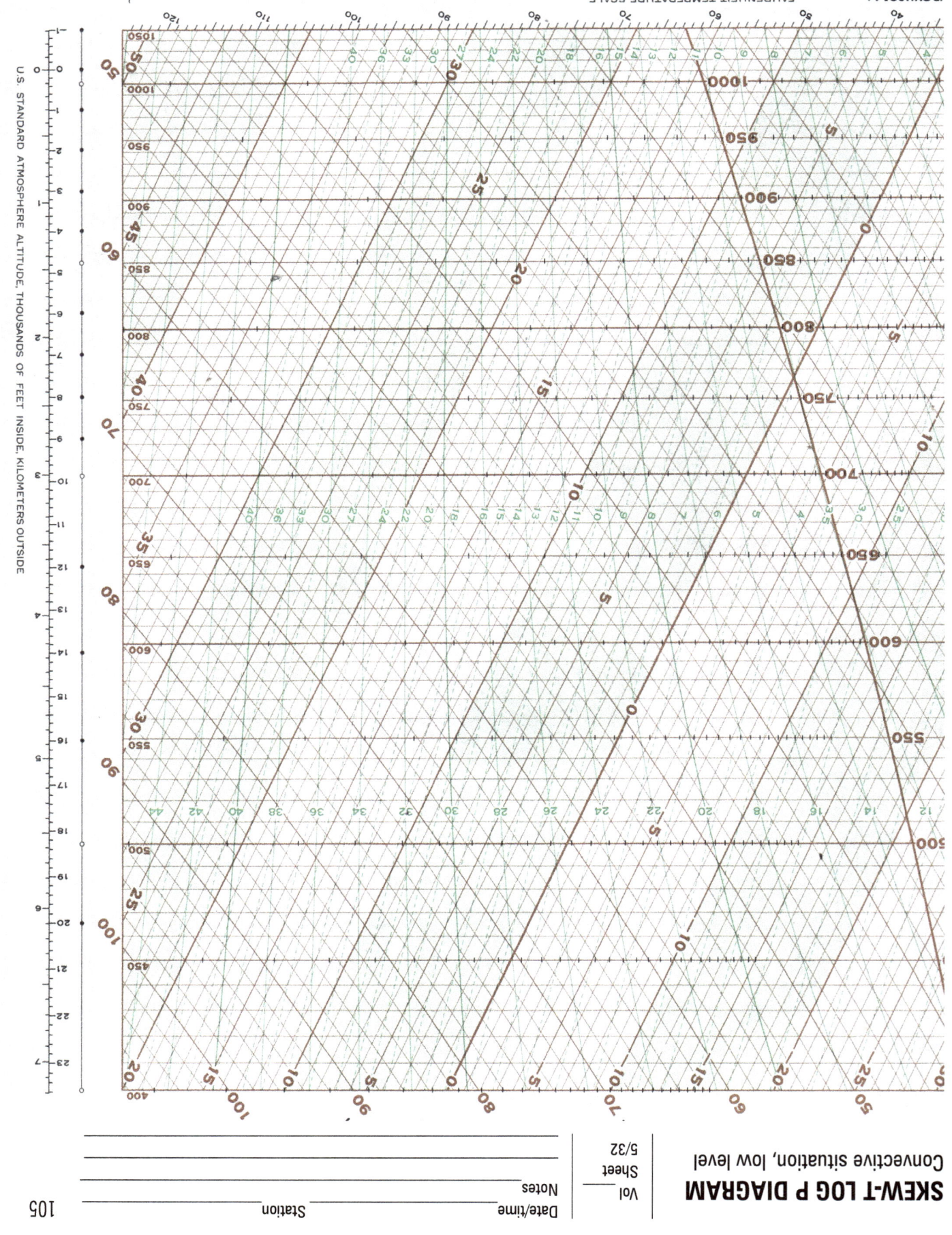

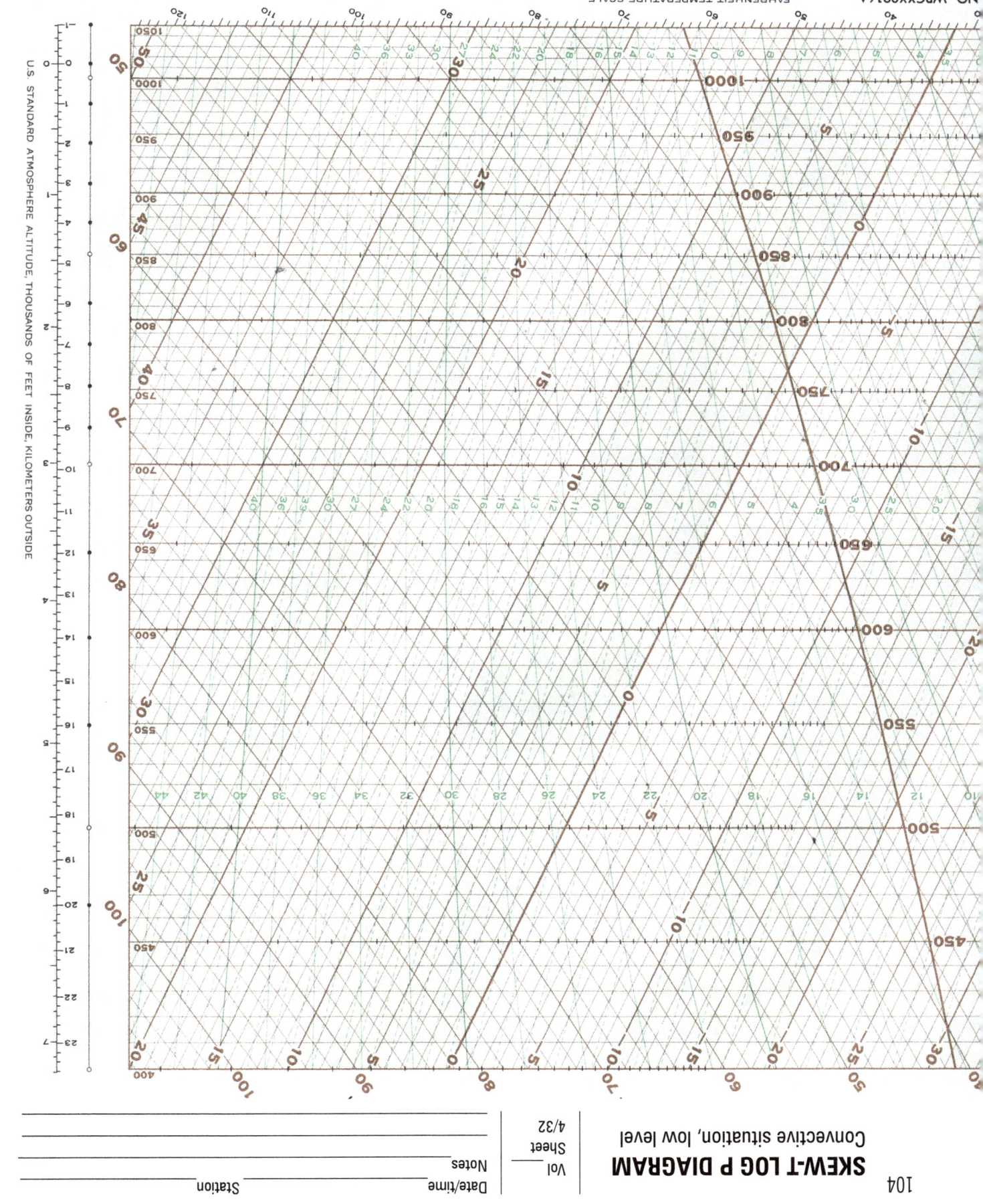

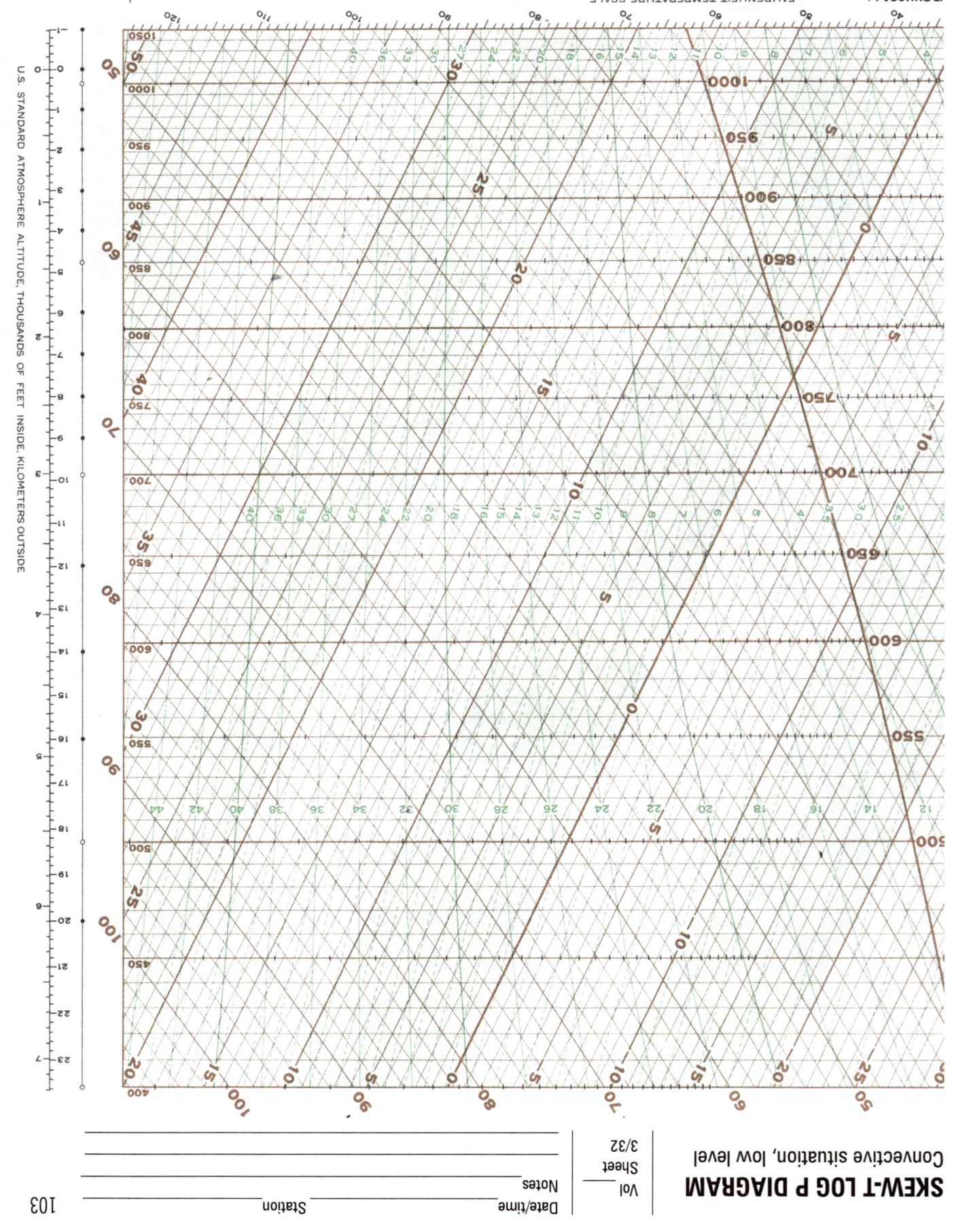

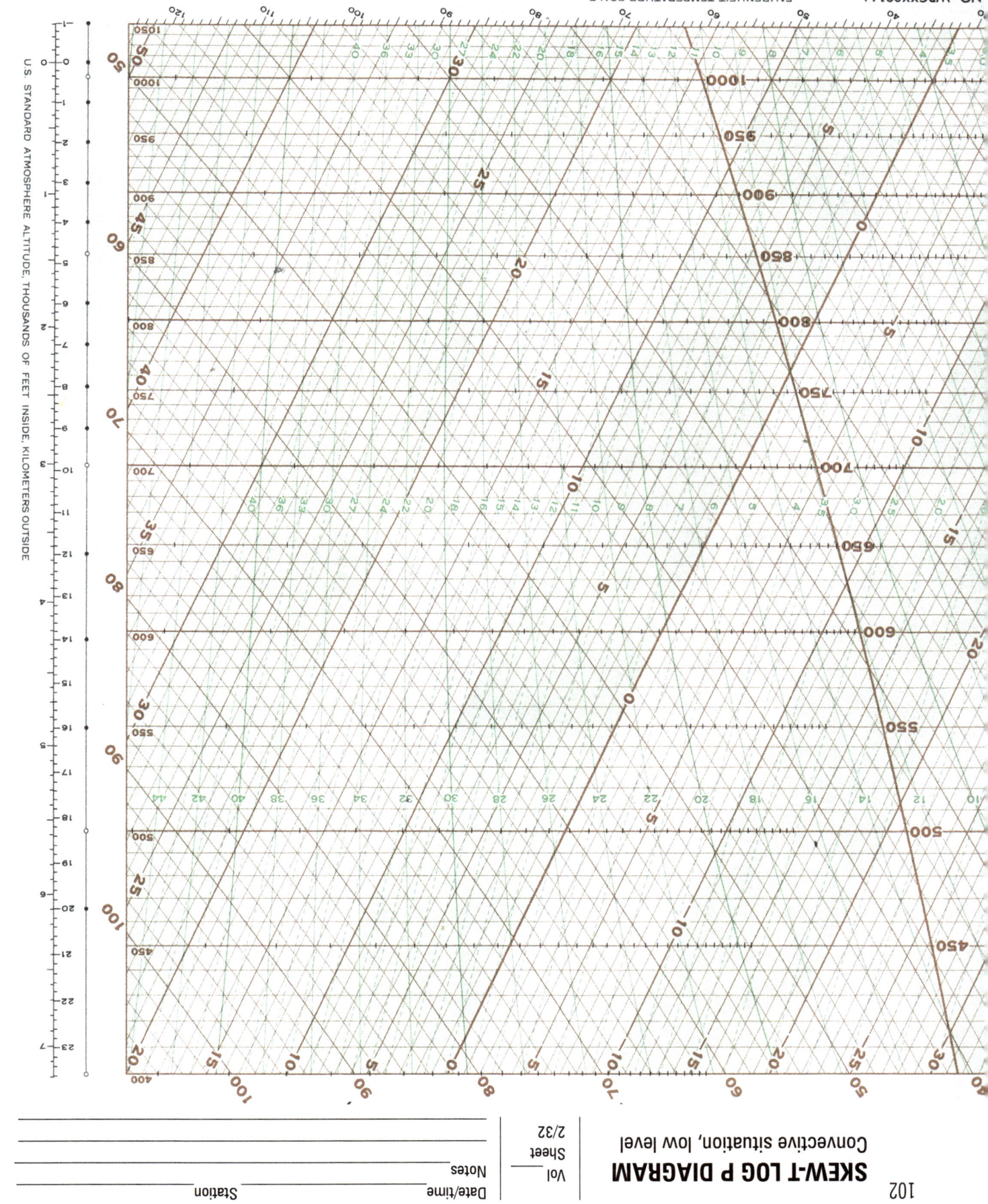

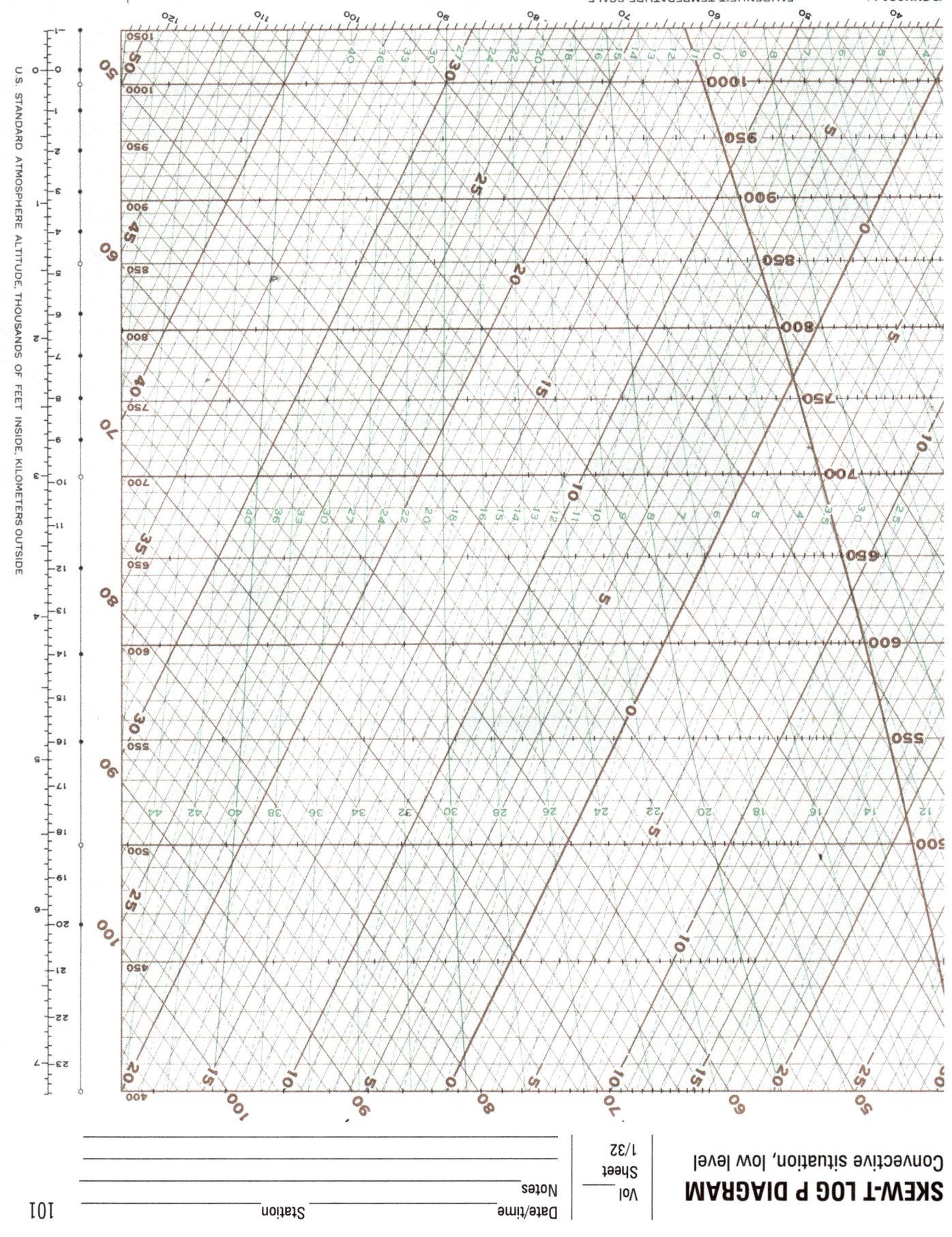

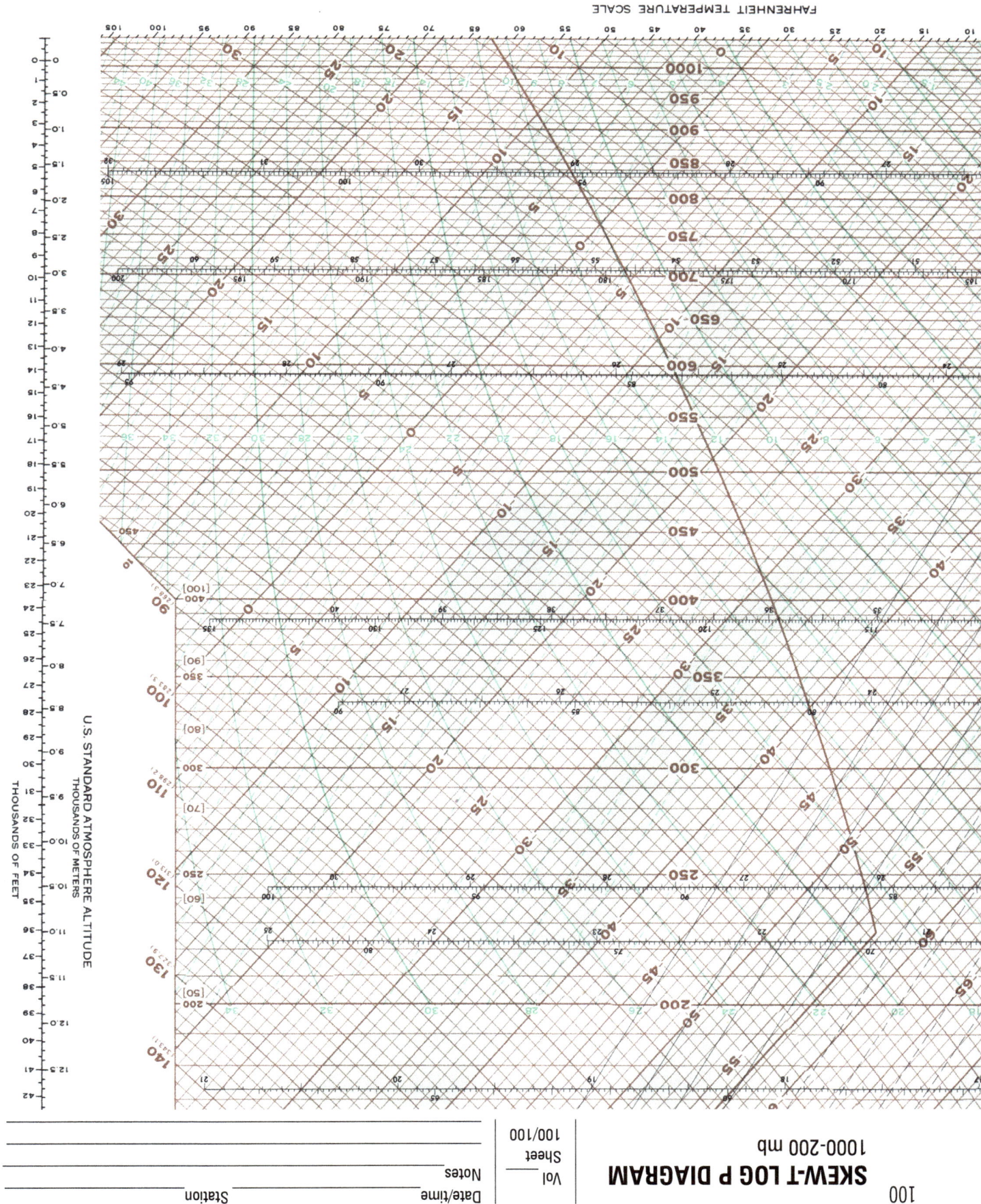

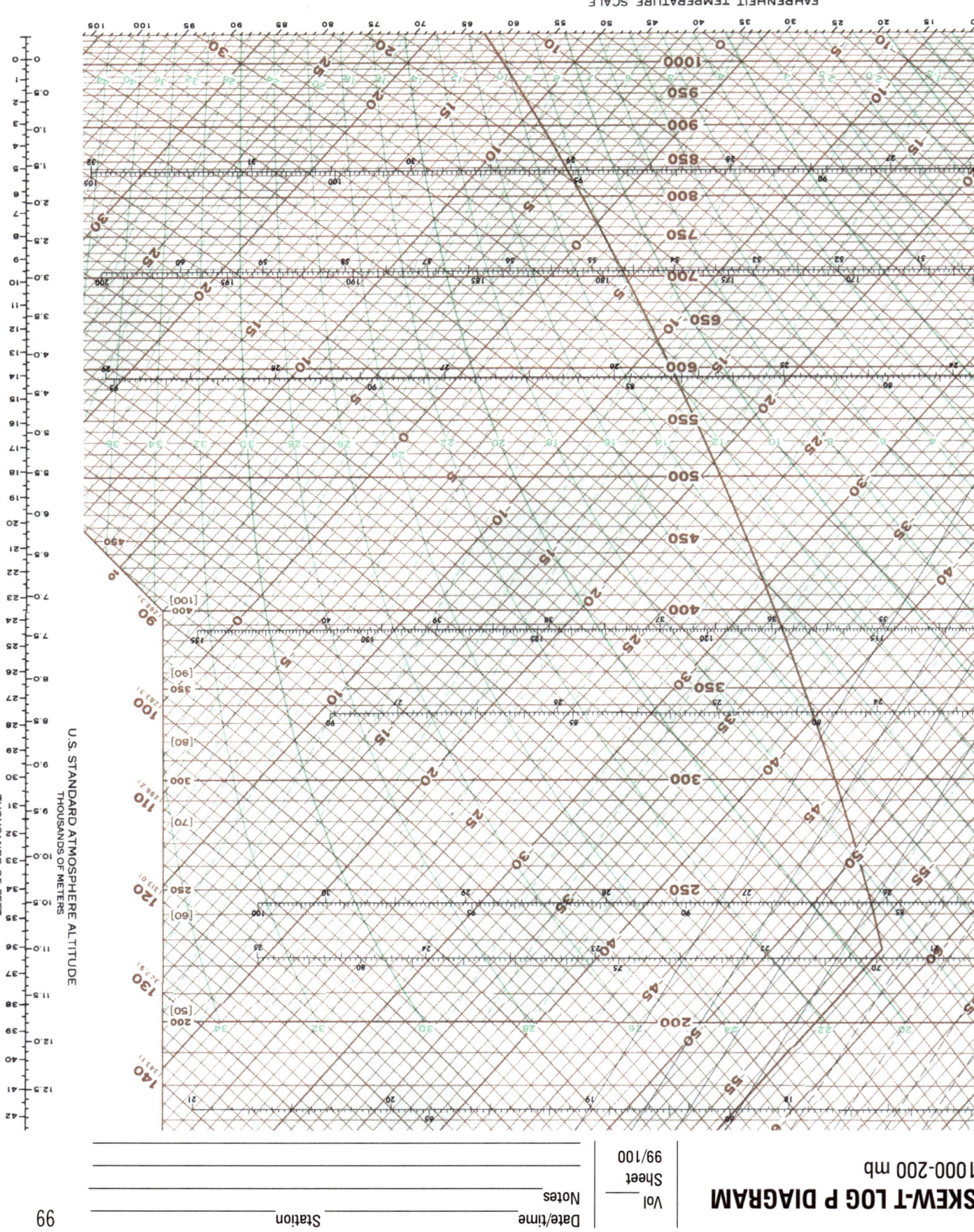

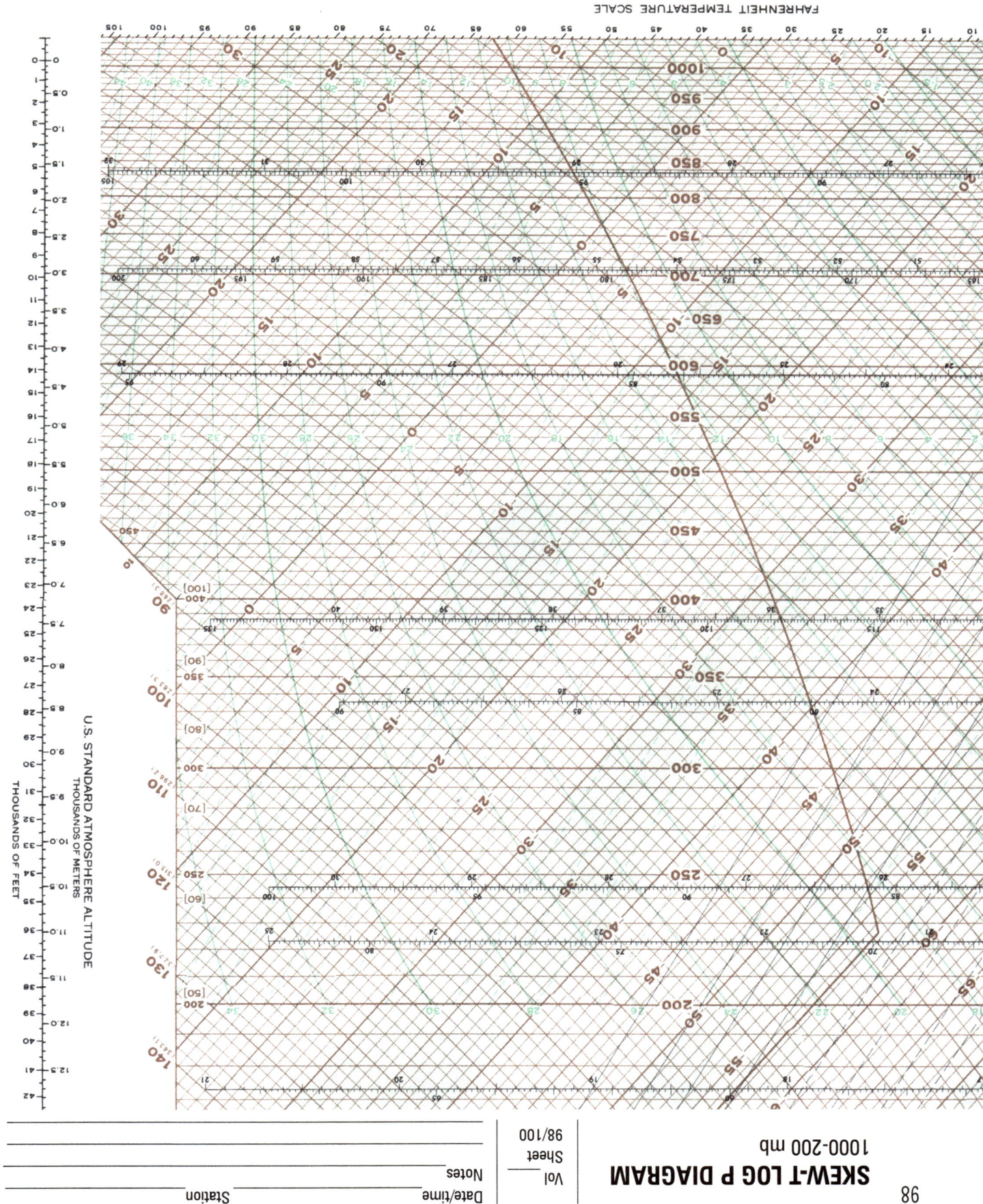

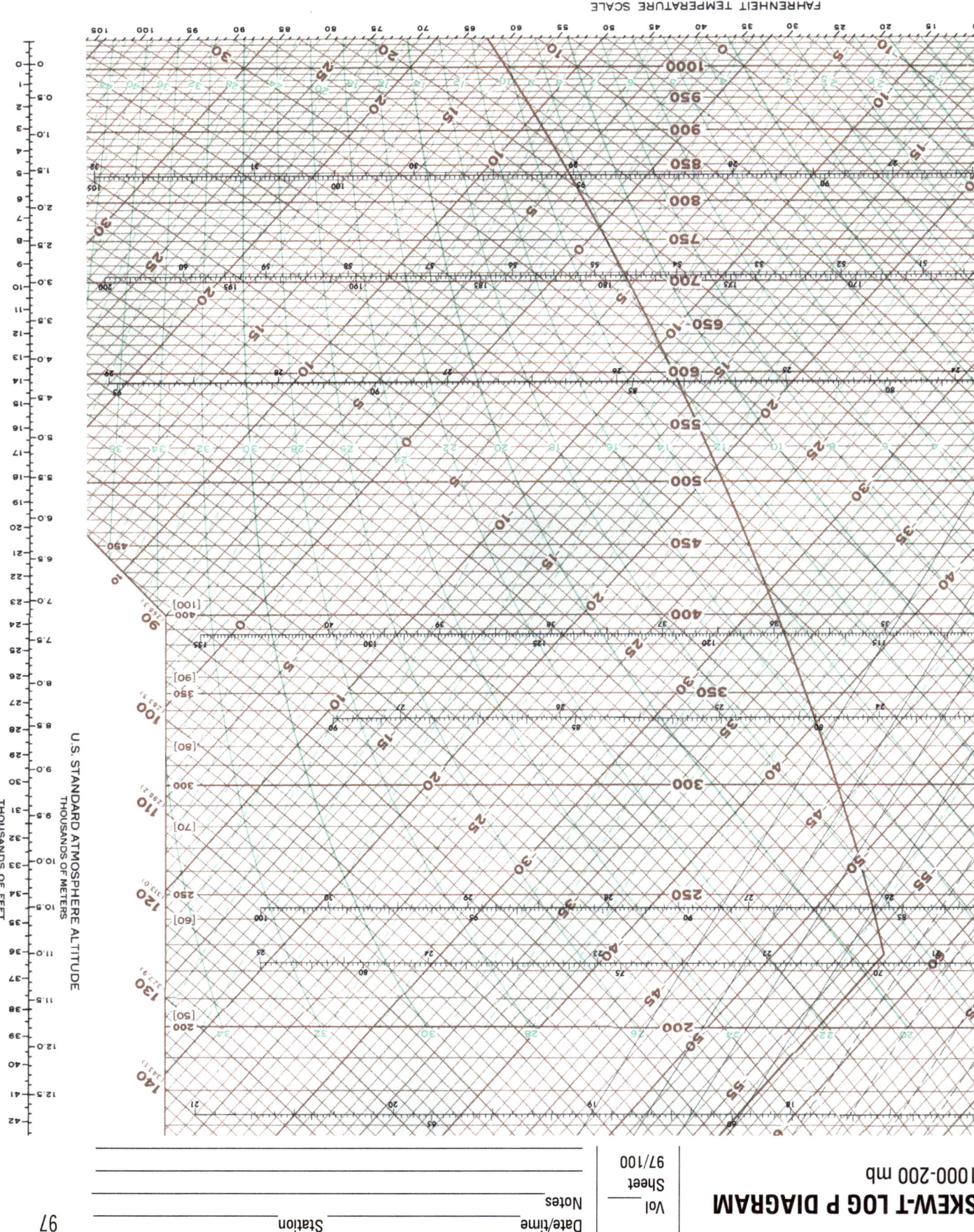

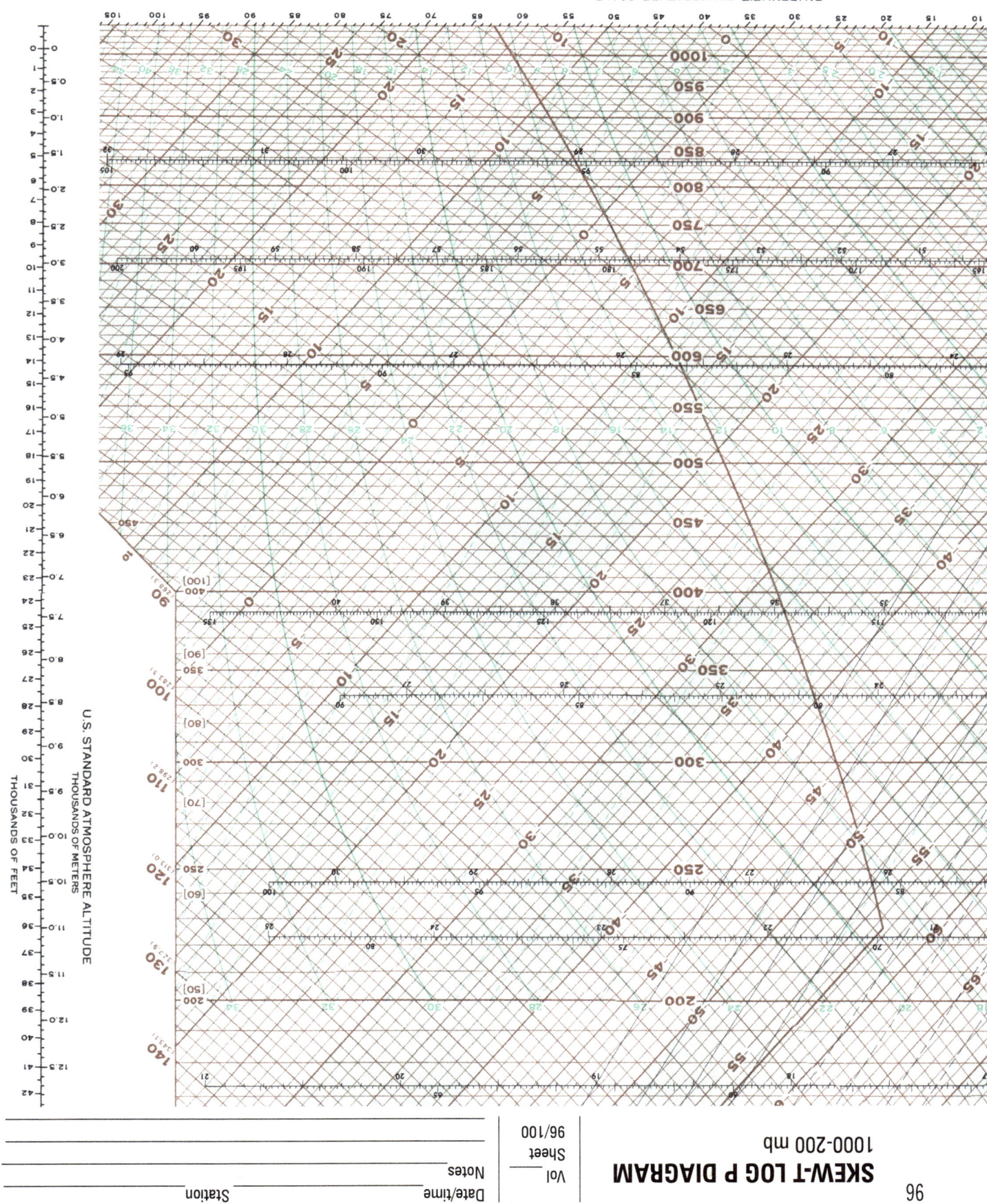

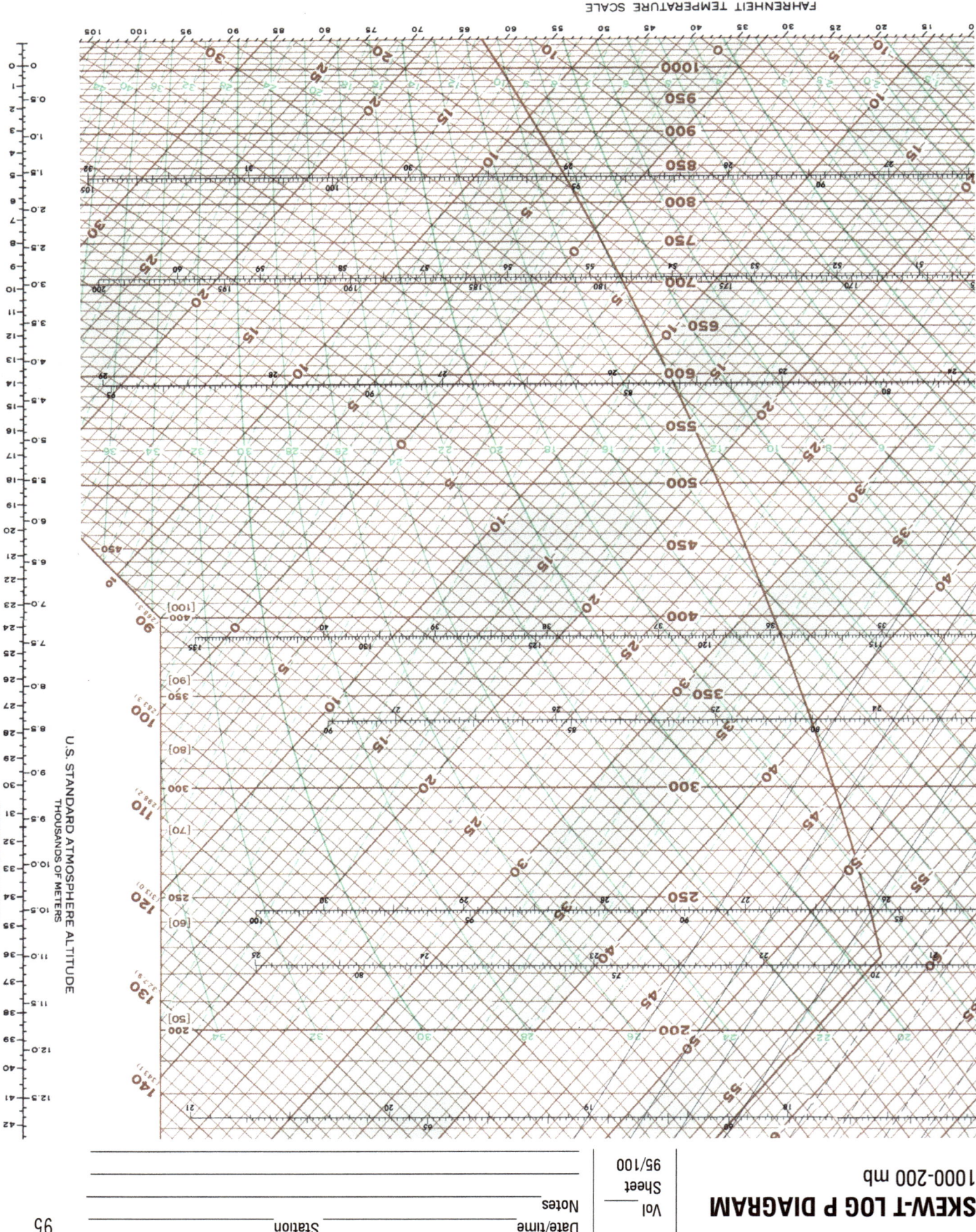

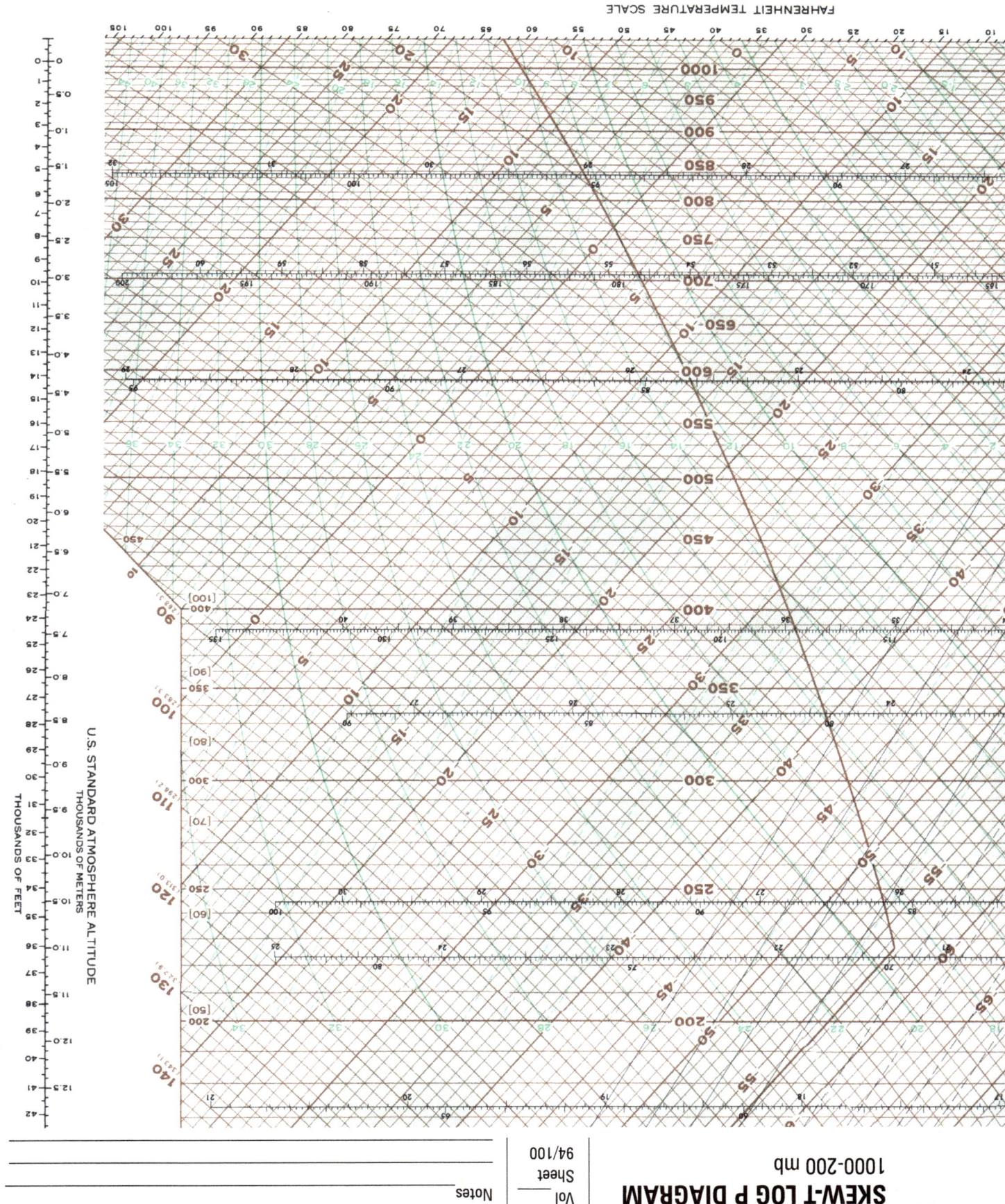

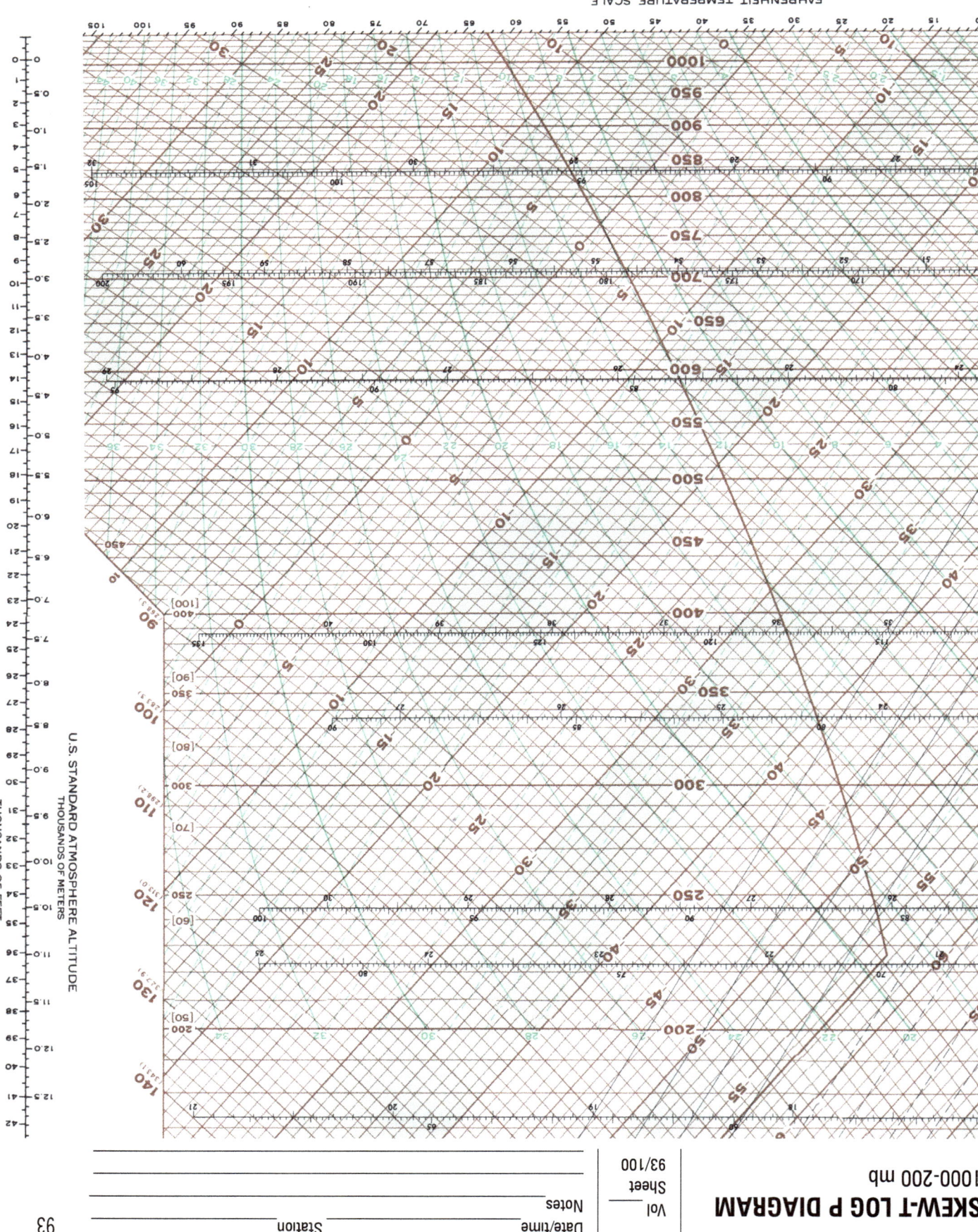

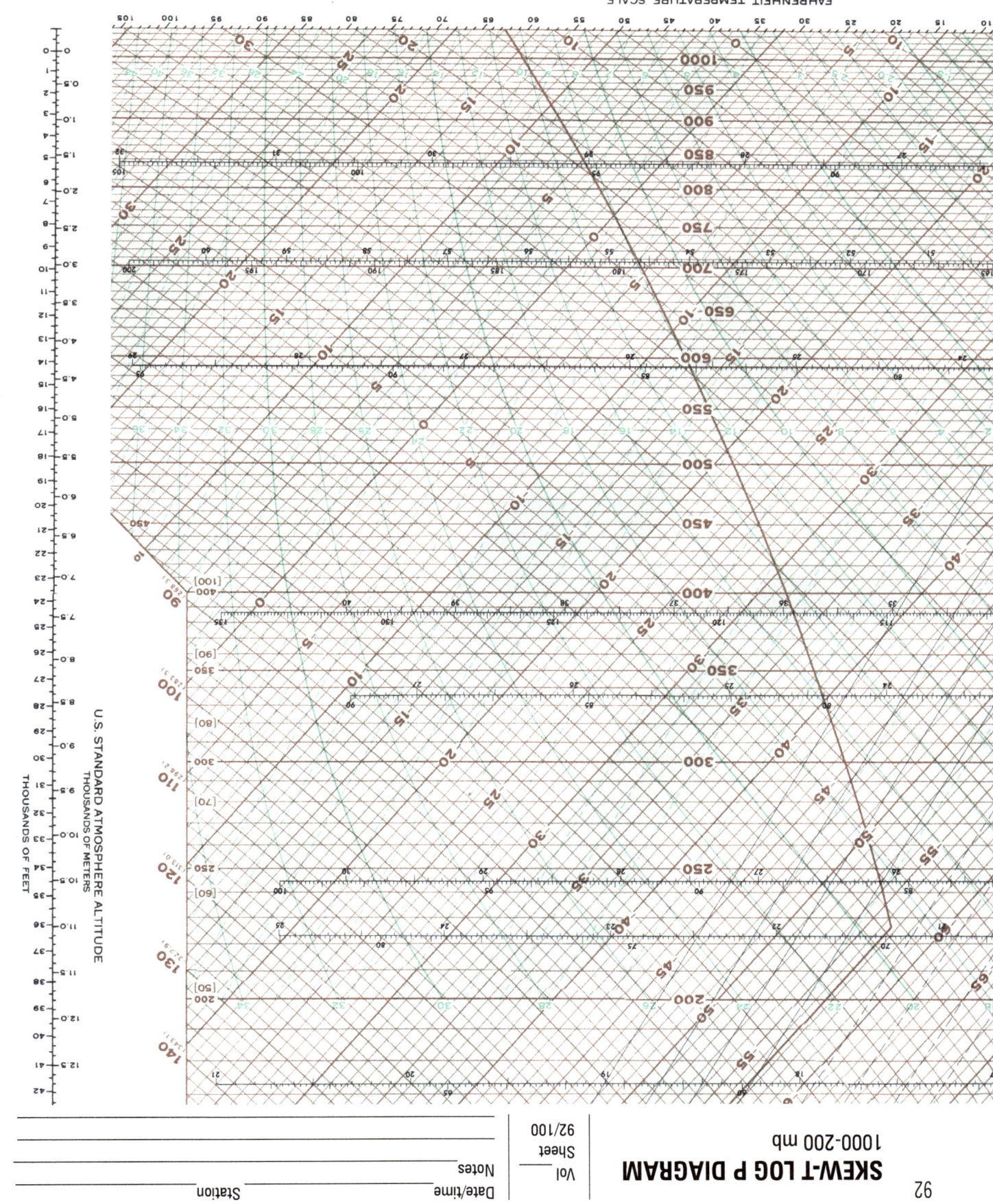

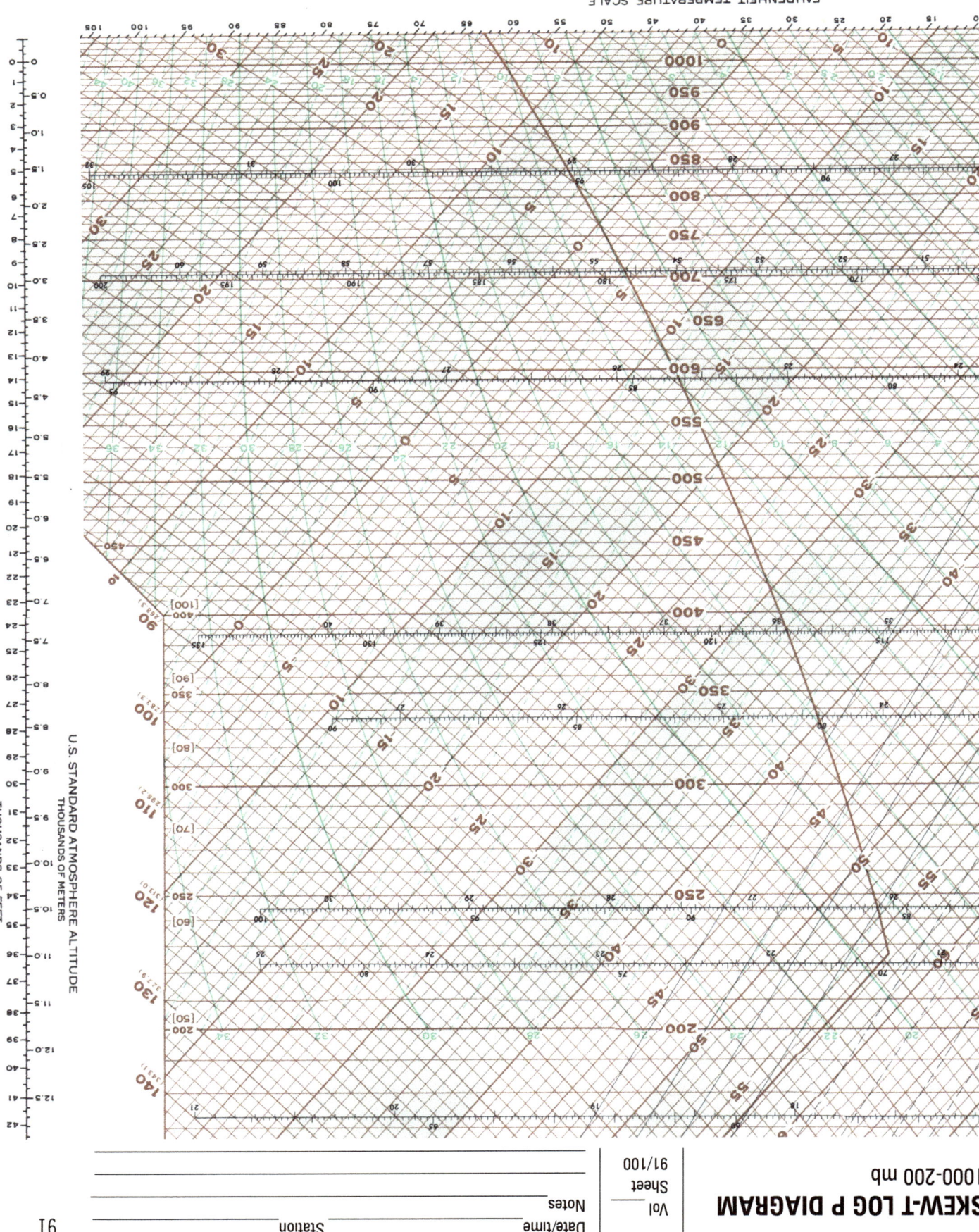

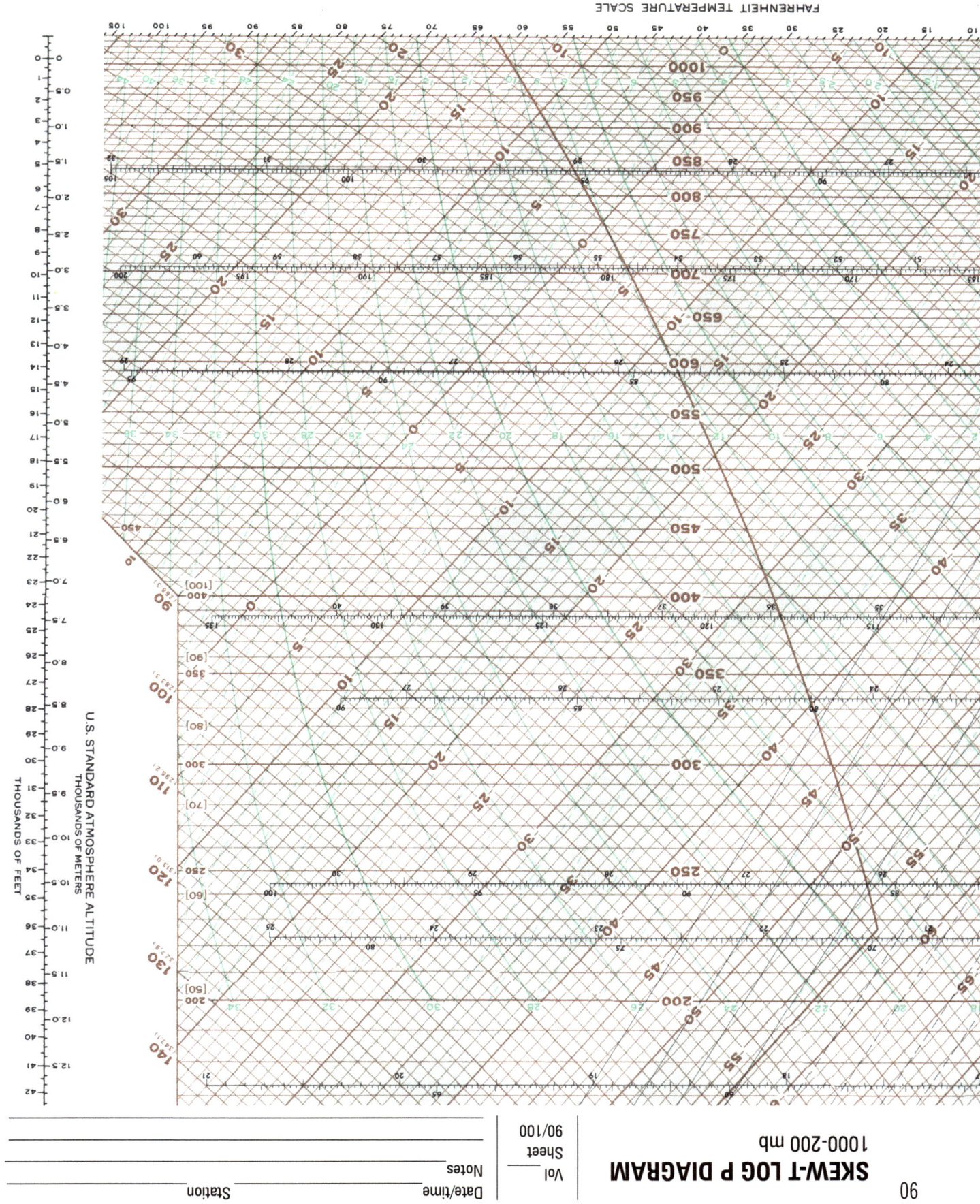

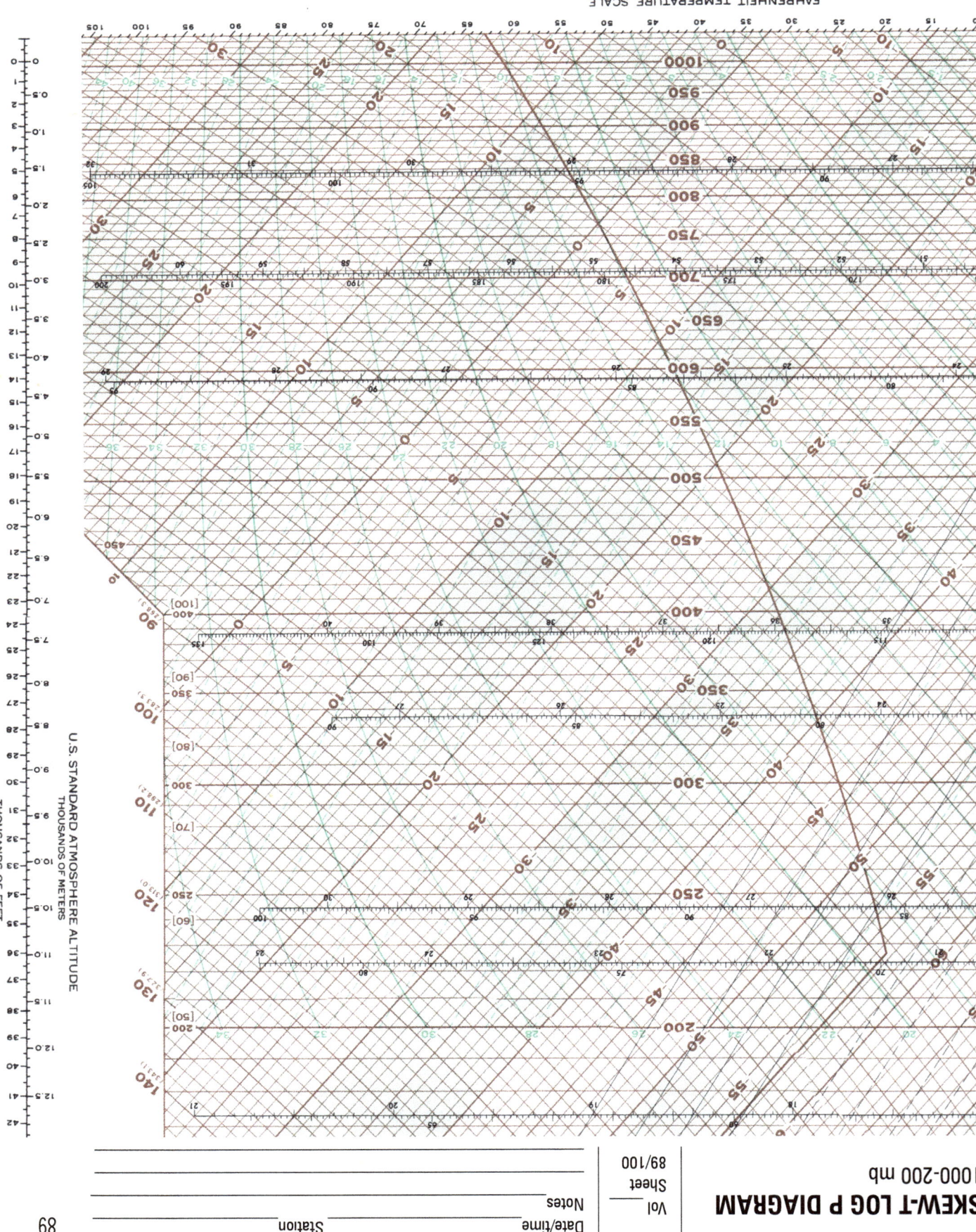

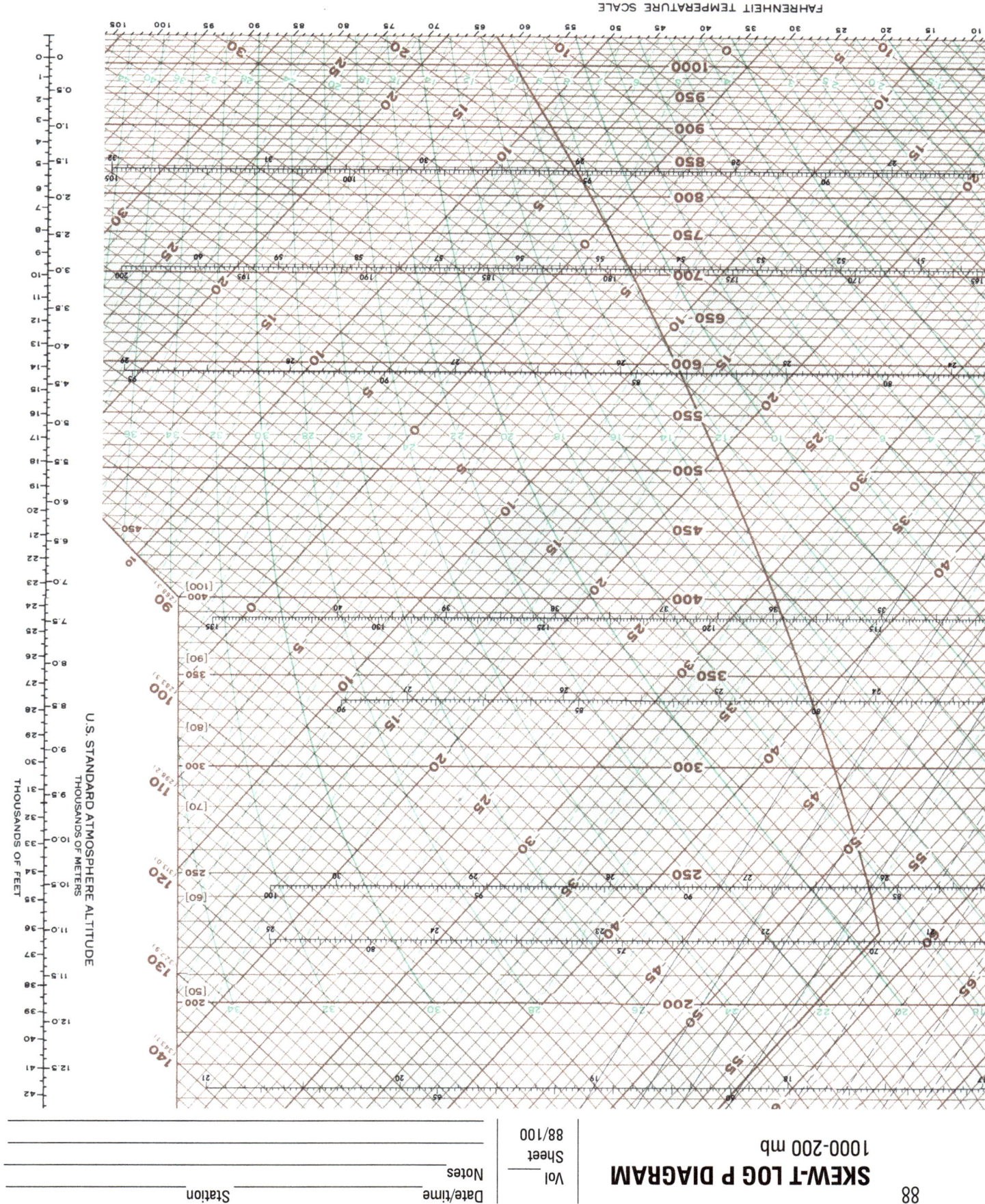

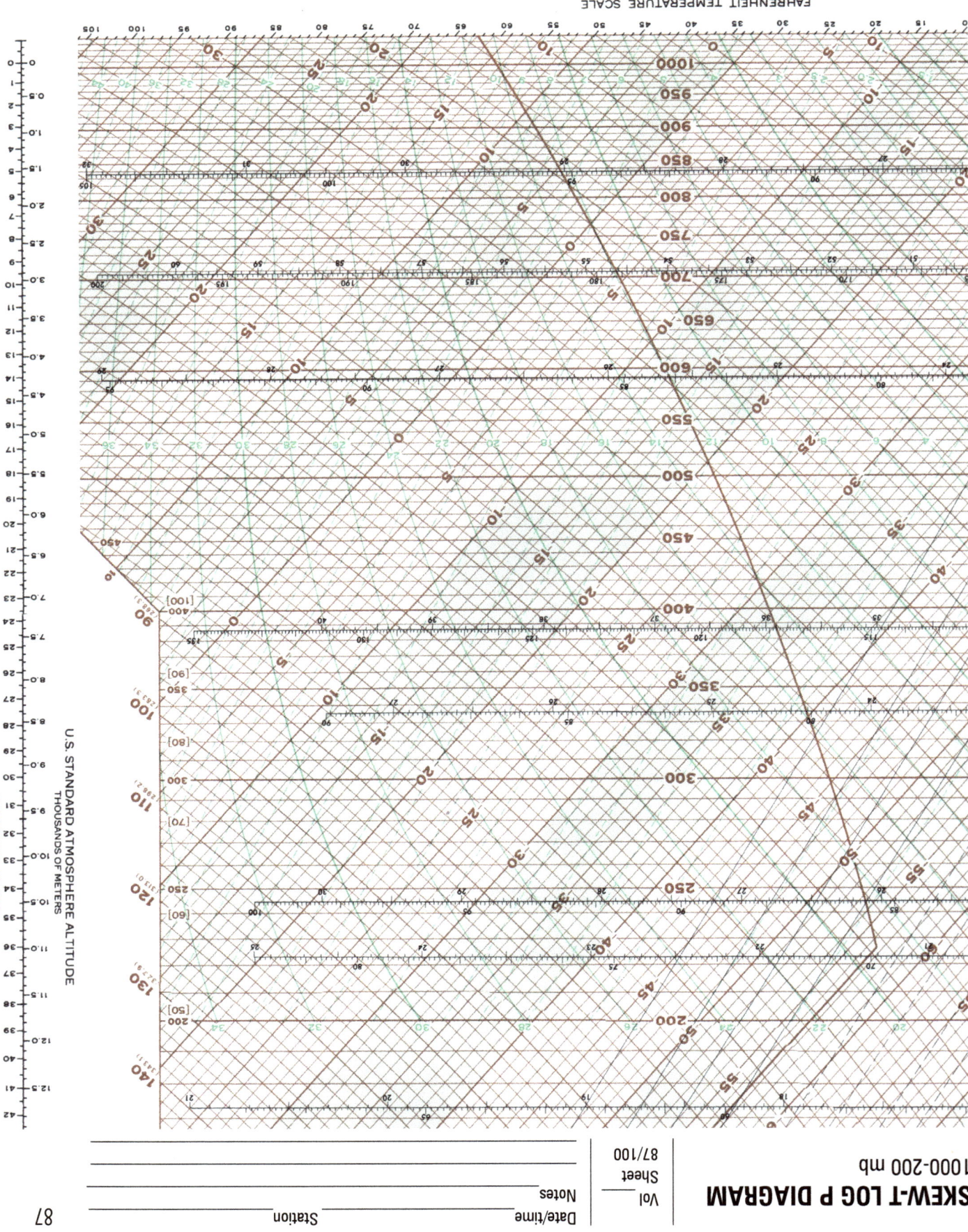

SKEW-T LOG P DIAGRAM
1000-200 mb

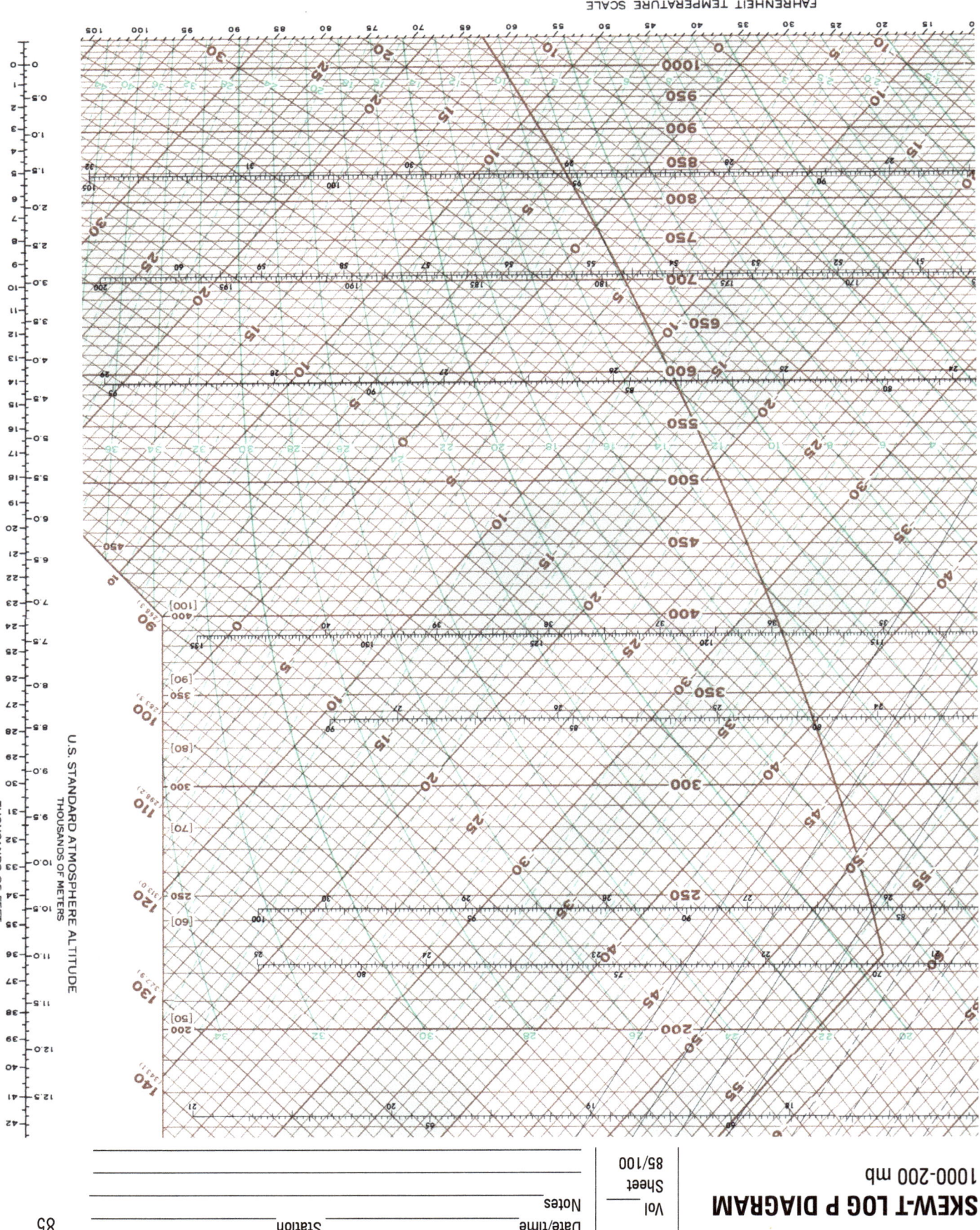

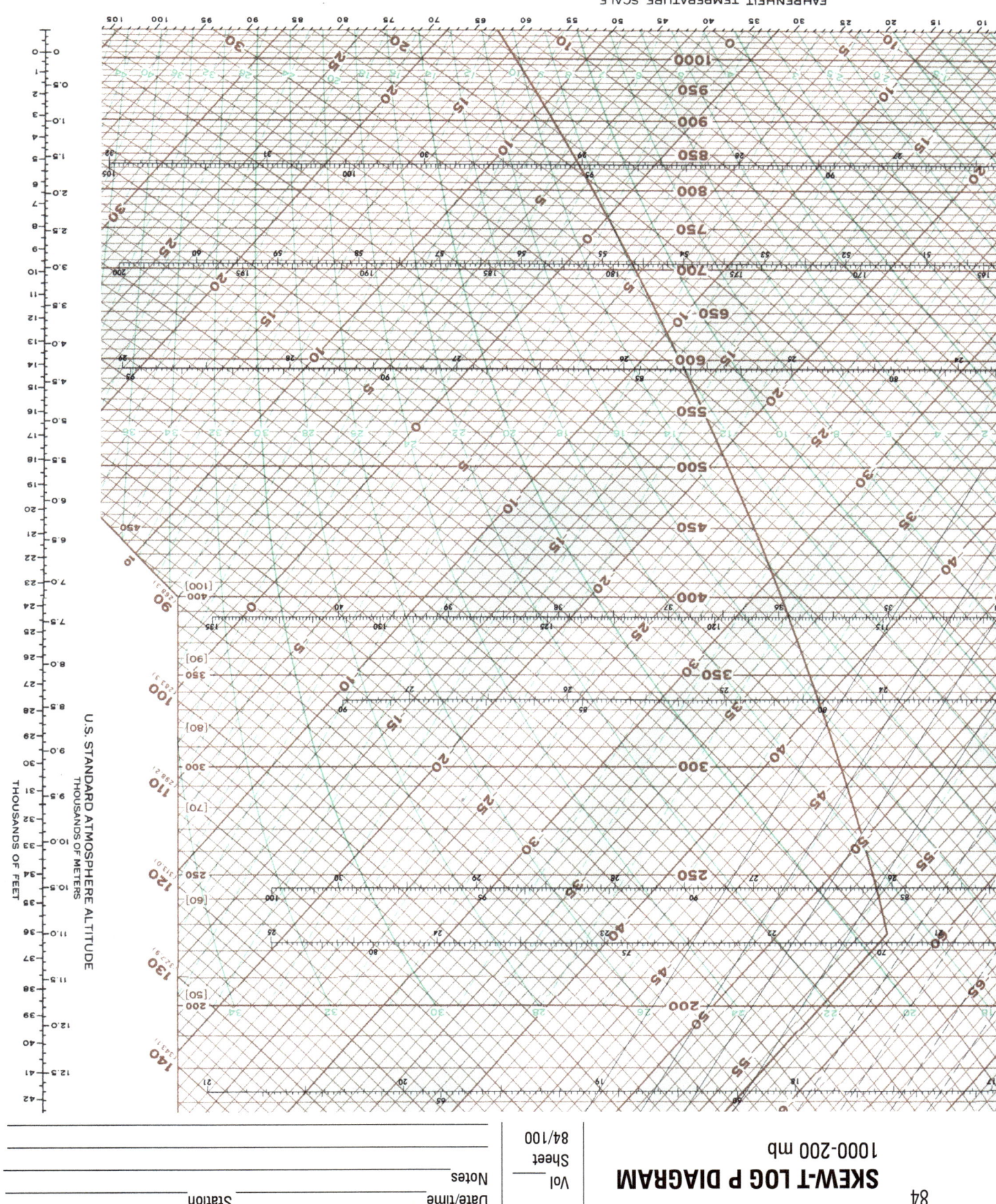

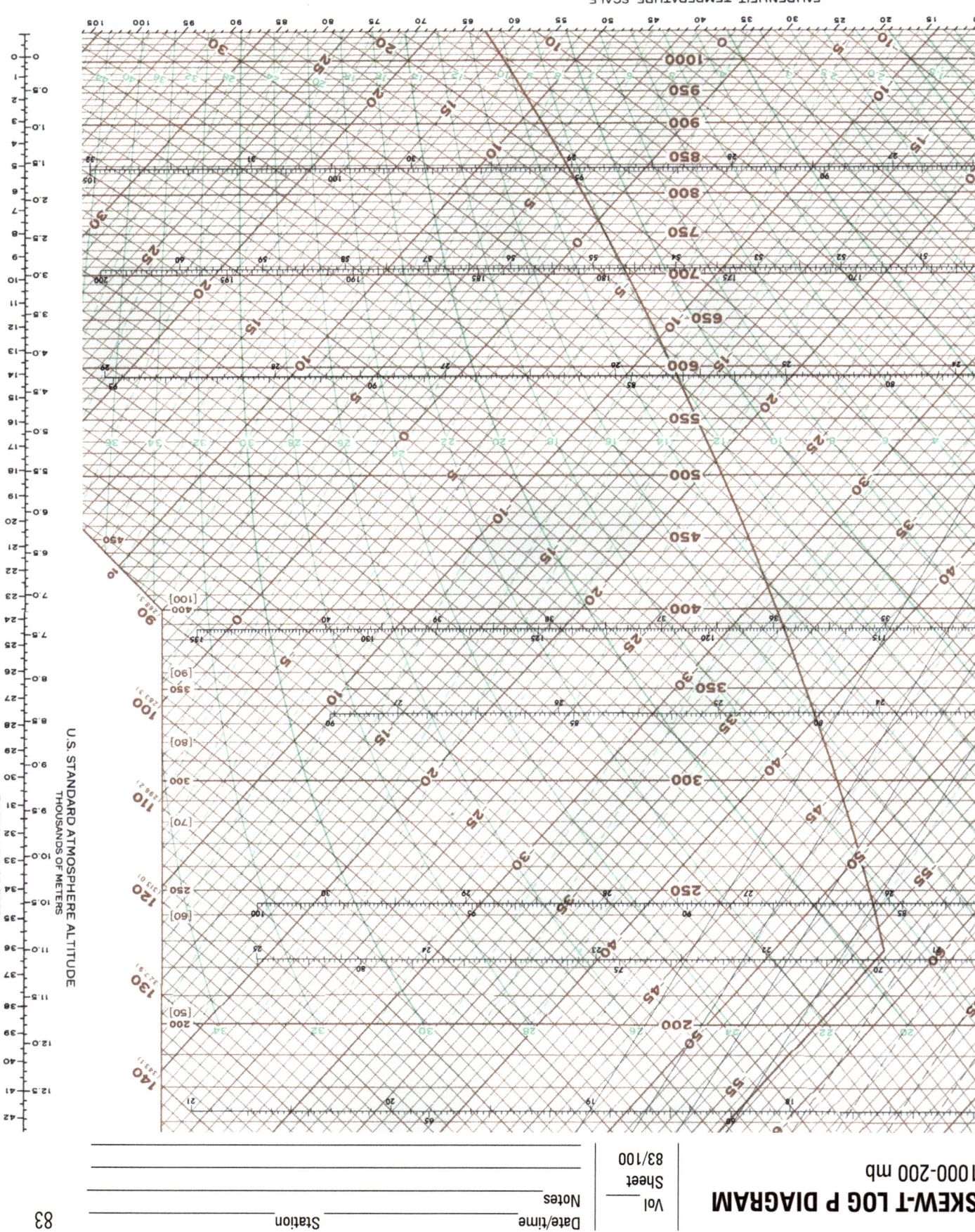

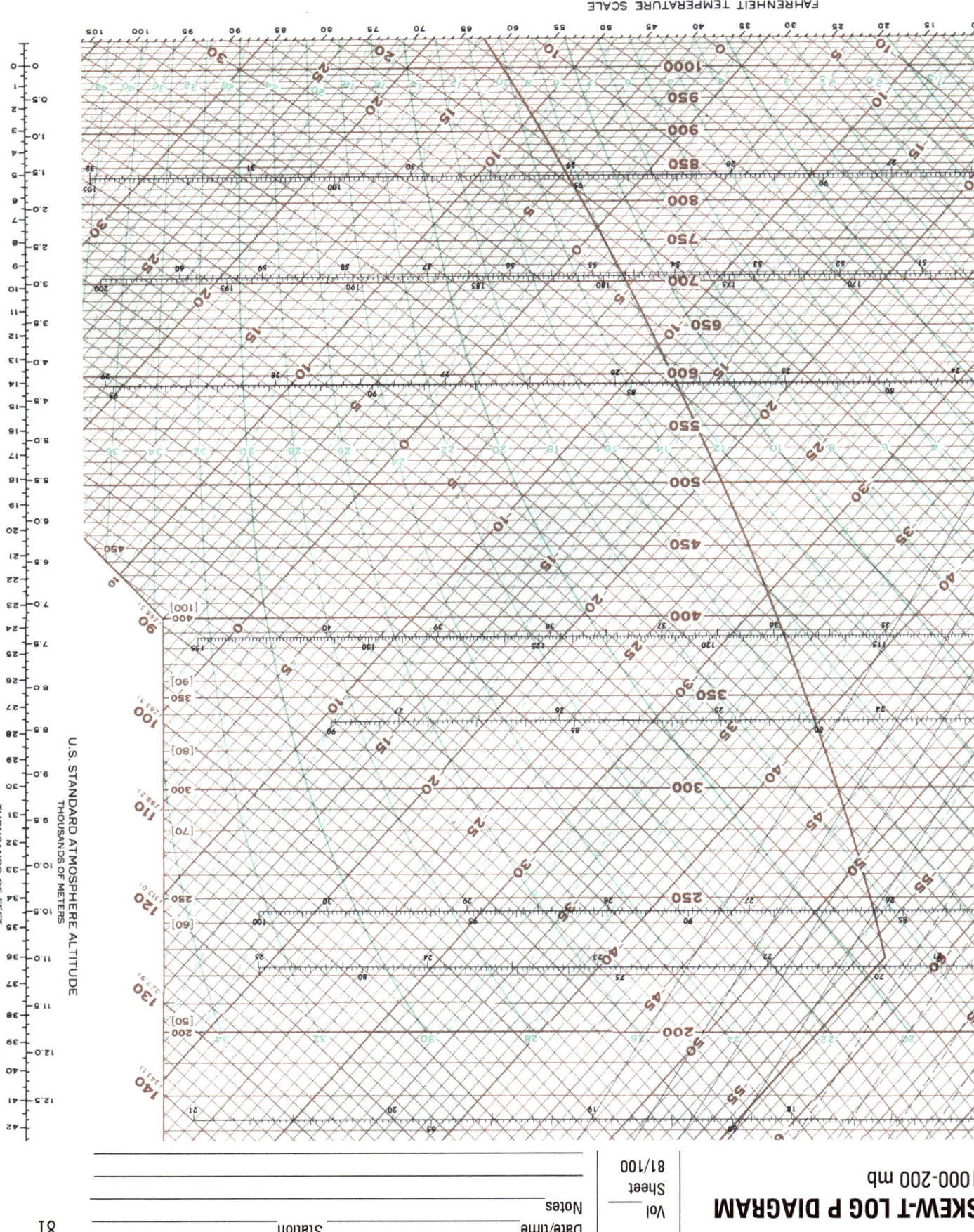

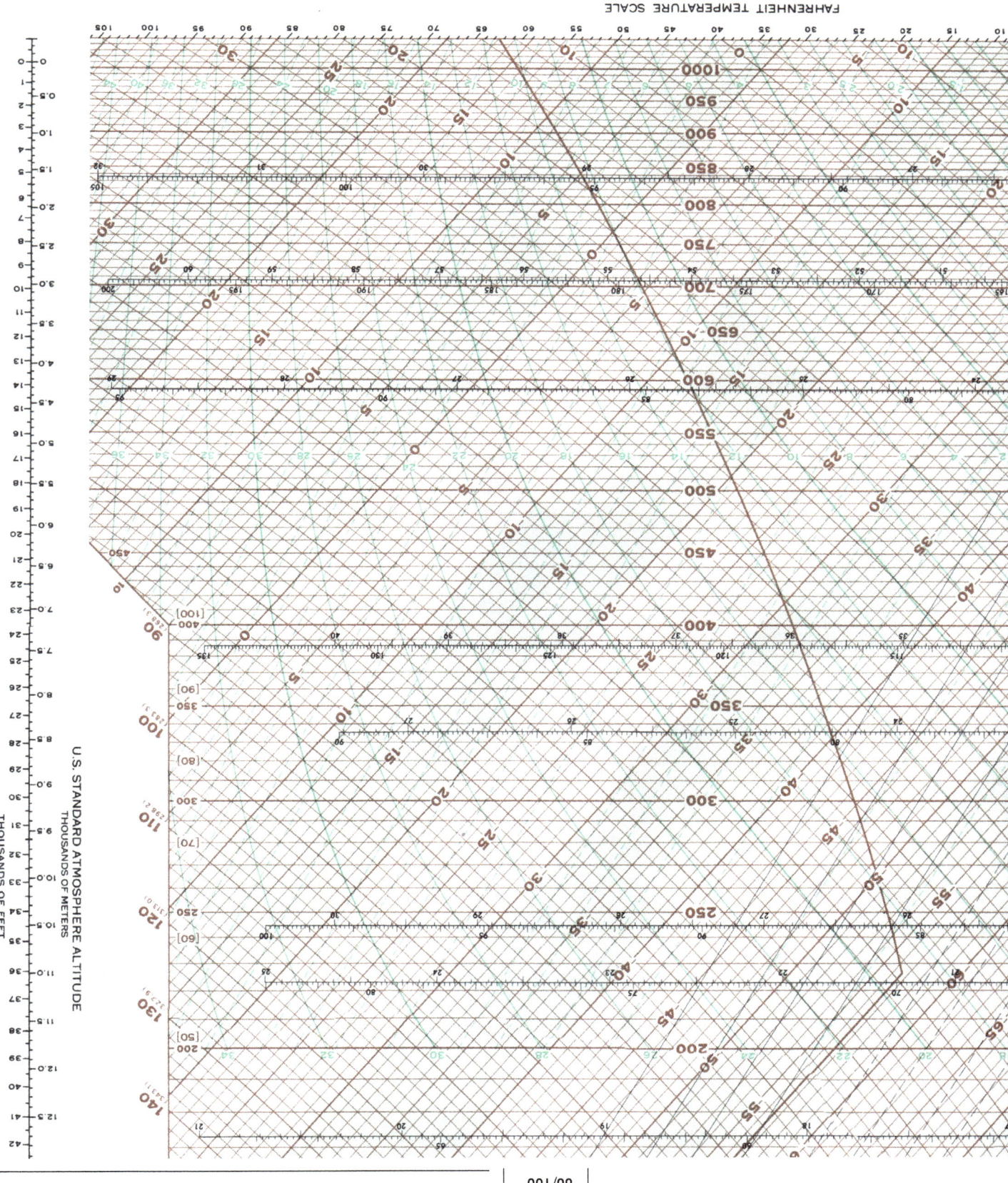

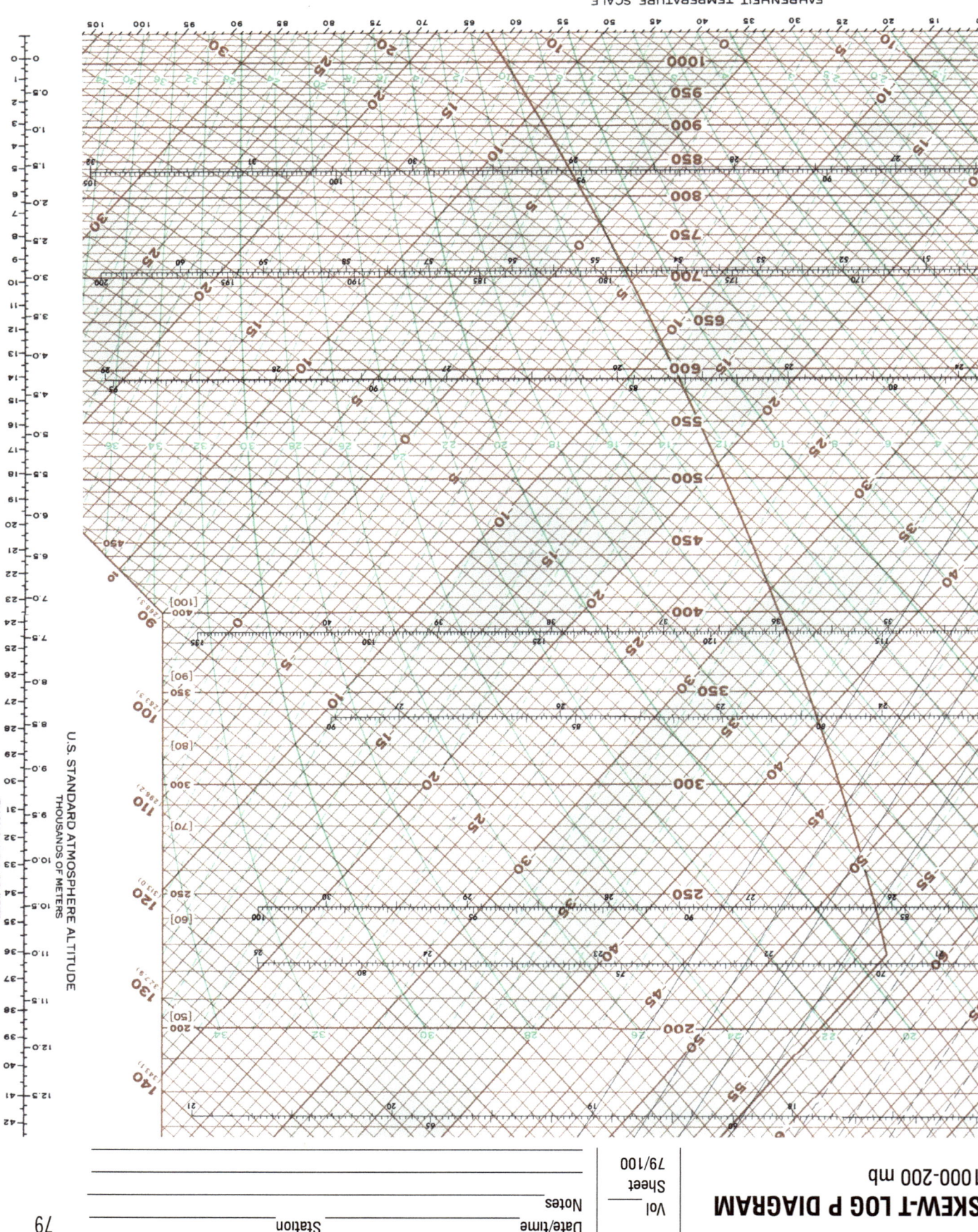

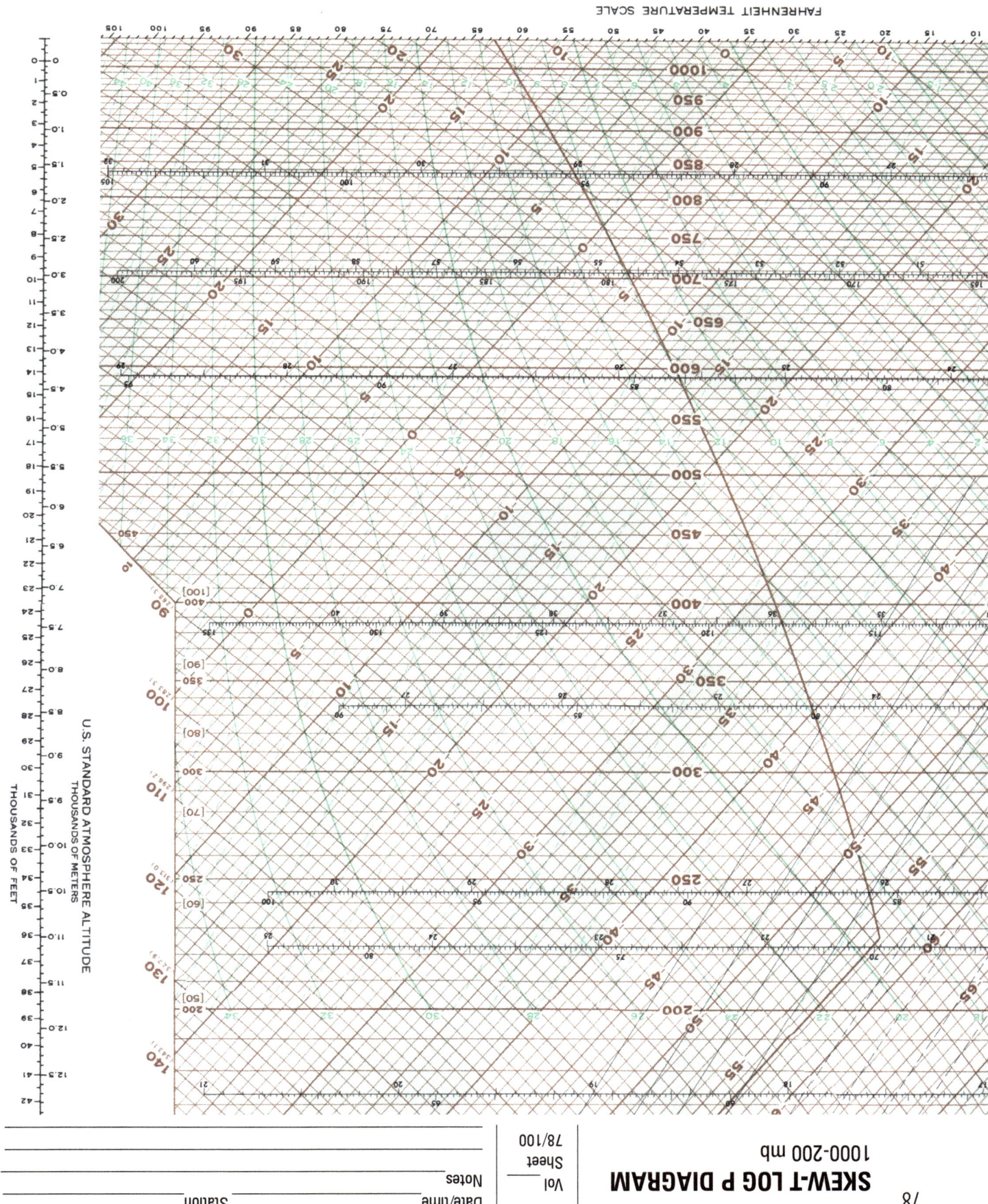

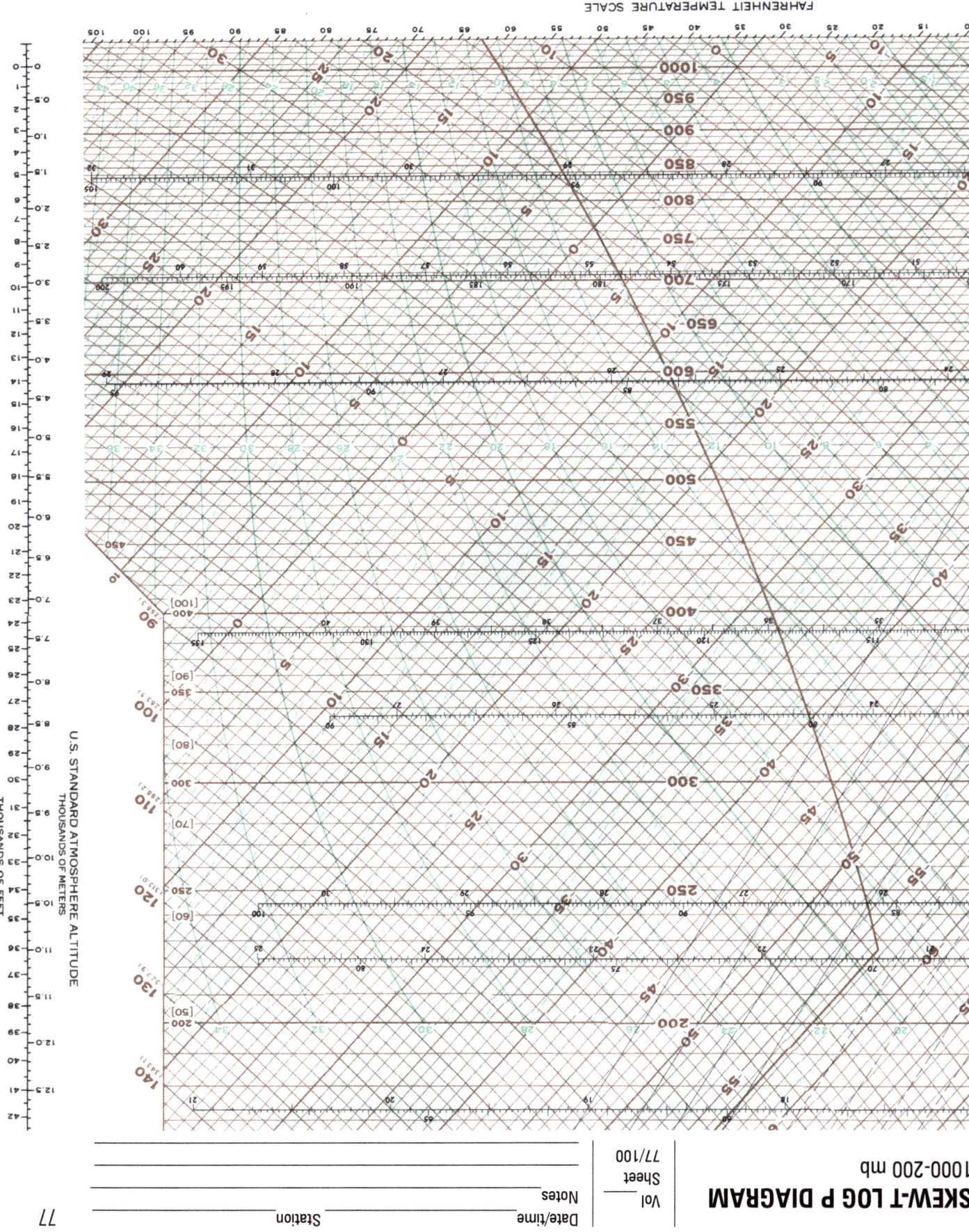

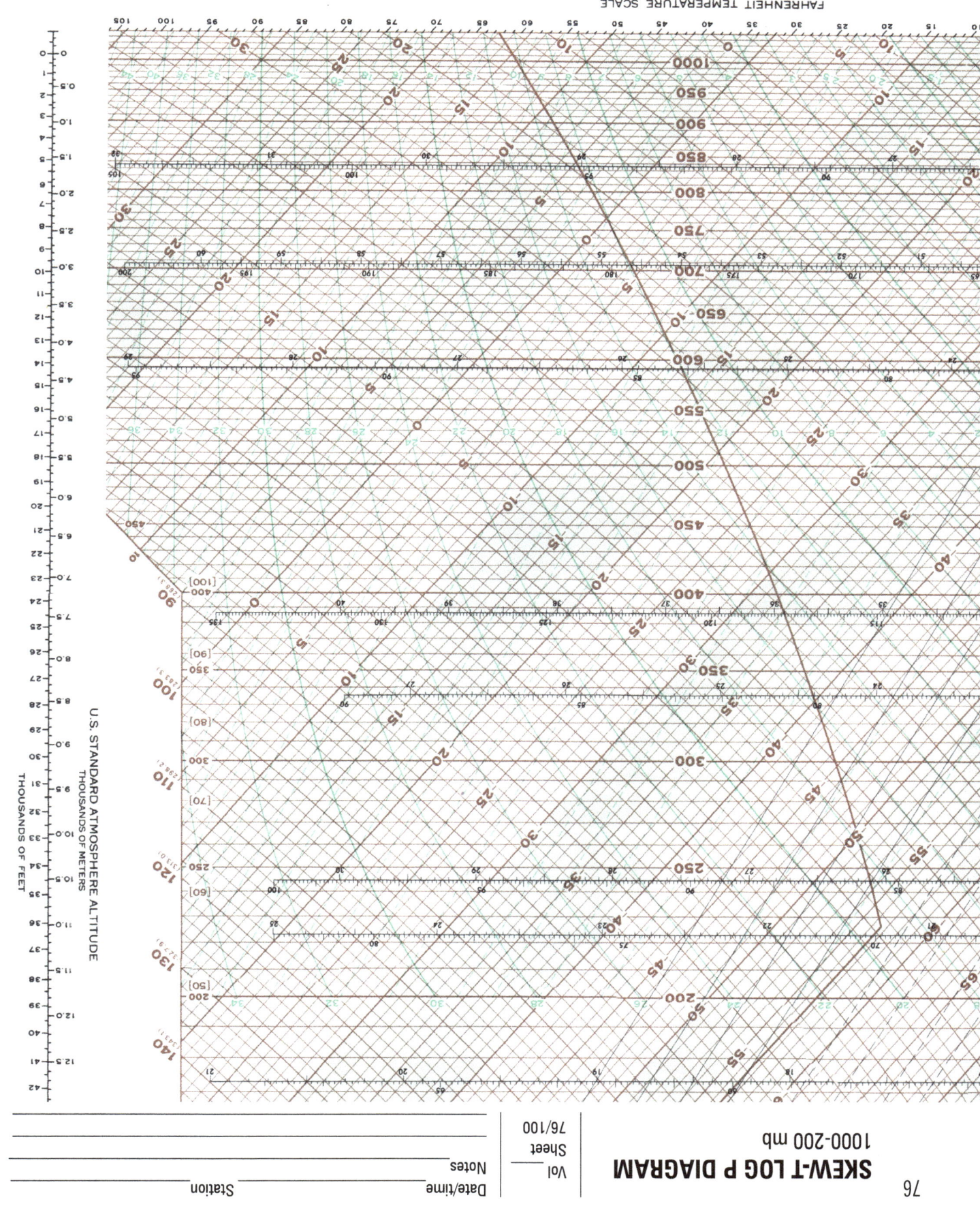

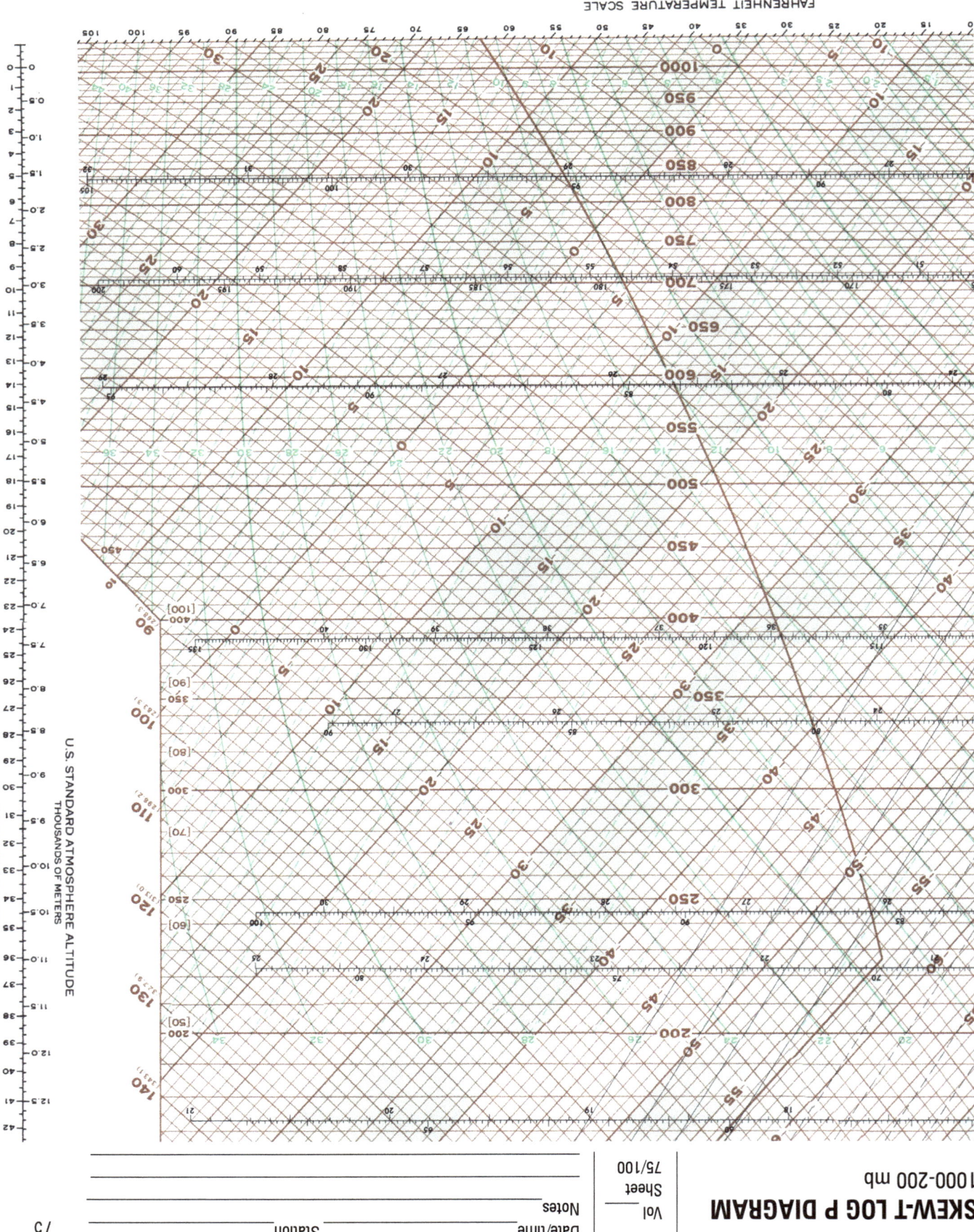

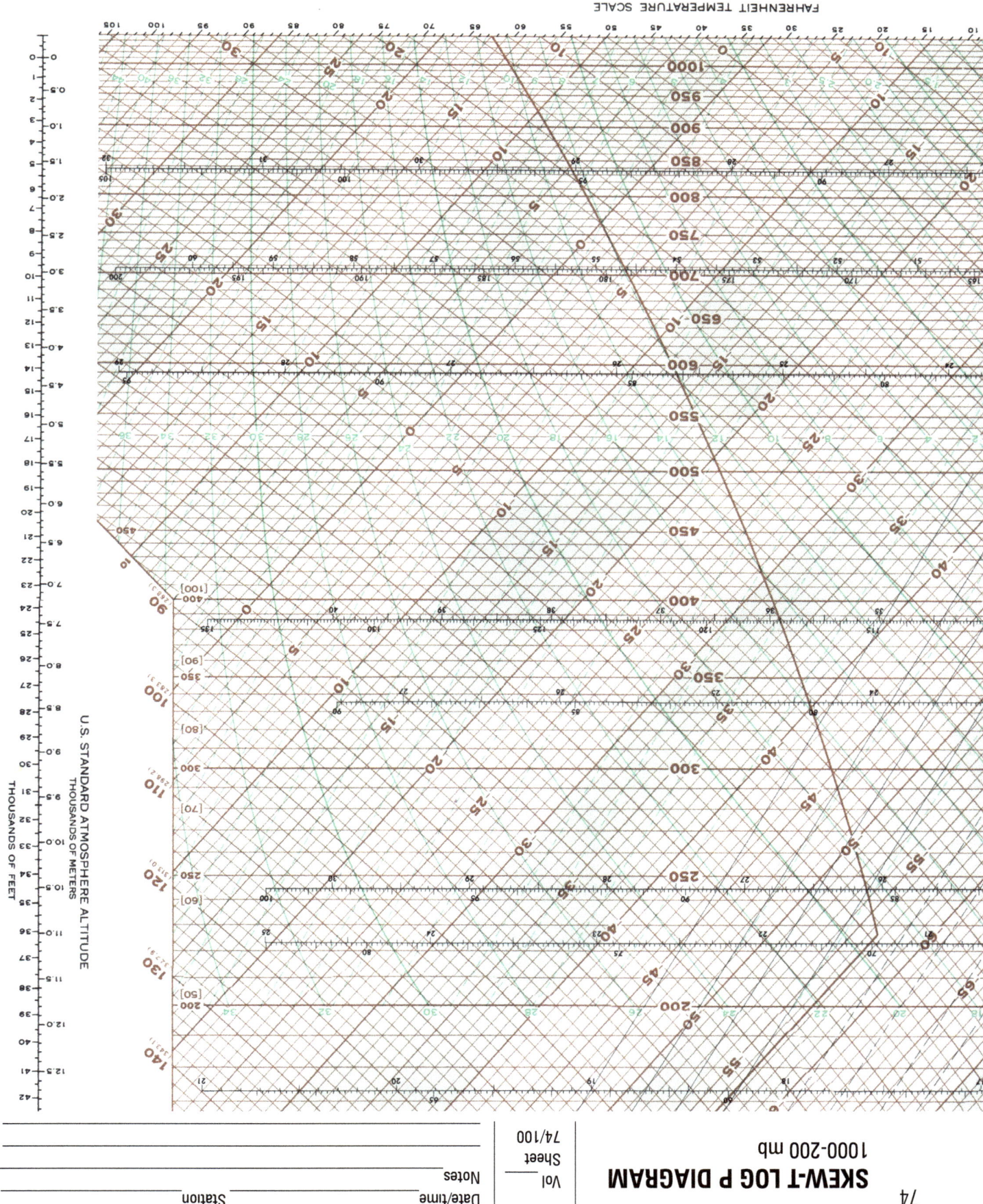

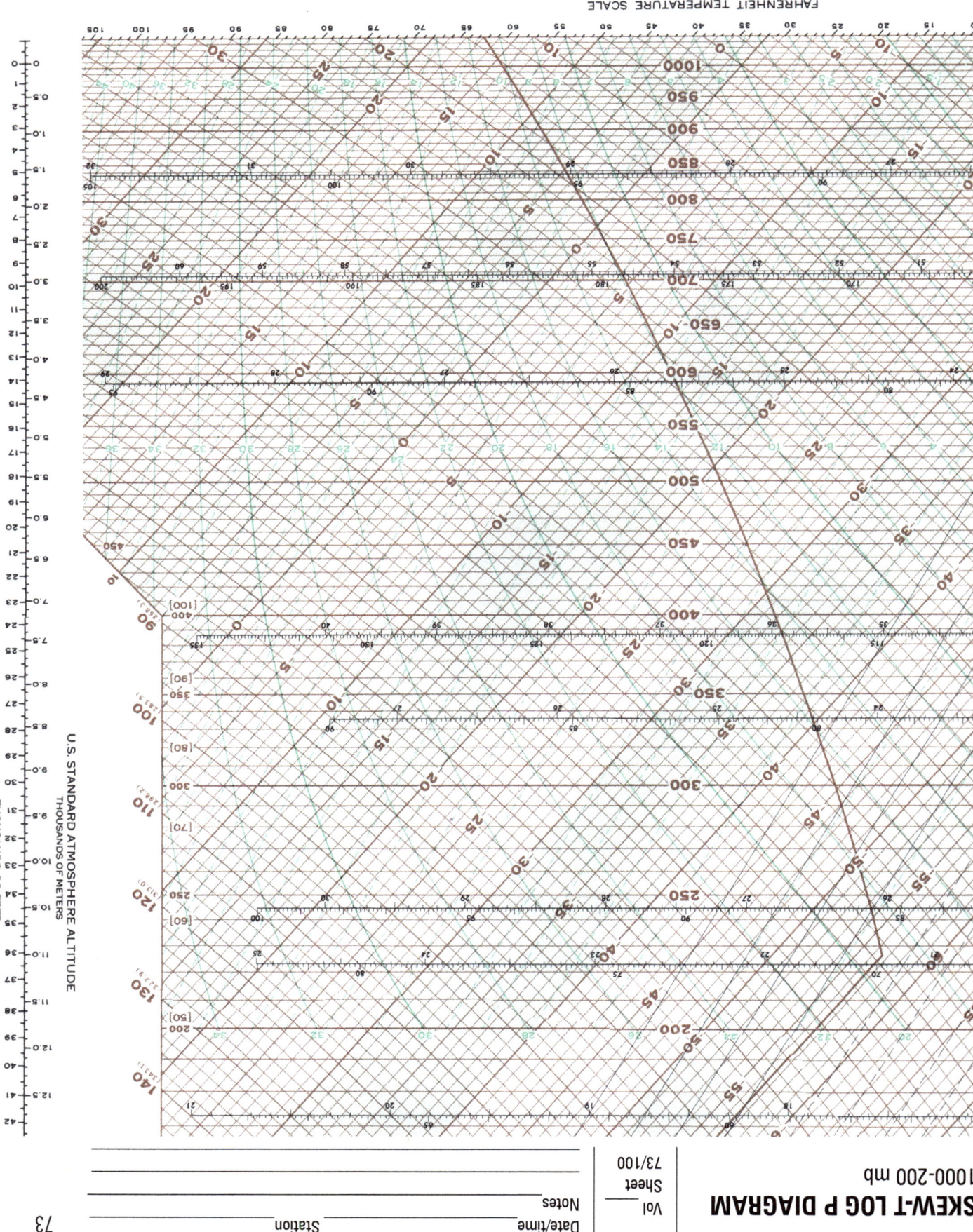

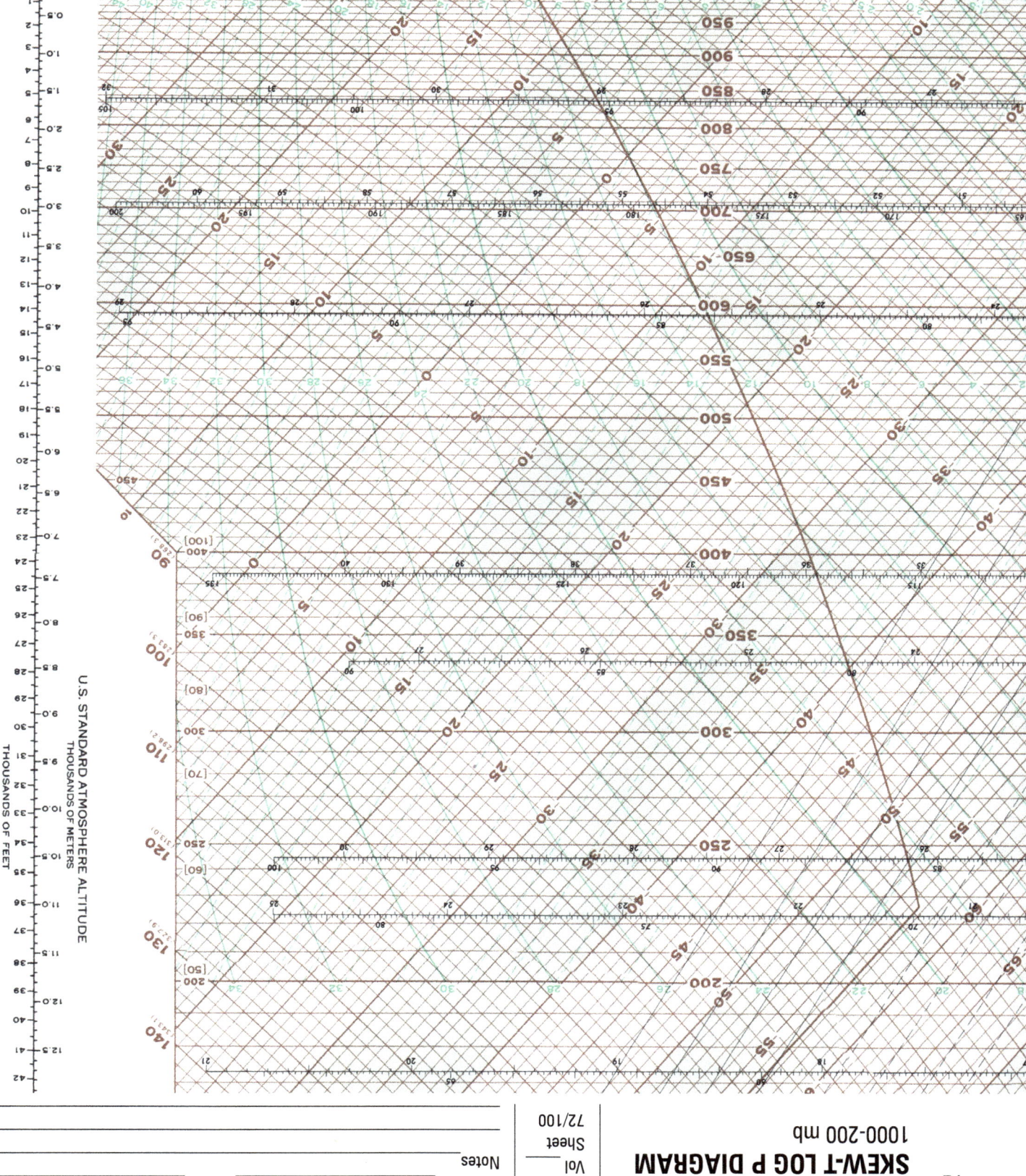

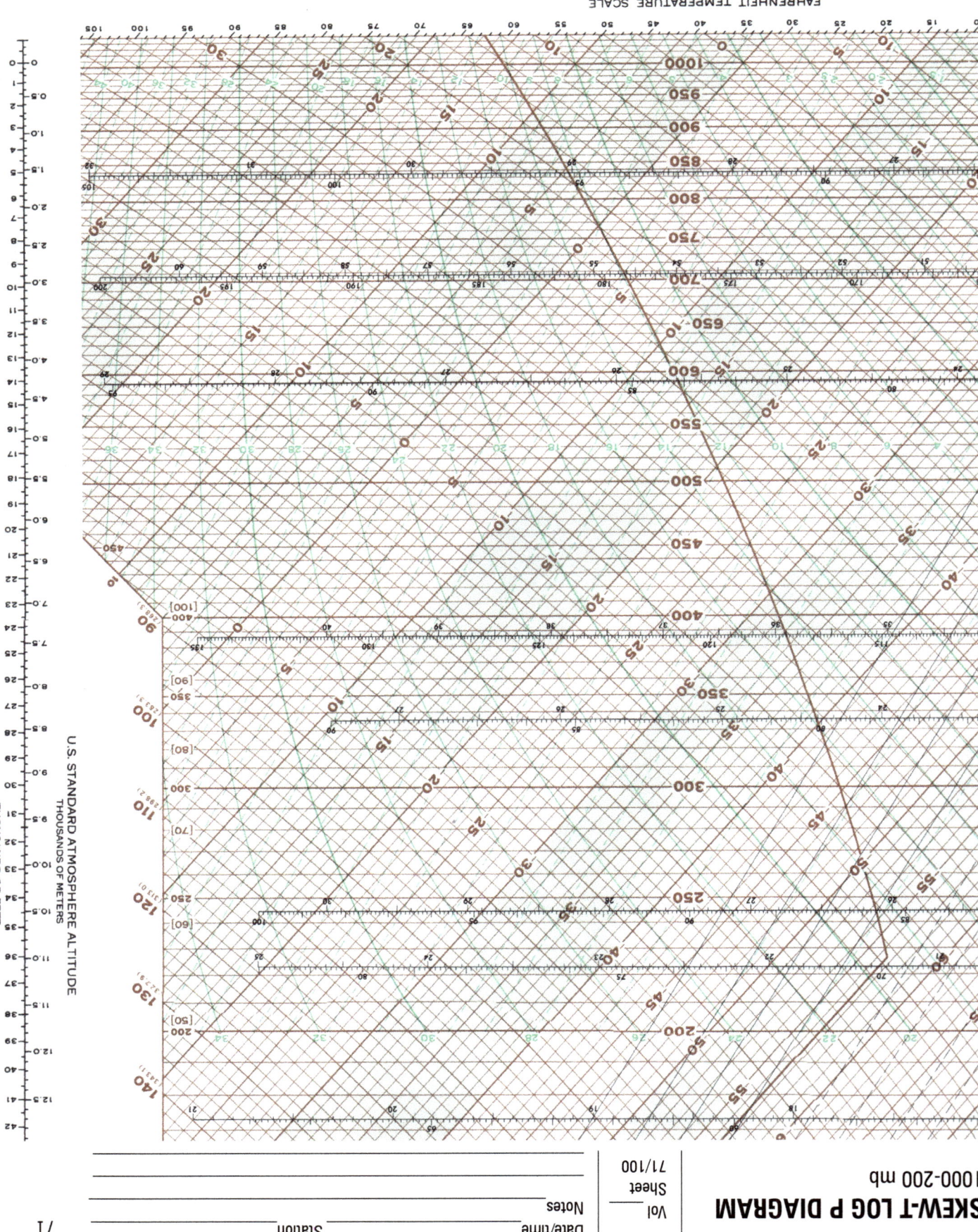

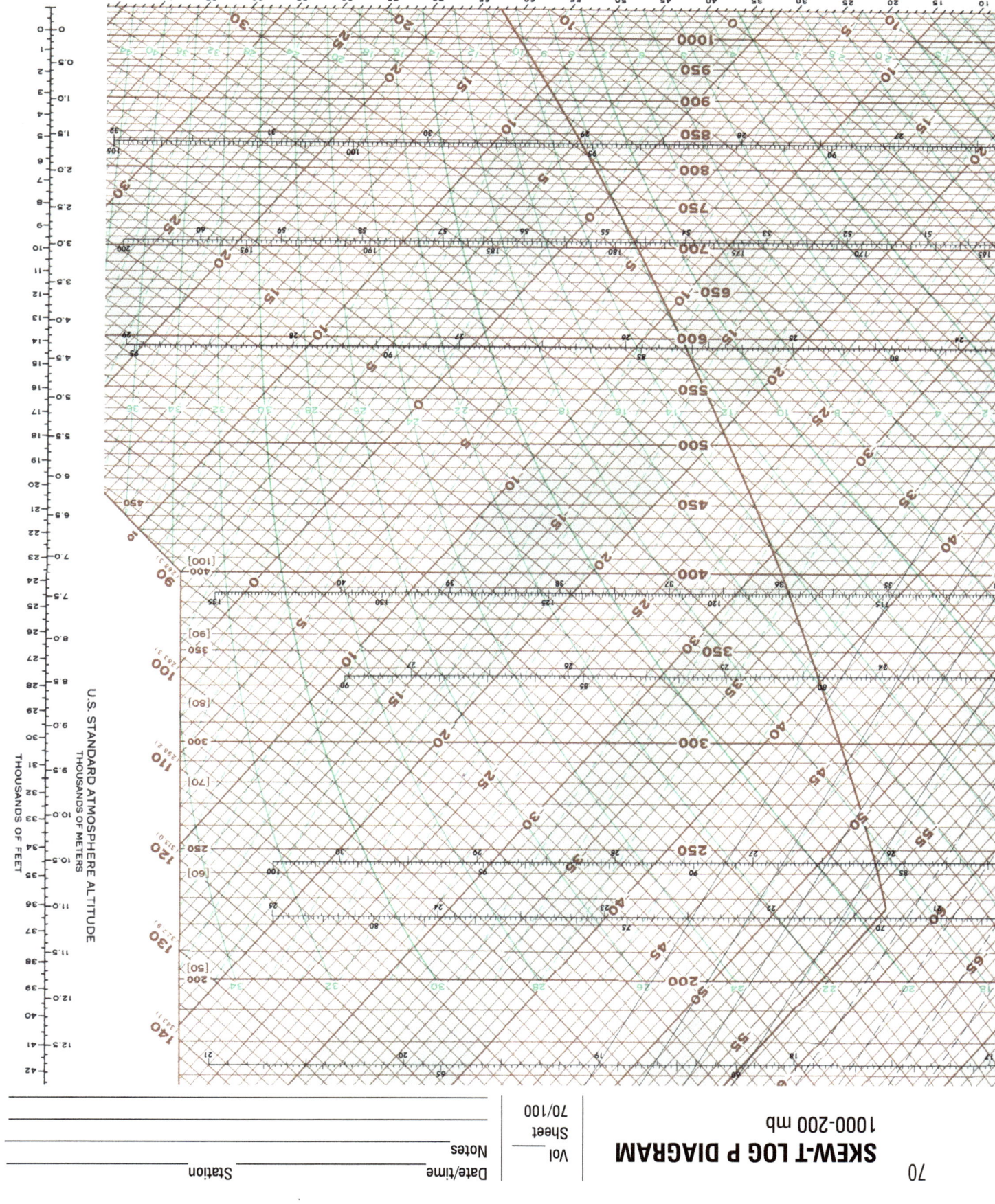

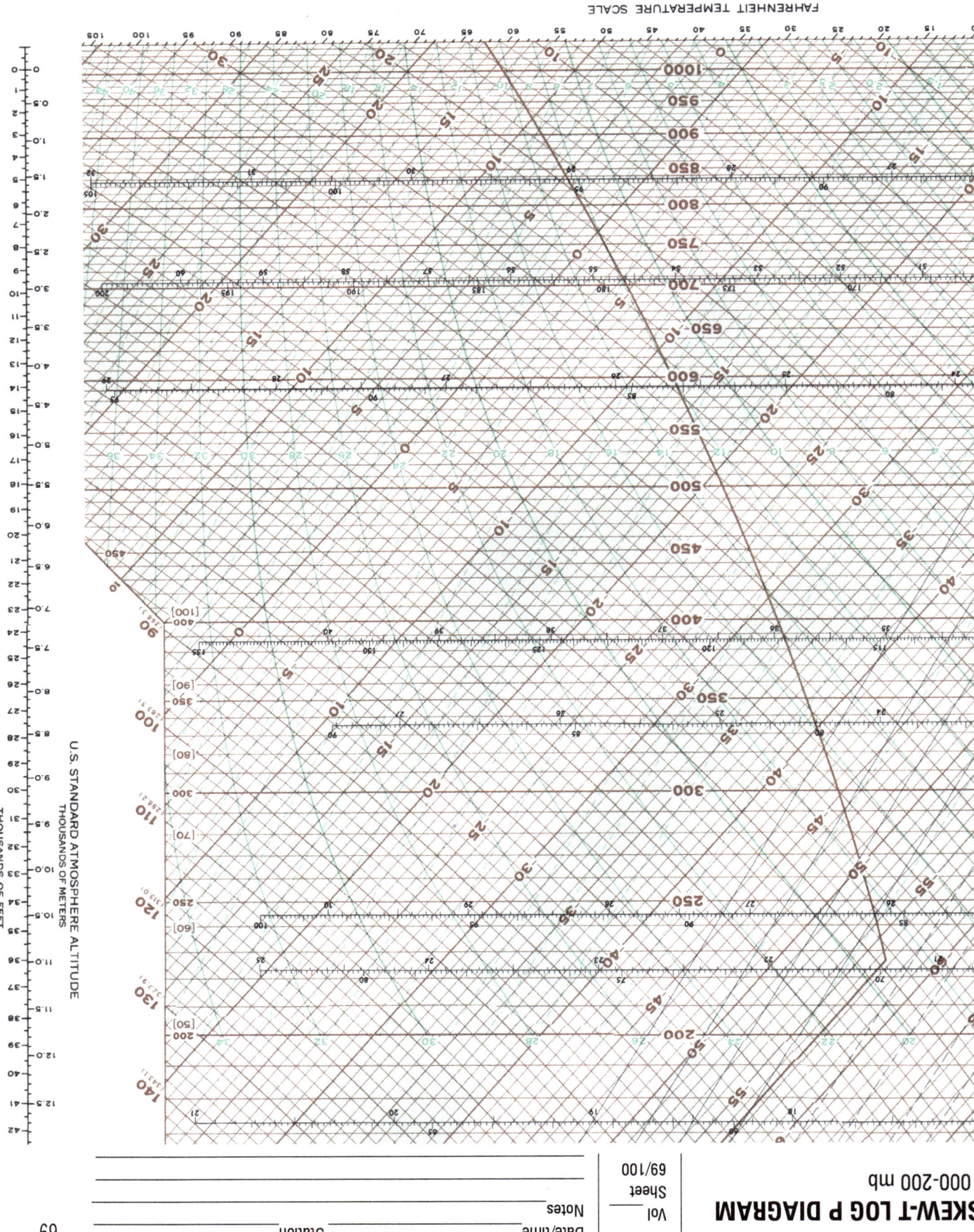

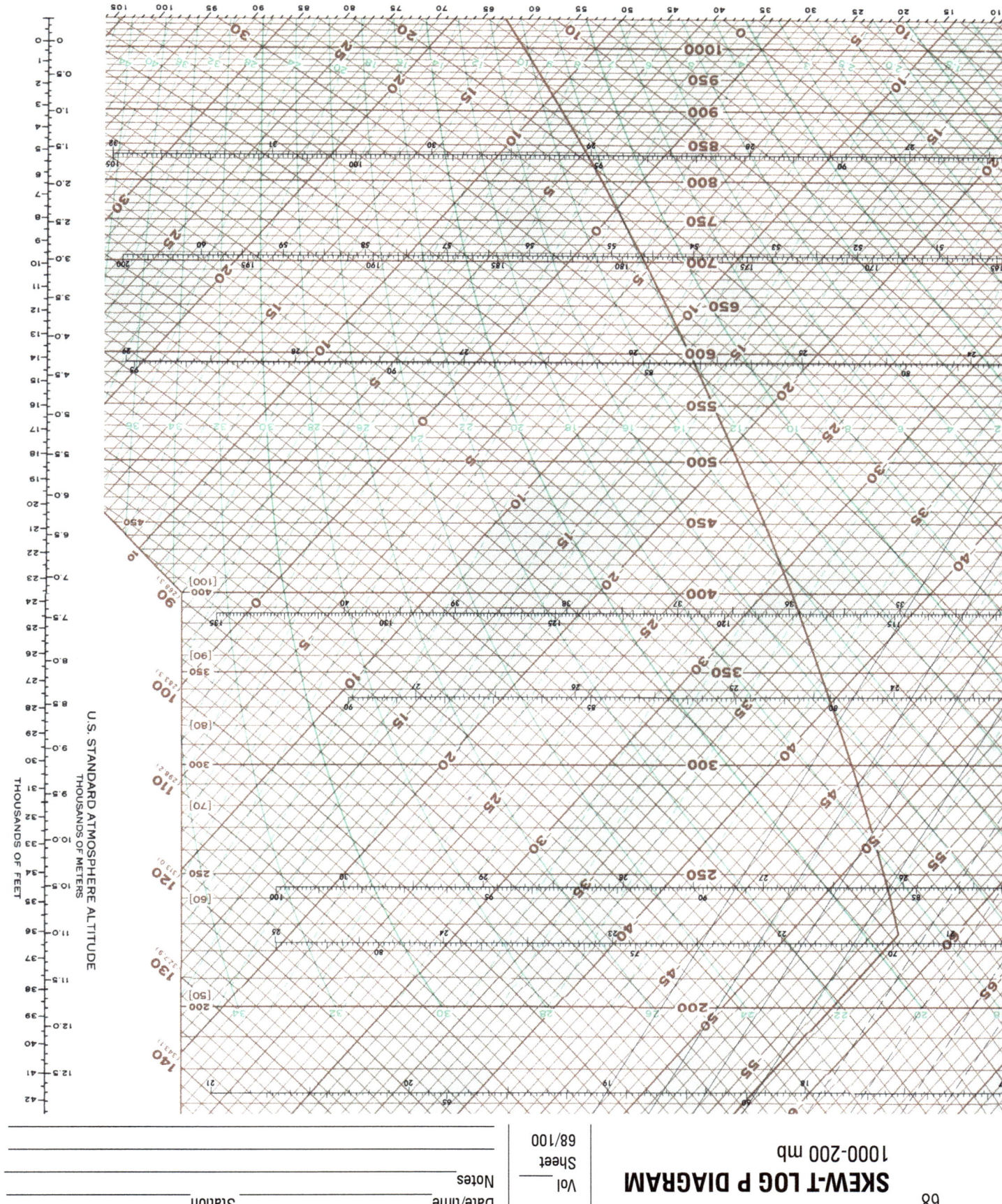

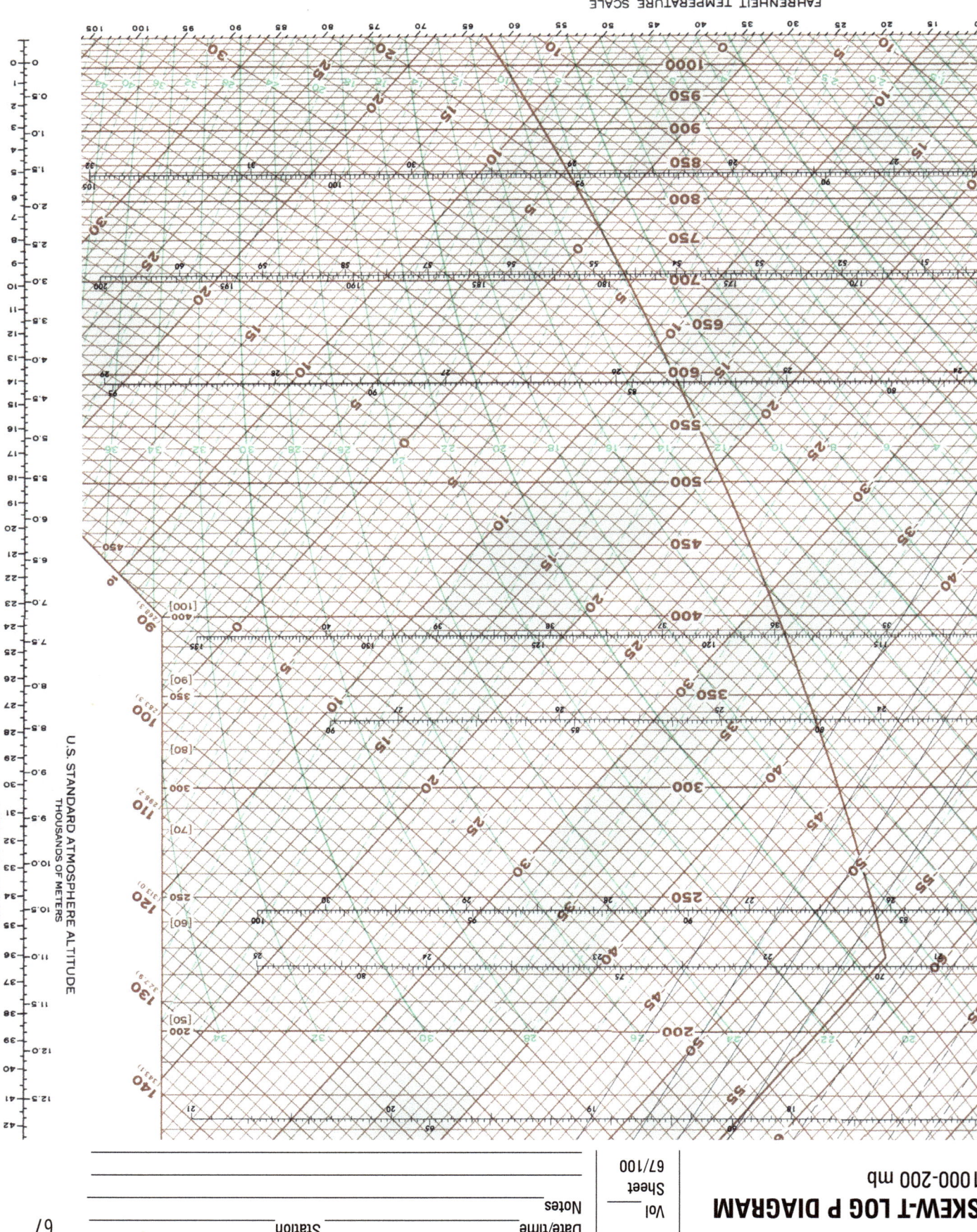

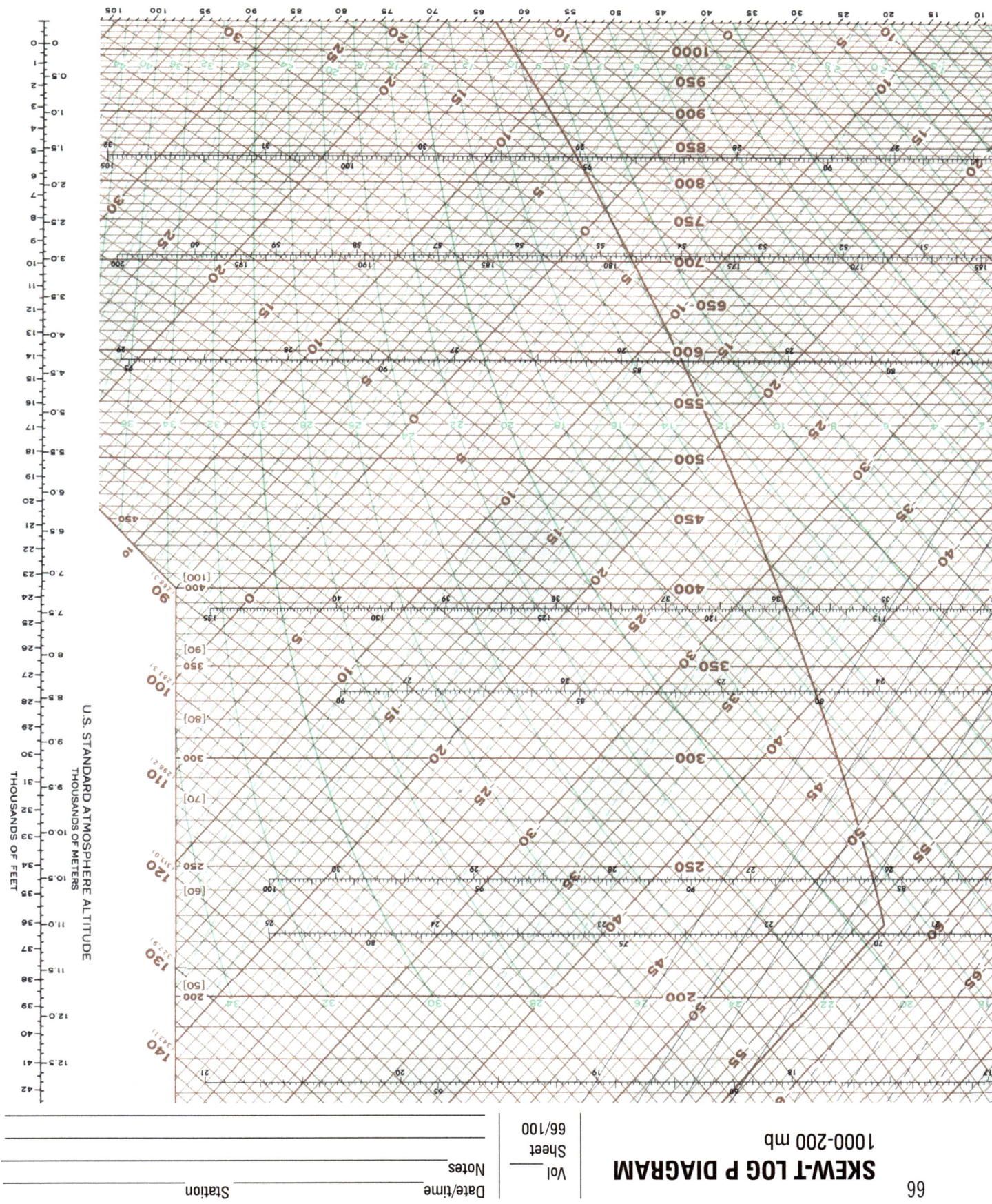

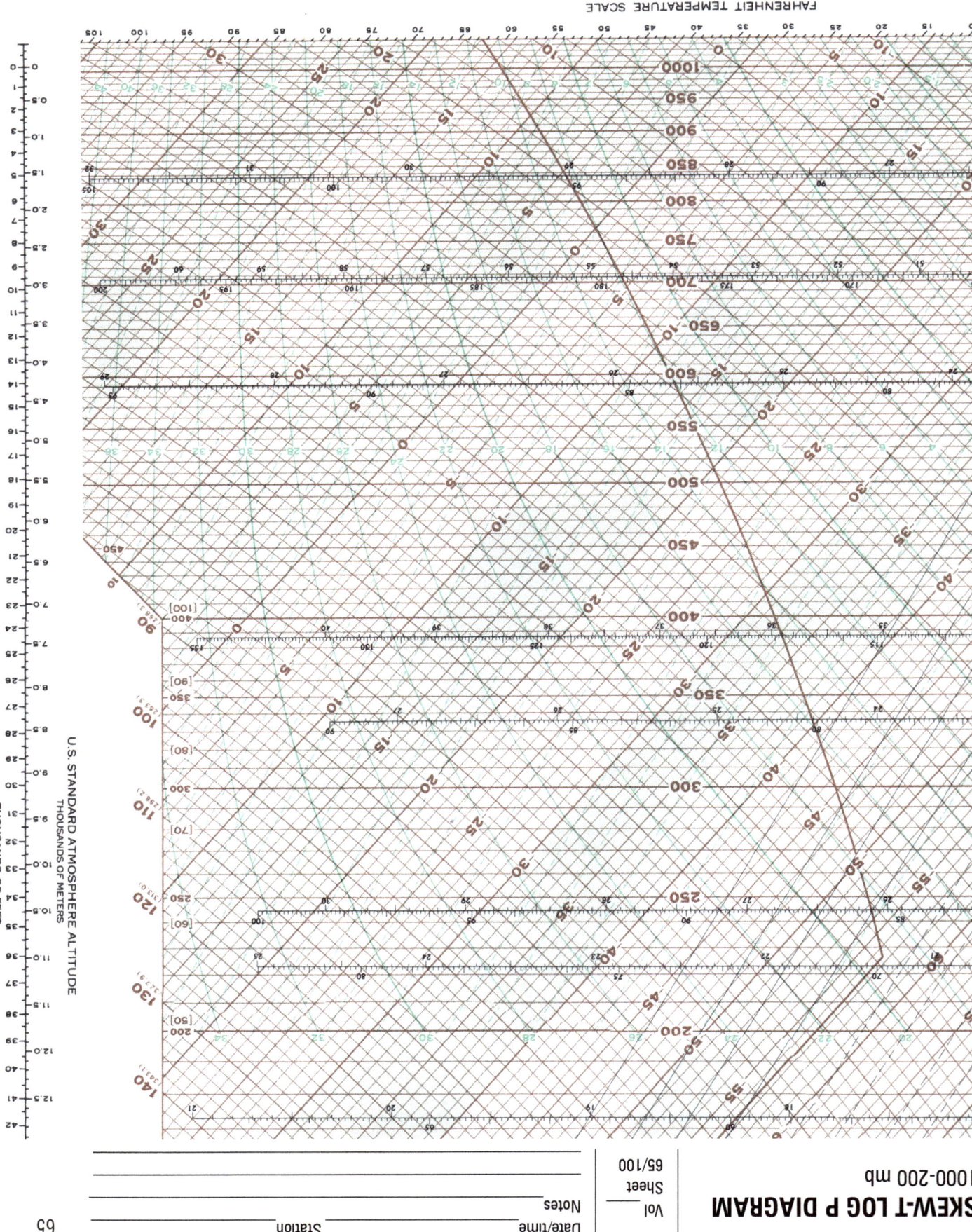

SKEW-T LOG P DIAGRAM
1000-200 mb

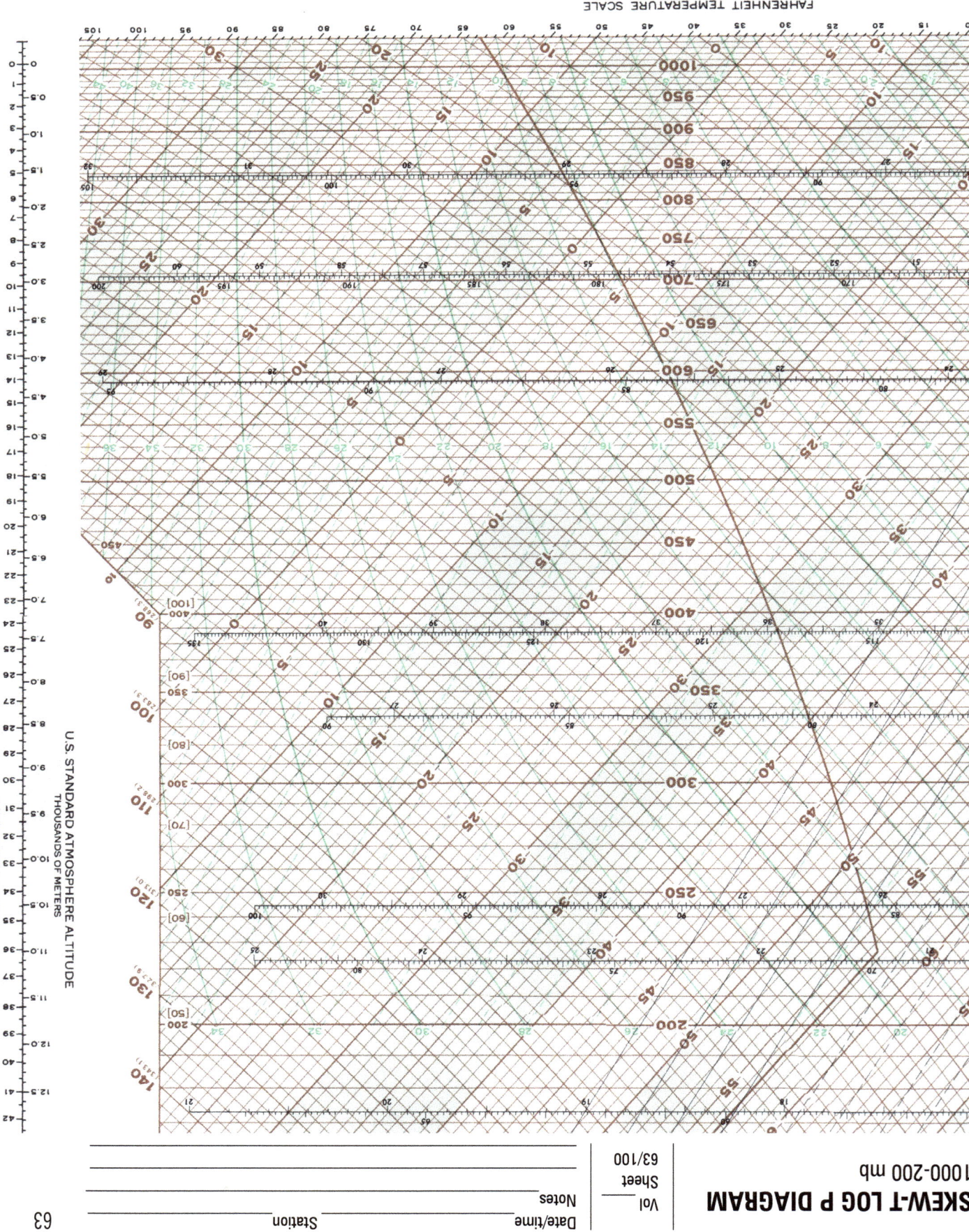

SKEW-T LOG P DIAGRAM
1000-200 mb

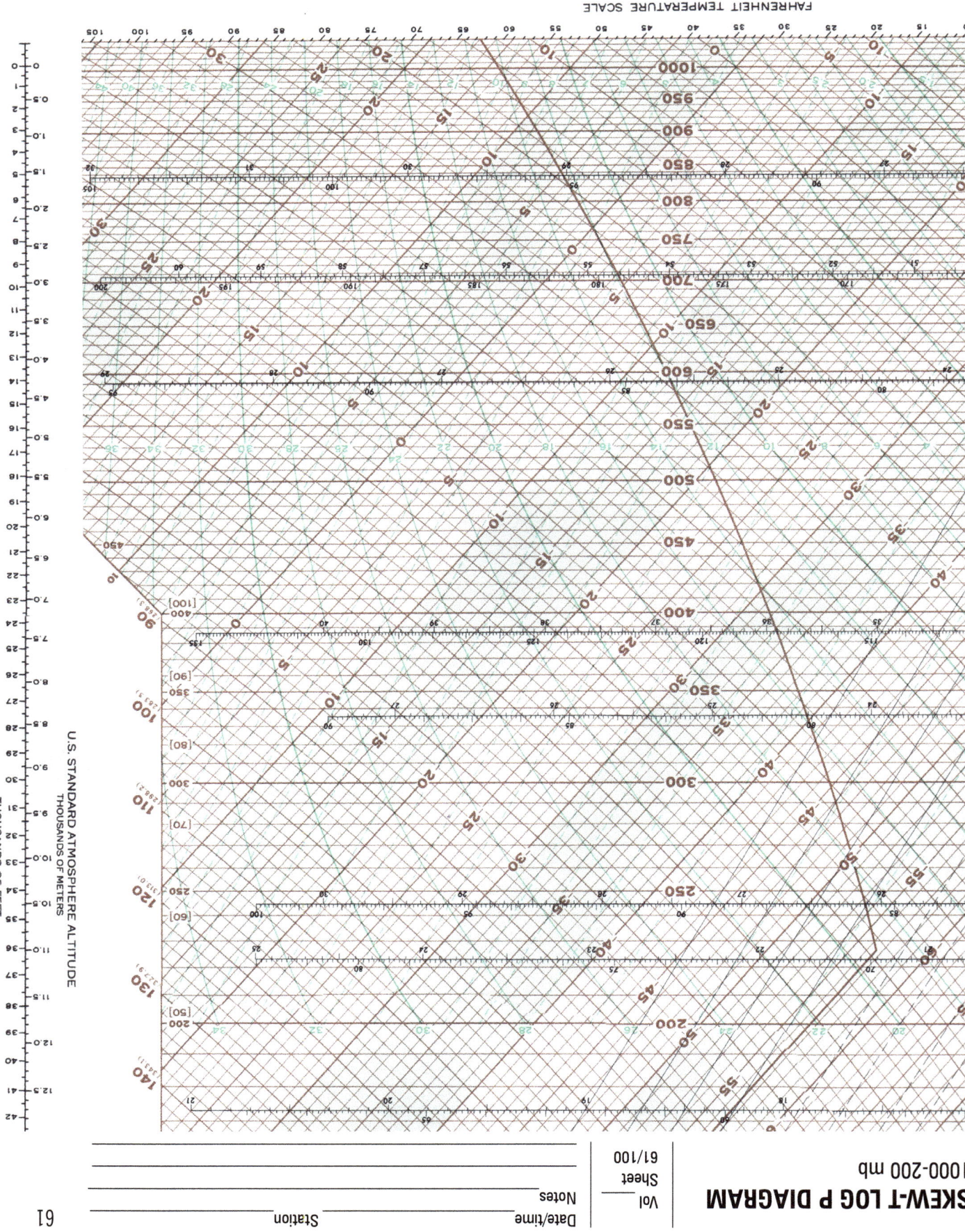

SKEW-T LOG P DIAGRAM
1000-200 mb

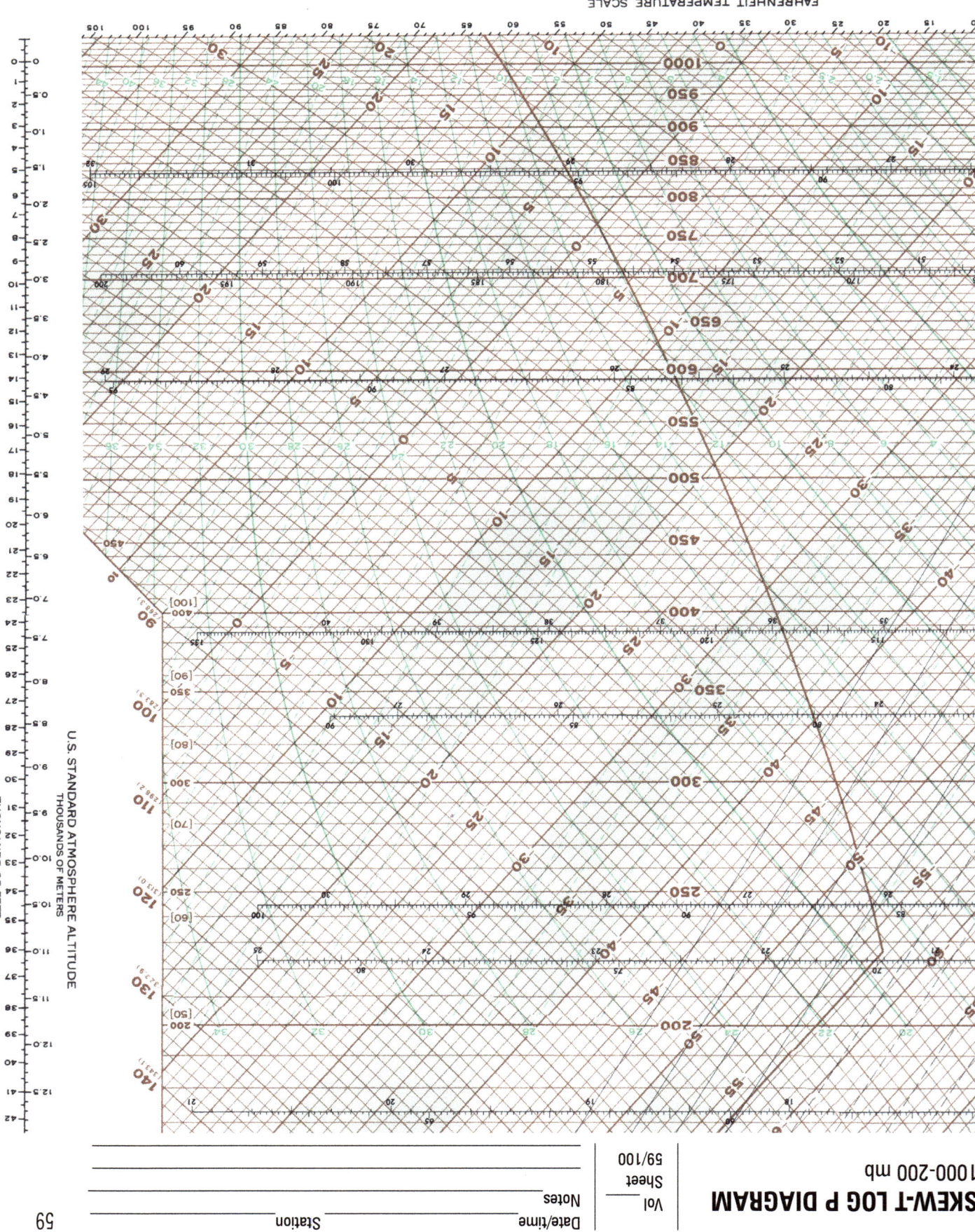

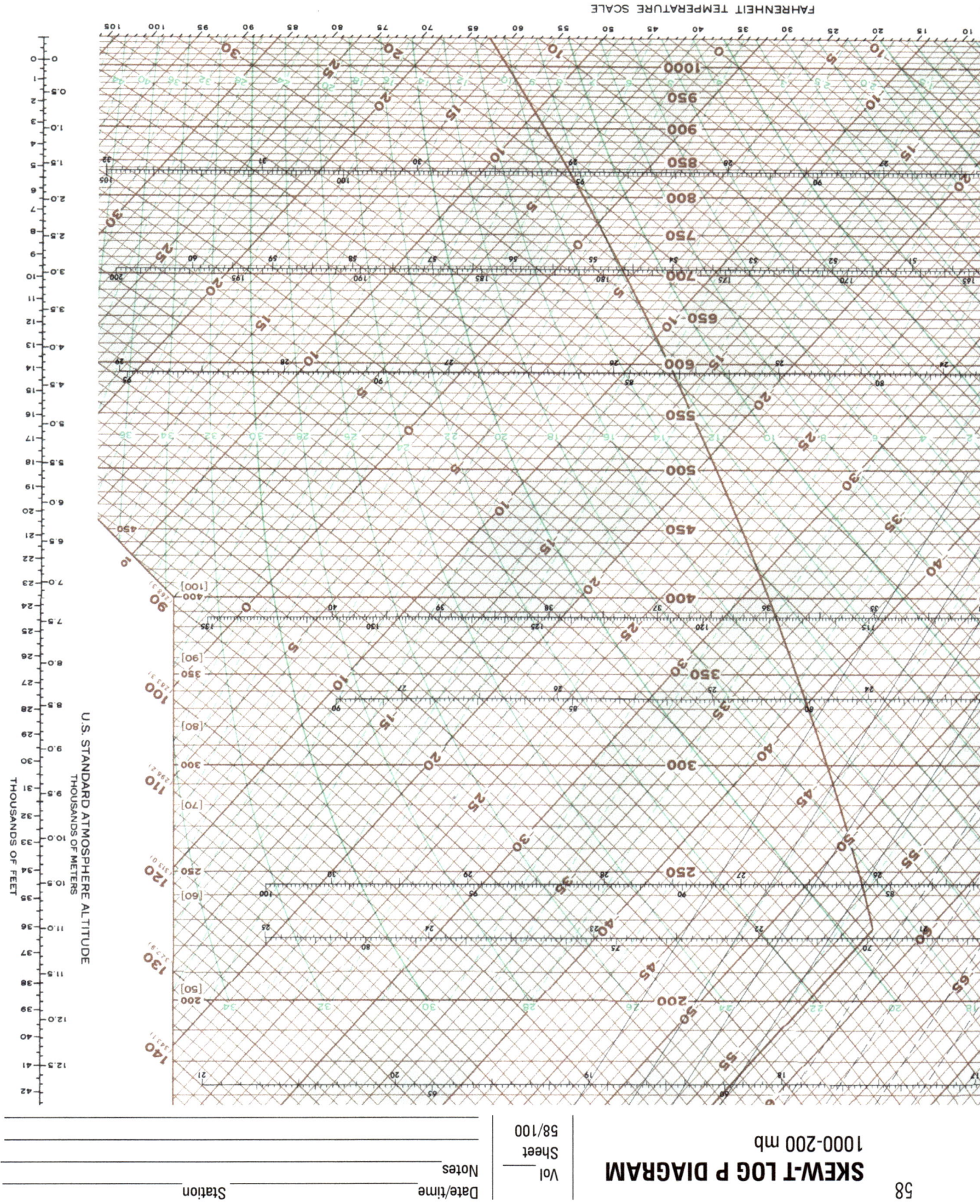

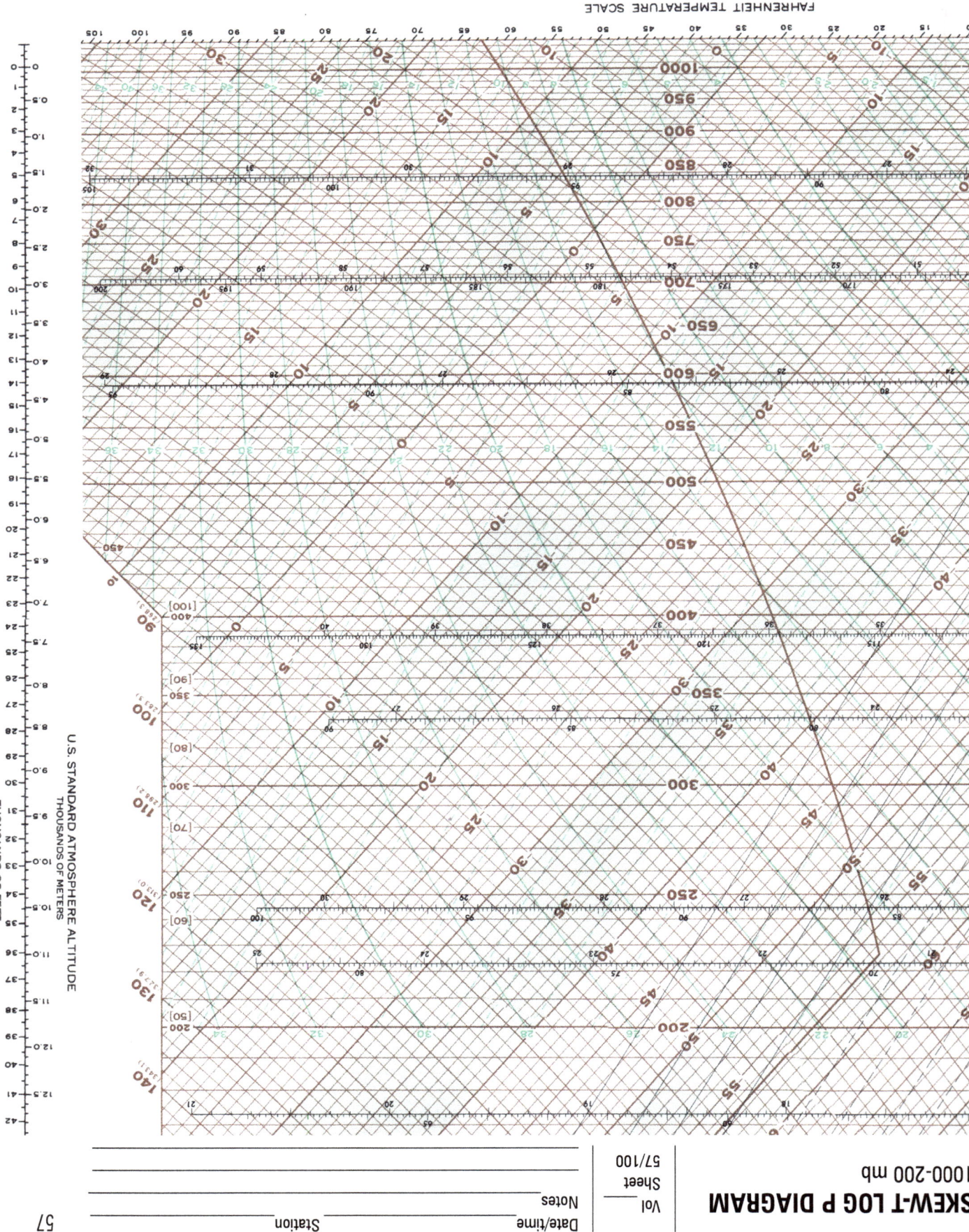

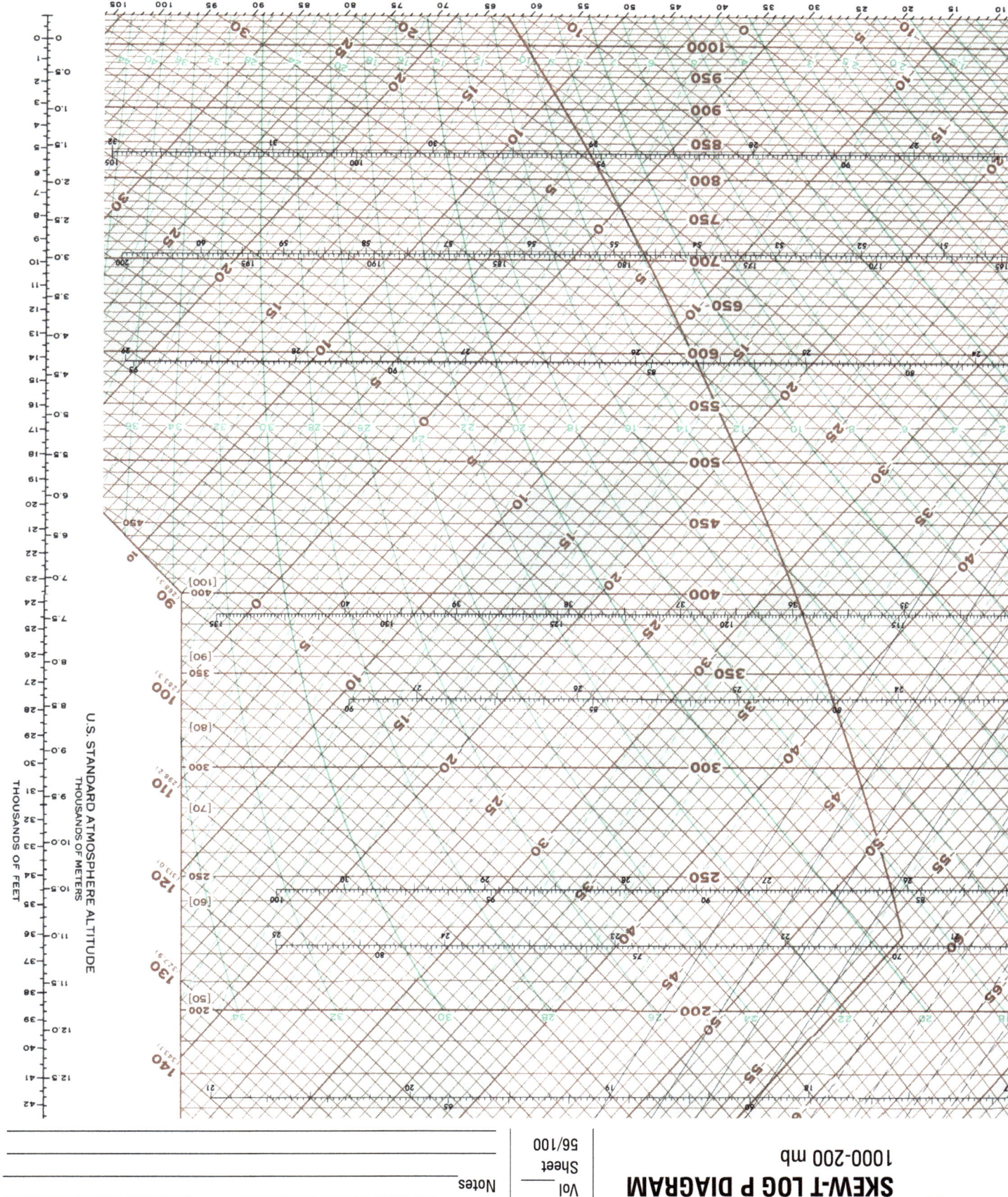

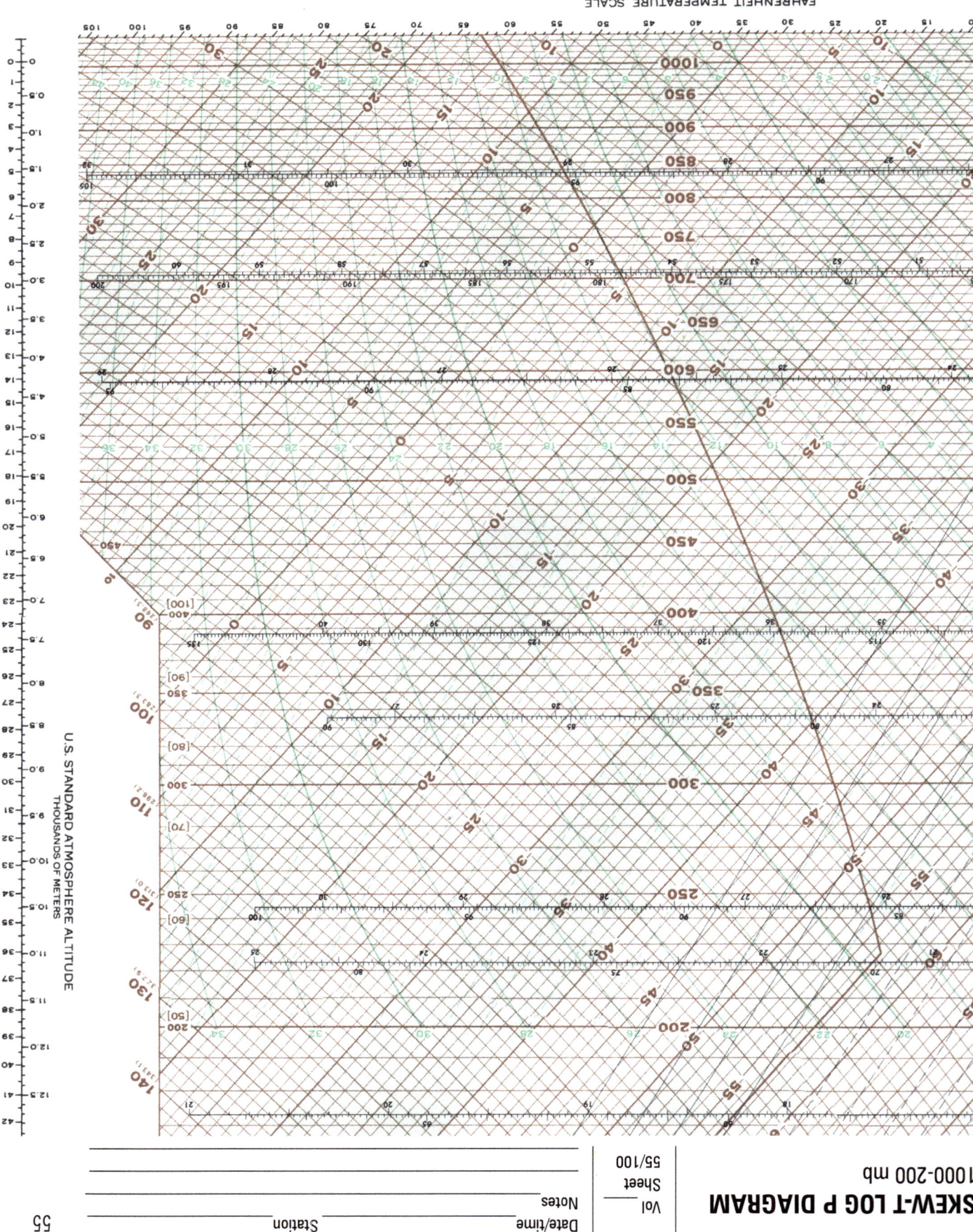

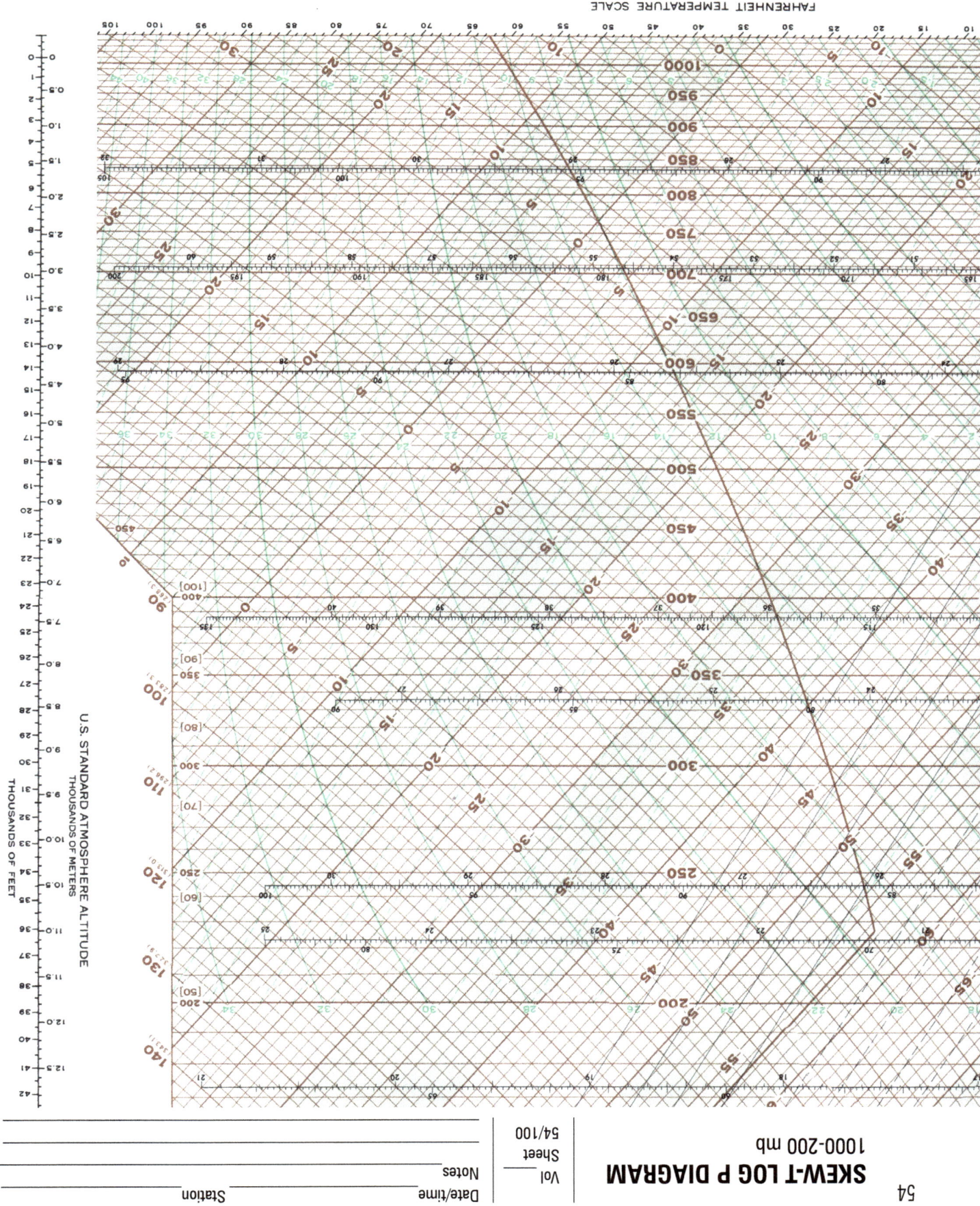

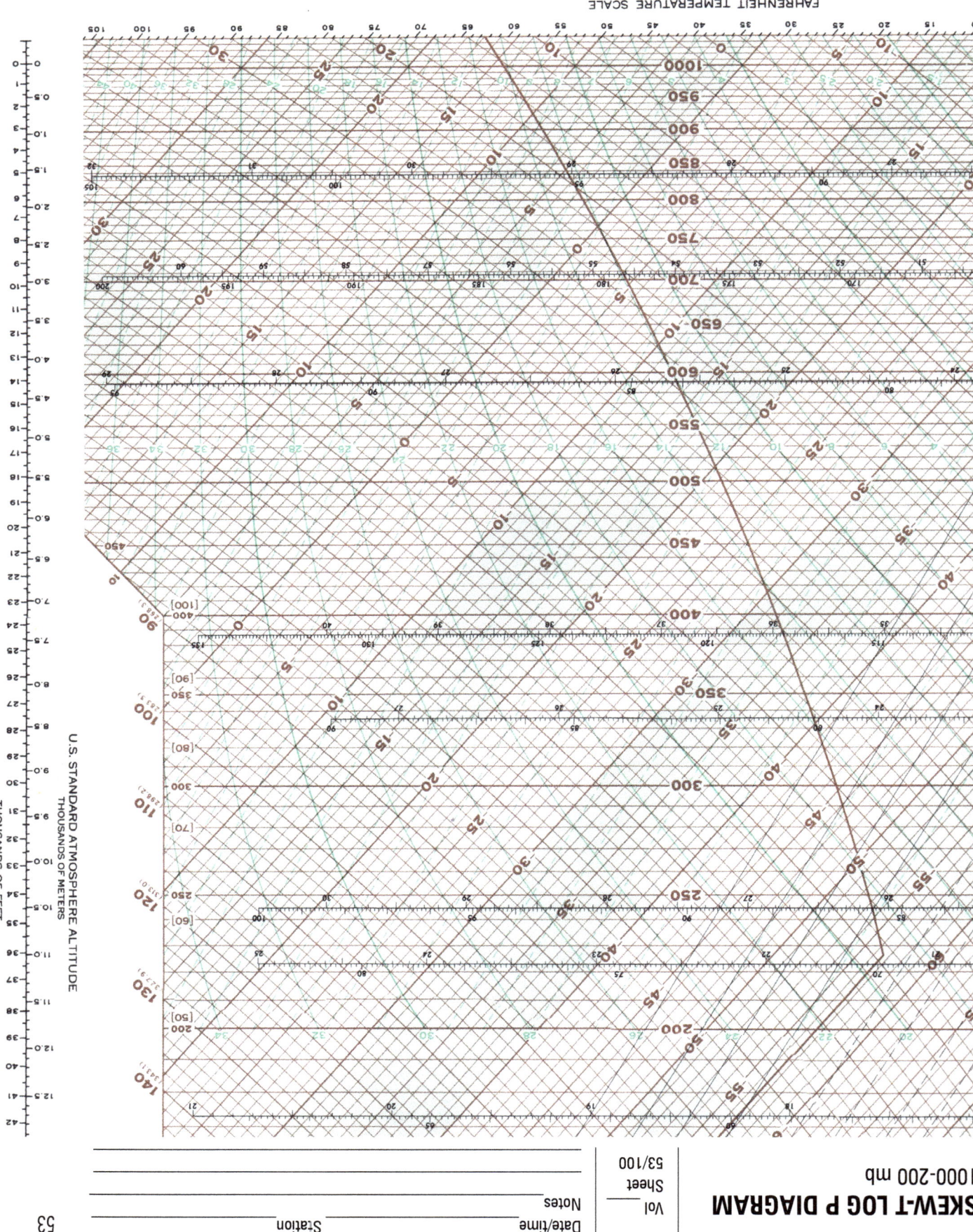

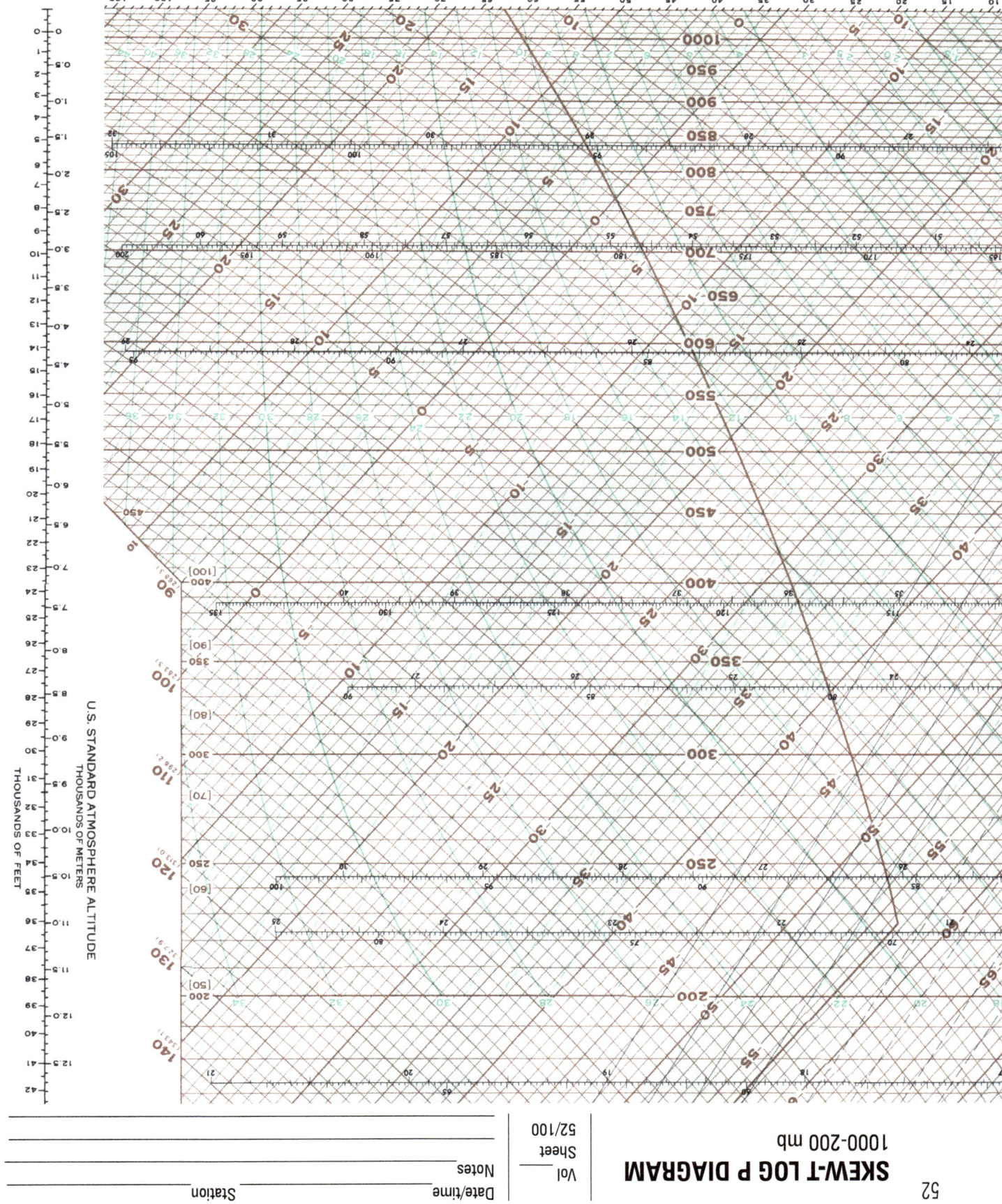

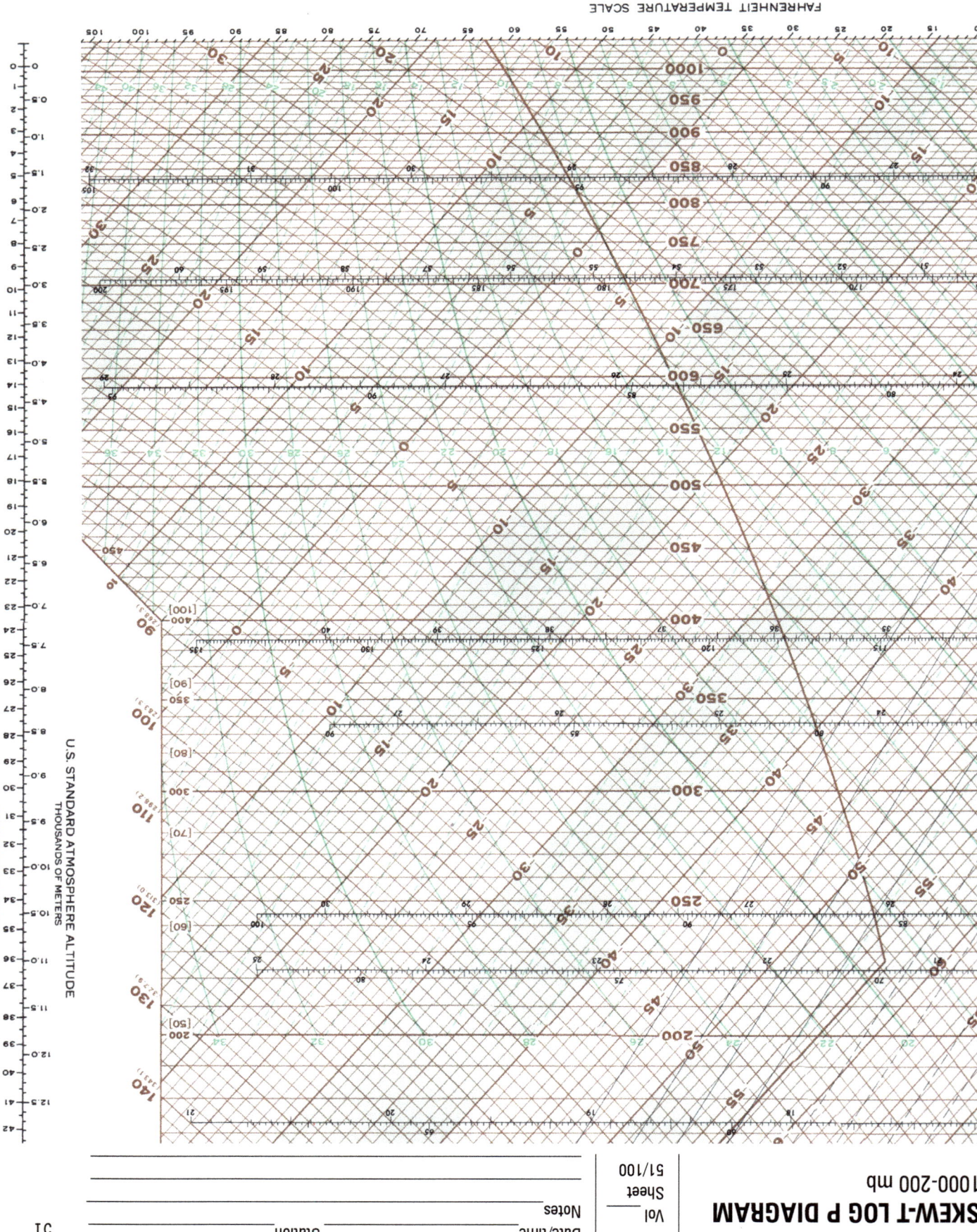

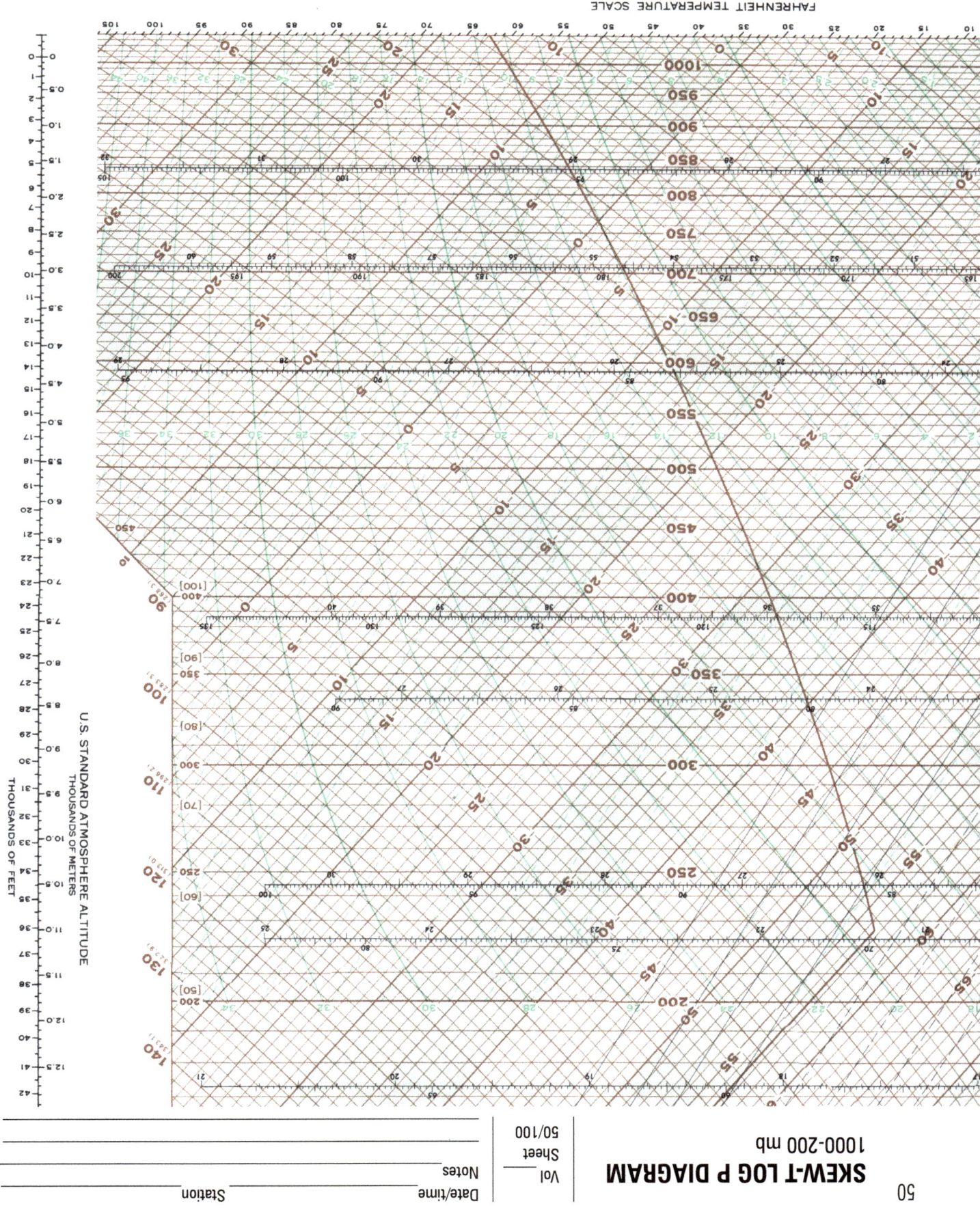

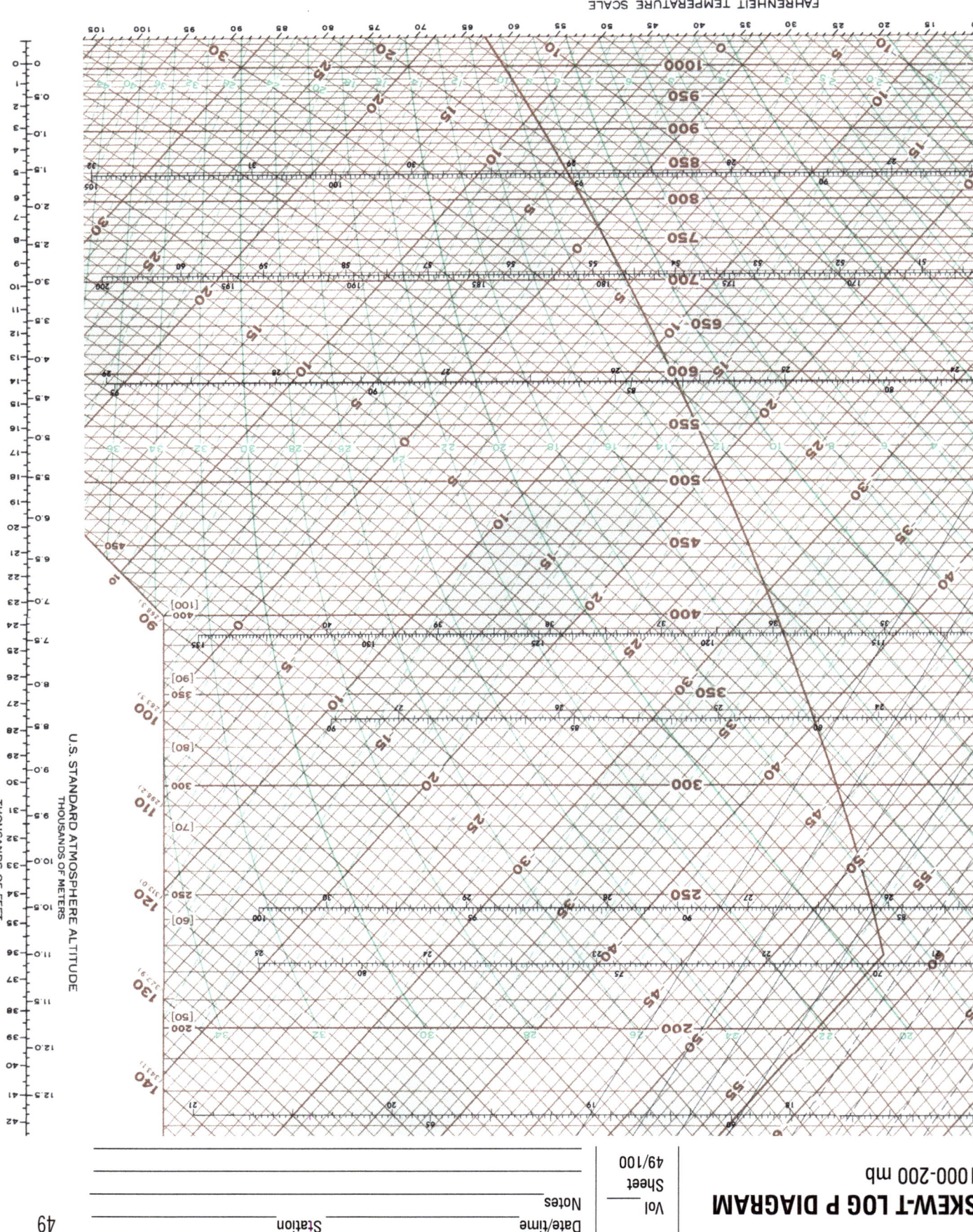

SKEW T, LOG P DIAGRAM
1000-200 mb

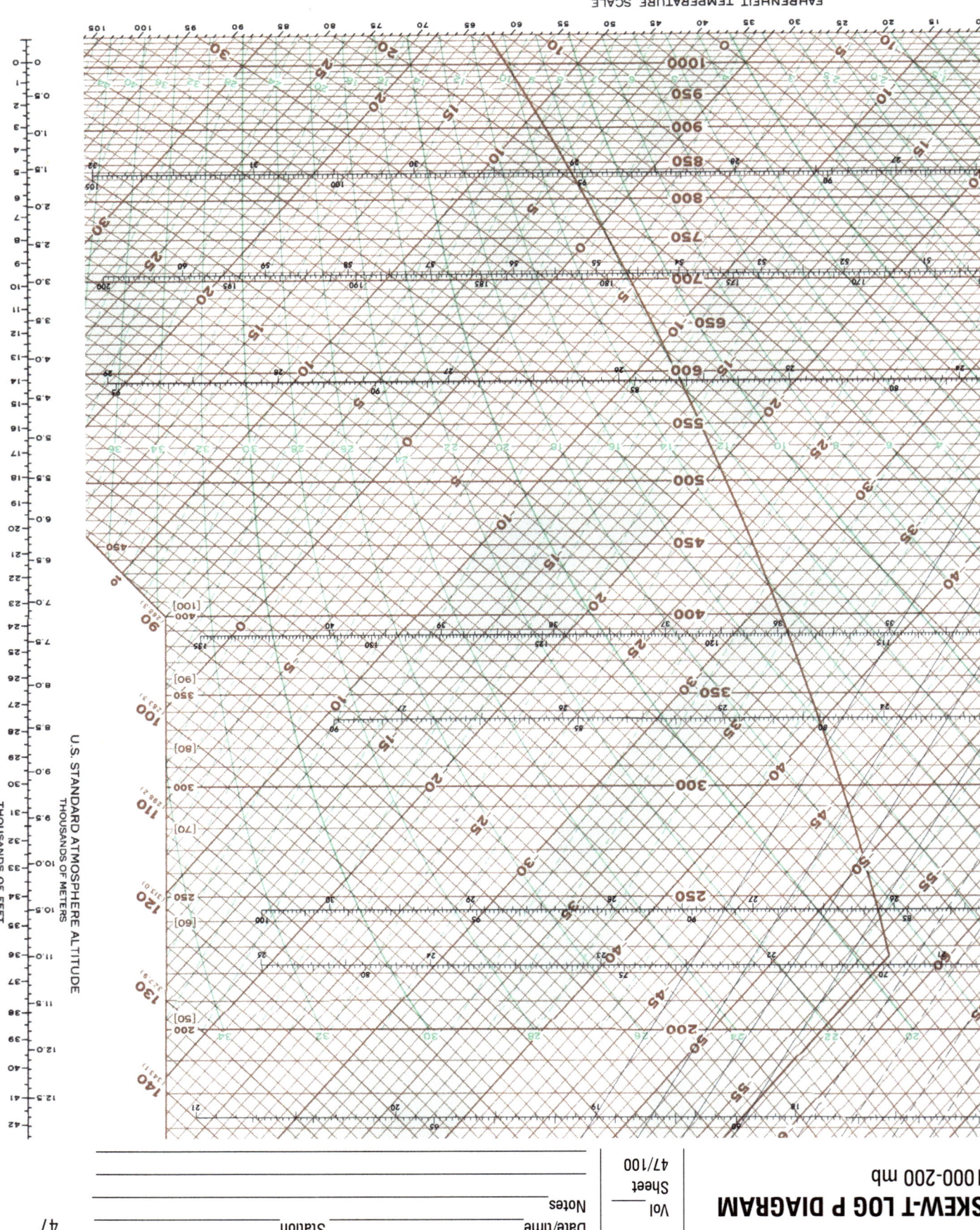

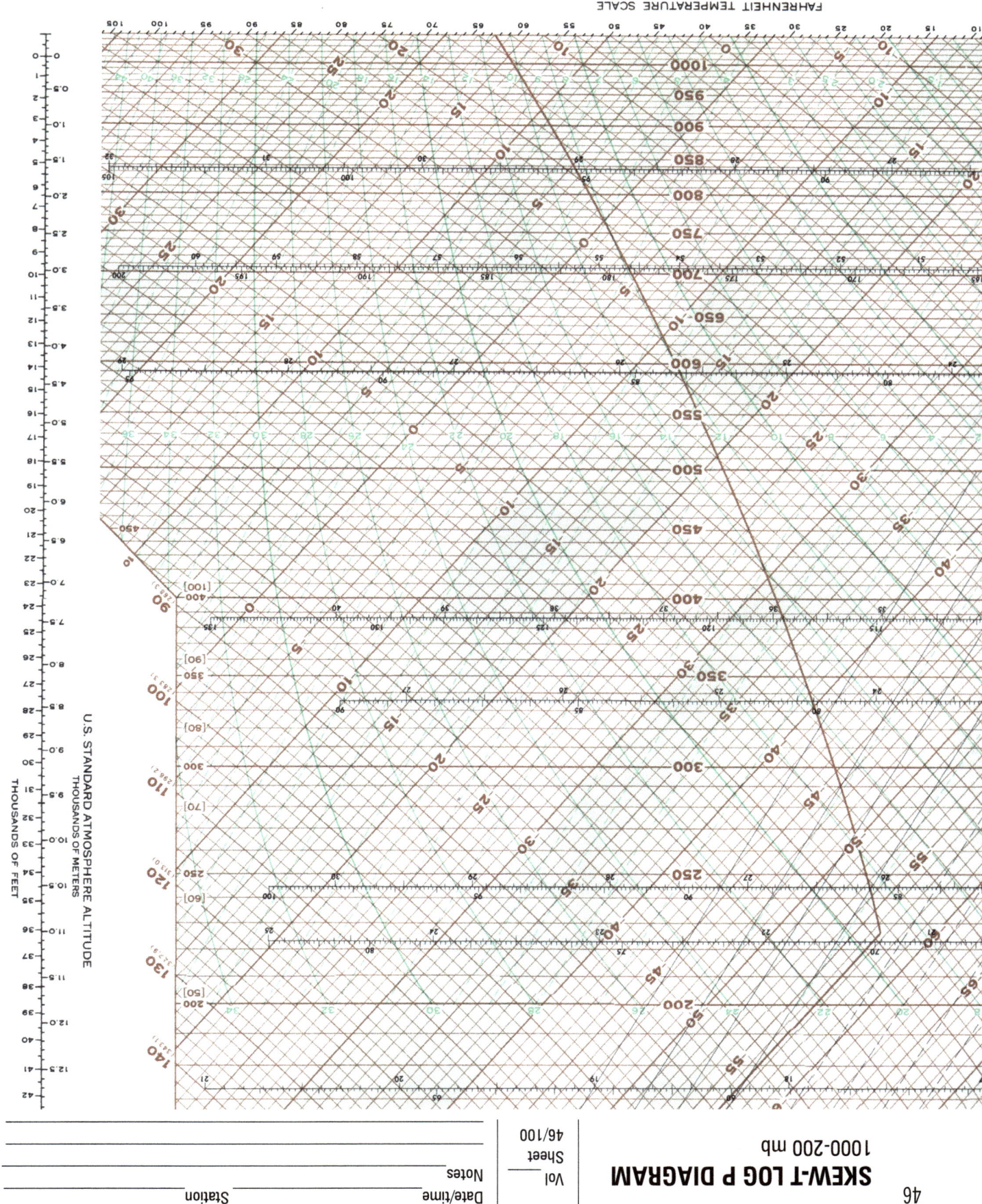

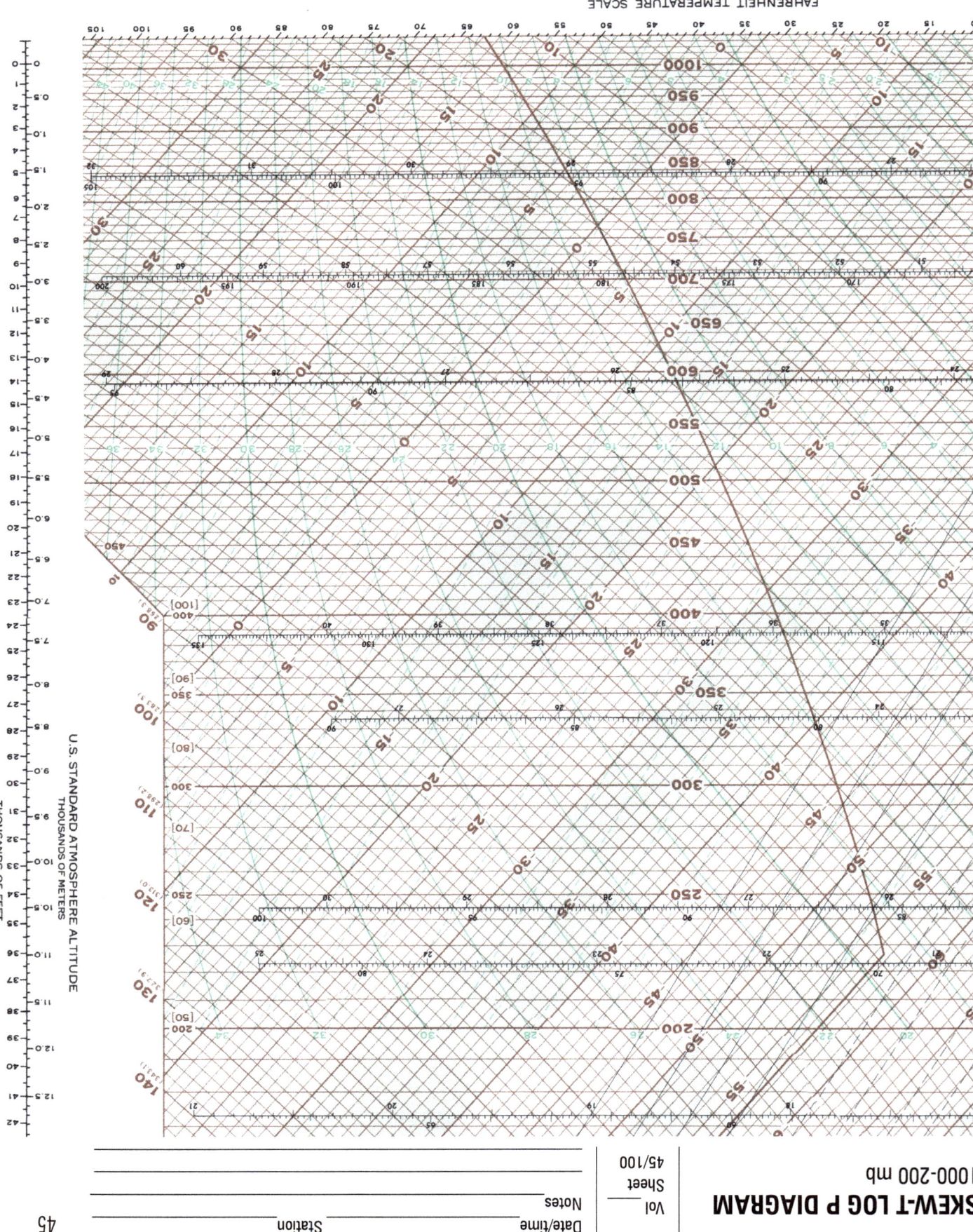

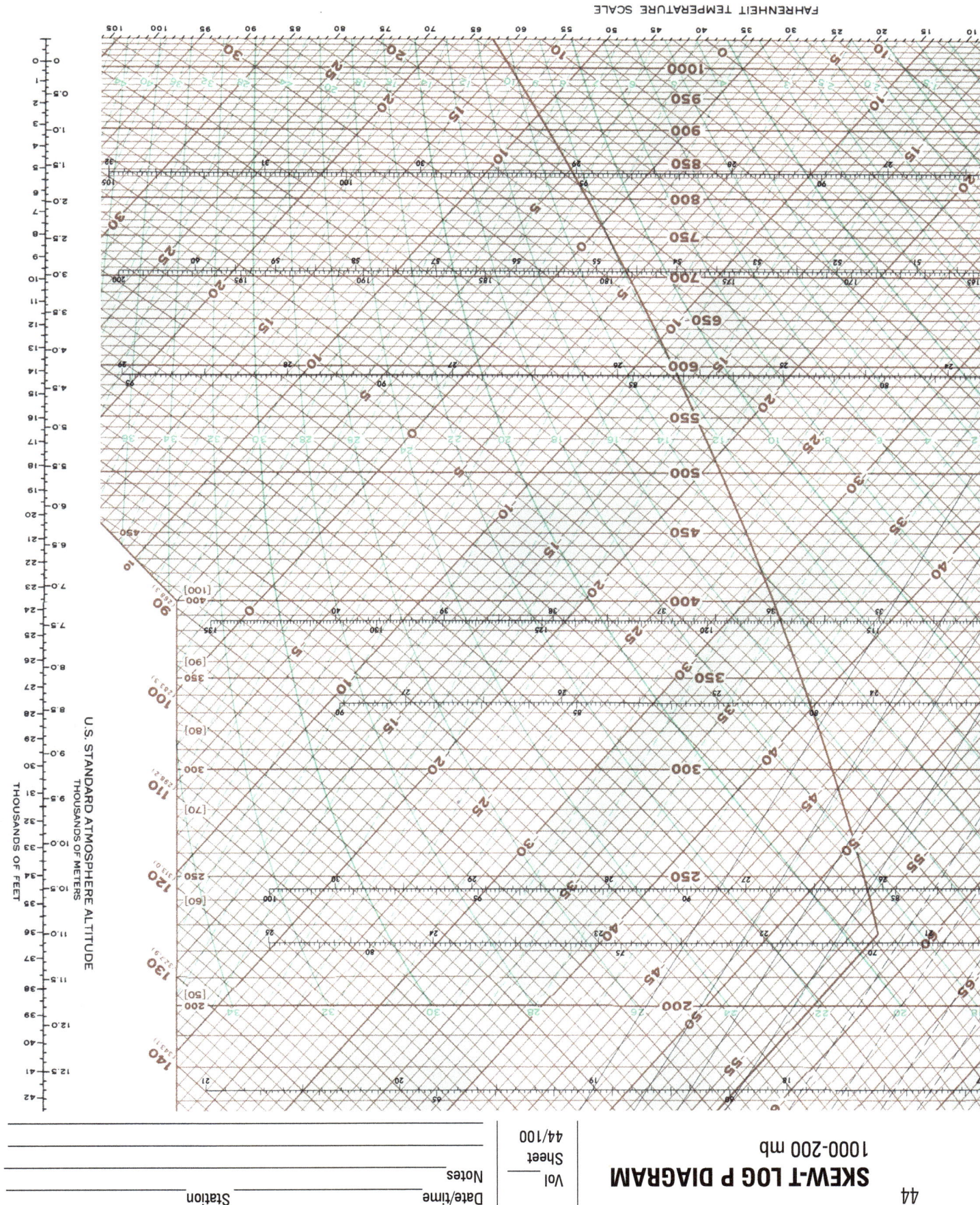

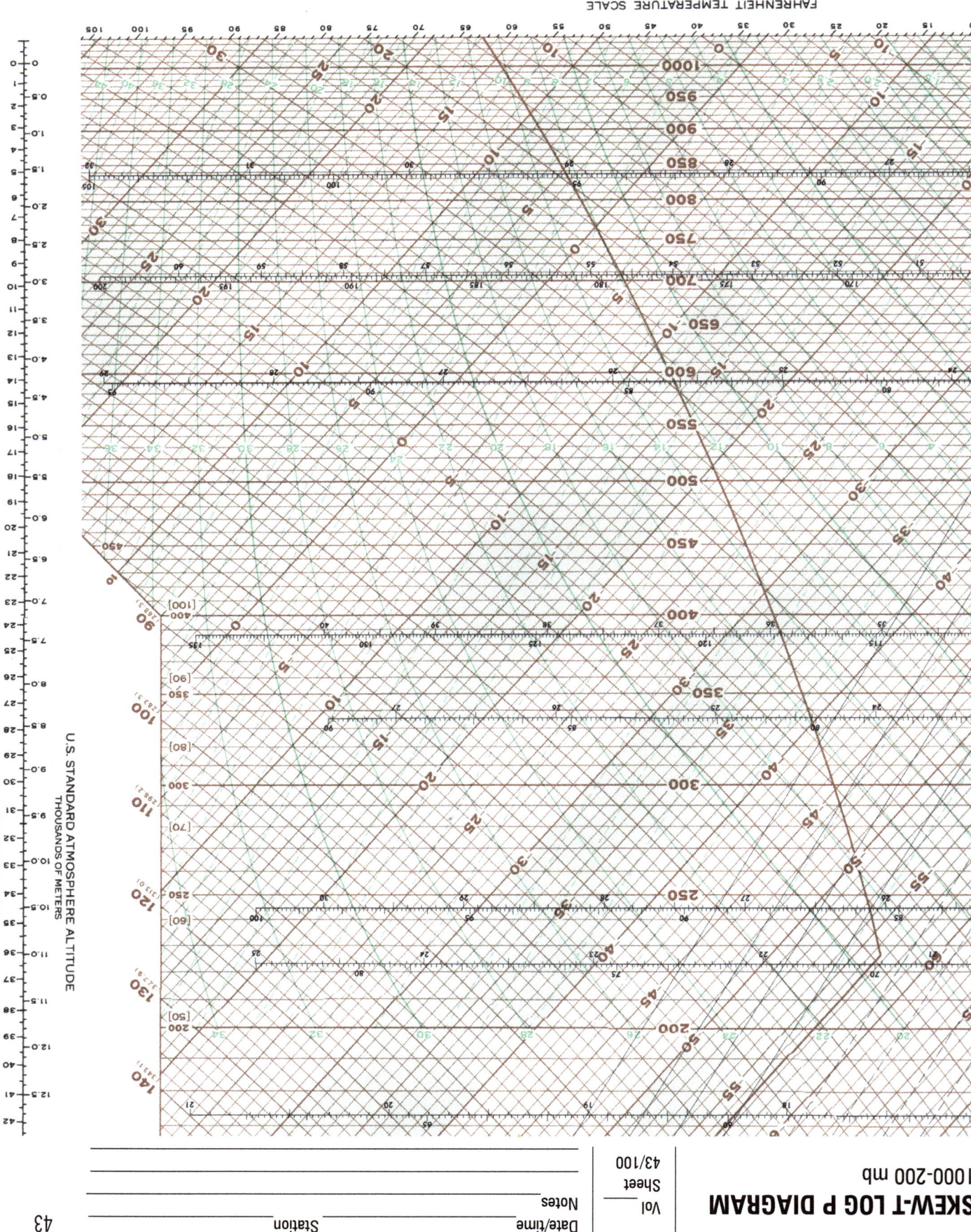

SKEW-T LOG P DIAGRAM
1000–200 mb

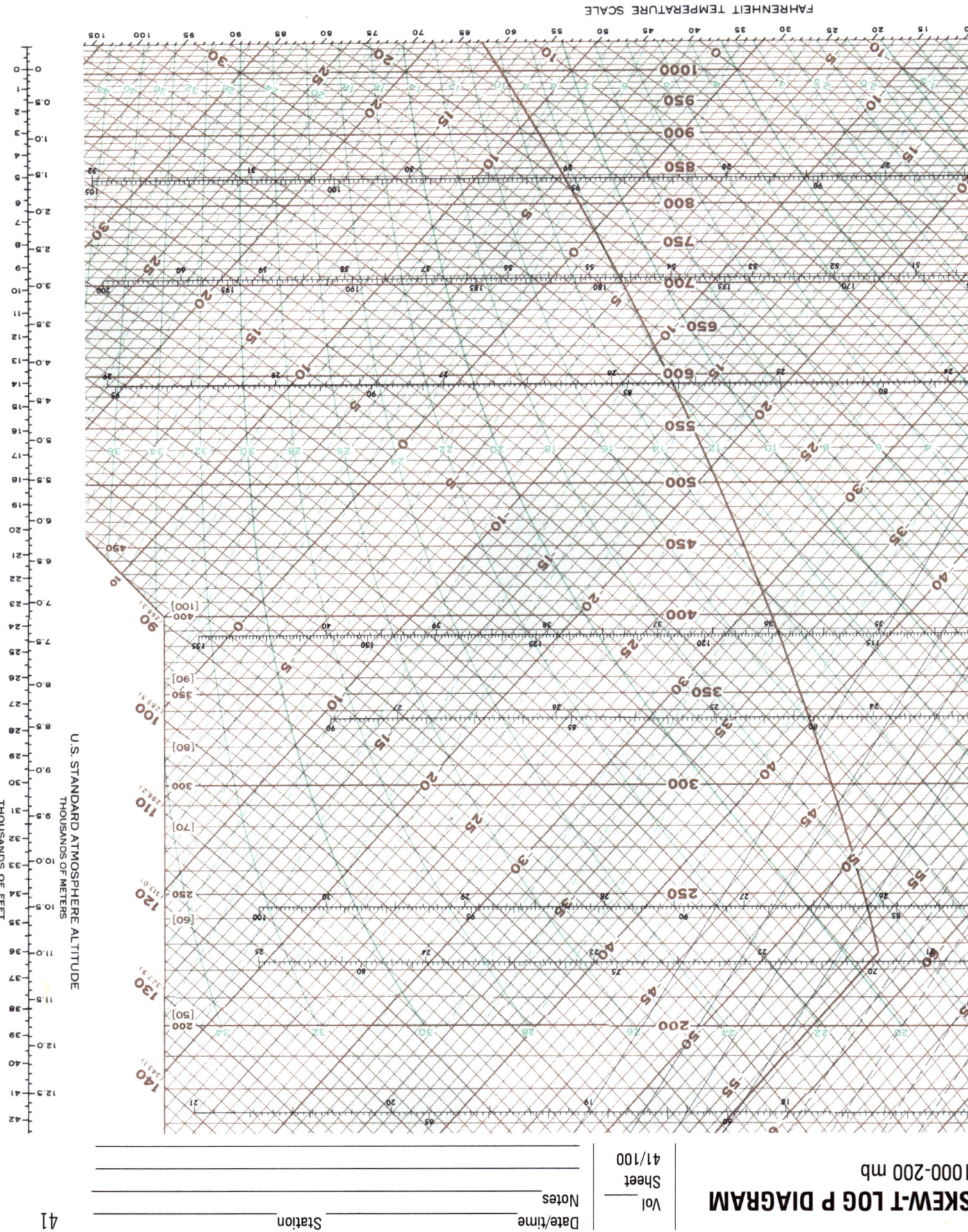

SKEW T-LOG P DIAGRAM
1000-200 mb

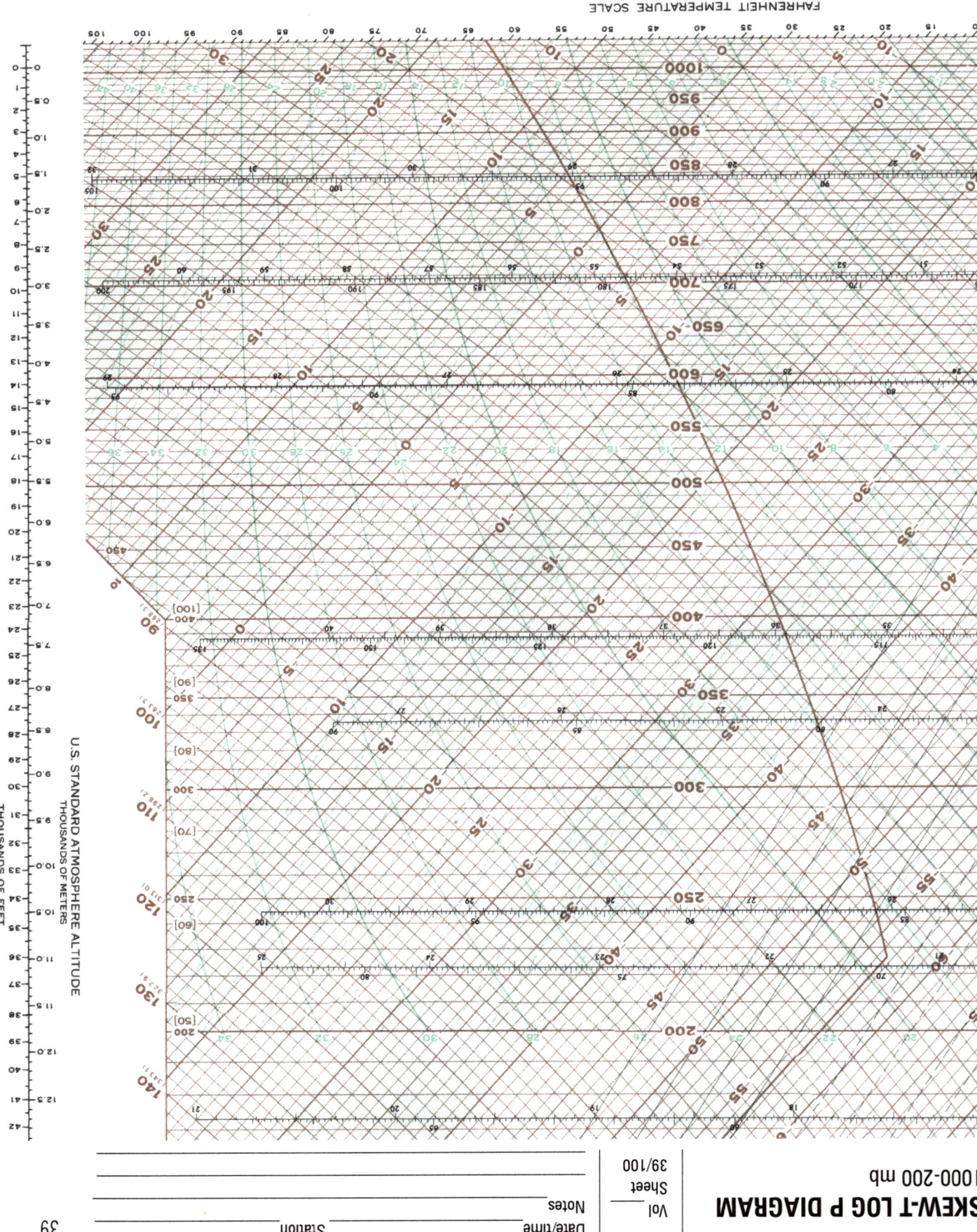

SKEW-T LOG P DIAGRAM
1000-200 mb

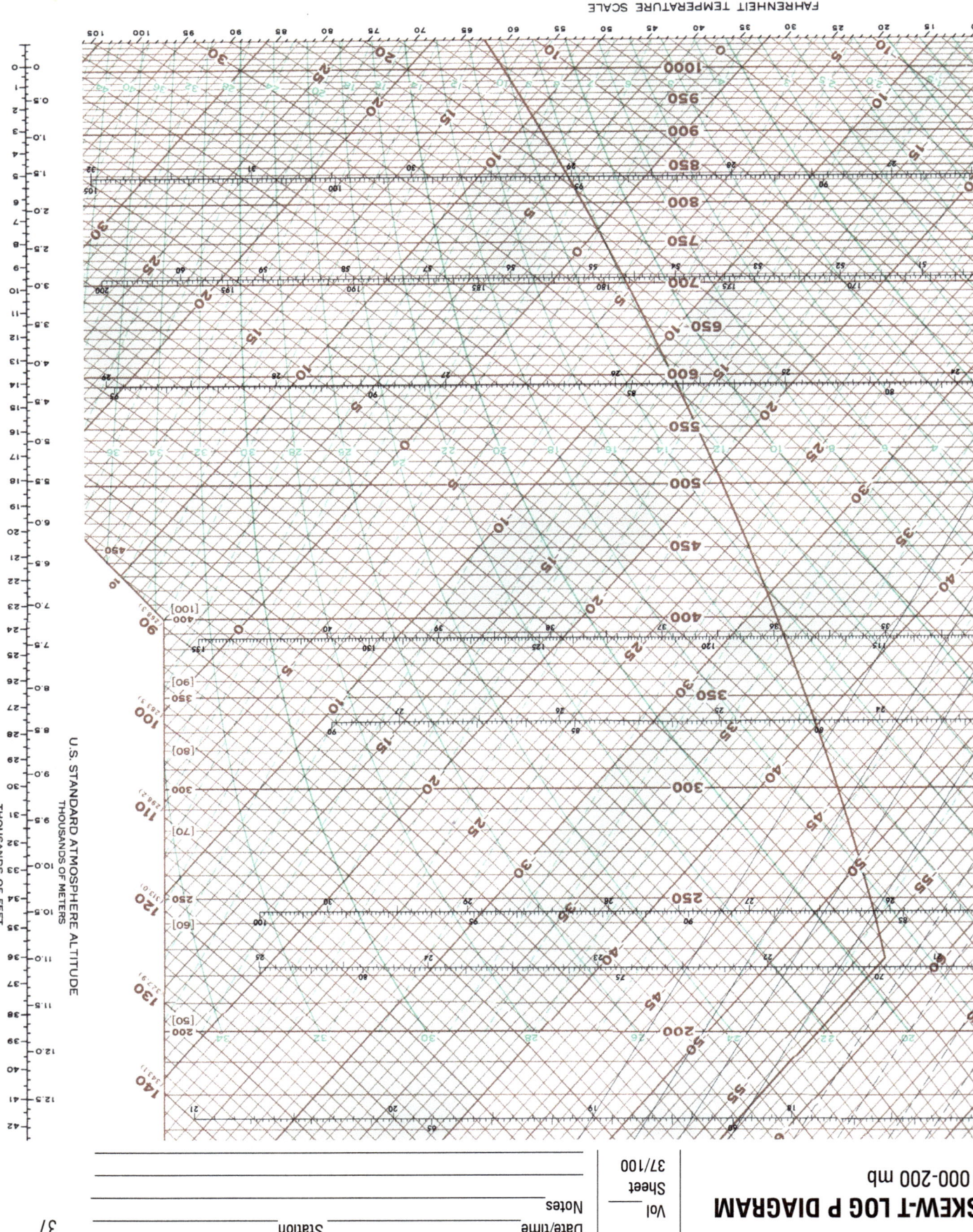

SKEW-T LOG P DIAGRAM
1000-200 mb

SKEW-T LOG P DIAGRAM
1000-200 mb

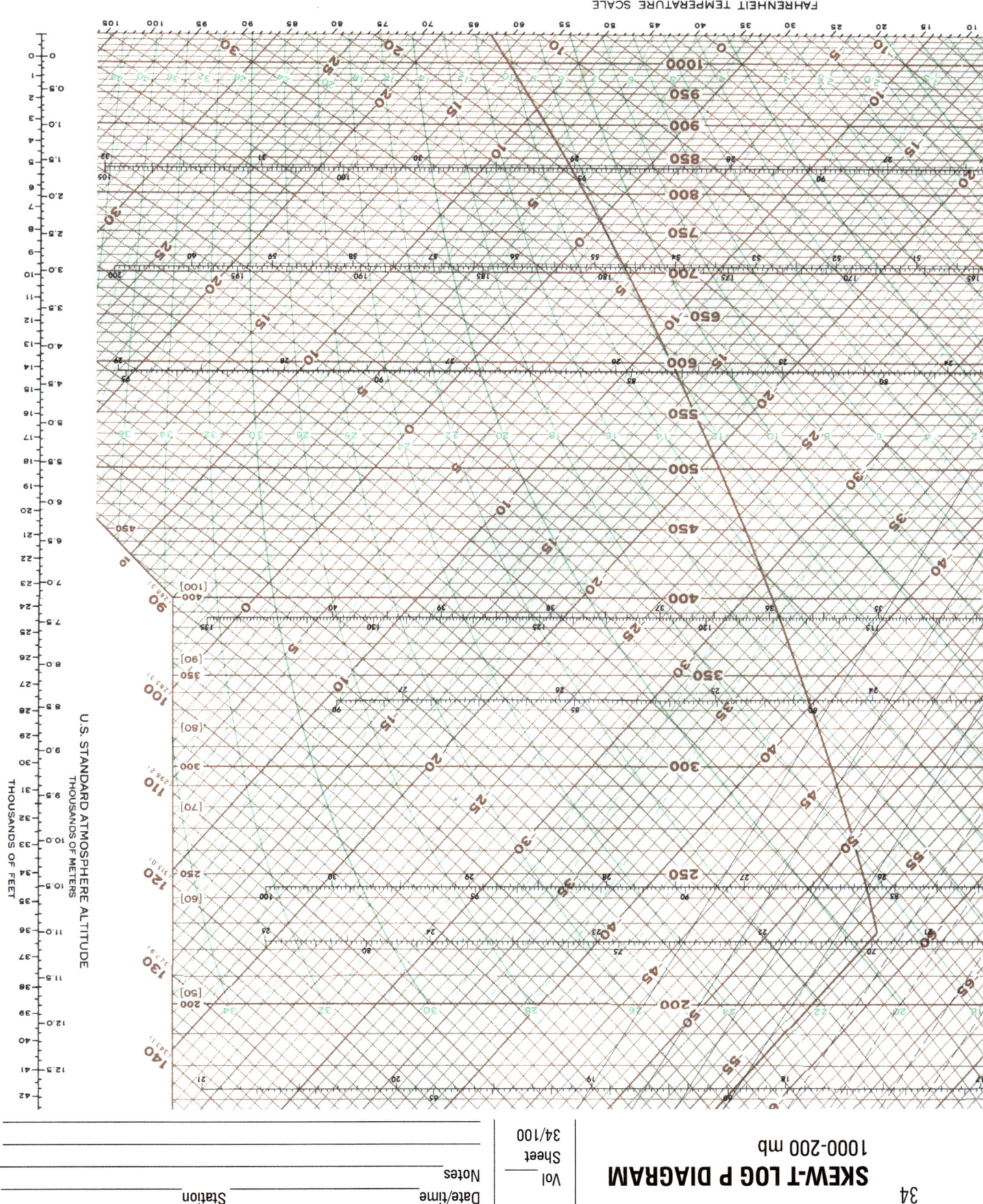

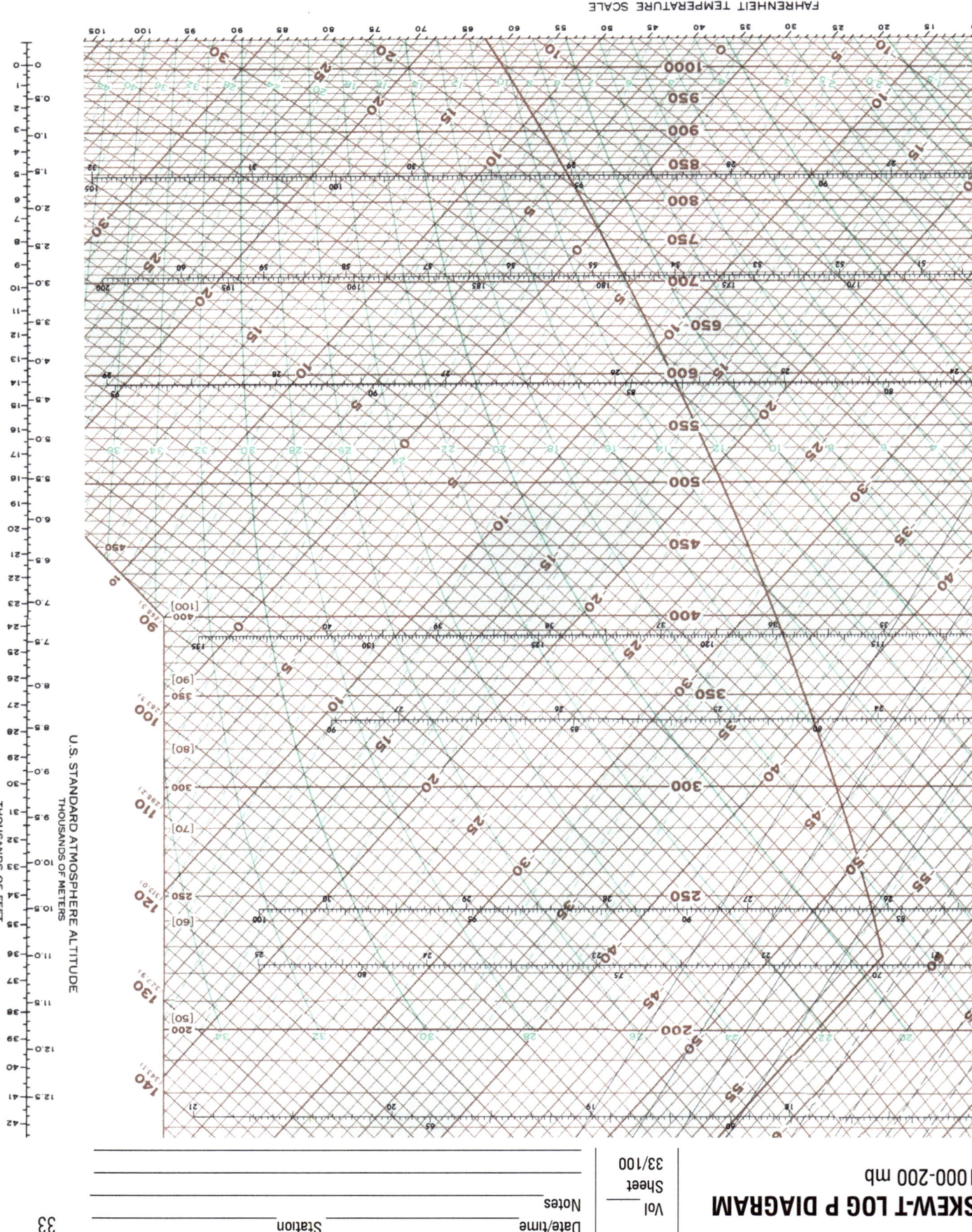

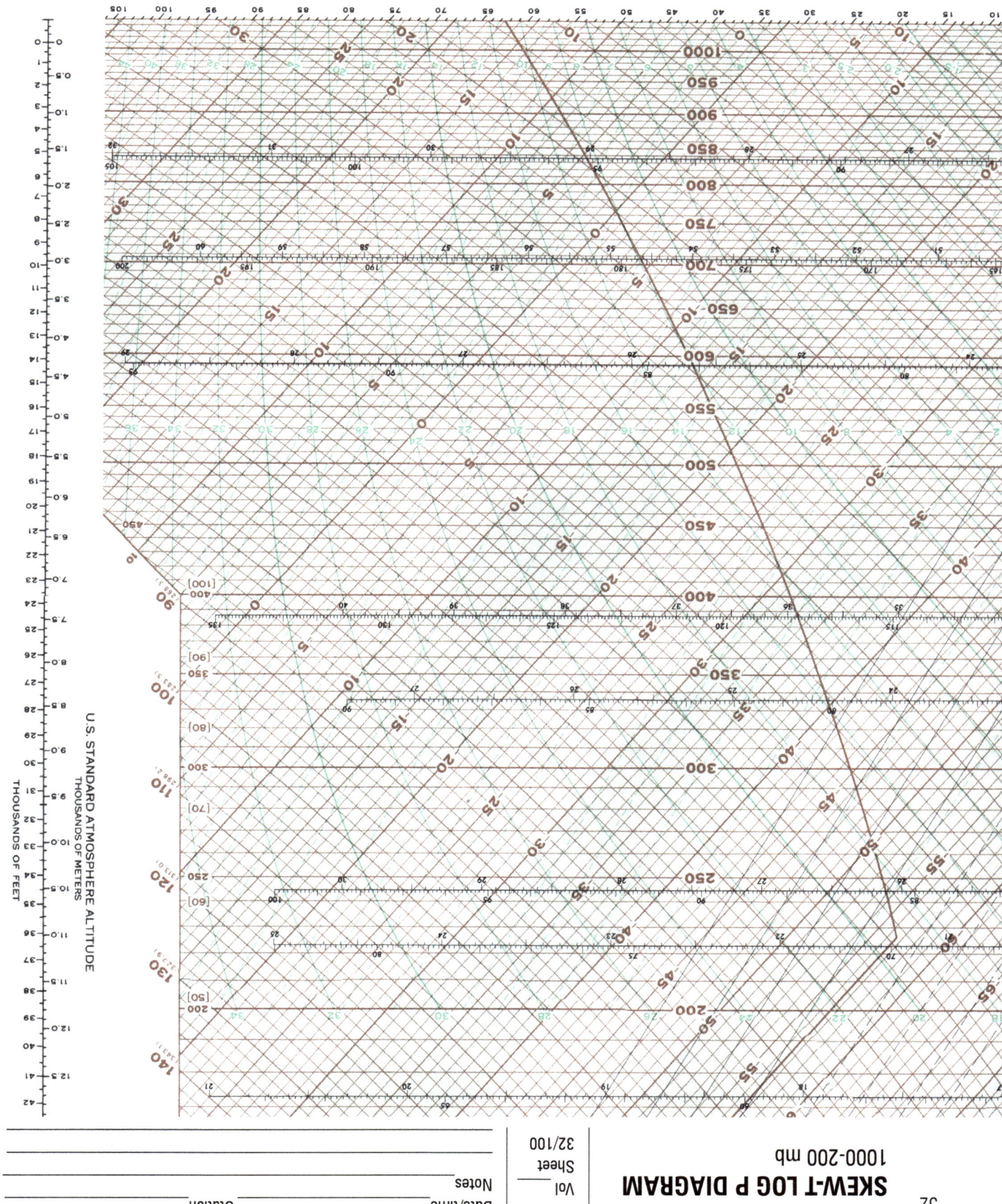

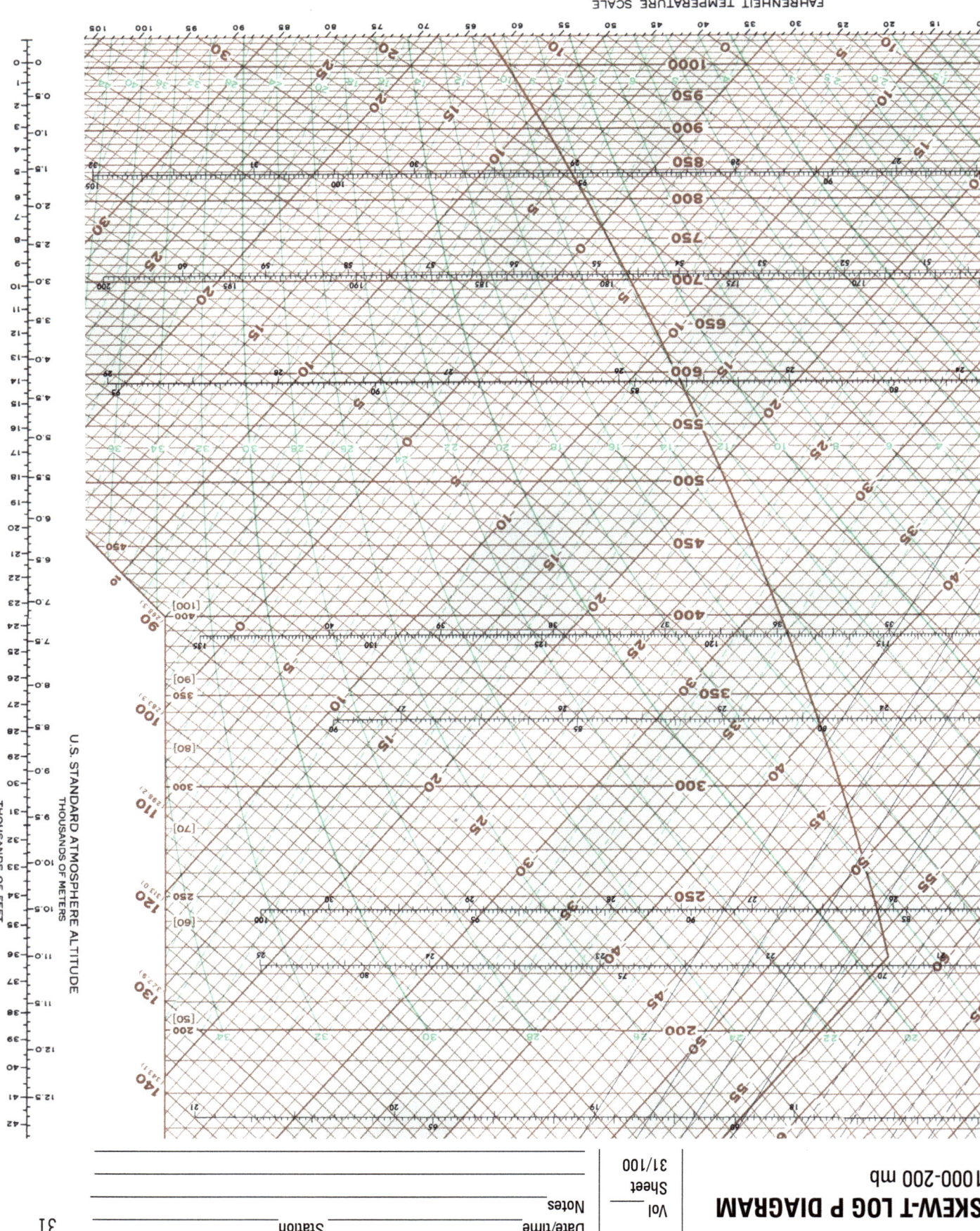

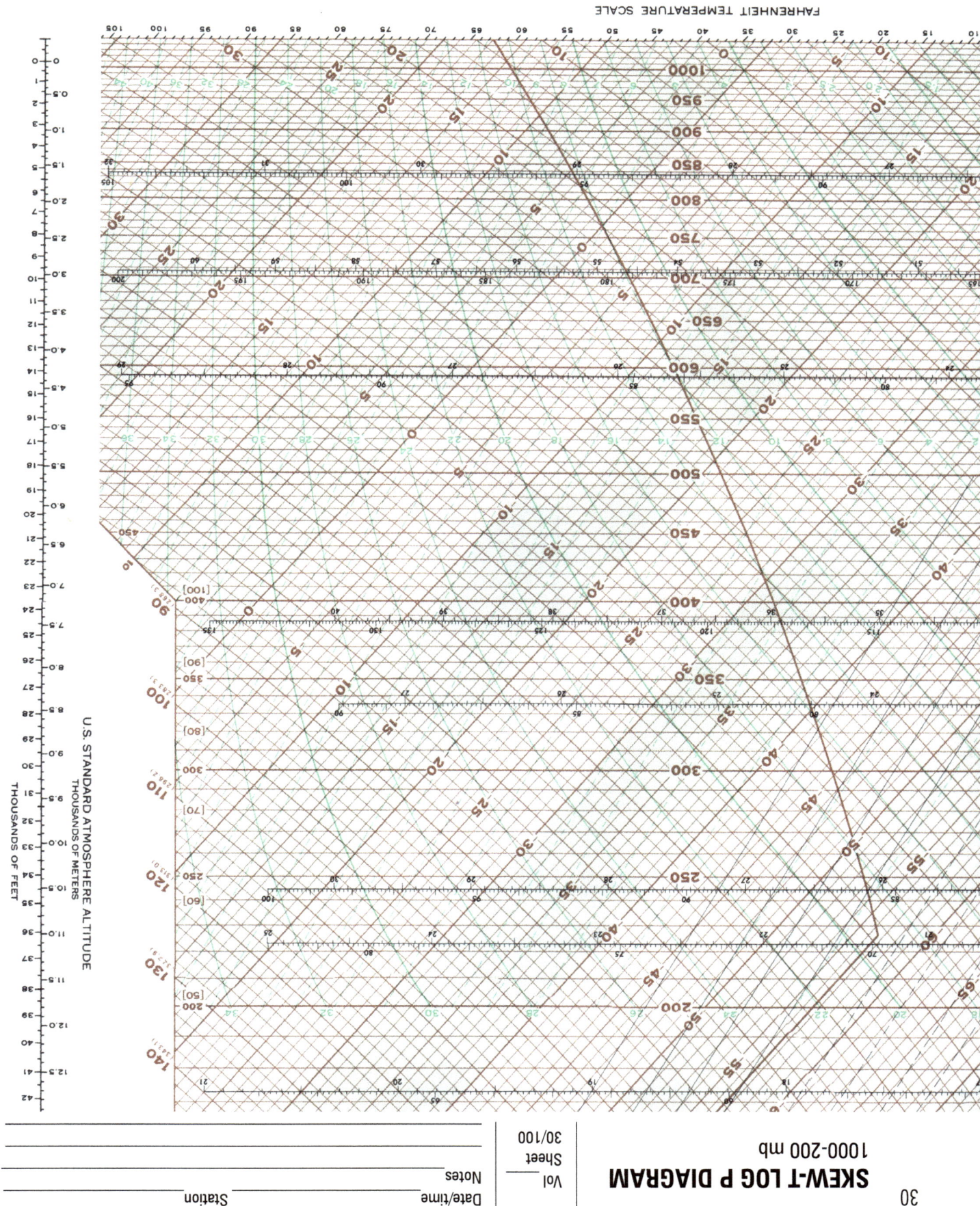

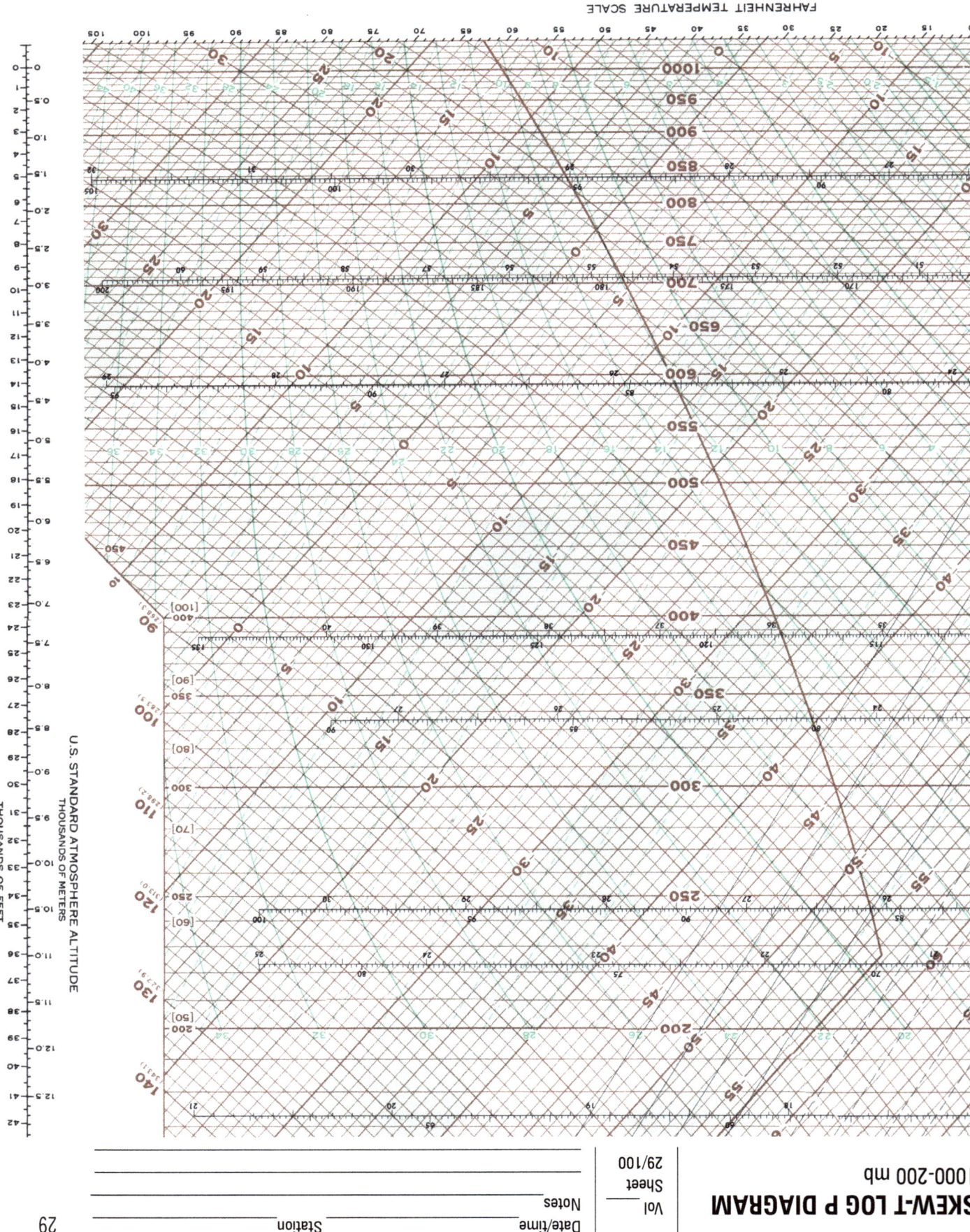

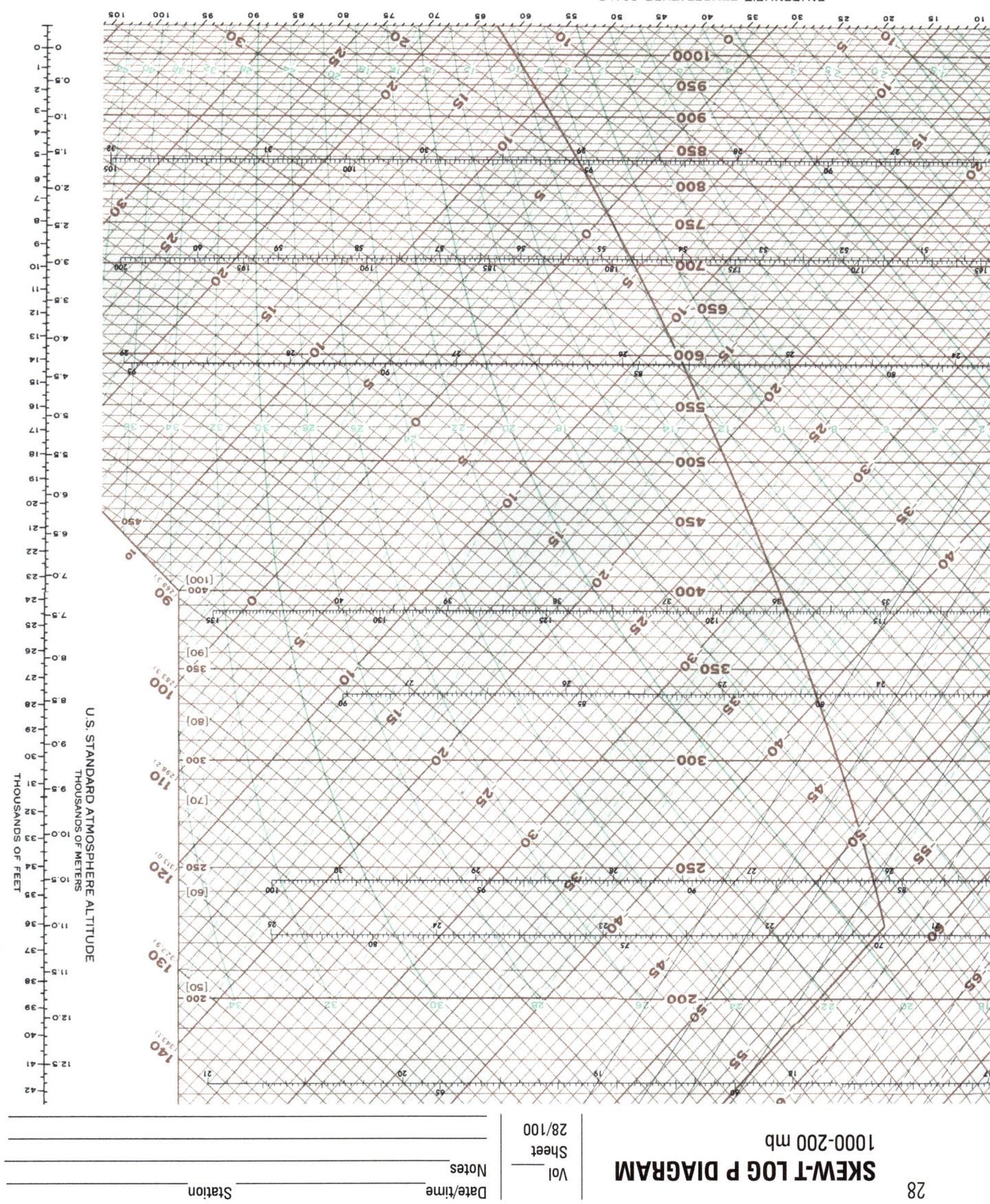

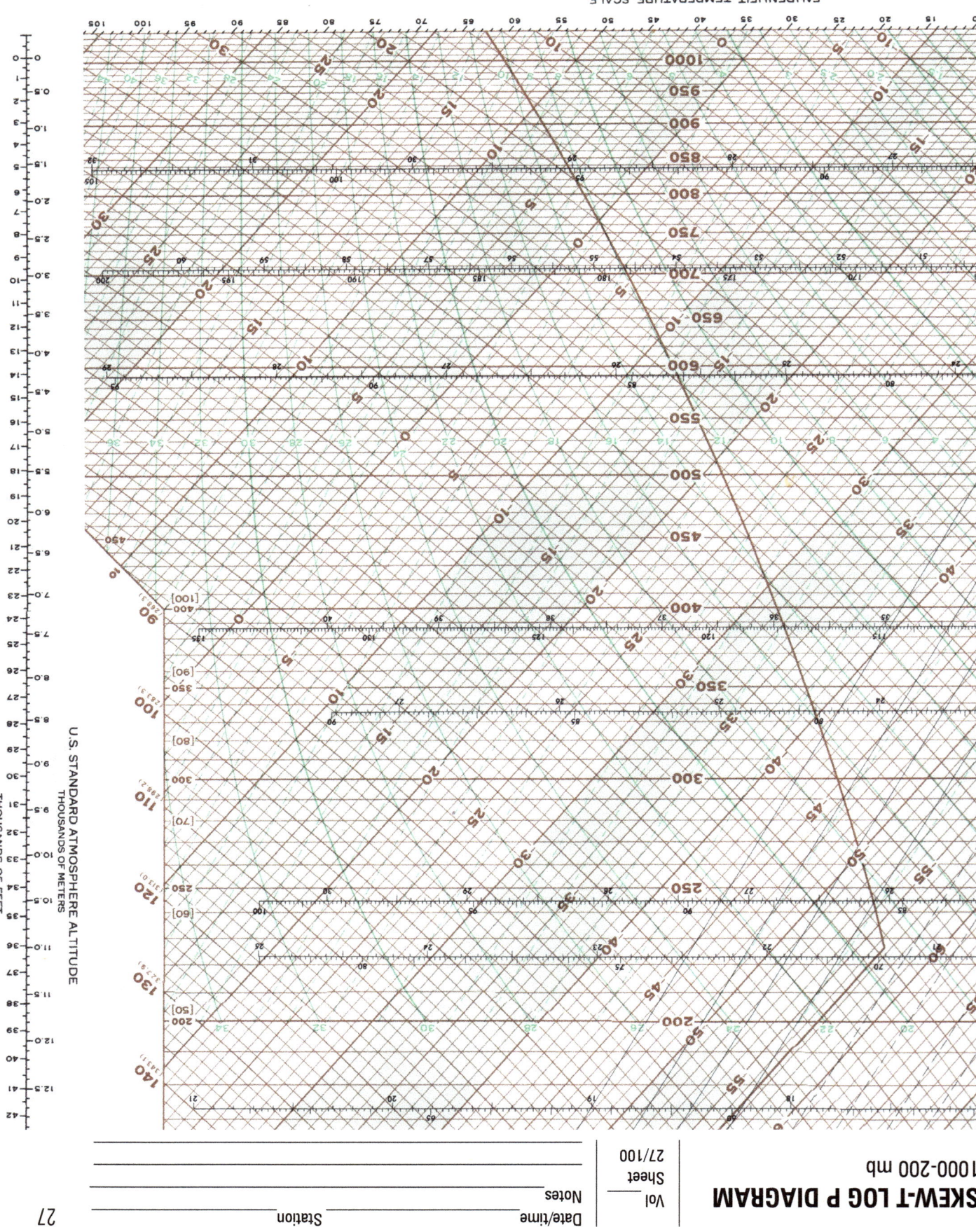

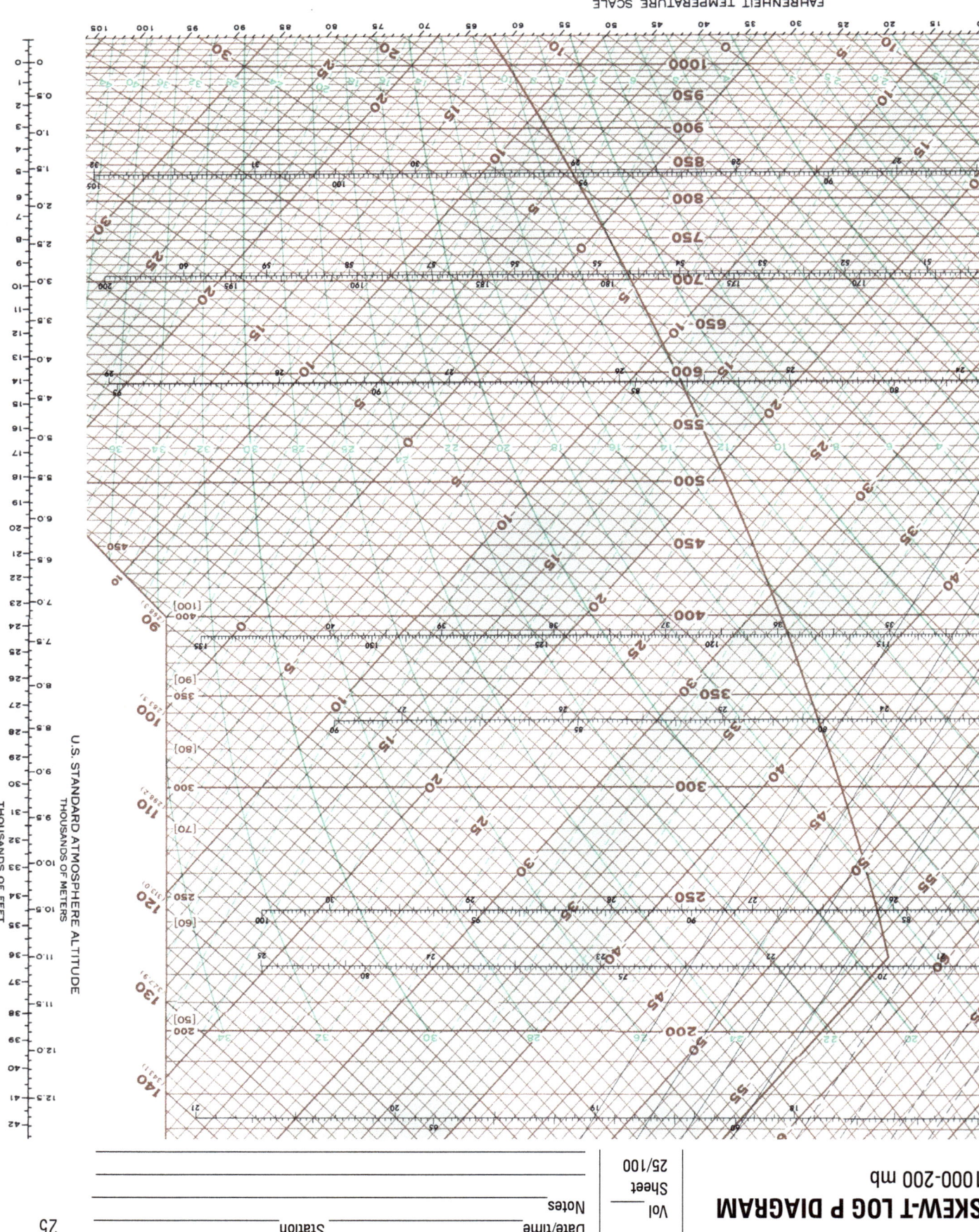

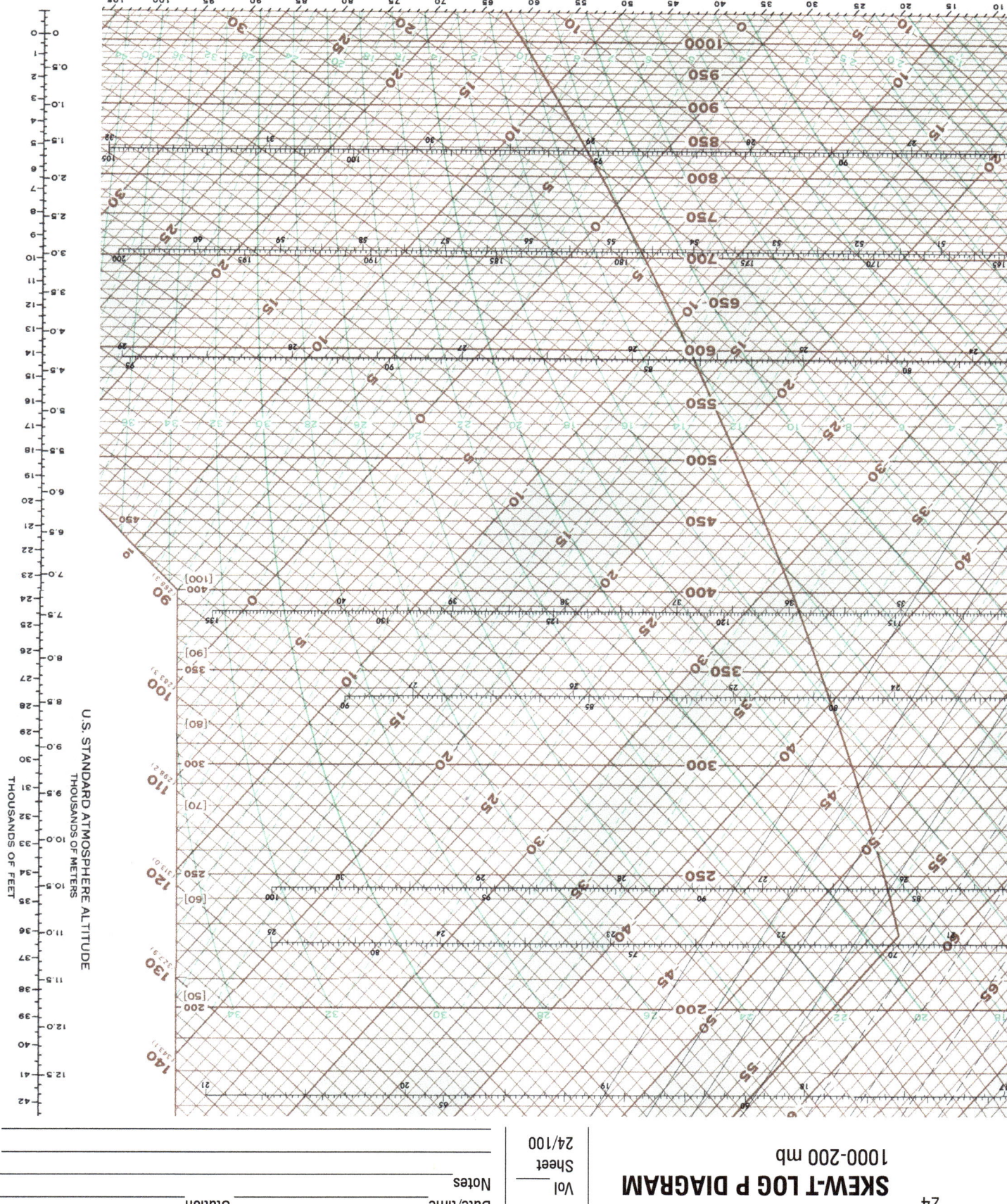

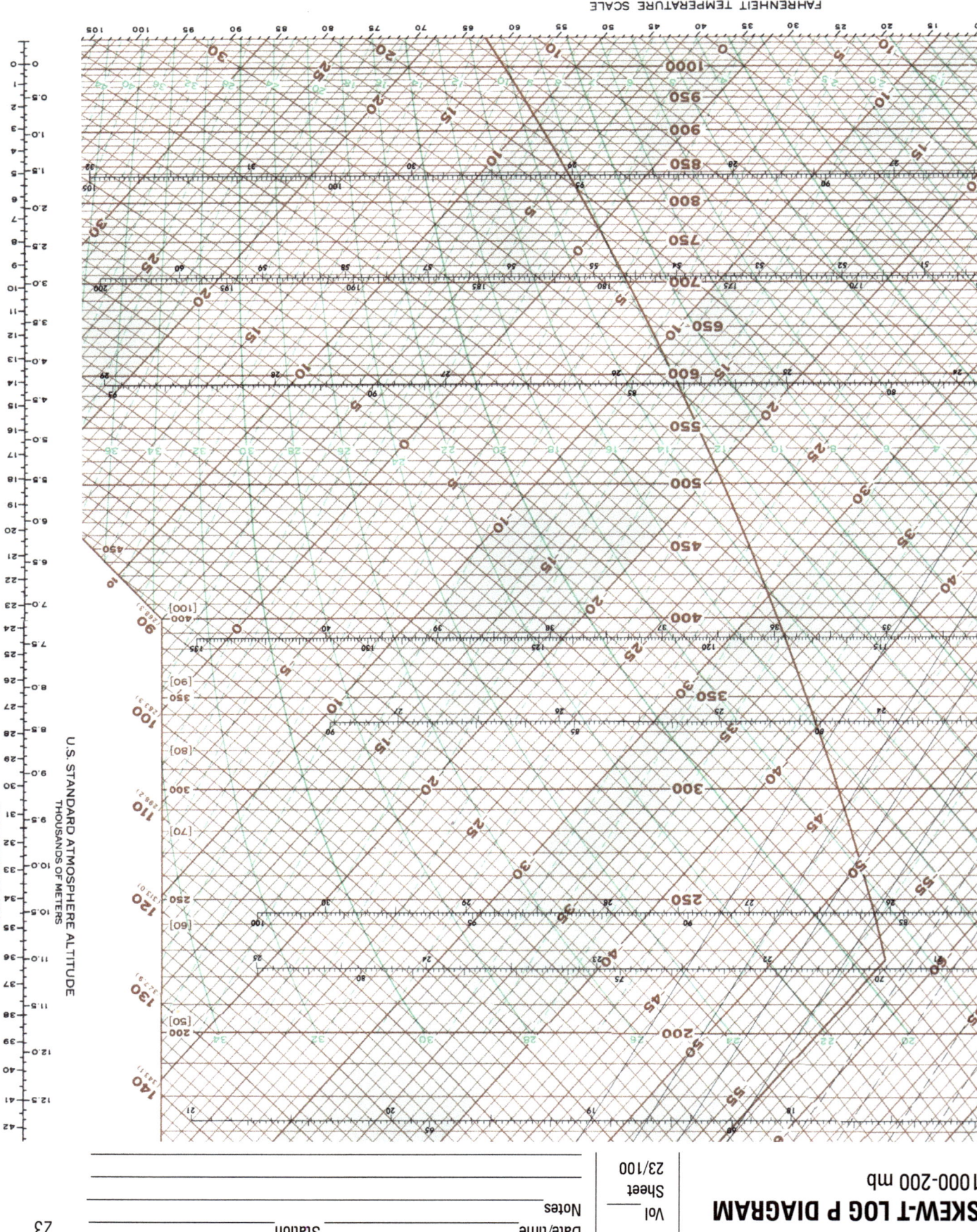

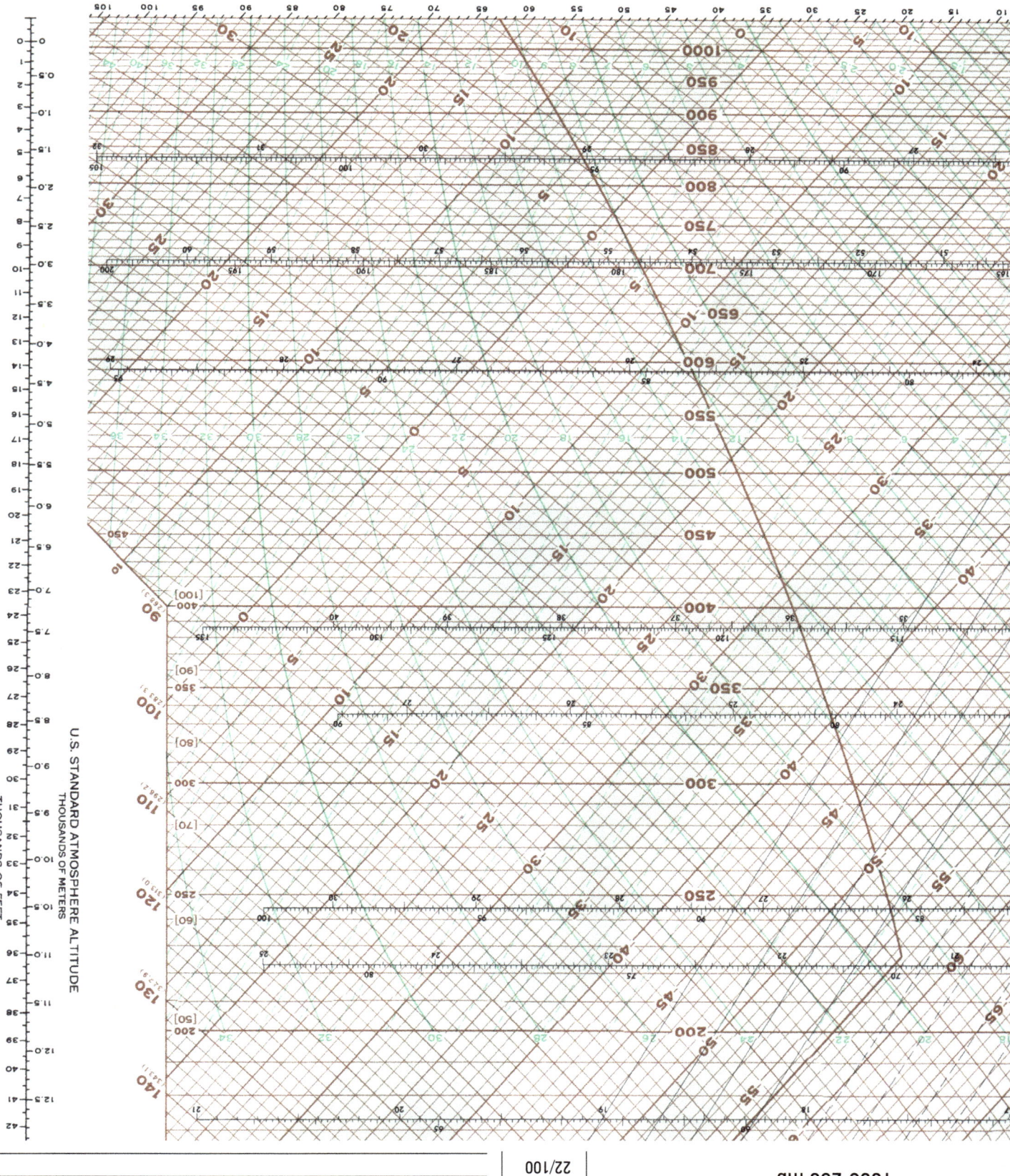

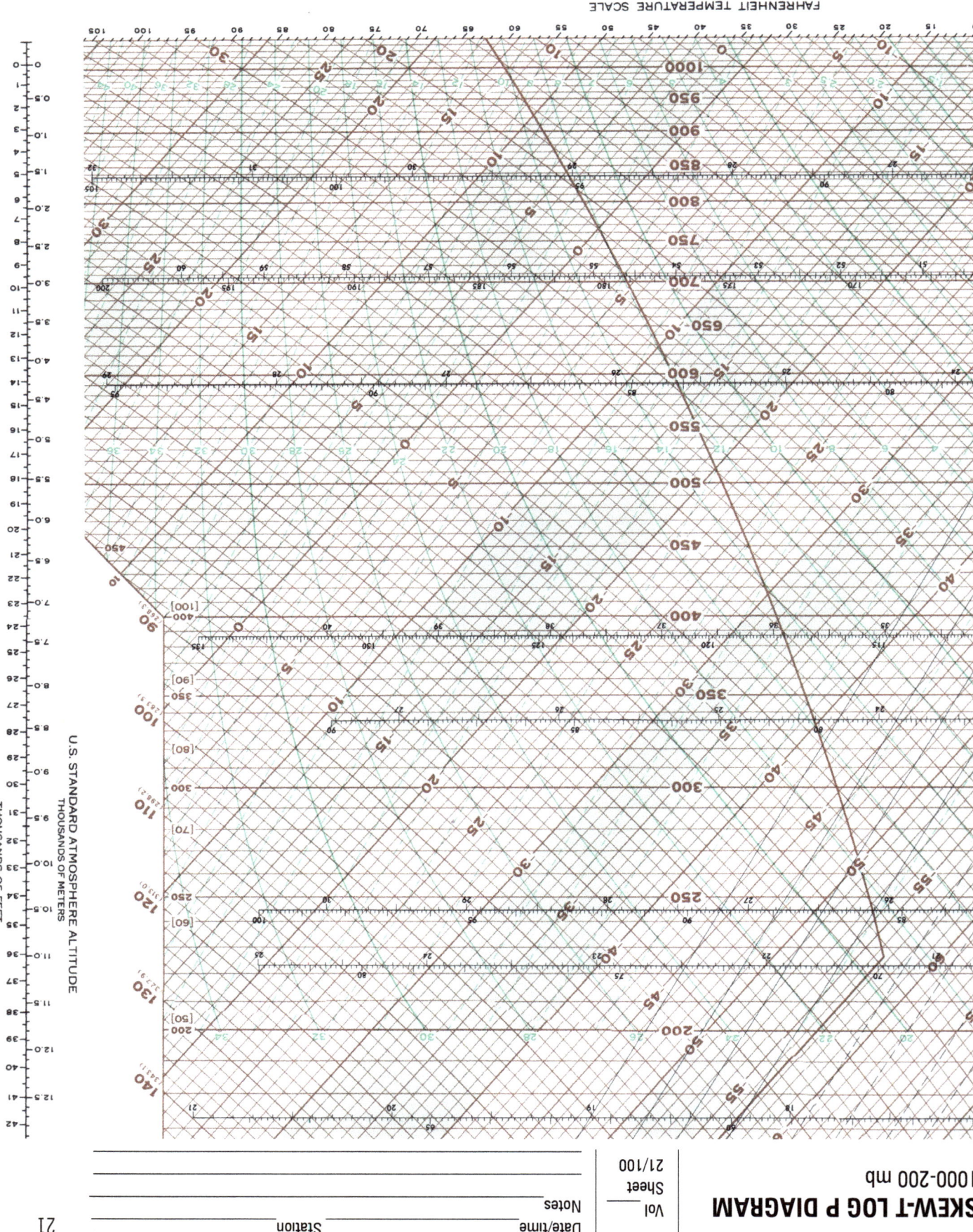

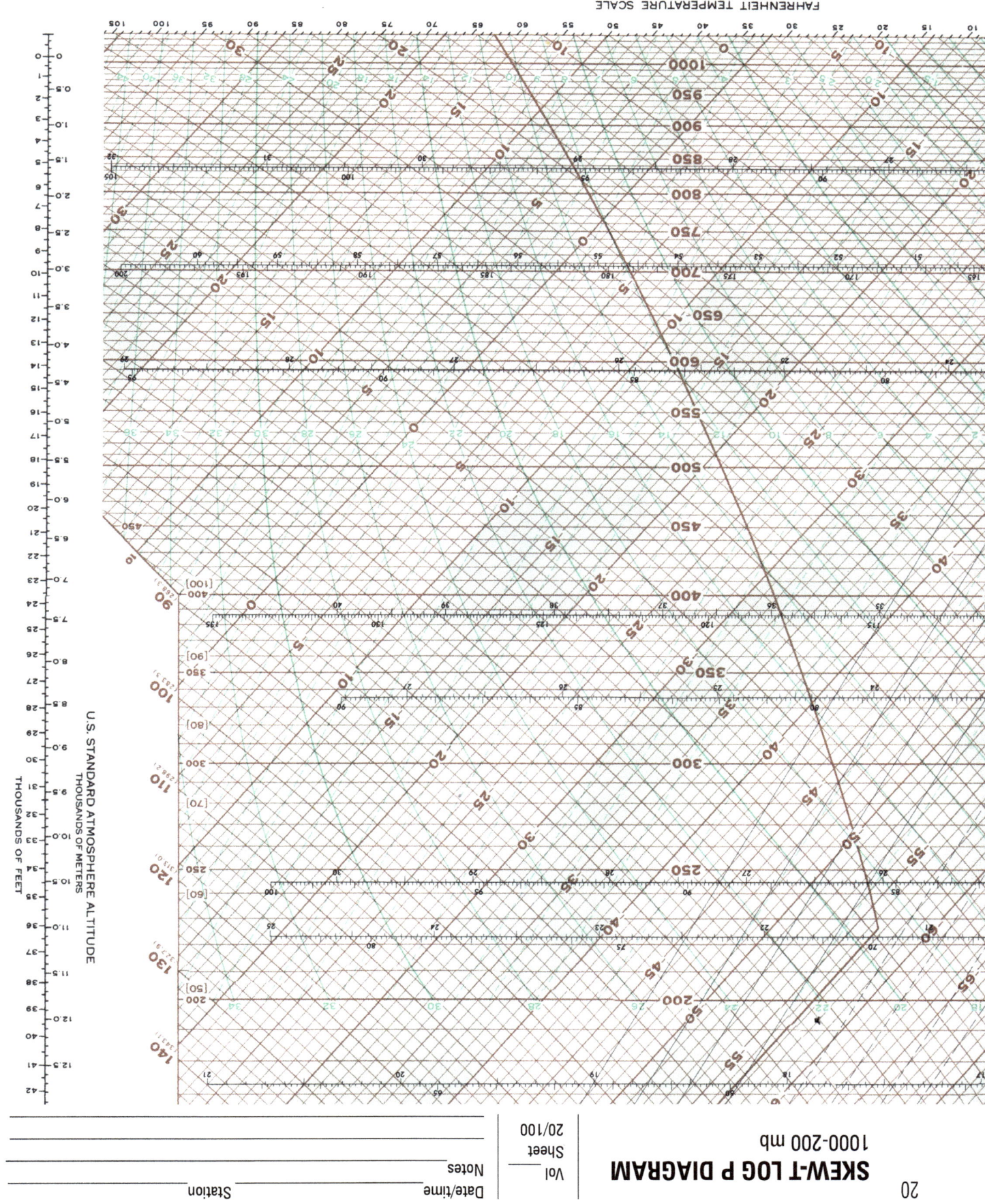

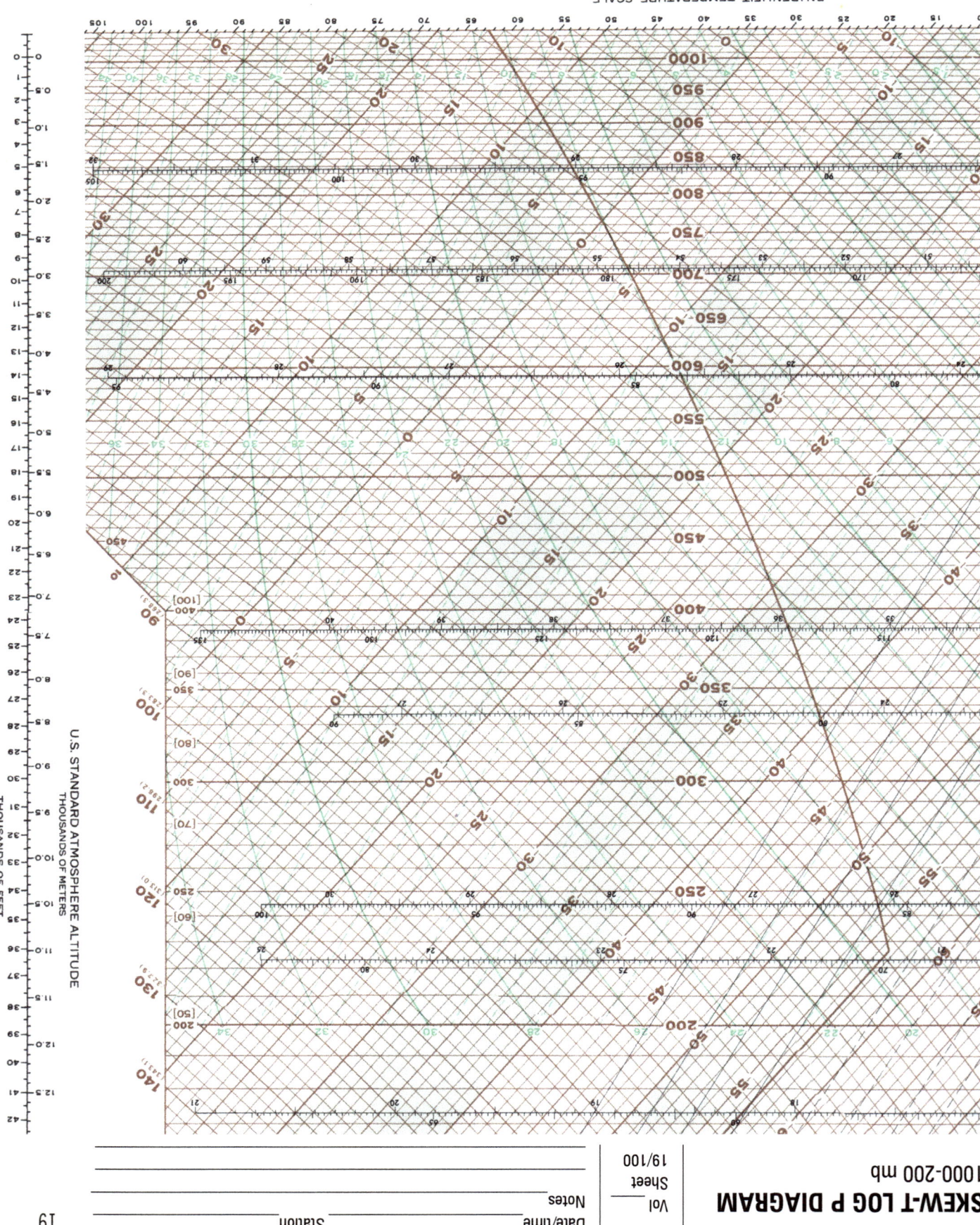

SKEW-T LOG P DIAGRAM
1000-200 mb

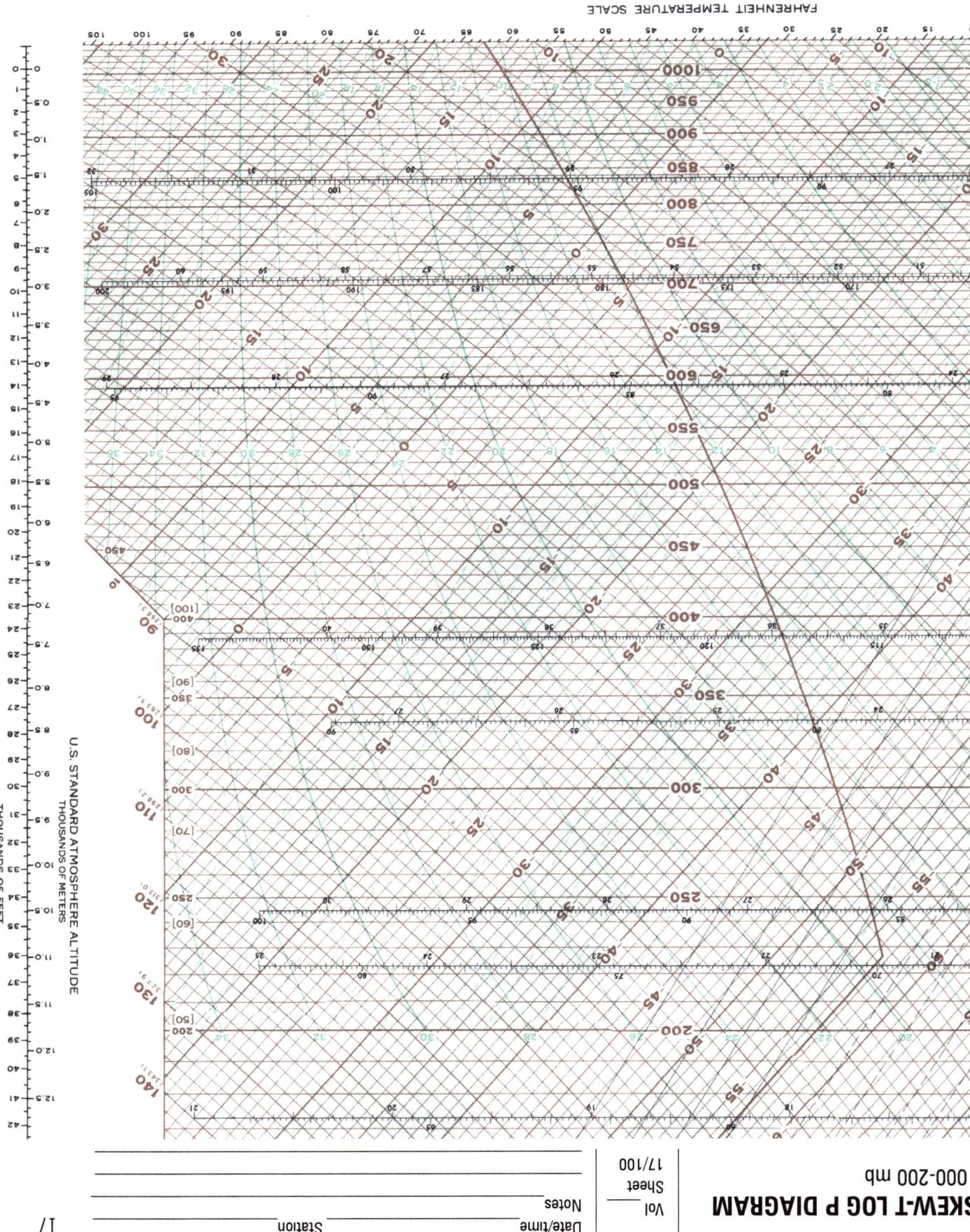

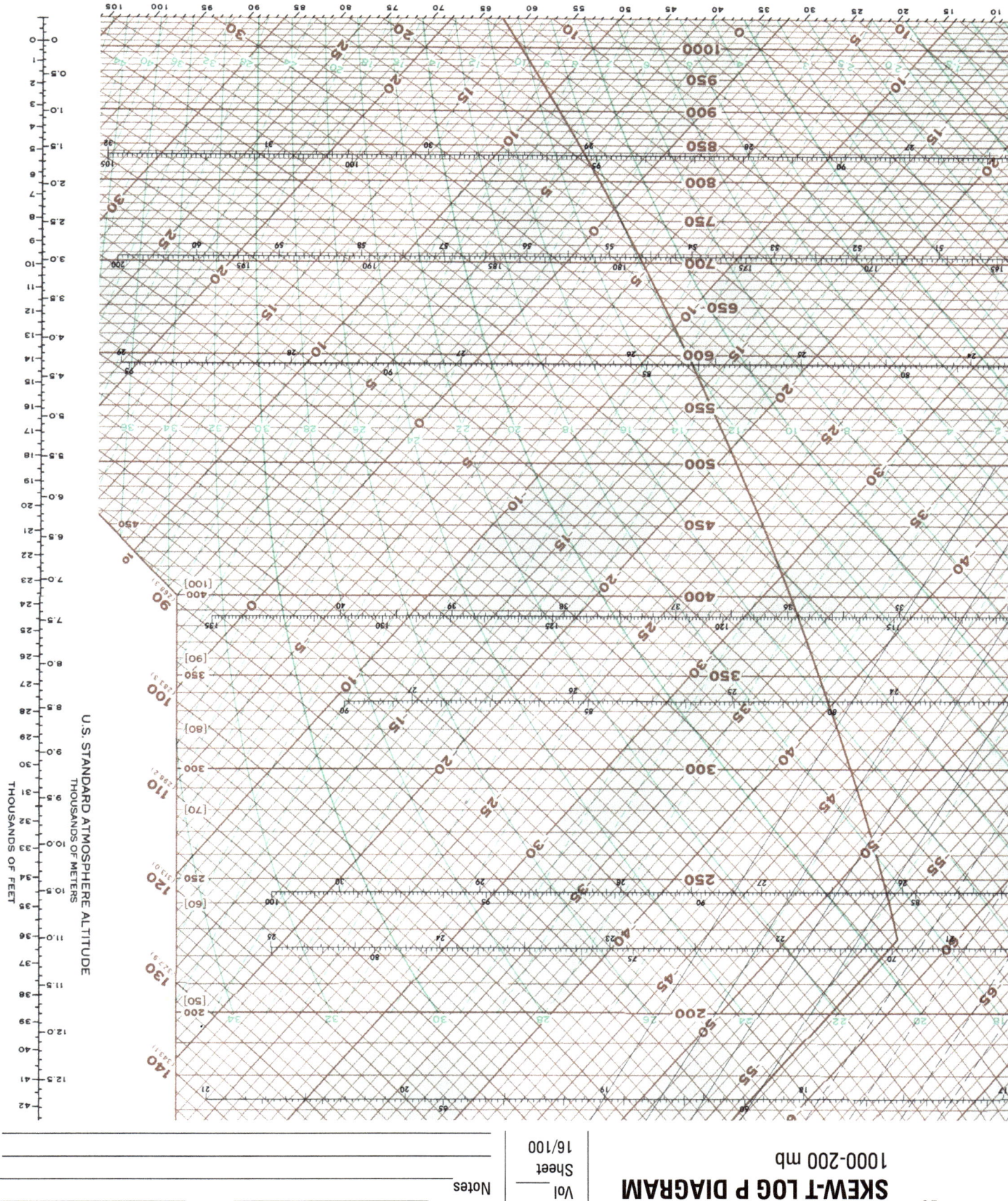

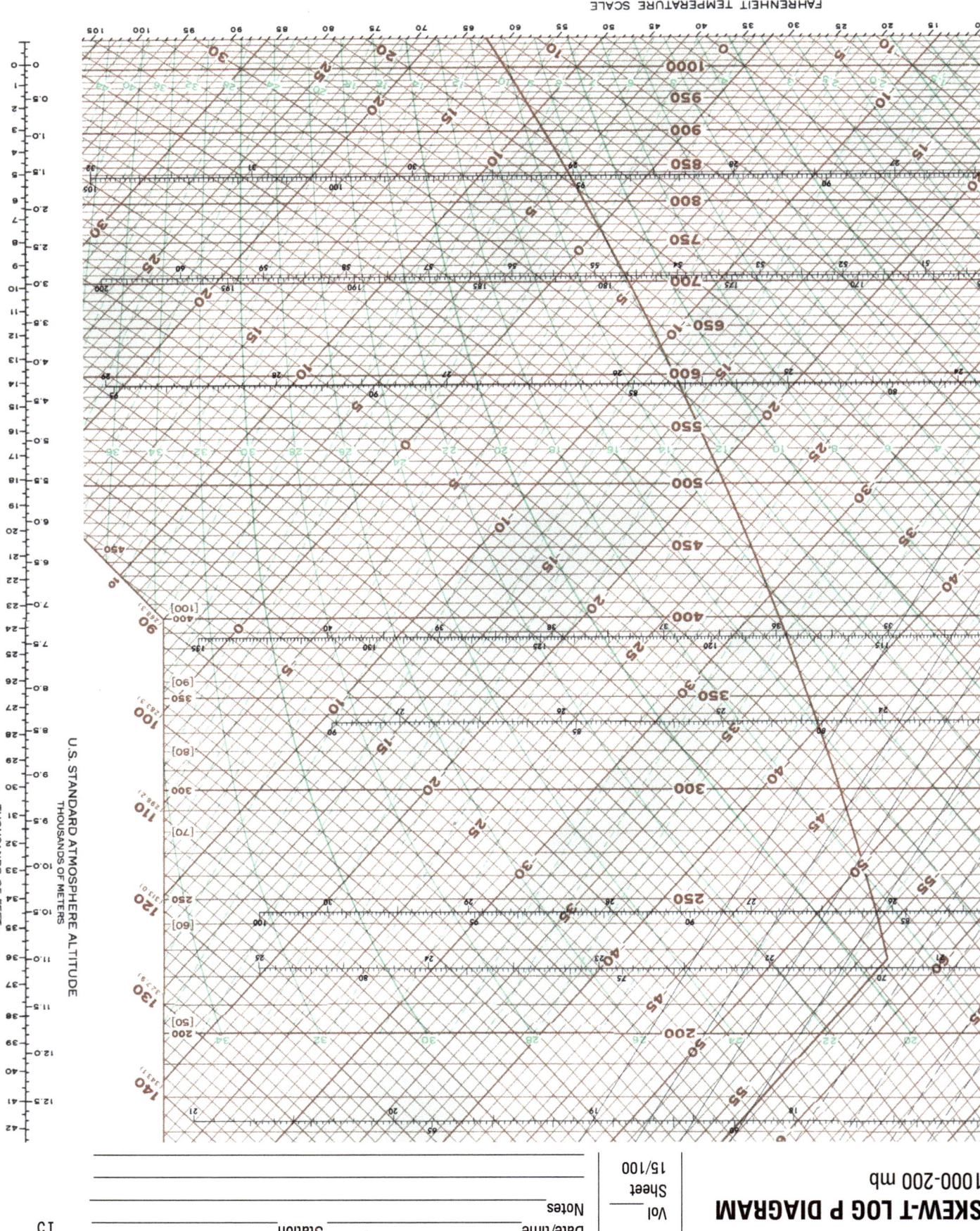

SKEW-T LOG P DIAGRAM
1000-200 mb

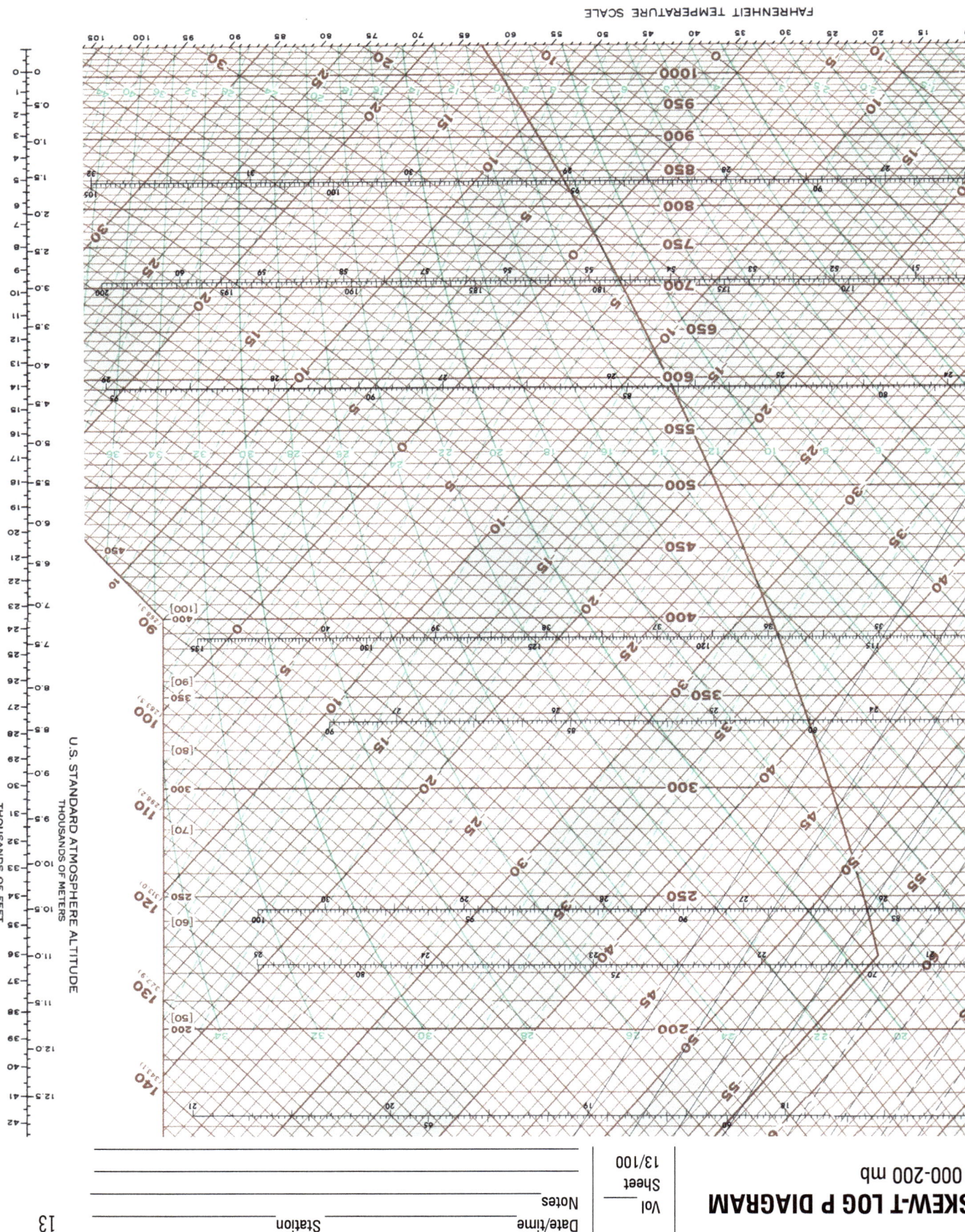

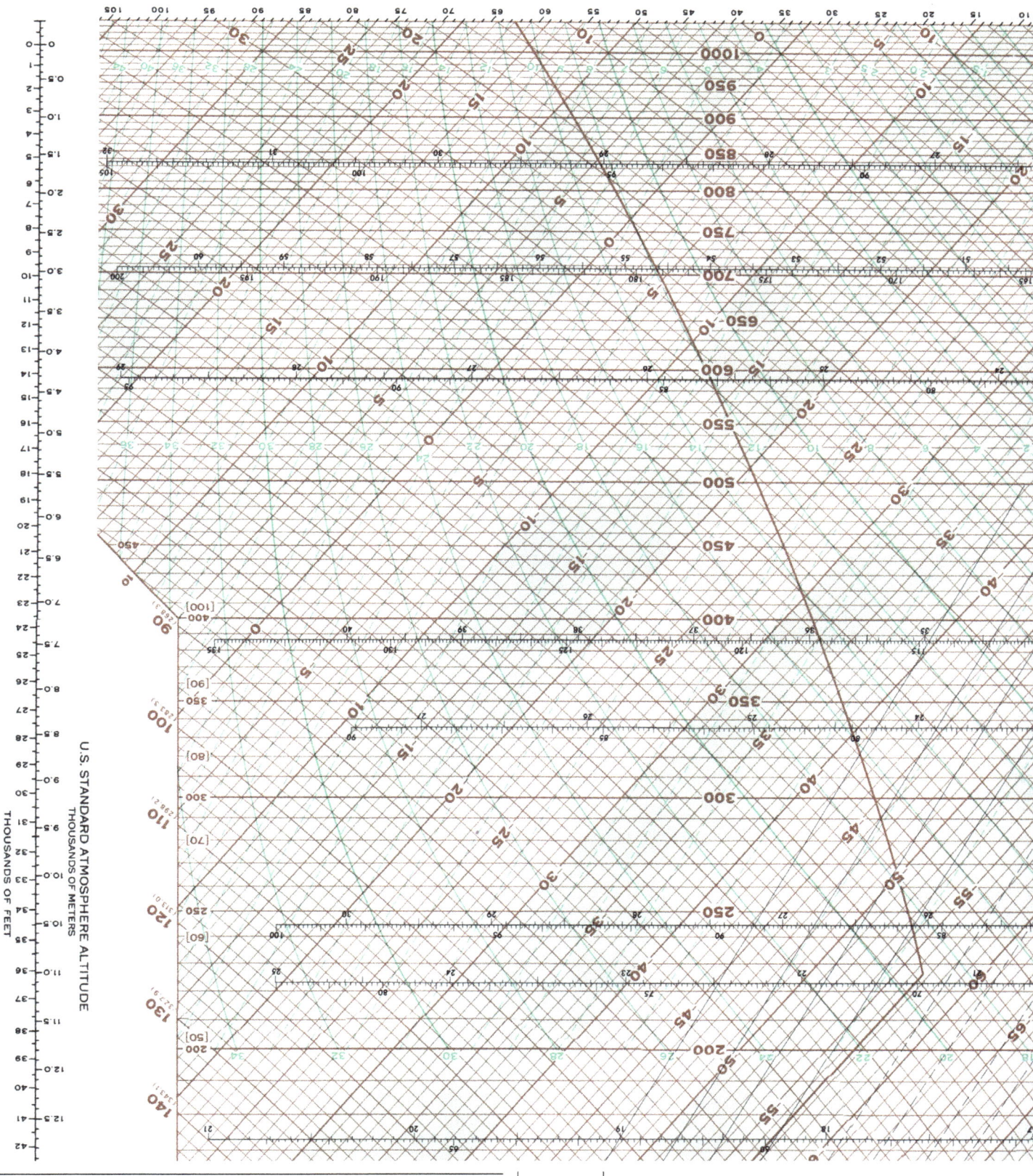

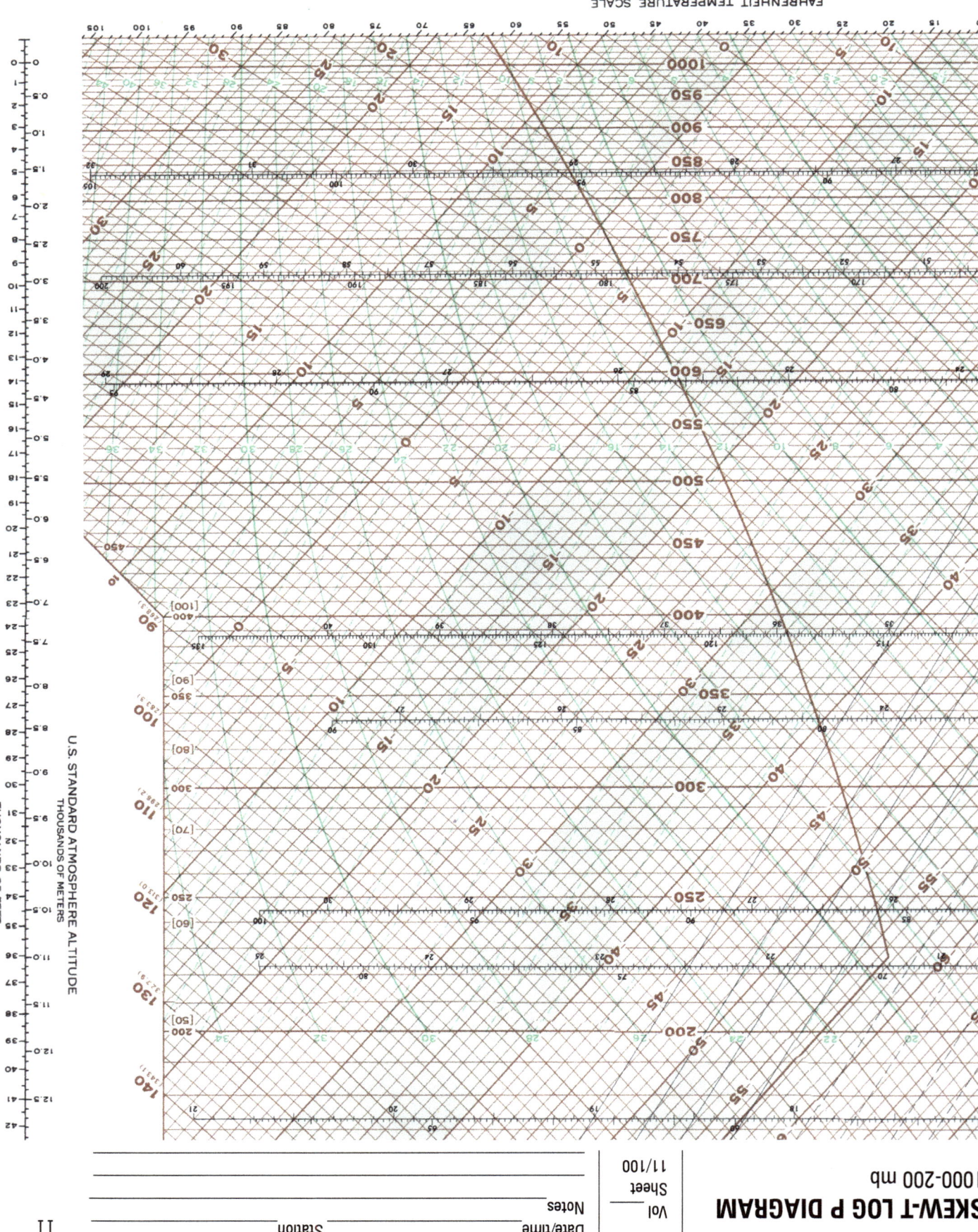

SKEW-T LOG P DIAGRAM
1000-200 mb

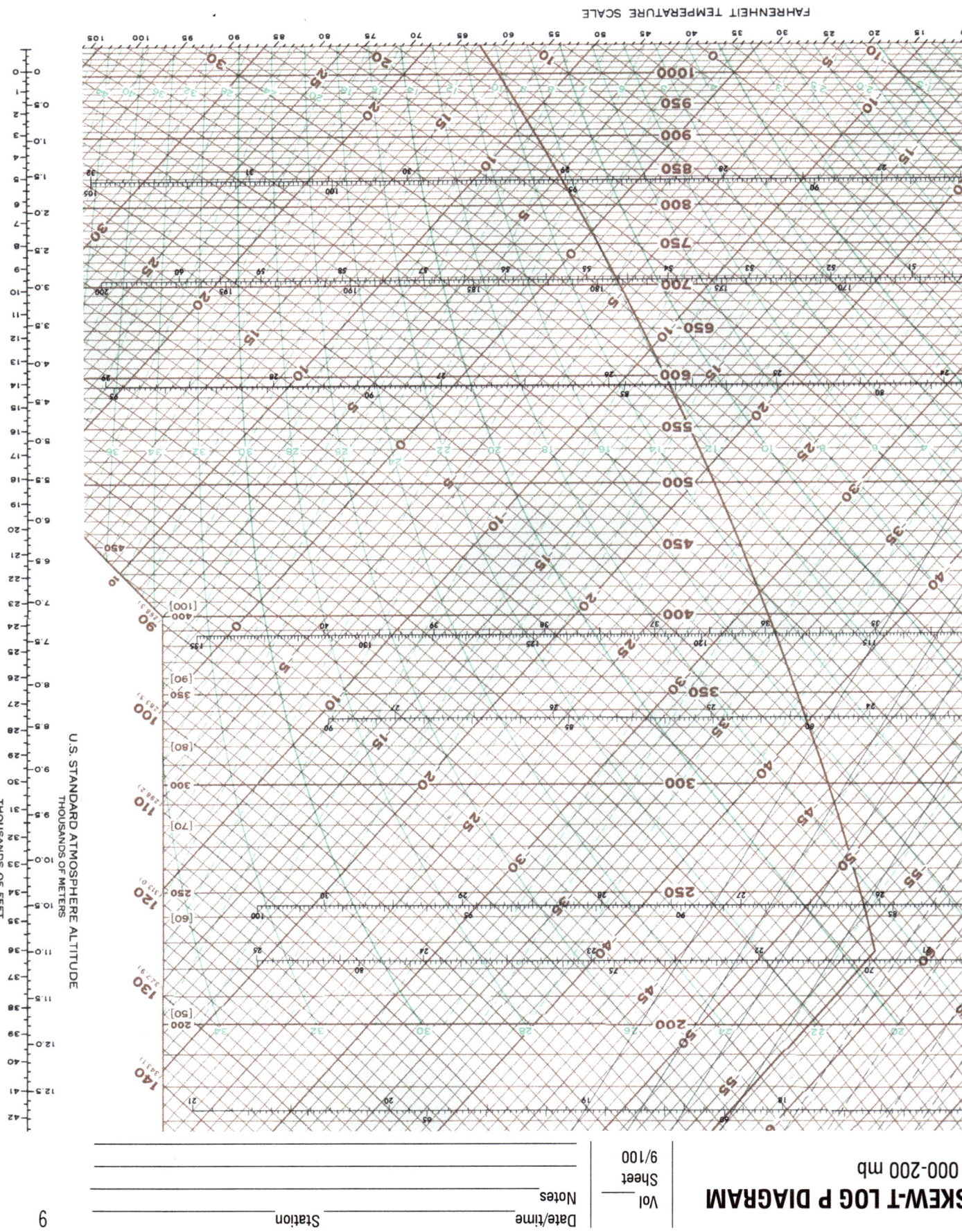

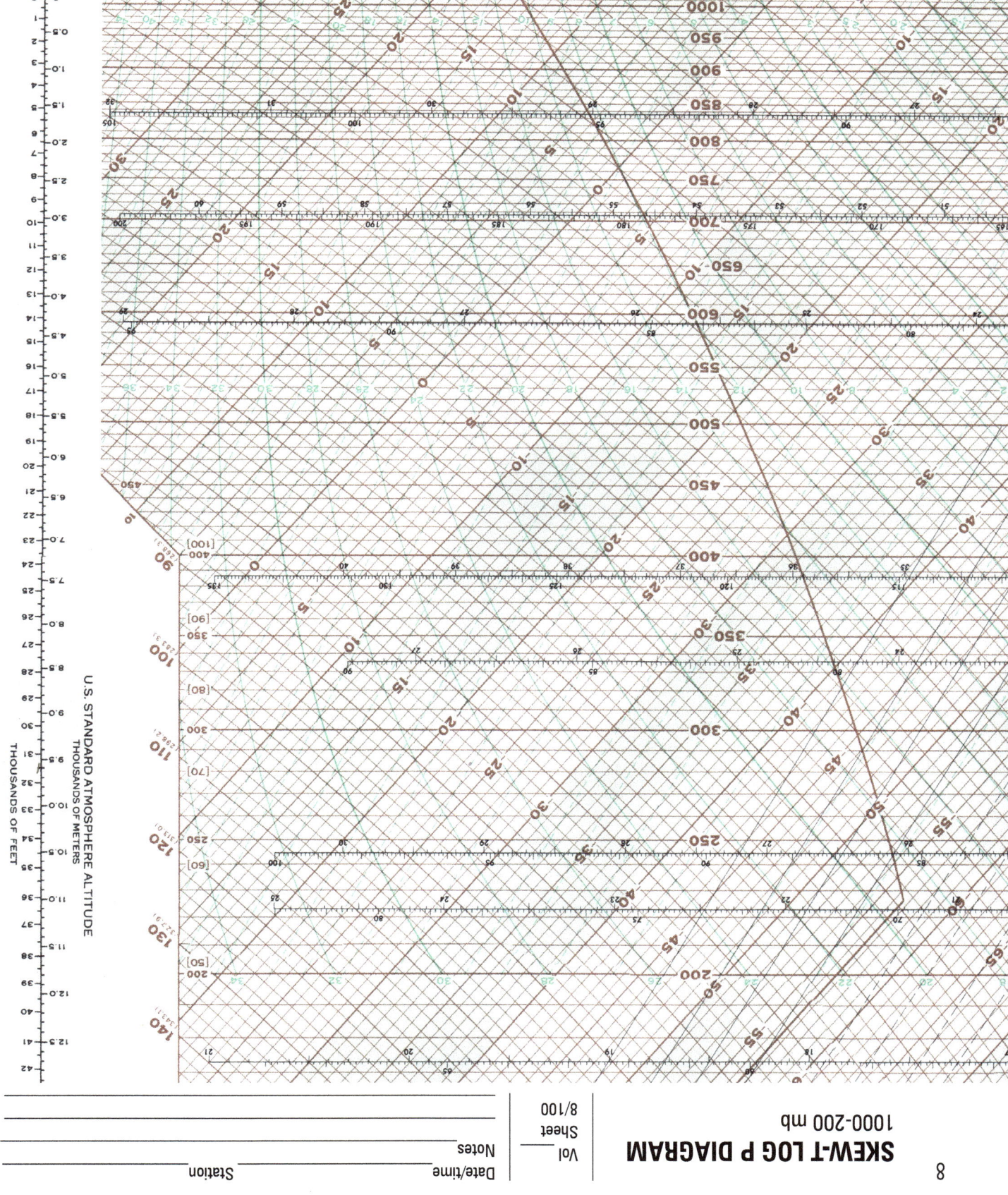

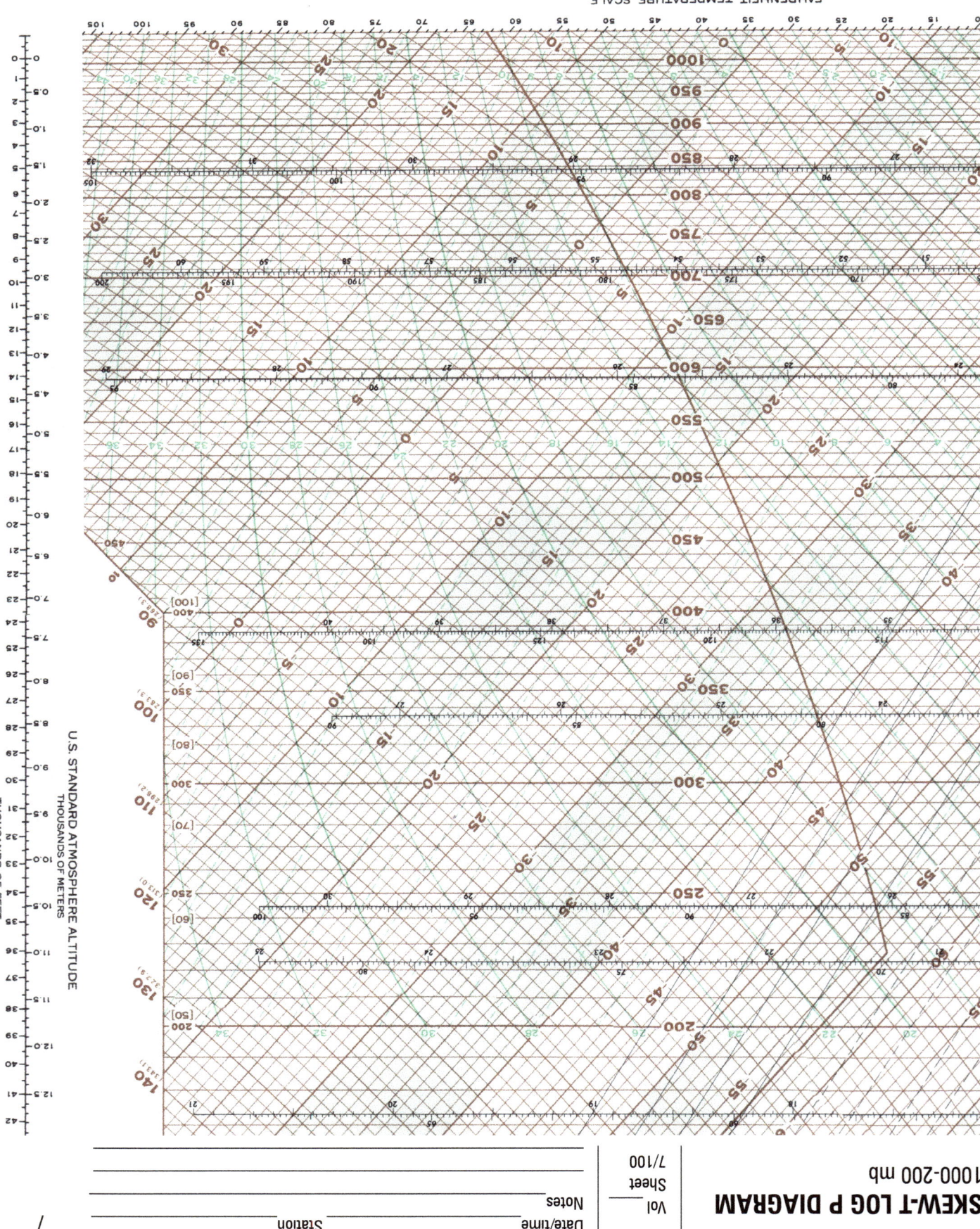

SKEW-T LOG P DIAGRAM
1000-200 mb

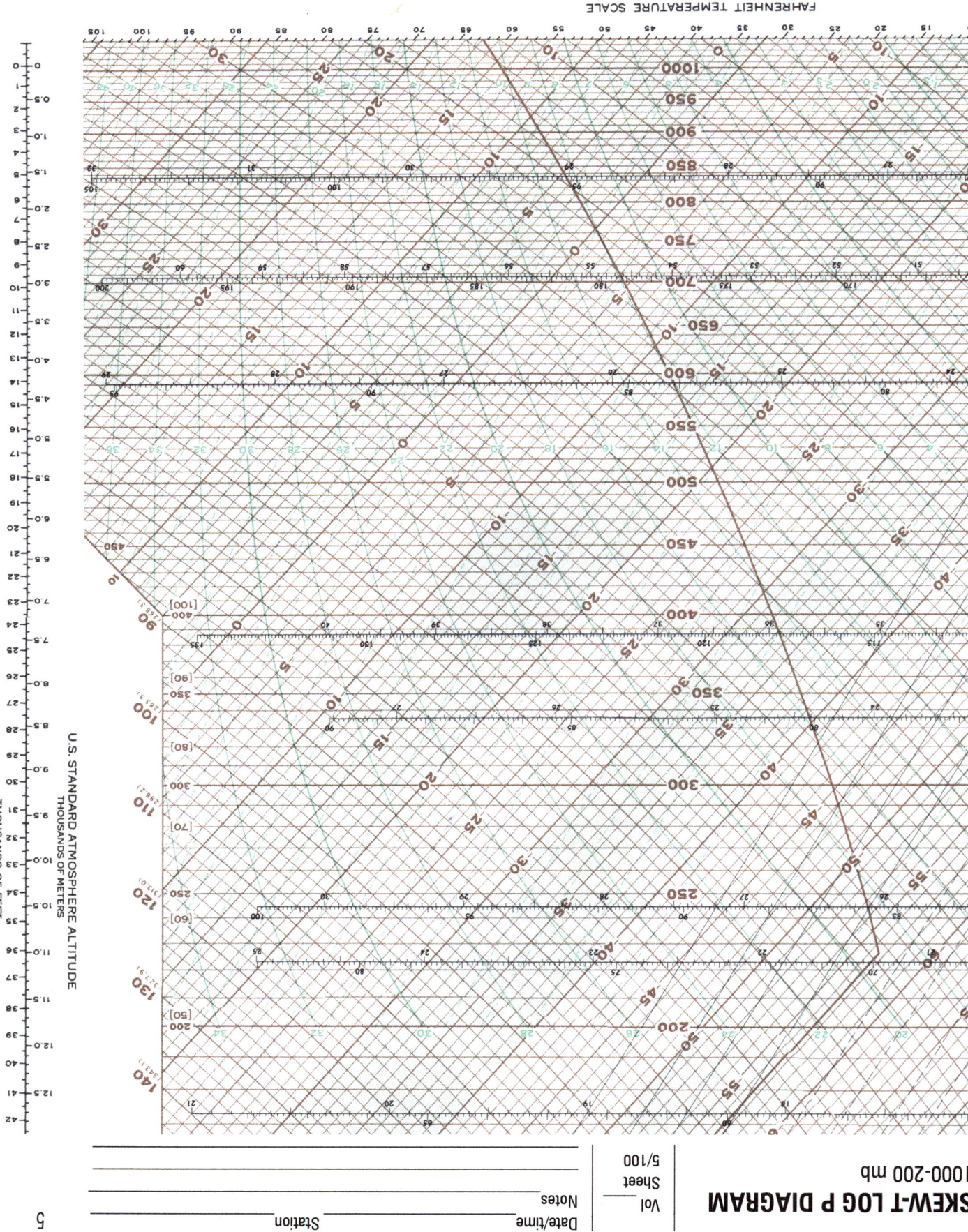

SKEW-T LOG P DIAGRAM
1000-200 mb

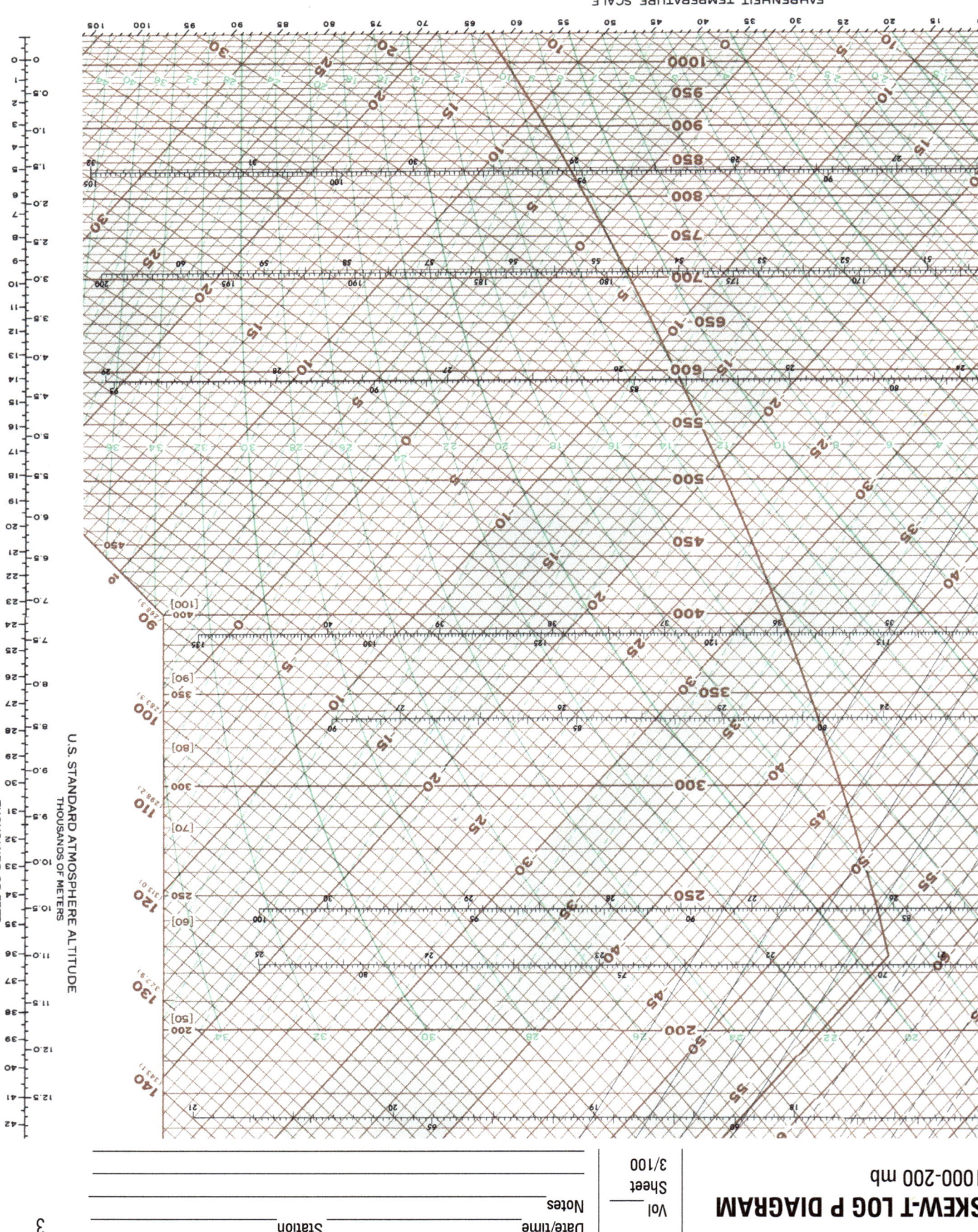

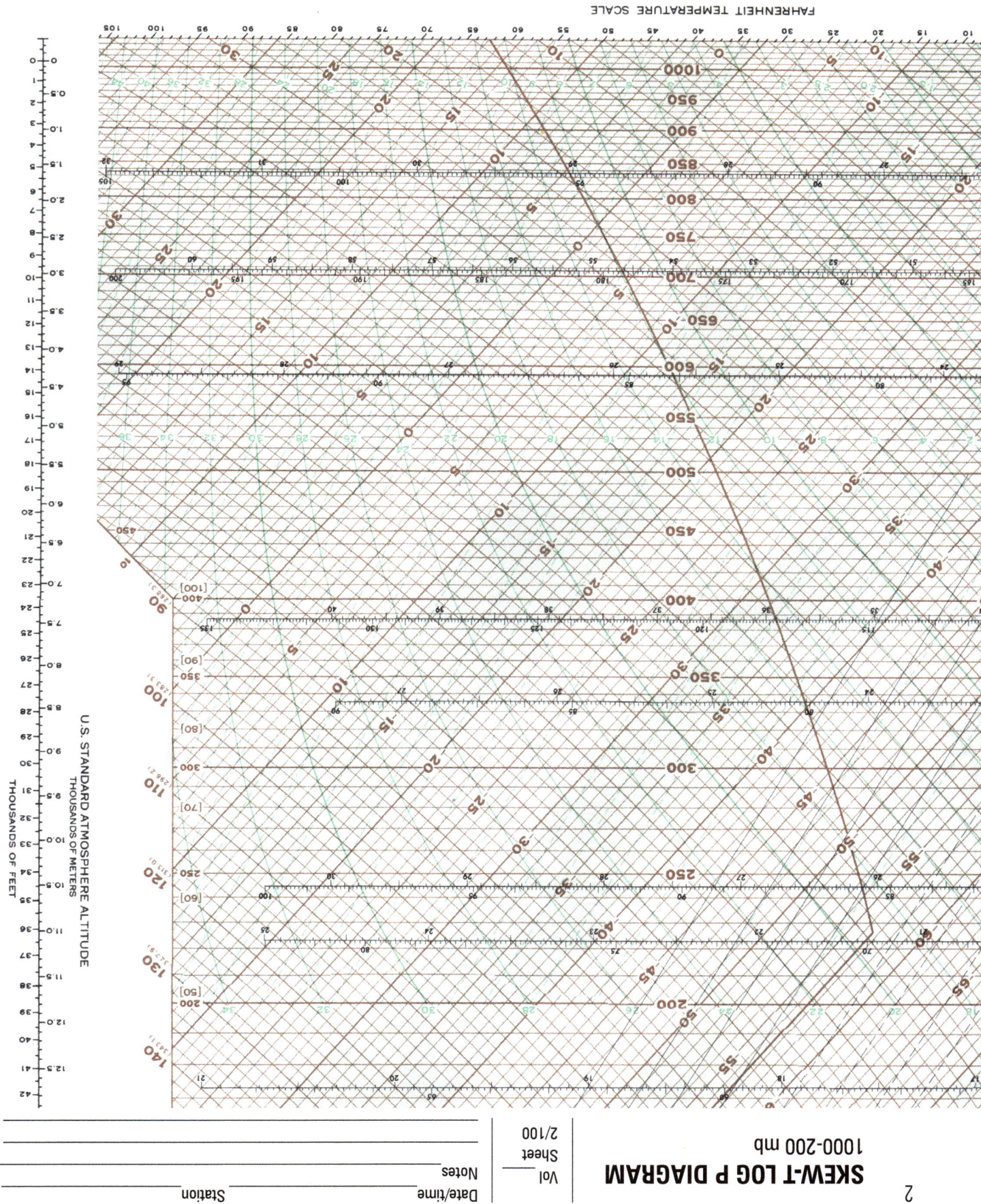

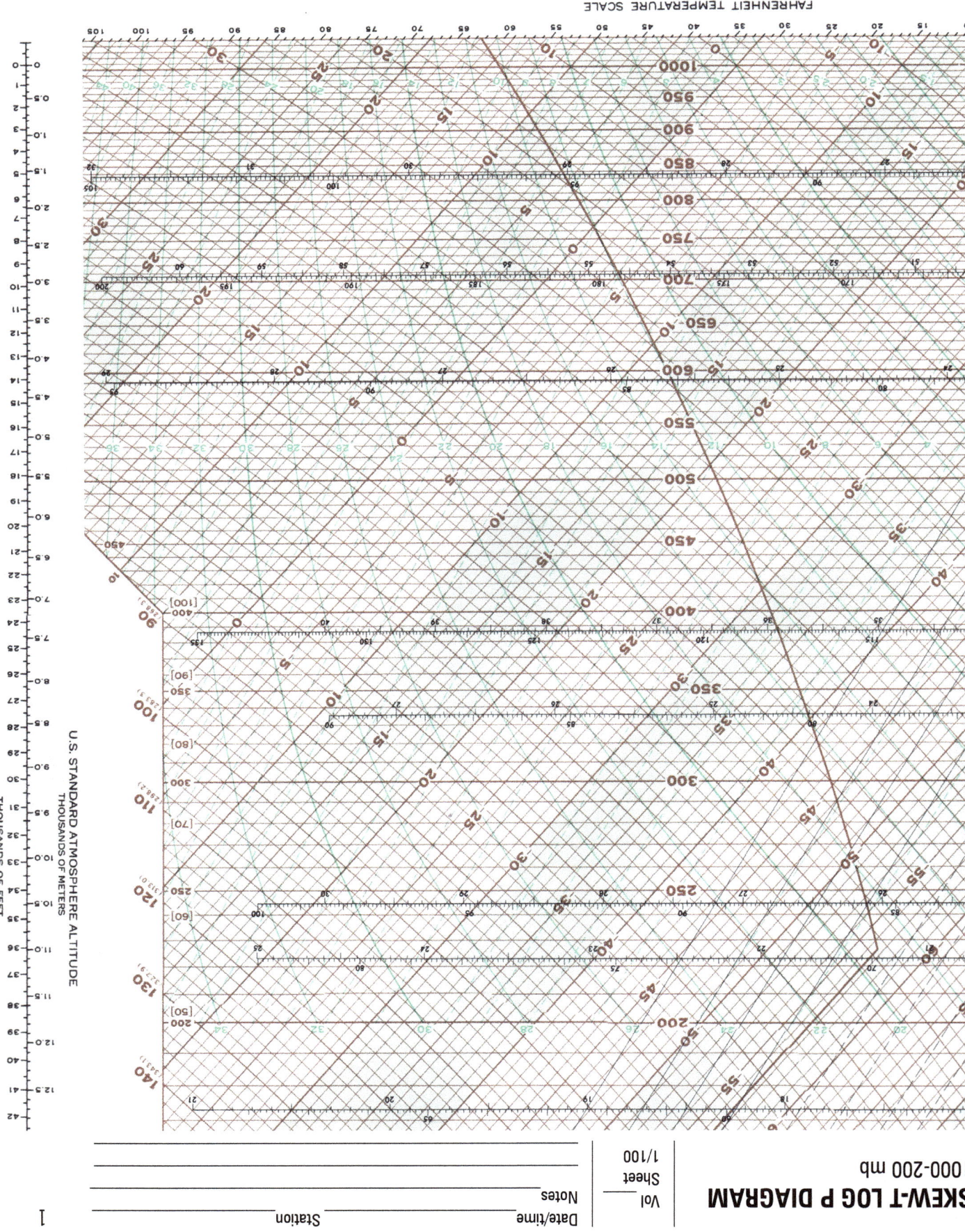

SOUNDINGS

INTRODUCTION

Around 2005, Defense Mapping Agency meteorological charts were discontinued, ending the only remaining source of bulk paper maps available to the public. For the next ten years I grew into a habit of scribbling notes and drawing on printed soundings and hodographs when working out predictions on storm days or forecasting for Chase Hotline clients.

It was at some point this year that it dawned on me that I (and other forecasters) could certainly use a disposable workchart book that's full of blank charts and which is designed to be written in and have its pages torn out or removed with an X-Acto knives. Even if you're just learning the ropes of forecasting, you'll eventually find many of these blank diagrams to be indisposable for trying your hand at a forecast or just scribbling on a diagram to understand better how a specific instability index works.

The workbook focuses mainly on soundings, using the diagram most commonly used at temperate latitudes. To avoid shrinking the diagrams excessively or adding to the size of the book, I have omitted forms of the sounding used in polar regions or near the stratosphere, placing only a few pages of the full version toward the rear of the book.

With that I remind you that even with the proliferation of computerized weather data, there are still many techniques that can't be done on computers (or can't be done efficiently) and depend on manual computation. In fact, manipulating and modifying soundings and hodographs is the very thing that connects forecasters in a qualitative sense to the forecast problem and gives them a true hands-on approach to completing the forecast. There are just some things that can't be done easily on the computer screen.

Keep in mind this book is intended to be disposable and is meant to help forecasters work out problems at the forecast desk. Tear out pages and write in it freely. You can always order more copies of the book.

Suggestions and feedback for making this workbook more useful are welcome.

TIM VASQUEZ
October 2016

table of contents

1 SOUNDINGS

1   Skew T log P diagram, 1000-200 mb (100 sheets)
101   Skew T log P diagram, convective situation, low level (1000-500 mb) (32 sheets)
133   Skew T log P diagram, full chart (1000-100 mb) (8 sheets)

141 HODOGRAPHS

142   Hodograph reference chart
143   Hodograph (32 sheets)

175 STORM MOTION

177   Gamma mesoscale (8 sheets)
185   Beta mesoscale (6 sheets)
191   Alpha mesoscale (6 sheets)

197 HORIZONTAL CHARTS

197   United States and Southern Canada (17 sheets)

215 RADAR

216   WSR-88D range/height diagram (5 sheets)

FORECASTERS REFERENCE WORKBOOK
First edition

October 2016

Copyright ©2016 Tim Vasquez
All rights reserved

For information about permission to reproduce selections from this book, write to Weather Graphics Technologies, P.O. Box 450211, Garland TX 75045 or servicedesk@weathergraphics.com. No part of this publication may be reproduced, stored in a retrieval system, or transmitted by any means without the express written permission of the publisher.

ISBN 978-0-9969423-2-4

Printed in the United States of America

Forecasters Reference Workbook

Tim Vasquez

2016